Advances in Chemical Reaction Dynamics

NATO ASI Series

Advanced Science Institutes Series

A series presenting the results of activities sponsored by the NATO Science Committee, which aims at the dissemination of advanced scientific and technological knowledge, with a view to strengthening links between scientific communities.

The series is published by an international board of publishers in conjunction with the NATO Scientific Affairs Division

A Life Sciences	Plenum Publishing Corporation
B Physics	London and New York
C Mathematical and Physical Sciences	D. Reidel Publishing Company Dordrecht, Boston, Lancaster and Tokyo
D Behavioural and Social Sciences E Engineering and Materials Sciences	Martinus Nijhoff Publishers The Hague, Boston and Lancaster
F Computer and Systems Sciences G Ecological Sciences	Springer-Verlag Berlin, Heidelberg, New York and Tokyo

Series C: Mathematical and Physical Sciences Vol. 184

Advances in
Chemical Reaction Dynamics

edited by

Peter M. Rentzepis
University of California, Department of Chemistry,
Irvine, California, U.S.A.

and

Christos Capellos
Energetics and Warheads Division,
Armament Engineering Directorate,
US Army Armament Research, Development and Engineering Center,
Dover, New Jersey, U.S.A.

D. Reidel Publishing Company

Dordrecht / Boston / Lancaster / Tokyo

Published in cooperation with NATO Scientific Affairs Division

Proceedings of the NATO Advanced Study Institute on
Advances in Chemical Reaction Dynamics
Iraklion, Crete, Greece
25 August-7 September 1985

Library of Congress Cataloging in Publication Data

NATO Advanced Study Institute on Advances in Chemical Reaction Dynamics (1985 , Crete)
   Advances in chemical reaction dynamics.
   (NATO ASI series. Series C, Mathematical and physical sciences; vol. 184)
   "Proceedings of the NATO Advanced Study Institute on Advances in Chemical Reaction
Dynamics, Iraklion, Crete, Greece, 25 August–7 September, 1985"—
   "Published in cooperation with NATO Scientific Affairs Division."
   Includes index.
   1. Chemical reaction, Rate of—Congresses. 2. Chemical reaction, Conditions and
Laws—Congresses. I. Rentzepis, Peter M., 1934– . II. Capellos, Christos. III. Title.
IV. Series: NATO ASI series. Series C, Mathematical and physical sciences; vol. 184.
   QD502.N365   1985   541.3'94   86–17441

ISBN-13: 978-94-010-8604-2   e-ISBN-13: 978-94-009-4734-4
DOI: 10.1007/978-94-009-4734-4

Published by D. Reidel Publishing Company
P.O. Box 17, 3300 AA Dordrecht, Holland

Sold and distributed in the U.S.A. and Canada
by Kluwer Academic Publishers,
101 Philip Drive, Assinippi Park, Norwell, MA 02061, U.S.A.

In all other countries, sold and distributed
by Kluwer Academic Publishers Group,
P.O. Box 322, 3300 AH Dordrecht, Holland

D. Reidel Publishing Company is a member of the Kluwer Academic Publishers Group

TABLE OF CONTENTS

PREFACE

This book contains the formal lectures and contributed papers presented at the NATO Advanced Study Institute on the Advances in Chemical Reaction Dynamics. The meeting convened at the city of Iraklion, Crete, Greece on 25 August 1985 and continued to 7 September 1985.

The material presented describes the fundamental and recent advances in experimental and theoretical aspects of reaction dynamics. A large section is devoted to electronically excited states, ionic species, and free radicals, relevant to chemical systems. In addition recent advances in gas phase polymerization, formation of clusters, and energy release processes in energetic materials were presented. Selected papers deal with topics such as the dynamics of electric field effects in low polar solutions, high electric field perturbations and relaxation of dipole equilibria, correlation in picosecond/laser pulse scattering, and applications to fast reaction dynamics. Picosecond transient Raman spectroscopy which has been used for the elucidation of reaction dynamics and structural changes occurring during the course of ultrafast chemical reactions; propagation of turbulent flames and detonations in gaseous energetic systems are also discussed in some detail. In addition a large portion of the program was devoted to current experimental and theoretical studies of the structure of the transition state as inferred from product state distributions; translational energy release in the photodissociation of aromatic molecules; intramolecular and intraionic dynamic processes. Dipolar contribution to the mechanism of triplet-triplet energy transfer in molecular solids; dynamics of excited states of atoms interacting with strong laser fields, as well as other important topics.

The purpose of the proceedings is to provide the reader with a rather broad perspective on the current theoretical aspects and recent experimental observations in the field of reaction dynamics. Furthermore, the material presented in the proceedings should make apparent the future trends for research and demonstrate the significance of collaborative research between theorists and experimentalist in areas of importance to the understanding of chemical reaction dynamics specially of large organic, energetic, and biological molecules.

The proceedings should be of interest to graduate and postgraduate students who are involved or starting research in these scientific areas, and to scientists who are actively pursuing research in reaction kinetics. Appropriate introductory overviews are provided in this text on the various aspects of reaction dynamics, which might be of help before continuing to more advanced concepts and the recent theoretical and experimental papers presented. An appreciable part of the information contained in the proceedings has not appeared yet in the textbooks on chemical reaction dynamics.

                                                   P.M. RENTZEPIS
                                                   C. CAPELLOS

ACKNOWLEDGMENTS

We express our appreciation to Professor P. Papagianakopoulos for his reliable assistance in organizing the NATO Advanced Study Institute. Special thanks are due to all lecturers and authors for their cooperation in preparing their manuscripts, excellent delivery of lectures, and dedication to the meeting.

Miss Irene Fedum provided invaluable secretarial assistance in preparation for the meeting, and Ms. Carol Novak for the expert typing of manuscripts.

Last, but not least, we should like to express our gratitude to the NATO Scientific Affairs Division, US Army Research Group (ERO) and the Energetic Materials Division (LCWSL-ARDC) for granting financial support for the meeting.

TRANSLATIONAL ENERGY RELEASE IN THE PHOTODISSOCIATION OF AROMATIC
MOLECULES

R. Bersohn
Department of Chemistry
Columbia University
New York, NY   10027

ABSTRACT. When a large amount of energy is suddenly placed in an
aromatic molecule, it can dissociate via a rather complex path.
Measurements of translational energy distributions of the
dissociated fragments of a variety of related molecules at
different wave lengths clarify the dissociation processes. The
reactions considered are
1)   aryl halides + h$\nu \rightarrow$aryl radical + halogen atom
2)   s-triazine + h$\nu \rightarrow$3 HCN
3)   styrene or cyclooctatetraene + h$\nu \rightarrow$benzene + acetylene
The first reaction may involve direct cleavage of the C-X bond
but when it does not there is an electronic energy transfer from
the delocalized π system to the C-X bond. The second and third
reactions involve dissociations from vibrationally hot electronic
ground states and comparatively little kinetic energy is
released. The use of isotopically labelled styrene shows that
rather extensive hydrogen atom migration precedes dissociation.

The spectroscopy of aromatic hydrocarbons is rather well
understood and even the spectroscopy of many substituted aromatic
hydrocarbons has been studied in detail. They are therefore an
interesting class of molecules for photodissociation studies. In
this review three types of processes will be discussed. The
first is the cleavage of a halogen atom from a haloaromatic: RX
+ h$\nu \rightarrow$R + X. The second is the explosion of a triazine (a
benzene three of whose CH groups have been replaced by nitrogen
atoms) into three HCN molecules. The third and the most
remarkable is the cleavage of a vinyl substituted benzene
(styrene) into benzene and acetylene, a process preceded by an
exchange of hydrogen atoms between the ring and the side chain.

1

P. M. Rentzepis and C. Capellos (eds.), Advances in Chemical Reaction Dynamics, 1–20.
© *1986 by D. Reidel Publishing Company.*

$$R-X + h\nu \longrightarrow R + X$$

The simple cleavage of a carbon halogen bond is not as simple as it would first appear. When methyl iodide absorbs a photon whose energy is in its first absorption band, the resulting dissociation is a direct process, as simple as one can imagine. The mere observation of methyl radicals and iodine atoms leaves unanswered questions about the internal vibrational state distribution of the methyl radical, the relative yield of $I(^2P_{3/2})$ and $I*(^2P_{1/2})$ atoms, and the distribution of relative translational energies. Even more subtle is the question of the anisotropy of the angular distribution of the fragments' relative velocity when not all directions of polarization of the dissociating light are equivalent. In the case of an aromatic iodide, the dissociation is indirect, that is, the first electronic surface to which the molecule is excited is not the surface from which dissociation takes place. Additional questions therefore arise such as the nature of the final surface and the mean time required for crossing from the initial to the final dissociated surface.

There are several lines of evidence that the dissociation of aryl halides is an indirect process. In the first place, the uv absorption spectrum of an aryl halide, RX, is very little different from that of its parent hydrocarbon RH. This is because the halogen atom states interact only very weakly with the π electron states of the aromatic rings. Moreover, the absorption of the π electron system is usually much stronger than that of the C-X bond so that the latter absorption is submerged. (An exception is the forbidden first absorption band of benzene near 250 nm; at this wavelength the absorption coefficient of iodobenzene is only about a factor of two greater than that of CH_3I as compared to the usual factor of 10-100.)

The above arguments while suggestive do not constitute a proof. Experiments using polarized light have proven that the transition is indirect. To understand these experiments one must first consider the alignment of the molecules dissociated with polarized light. The following is a classical derivation of this alignment. The molecular absorption coefficient is proportional to $|\vec{\mu}_{if} \cdot \vec{\mathcal{E}}|^2 = \mu_{if}^2 \mathcal{E}^2 \cos^2\theta$ where $\mathcal{E}$ is the electric vector of the light wave, μ_{if} is the matrix element of the dipole moment operator between the initial and final states and θ is the angle between $\vec{\mathcal{E}}$ and $\vec{\mu}_{if}$. The anisotropy of molecular absorption is the ultimate source of the alignment. The distribution of transition dipole axes with respect to the electric vector i.e. $\cos^2\theta$ is not what is observed. The observed distribution is that of dissociation axes at the instant of dissociation with respect to the electric vector (see Fig. 1). To obtain this distribution we must average $\cos^2\theta$ over the cone of angles χ between the dissociation direction and $\vec{\mu}_{if}$.

Using the identity $\cos^2\theta = \frac{1}{3}(1+2P_2(\theta))$,

$$\langle\cos^2\theta\rangle = \frac{1}{3}(1+2\langle P_2(\theta)\rangle) = \frac{1}{3}(1+2P_2(\chi)P_2(\theta_\epsilon) = \frac{1}{3}(1+\beta P_2(\theta_\epsilon)$$

where the Legendre polynomial addition theorem has been used to carry out the averaging. We obtain an asymmetry parameter $\beta = 3\cos^2\chi - 1 = 2P_2(\chi)$ which varies from 2 to -1.

As an experimental example[1], consider the isomers 1- and 2-iodonaphthalene. When naphthalene is excited in the uv region 240-280nm, the symmetry of this $S_2 \leftarrow S_0$ transition is such that the transition is polarized along the short (9,10) axis. If the transition responsible for the photodissociation is simililarly polarized, then the angle χ will be very different for the two molecules when they are excited in this uv region:

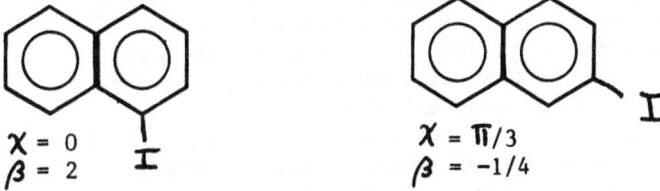

On the other hand, if the transition dipole is fixed with respect to the C-I bond then χ and hence the asymmetry parameter β will be the same. Figs. 2 and 3 show the flux of iodine atom fragments emerging from 1- and 2-iodonaphthalene. These curves prove that the dissociation starts from the stable delocalized $\pi,\pi*$ state and ends at an unstable antibonding state largely localized on the C-I bond.

So far the discussion has centered on spectroscopy and on electronic pathways, but nothing has been said about the rate at which the dissociation takes place. Nature has provided us with a clock which is the rotational period of the molecule. Let us consider for the sake of simplicity a linear molecule which is rotating at an angular velocity ω. The faster the molecule rotates relative to the rate at which it dissociates, the more anisotropy is lost. One can prove[8,9] that

$$\beta = 2P_2(\chi)(1+\omega^2\tau^2)/(1+4\omega^2\tau^2) \tag{1}$$

where τ is the average lifetime. The angular velocity depends on the rotational state, but the square of the angular velocity averaged over the Boltzmann distribution is given by $I\omega^2 = kT$. At room temperature a typical molecular rotation period ($2\pi/\omega$) is about one picosecond. When the product $\omega\tau \to 0$, Eq. (1) reduces to that previously derived whereas when $\omega\tau \to \infty$, only one-fourth of the original anisotropy remains. This is contrary to one's intuition because one would assume that fast rotation should destroy all the anisotropy. The explanation is that, in the absence of collisions, the molecule rotates around a unique axis fixed in space; in the presence of collisions as in solution all anisotropy would disappear.

Eq. (1) implies that if one knows χ, measures β and can average ω^2 properly, the lifetime τ can be extracted from the measured β. Using a more complex version of Eq. (1) one can show that the experimental value of $\beta = 1.18 \pm 0.13$ for 1-iodonaphthalene implies a lifetime for dissociation of 1.0 ± 0.2ps. This is an example of a picosecond measurement made with a dc lamp.

Methyl iodide directly dissociates in a time much less than the rotational correlation time. 1-naphthyl iodide dissociates in a time slightly less than the rotational correlation time. Fig. 2 shows the Br atom signal versus polarization angle. The very small angular dependence is consistent with an average dissociation time longer than the rotational correlation time. Similar results were found for 4-bromobiphenyl and 9-bromoanthracene. These results are consistent with the fact that aryl bromides fluoresce but with small quantum yield, whereas aryl iodides are not observed to fluoresce. Indeed, Huppert et al[10] have shown that the fluorescence lifetimes of a number of aryl bromides were in the 30-150ps range, whereas the fluorescence lifetime of aryl iodides was immeasurably short. The reason that aryl bromides decompose more slowly than aryl iodides may be that the spin-orbit coupling which mixes the singlet and triplet manifolds is weaker in bromine atoms than in iodine atoms. We assume that the crossing is from the S_1 (π,π^*) state not to T_1 but to a higher triplet (n,π^*) state localized in the C-I bond.

The variation of kinetic energy release with photon energy sometimes provides a clue as to the mechanism of dissociation. For example, when CdI_2 is dissociated at 278 nm, its fragments have, on the average, more kinetic energy than when 300 nm is used (Fig. 4). This is just what one would expect from a direct dissociation. When 4,4'-diiodobiphenyl is dissociated by 265, 279 and 299 nm light, exactly the same distribution of kinetic energies is obtained in each case (Fig. 5). Evidently the excess energy above the electronic origin is used in vibrationally exciting the 4-iodobiphenyl iodical and plays no role in the dissociation process. In some cases when the photon energy is increased, the kinetic energy even decreases. Fig. 6 shows that

when iodobenzene is dissociated at 249 nm, fairly fast fragments
result, but at 193 nm there are slower fragments as well[5].
Indeed, the distribution at 193 nm appears to be a superposition
of two distributions: one identical with the "fast" peak seen at
248 nm and the other a "slow" peak. We tentatively assume that
the fast peak is due to a rapid dissociation from the S_1 state,
whereas the slow peak is due to a slow dissociation from a
vibrationally hot electronic ground state (Fig. 7). Fortunately
this assumption can be tested by the use of polarized light. The
rapidly dissociating fragments will exhibit strong angular
dependence, whereas the slowly dissociating fragments will be
almost isotropically distributed. Fig. 8 shows exactly this
behavior. When the light is polarized parallel to the detection
direction, the faster fragments are more abundant; whereas for
the perpendicular polarization the slower fragments are more
abundant.

The dissociation of benzene into three acetylenes is highly
endothermic, but as nitrogen atoms successively replace the
isoelectronic C-H groups, the rings become progressively weaker.
Consider the following series of dissociations:

$$\Delta H^{\circ}(kcal/mole)$$

		ΔH°
	$3C_2H_2$	+151.3
	$2C_2H_2 + HCN$	+117.1
	$C_2H_2 + 2HCN$	+ 84.1
	$3HCN$	+ 43.2
	$2HCN + N_2$	⌒10

A ring with 5 or 6 nitrogen atoms is unstable with respect to
dissociation into three molecules. All of the above molecules
when excited by a photon with an energy somewhat in excess of
that needed thermodynamically to dissociate them are perfectly
stable and have sharp spectra. The explanation of this anomaly
follows from correlation considerations. The ground state of
each of these aromatic molecules is a non-degenerate singlet and
so are the ground states of the small molecules into which they
dissociate. Any electronically excited state of the parent
molecule must therefore dissociate into three small fragment
molecules, at least one of which must be electronically excited.
In the experiments to be discussed[6,7], which were carried out at
248 and 193 nm, not enough energy was available for electronic
excitation of any fragment. Dissociation had therefore to take
place from the ground state of the parent molecule. Thus we can
distinguish between two threshold energies for dissociation. One
is the amount of electronic energy which must be supplied before
the molecule can rapidly cross to the ground state; the other is
the minimum amount of vibrational energy which must be supplied
to the ground state molecule to produce dissociation. The
former, for s-triazine is about 100 kcal/mole; the latter would
be the activation energy for thermally dissociating the molecule
and might be only slightly greater than the dissociation energy
of 43 kcal/mole.

Fig. 9 shows the distribution of times of flight of HCN
fragments from the point where they are formed at the
intersection of the molecular beam and the laser beam to the
ionizer of the mass spectrometer which detects them. The
fragments generated by light at 248 nm, paradoxically, move
faster than those generated at 193 nm. The translational energy
distribution shown in Fig. 10 is calculated from the time of
flight distribution assuming that each HCN travels with the same
speed.

Table I shows the average partitioning of energy at the two
photon energies. The work of Goates, Chu and Flynn[7] enables us
to analyze $\langle E_{INT} \rangle$ in more detail. They also dissociated triazine
vapor with 193 nm light and observed subsequent ir emission. By
observing the emission through different filters, it was found
that there are about 70 times as many quanta in the bending modes
as in the C-H stretch. (The CN stretch could not be investigated
because of the very small dipole derivative associated with this
mode.)

Table I Average energy partitioning in HCN fragments

λ(nm)	$h\nu$	E_{AVAL}	$\langle E_T \rangle/3$	$\langle E_{INT} \rangle/3$
193	148	105	2	33
248	115	72	10	14

E_{AVAL} = $h\nu$-dissociation energy. $\langle E_T \rangle$ and $\langle E_{INT} \rangle$ are the average total translational and internal energy, respectively. All energies are in units of kcal/mole.

Once the surface crossing is accomplished, the dissociation is so quick that there is no time for vibrational equilibration. The evidence for this is as follows. The fact that the vibrational energy is greater and the translational energy is less at the higher photon energy suggests that the surface crossing occurs at a different region in configuration space corresponding to a greater distance between HCN molecules. If vibrational redistribution were rapid on the time scale of dissociation, we would expect more translation energy at the higher photon energy. Moreover, we would not find a large imbalance of populations in the bend and C-H stretch modes. As a mnemonic this extremely rapid separation of the three HCN molecules can be called an "explosion."

$$C_8H_8 + h\nu \rightarrow C_6H_6 + C_2H_2$$

This section describes a remarkable series of intramolecular hydrogen atom transfers occurring in styrene after it is electronically excited. To begin with, a more energy rich isomer of styrene, namely cyclooctatetraene (COT) has a complex photochemistry[11] when irradiated in its first weak $^1B \leftarrow ^1A_1$ absorption band near 250 nm. Dissociation into benzene and acetylene occurs as well as isomerization to a variety of compounds including bicyclo-[4,2,0]-octa-2,4,7-triene (BOT), tricyclo-[3,3,0,0^{2,8}]-octa-3,6-diene (semibulvalene), 1,5-dihydropentalene and styrene. COT has a very strong $^1A_1 \leftarrow ^1A_1$ absorption peaking at 193 nm which made it an attractive candidate for a molecular beam study of its fragmentation by a 193 nm ArF laser. Consider the energetics of the various C_8H_8 isomers shown in Fig. 9. Only 3 kcal/mole are required to dissociate COT into benzene and acetylene; at 193 nm, 145 kcal/mole would be available to keep the molecule overcome any barrier to dissociation. How would this large amount of available energy be distributed between translation and internal energy of the fragments? In the past, such a question was seldom

asked for such a complex dissociation because no experimental
answer could be given. Today, with the advent of molecular beams
and lasers, such a question can be answered.

When photodissociated in a molecular beam at 193 nm
cyclooctatetraene dissociates exclusively into acetylene and
benzene[12]. The total average translational energy release is 18
kcal/mole which is 12% of the 145 kcal/mole available energy.
When styrene is dissociated under the same conditions, again only
acetylene and benzene result. The average total translation
energy release is 13 kcal/mole which is again just 12% of the
109 kcal/mole available energy. Had every final state been
equally occupied, given that there are 37 vibrational modes in
the products, only 4% of the available energy would have been
released as translational energy. Another way of stating the
results is that the reaction coordinate along which dissociation
takes place largely involves distortion of internal coordinates
of the benzene and acetylene molecules. As the separation takes
place the energy does not flow out of the vibrations and into
translation but remains in the internal vibrations.

What is the mechanism of dissociation of COT into benzene
and acetylene? It is likely that the acetylene is derived from
adjacent carbon atoms of the eight membered ring so that the
activation energy for the process is that required to bring
together, let us say, carbon atoms 1 and 4 thus squeezing out an
acetylene moiety from the two intervening CH groups:

COT   BOT

The postulated intermediate is in fact an energy rich BOT
molecule.

A more difficult question is the mechanism of dissociation
of styrene. The side chain contains three hydrogen atoms and one
of them must be removed in order to produce the C_2H_2 product.
(No vinyl radicals were observed.) As the kinetic energy release
for the two dissociations is similar, it is tempting to
postulate an identical intermediate, namely BOT. Another
possibility is the reformation of COT. Isotopic studies show
that the situation is in fact more complex than is implied by
these straightforward questions. An experiment with $C_6H_5CD=CH_2$
would at first sight appear to provide a definitive answer. If
the α-deuteron is transferred to the adjacent carbon atom, then
only mass 26 i.e. light C_2H_2 will be found. (If vinylidene,
$C=CH_2$ is formed it will isomerize to acetylene but its mass is

still 26.) If the β-proton is transferred to the ring carbon atom by a 1,3 jump, then only mass 27, C_2HD will be found. In fact both masses are observed, the ratio of 26/27 being 1.46±0.10. Evidently both α and β hydrogen atoms can be transferred to the ring. An experiment on $C_6D_5CH=CH_2$ would yield only light acetylene if the reverse jump, from ring to side chain were forbidden. Instead what was found was a ratio of 2.29±0.10 for mass 26/mass 27. Dissociation via an 8-membered ring is clearly excluded by the measurements.

A unified explanation of these observations was given by N.J. Turro who proposed a BOT intermediate across whose square ring 1,3 hydrogen atom jumps take place easily. Fig. 11 shows the various jumps which are possible. The model assumes a rate constant k_f for 1,3 jumps to the ring and k_b for 1,3 jumps from the ring. k_H and k_D are the rate constants for the dissociation into light and heavy acetylene, respectively. If the H atom jump rates k_f and k_b were much faster that the rates of dissociation k_H and k_D, then the ratio of mass 26/mass 27 would be k_H/k_D for both $C_6H_5CD=CH_2$ and $C_6D_5CH=CH_2$. The ratio is clearly different which rules out this possibility. If the H atom jump rates were sufficiently slower than the rates of dissociation, then $C_6H_5CD=CH_2$ would give only heavy acetylene and $C_6D_5CH=CH_2$ would give only light acetylene in contradiction to the experiment. We conclude that the rates are of comparable order of magnitude. When the seven rate equations implied by Fig. ? are integrated, one obtains good agreement with the experiment using the values $k_b/k_f=2.20$, $k_H/k_f=0.40$ and $k_D/k_f=0.20$. A plausible reason that BOT dissociates faster to form C_2H_2 than C_2HD is the Franck-Condon factors associated with the bent acetylene as it dissociates. These factors will be smaller for HCCD which for the same configurational energy will be excited into states of higher vibrational quantum numbers than HCCH.

Acknowledgment: This work was supported by the U.S. National Science Foundation

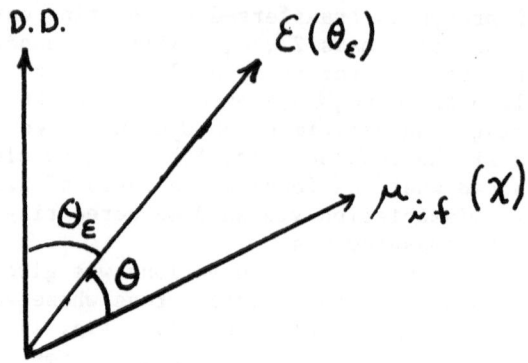

Fig. 1 → The electric vector ($\vec{\mathcal{E}}$), the transition dipole moment vector (μ_{if}) and the dissociation direction D.D.

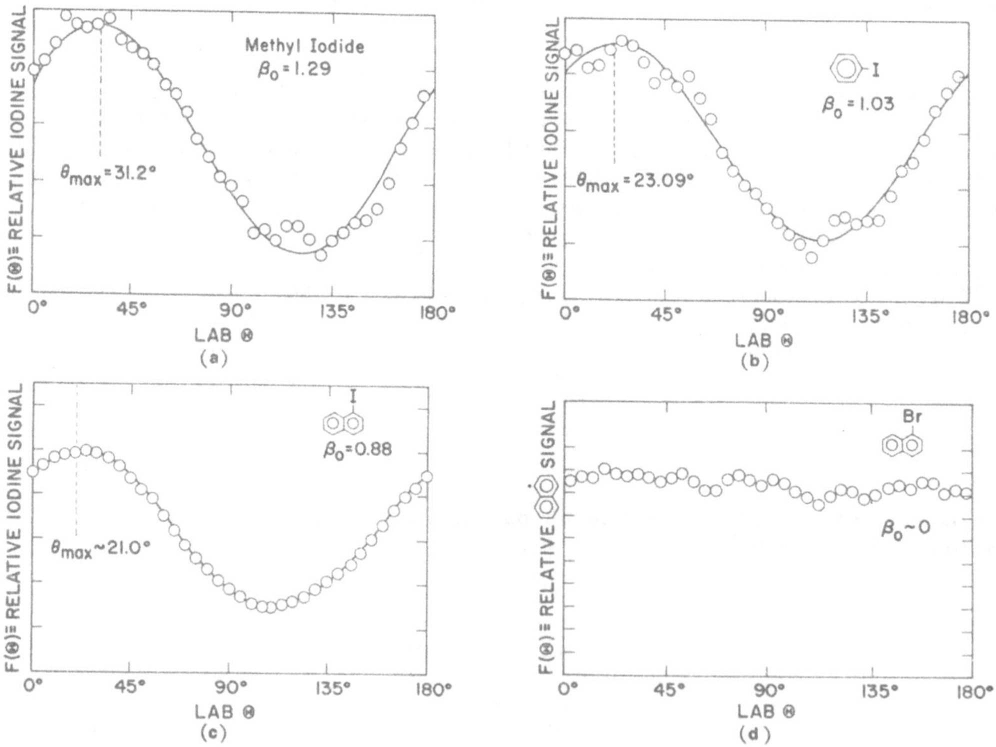

Fig. 2   Photofragment intensities as a function of angle between detector axis and polarization of light; β_0 values have not been corrected for imperfect polarization of the incident light.

Fig. 3   Iodine atom signal as a function of angle between detector axis and polarization of light.

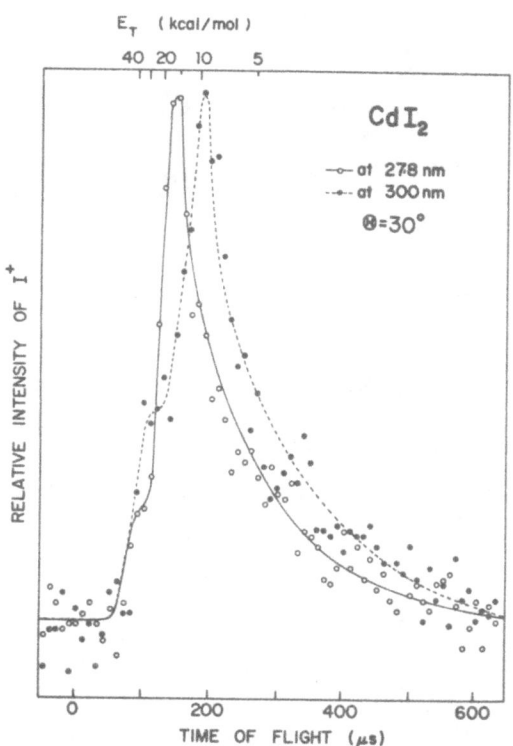

Fig. 4   Relative intensity of I^+ mass spectrometer signal as a function of translational energy when CdI_2 is dissociated by 278 and 300 nm light.

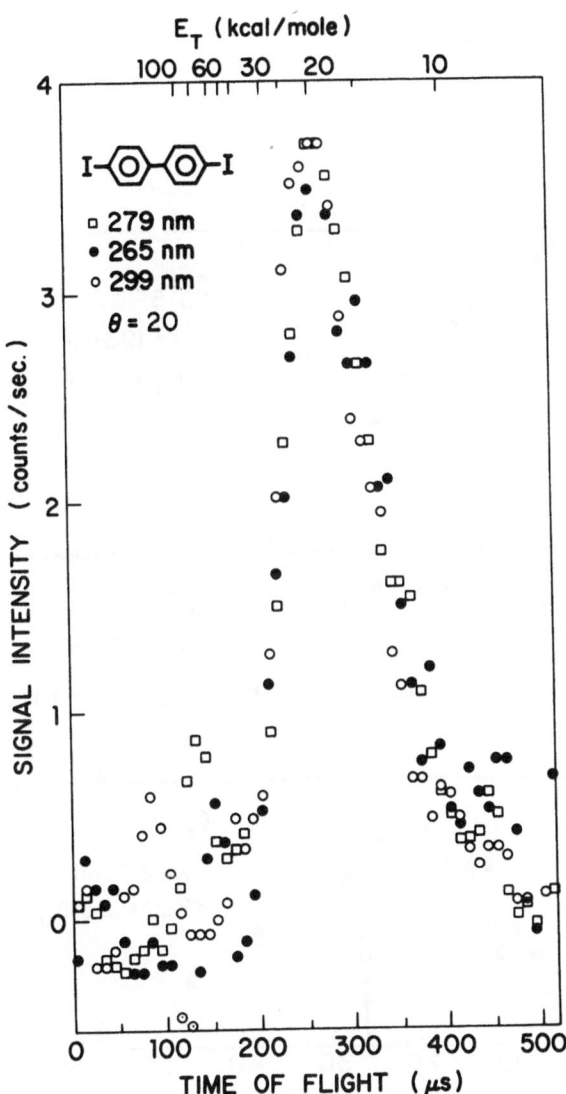

Fig. 5   Time of flight distribution of iodobiphenyl fragments (detected as mass 150 ions) from the photolysis of 4,4'-diiodobiphenyl at mass 265, 279 and 299 nm. Signals at different wavelengths have been normalized to the same maximum value.

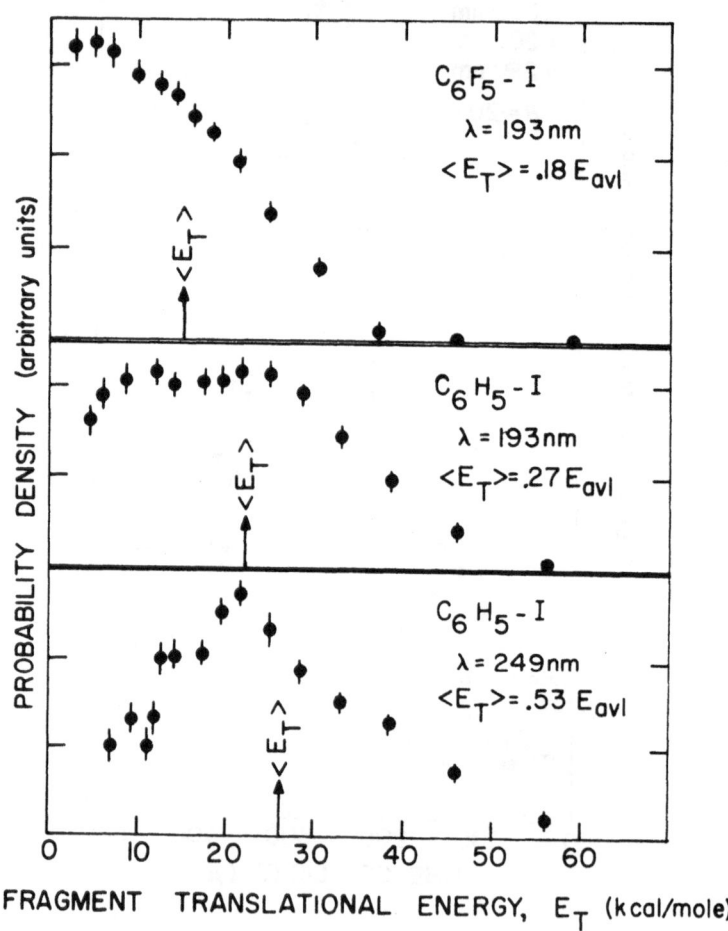

Fig. 6   Fragment translational energy distributions

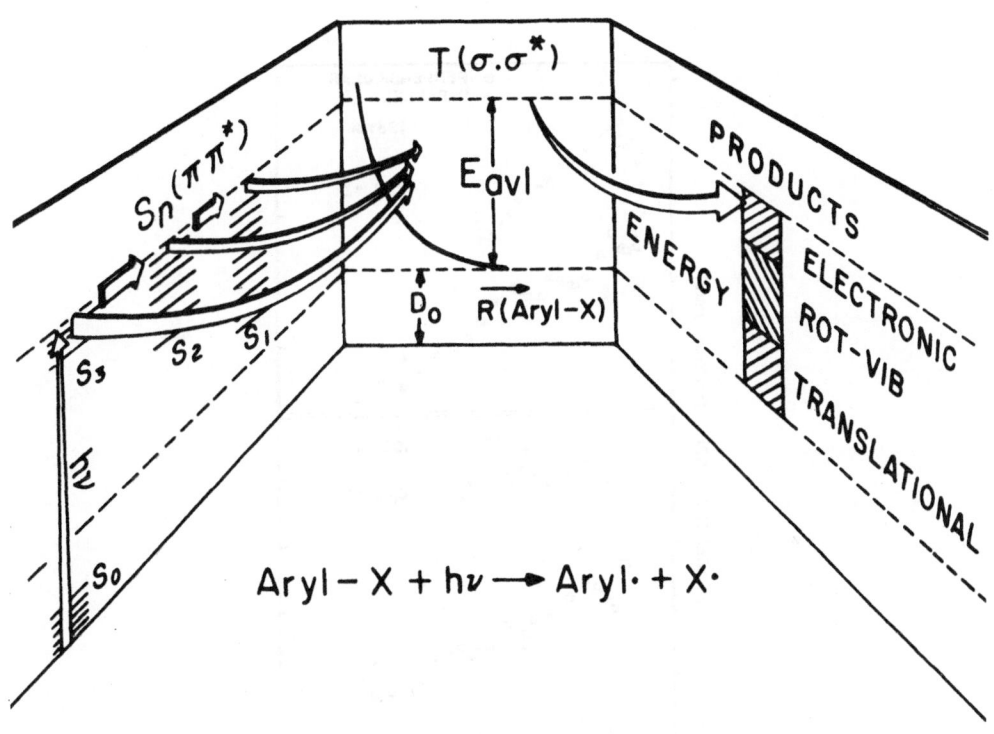

Fig. 7   Schematic representation of dissociation process of aryl halide molecules when excited to upper singlet levels.

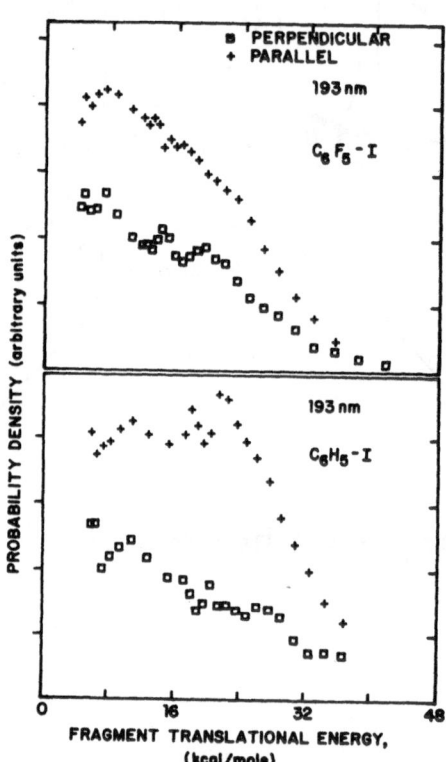

Fig. 8   Fragment translation energy distributions resulting from linearly polarized light. Fragment density is higher with E vector of the light parallel (+) to the flight path to the mass spectrometer than with E perpendicular (☐).

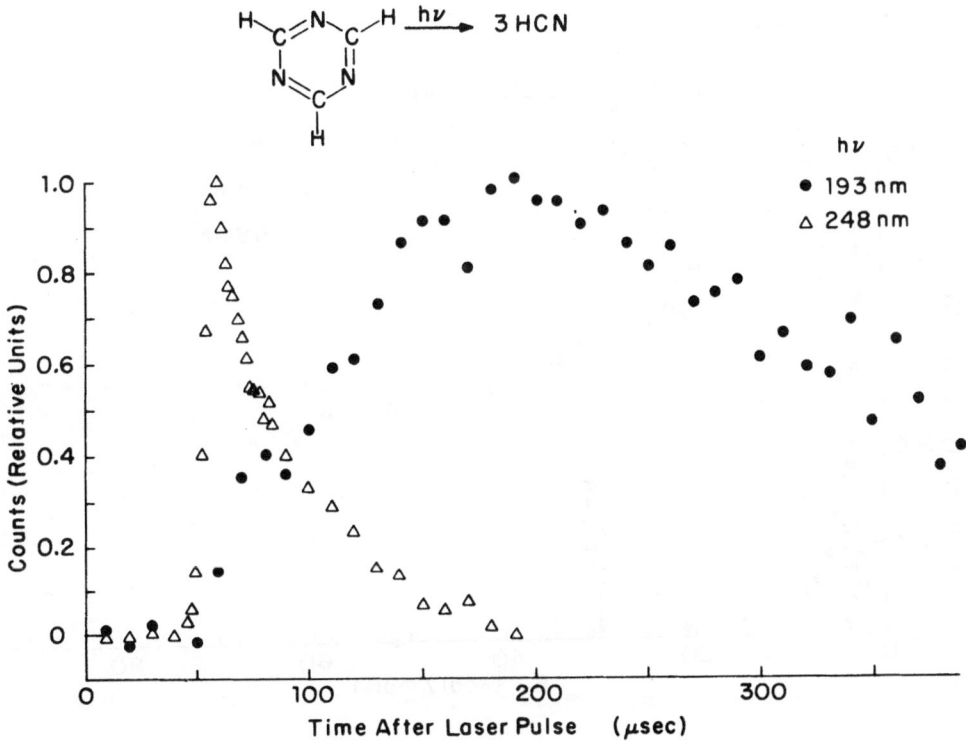

Fig. 9 Distribution of times of flight of HCN fragments from the photodissociation of s-triazine

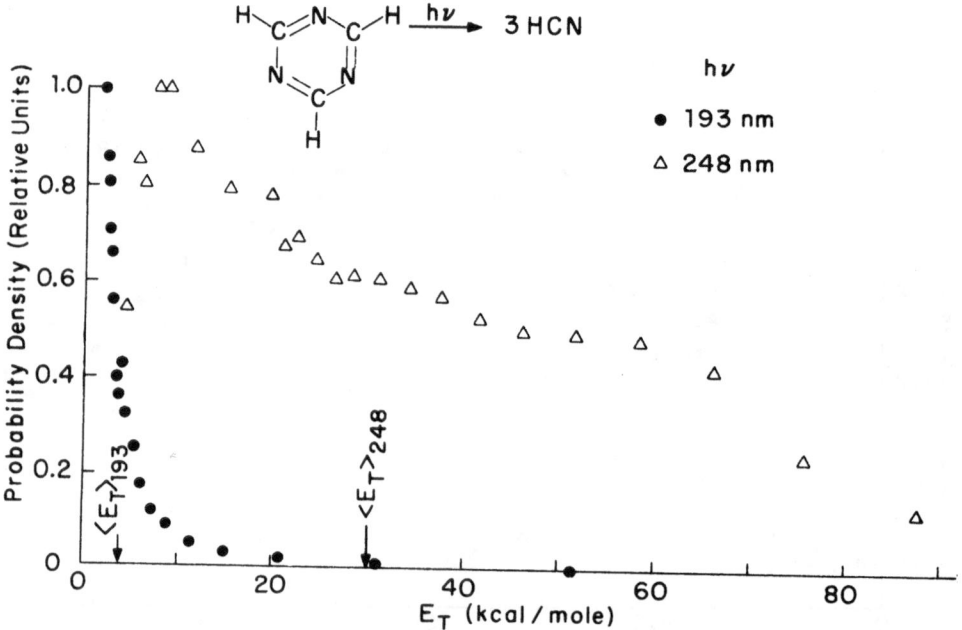

Fig. 10   Distribution of total translational energies E_T of the HCN fragments from the photodissociation of s-triazine. The total translational energy has been assumed to be three times measured individual kinetic energies.

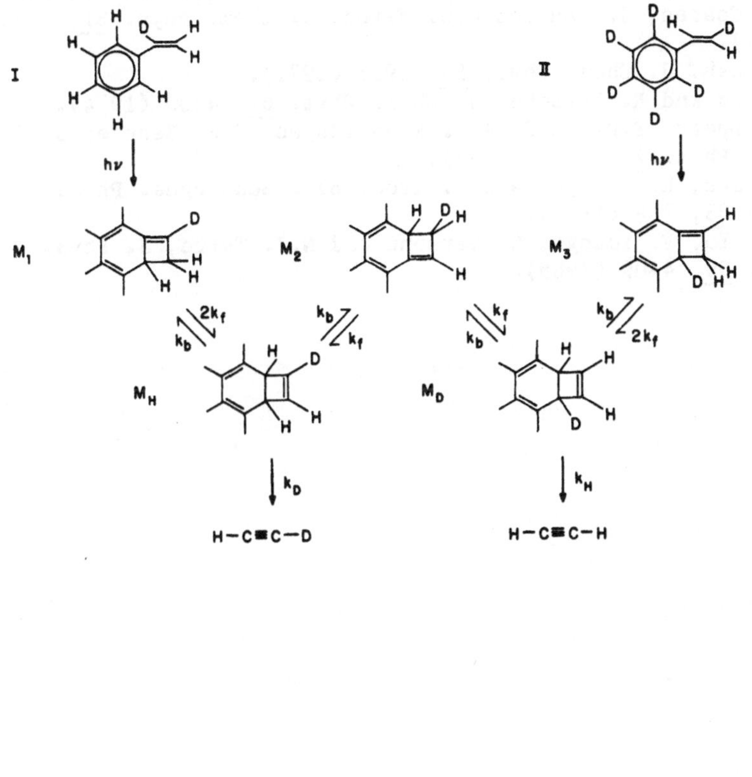

Fig. 11   Mechanistic steps assumed in the photodissociation of
deuterium labelled styrene. Open-ended bonds around six-membered
rings stand for -H if starting with compound I, for   -D if
starting with compound II.

References

1. M. Dzvonik, S.C. Yang and R. Bersohn, J. Chem. Phys. $\underline{61}$, 4408 (1974).
2. R. Bersohn, Israel J. Chem. $\underline{11}$, 675 (1973); $\underline{14}$, 111 (1975).
3. M. Kawasaki, S.J. Lee and R. Bersohn, J. Chem. Phys. $\underline{71}$, 1235 (1979).
4. M. Kawasaki, S.J. Lee and R. Bersohn, J. Chem. Phys. $\underline{66}$, 2647 (1979).
5. A. Freedman, S.C. Yang, M. Kawasaki and R. Bersohn, J. Chem. Phys. $\underline{72}$, 1028 (1980).
6. G.S. Ondrey and R. Bersohn, J. Chem. Phys. $\underline{81}$, 4517 (1984).
7. S.R. Goates, J. Chu and G.W. Flynn, J. Chem. Phys. $\underline{81}$, 4521 (1984).
8. C. Jonah, J. Chem. Phys. $\underline{55}$, 1915 (1971).
9. S. Yang and R. Bersohn, J. Chem. Phys. $\underline{61}$, 4400 (1974).
10. D. Huppert, S.D. Rand, A.H. Reynolds and P.M. Rentzepis, J. Chem. Phys. $\underline{77}$, 1214 (1982).
11. D. Dudek, K. Glanger and J. Troe, Ber. Bunsenges. Phys. Chem. $\underline{83}$, 788 (1979).
12. C.-F. Yu, F. Youngs, R. Bersohn and N.J. Turro, J. Phys. Chem. $\underline{89}$, 4409 (1985).

THE STRUCTURE OF THE TRANSITION STATE AS INFERRED FROM PRODUCT STATE DISTRIBUTIONS

Richard Bersohn
Department of Chemistry
Columbia University
New York, NY 10027

ABSTRACT.   Three reactions are considered:

$$S(^1D) + OCS = S_2 + CO$$
$$F + CH_3I = CH_3F + I$$
$$A + HD = AH(D) + D(H) \qquad A = O(^1D), S(^1D), F$$

In each case laser induced fluorescence measurements on the nascent
products permits a strong inference about the nature of the transition
state.

When $S(^1D)$ reacts with OCS, S_2 is produced both in the
$X^3\Sigma$ and $a^1\Delta$ states. In spite of the large exoergicity of both
reactions, the $S_2(X^3\Sigma)$ is rotationally cold whereas the $a^1\Delta$ state is
rotationally excited with a maximum population at $J = 65$. An
explanation is that the non-adiabatic process which produces $S_2(X^3\Sigma)$
proceeds via a C_{2v} intermediate OCS_2. On the other hand the moderately
narrow J distribution of $S_2(a^1\Delta)$ implies a limited range of impact
parameters in an abstraction process.

When fluorine atoms react with methyl, ethyl, isopropyl and
t-butyl iodide, IF is produced in comparable amounts. On the other hand
I atoms are also products of the reaction with CH_3I but the yield
rapidly diminishes as the bulk of the alkyl group increases. This is
evidence for a Walden inversion.

Reactions of the type $A + H_2$ fall into two groups. When
AH_2 is not a stable molecule, (H_3, FH_2) the reaction proceeds by
abstraction whereas if AH_2 is stable (H_2O, H_2S, CH_2), the reaction
involves an insertion. Physical arguments are given that the H/D atom
ratio in reactions $A + HD$ will be less than one for abstractions and
greater than one for insertions. Experimental results for $F, O(^1D)$ and
$S(^1D)$ are in agreement.

P. M. Rentzepis and C. Capellos (eds.), Advances in Chemical Reaction Dynamics, 21–39.
© 1986 by D. Reidel Publishing Company.

$$A + H_2 = AH + H$$

$$H + D_2 = HD + D$$

The H + H_2 reaction is the most central reaction in chemical dynamics. However unless one observes directly a para-ortho transition on collision it is hard to observe reactive collisions in this system. The next best thing, an isotopic variation of the reaction has recently been studied by laser methods.[1,2] The H atoms were obtained by photolysis of HI at 266 nm. The photolysis produces two kinds of iodine atoms in a ratio of 1:2 , those with and without 7603 cm^{-1} (21.7kcal/mole) of electronic energy. Correspondingly there are slow and fast hydrogen atoms produced with kinetic energies of 15.2 and 36.8 kcal/mole in the laboratory system. If the hydrogen atoms are colliding with D_2 molecules which are virtually at rest, then the kinetic energy of relative motion of the H and D_2 is 4/5 of the hydrogen atom kinetic energy in the lab system. The balance of the energy is the translational energy of the center of mass. Thus the kinetic energies of relative motion are 12.2 and 29.4 kcal/mole.

An extensive calculation of the potential surface of three hydrogen atoms[3] shows that the threshold energy for this reaction is in the collinear approach at 9.7 kcal/mole. Thus in principle some of the minority slow hydrogen atoms could be reactive. However the barrier rises rapidly as the HHH angle increases from 0 degrees. For a perpendicular approach the barrier to reaction is 62 kcal/mole.[4] Thus we can safely neglect the slow hydrogen atoms which can only react when their relative velocities make only a small angle with respect to the D_2 molecule. Even the fast hydrogen atom is restricted in its reaction possibilities. When the H atom approaches at an HDD angle of greater than 90 degrees, the barrier will be higher than 29.4 kcal/mole and reaction will be impossible. In other words sidewise collisions can not be reactive. If a H atom travelling with the speed attained by dissociation of HI at 266 nm were to strike a D_2 molecule perpendicularly at its center and were it able to react, the HD product would have a rotational quantum number J = 8. In fact a maximum in the angular momentum distribution of the HD product molecules occurs for J < 4 when v = 0, J = 4 when v = 1 and J = 2 when v = 2.[1,2] This is an experimental proof of the theoretical prediction that end on collisions are more likely to be reactive than sidewise collisions.

$F + H_2 = HF + H$

The reaction of fluorine atoms with hydrogen molecules has probably been more extensively studied by chemical physicists than any other reaction. The central attraction has been that the reaction exothermicity, 31.7 kcal/mole is mainly channelled into vibrational energy of the HF product. The population inversion is so great that the system can be made to lase in the ir at the HF vibrational frequency. Also important is its theoretical simplicity approaching that of $H + D_2$. An ab initio calculation has been made of the potential energy surface.[5,6] Infrared chemiluminescence studies have revealed the distribution over v and J states (except for v = 0) of the HF product.[7] Rate constants have been measured as a function of temperature[8,9] and isotopic substitutions i.e. D_2 or HD for H_2.[10] Extensive crossed molecular beam studies have been reported.[11] Classical trajectory calculations have been carried out on a variety of surfaces.[12,13]

The potential surface for the $F + H_2$ reaction has a barrier of 1.6 kcal/mole for the collinear approach and 13.6 kcal/mole for a symmetric sidewise approach. This theoretical preference for the end on collision is confirmed by the observation that the HF products are formed in relatively low J states. Fig.1 contains two triangular plots of the fractions of the reaction exoergicity released as translation(T),vibration(V) and rotation(R). For $H+D_2$ the energy is mainly released as T whereas for $F+H_2$ it mainly appears as V. In both cases the rotational energy release is small.

$O(^1D) + H_2 = OH + H$

The potential surface for the $O(^1D)$ reaction has a slight barrier of 0.8 kcal/mole in the collinear system but is strongly attractive for the sidewise approach. The OH radicals produced in this reaction are strongly rotationally and vibrationally excited.

There thus appear to be two classes of potential surfaces for reactions of the type $A+H_2$. There are those such as $H + H_2$ and $F + H_2$ for which the potential barrier rises steeply as the approach of the H atom deviates from collinearity; this kind of surface favors the collinear reaction which is characterized by a backward motion of AH in the center of mass system. The atom A "abstracts" an H atom from the H_2 molecule. The other class of potential surface is characterized not only by the absence of any barrier in the perpendicular approach but rather a deep well. This well may capture transiently the H_2 molecule forming an AH_2 molecule which will bend violently, undergoing inversion and ultimately dissociate into AH and H. Such a reaction is

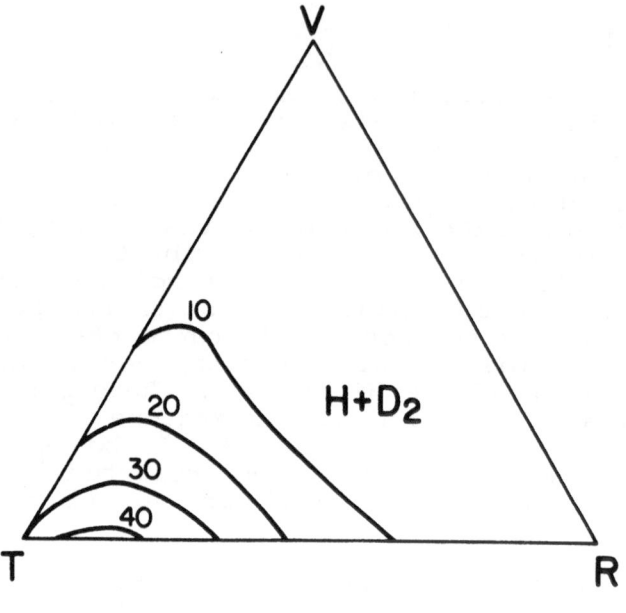

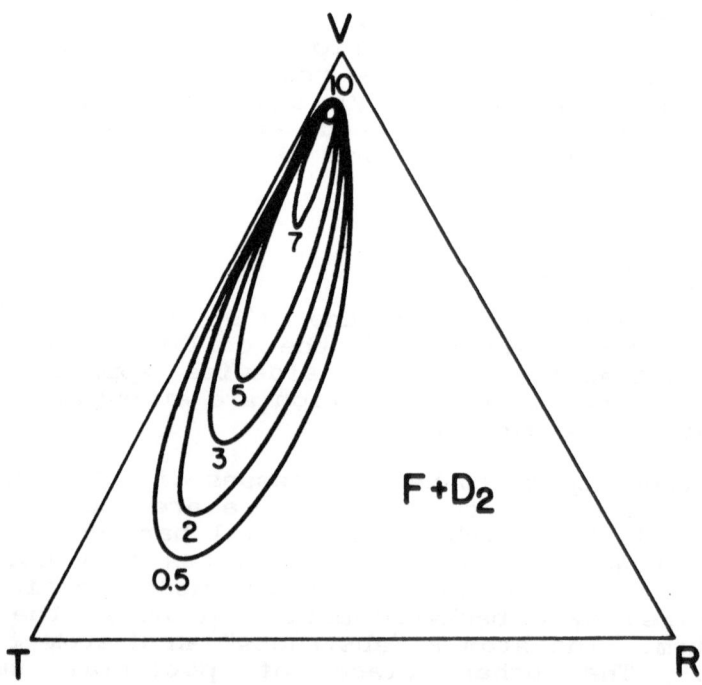

Fig.1 Triangular plots of the energy release in the reactions $H + D_2$ and $F + D_2$. The "contours" of equal probability are interpolated between discrete quantum states.

called an insertion reaction. The reason for this difference of behavior is clear on chemical grounds. The abstraction reactions of H_2 with an atom A having a single unpaired electron take place on an electronic surface which has no potential minimum at finite distance e.g. H_3 and H_2F are not stable molecules. On the other hand when the atom A has two spin paired electrons each in a different spatial orbital the reaction takes place on a surface with a finite minimum corresponding to a stable molecule e.g. H_2O. In this case, insertion to form the transient molecule is the typical event.

An observable which usually distinguishes between a direct and an indirect (intermediate complex) reaction is the differential reactive cross section. In a direct recoil reaction the heavy product is scattered into the backward hemisphere, i.e., opposite to the initial velocity of the heavy atom. When a reaction complex is formed which lasts at least as long as a rotational period, the differential cross section is symmetric[15] between forward and backward directions. As Buss et al.[15] have remarked, the converse is not necessarily true. That is, when an atom makes a lateral attack on a homonuclear diatomic molecule, there will be forward backward symmetry in the differential cross section regardless of the duration of the reaction complex. The observation of this symmetry as in the $O(^1D) + H_2$ reaction[15] is nevertheless a valuable guide to the structure of the transition state.

Another aspect of these fundamental reactions to which relatively less attention has been directed is the isotopic branching ratio of the reaction
$$A + HD = AH(D) + D(H)$$
Laser techniques have provided the means to measure the H/D ratio directly at a time so early that the product hydrogen atom does not have time to collide after it is formed. Recently, Wallenstein[16] has shown that Lyman alpha radiation at 121.6 nm can be generated by focusing 364.8 nm light in krypton gas. It is now convenient to use a laser induced fluorescence (LIF) or multiphoton ionization scheme to measure the relative abundance of H and D atoms whose absorption lines are only 22 cm^{-1} apart. Typical spectra are shown in Fig.2. The H/D ratios for the reactions of $F, O(^1D)$ and $S(^1D)$ with HD are $0.66\pm0.10, 1.13\pm0.08$ and 1.91 ± 0.10 respectively[40].

Muckerman has shown by classical trajectory studies on semiempirical potential energy surfaces that at low energies fluorine atoms form HF in approximately collinear collisions and the product HF recoils backward in the center of mass system[12] On the other hand, for ^1D atoms with deep potential wells, at low energies it is found that most of the reactive collisions take place by the entrance of the reactants into a strongly attracting region[14]. The bending vibration of

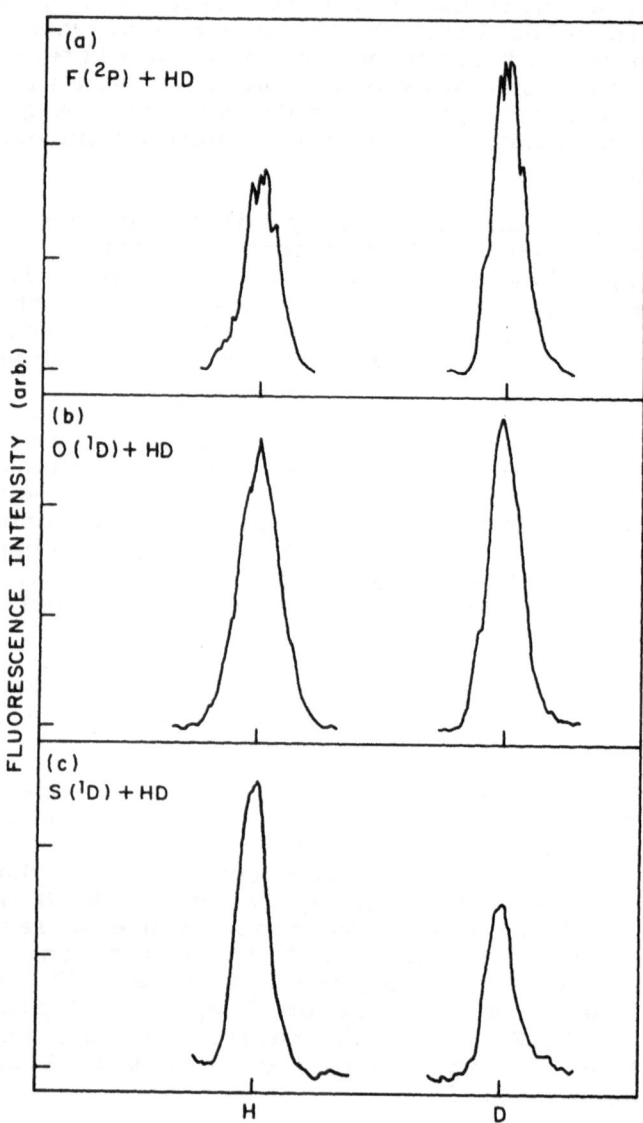

Fig.2 Fluorescence excitation spectra of the H and D atoms products of reactions A + HD. The H and D peaks are 22cm^{-1} apart.

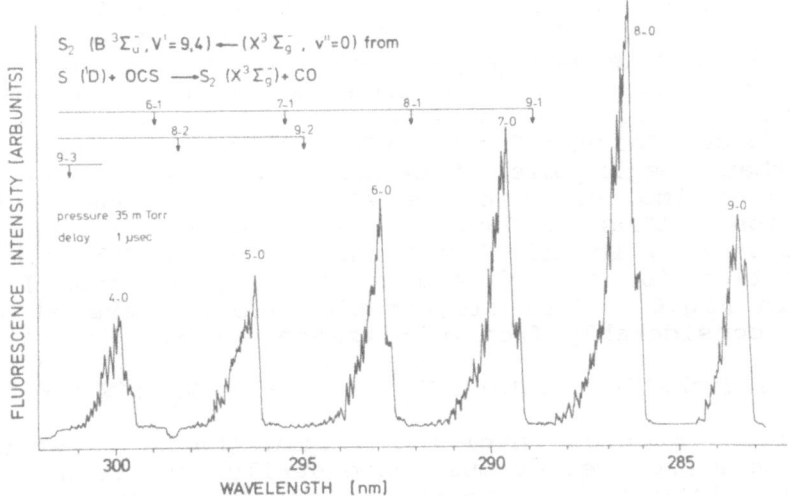

Fig.3 LIF spectrum of $S_2(X^3\Sigma_g^-)$ molecules one microsecond after the production of the $S(^1D)$ atoms. The individual transitions(v',v'') are indicated at the bandheads. The arrows show the bandhead positions of transitions from $v'' = 1,2,3$ which are obviously absent.

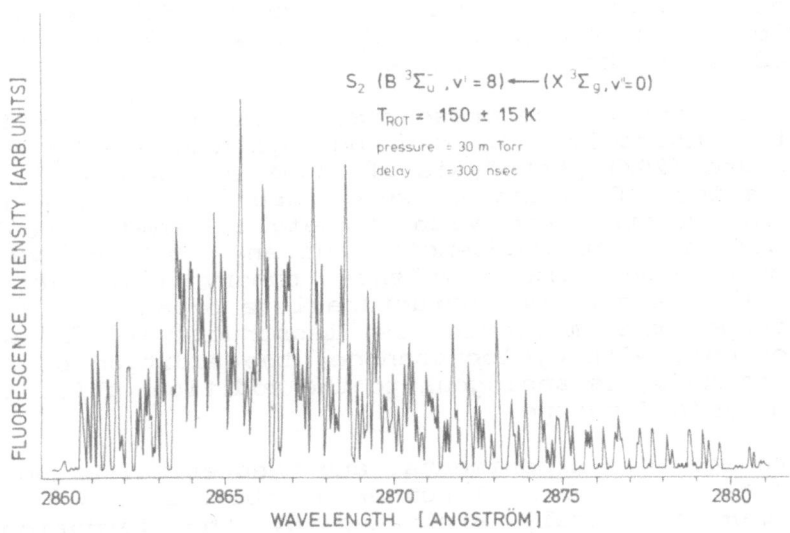

Fig.4 LIF spectrum of the $^3\Sigma_u^-(v'=8)$ -- $^3\Sigma_g^-(v''=0)$ transition in S_2 obtained under single collision conditions.

between the upper(f) and lower(a) states, the Franck-Condon factors of lines originating from the ground state are all very small in the wavelength region which was accessible in the experiment. The (5,1) line is the most intense line originating from v"=1 and was studied in detail. The Franck-Condon factors for the v"=2 states are larger and the fact that negligible intensity was seen for these transitions implies that the v" = 2 state has negligible population. Thus we conclude that there is appreciable population only in the v" = 0 and 1 states. The rotational distribution for the v" = 1 state inferred from Fig.5 is shown in Fig.6. This distribution which peaks at J = 65 differs considerably from a Boltzmann curve.

The diabatic reaction $S(^1D) + OCS = S_2(X^3\Sigma^-) + CO$

A chemical reaction involving a spin change in which only light atoms are present must necessarily involve a collision complex. This rather sweeping generalization follows from the specific example discussed by Tully[23] of the large rate constant for the process $O(^1D) + N_2 = O(^3P) + N_2$. A simple atom-atom like collision does not provide enough time for the comparatively weak spin-orbit interaction to effect a change of state. Otherwise put, when the system passes only twice through the intersection of two potential surfaces, there is a much lower probability of surface crossing than when there are many passages. This will happen when the trajectories of the reactants become highly tangled so that the collision lasts longer. Collision complexes will be formed if there is a deep potential well between the reactants. Thus, in Tully's example, N_2O is a stable molecule and in the present case we will argue by analogy that COS_2 is a stable molecule.

Most germane to the present study is the observation[24,25] that the molecule CO_3 can be synthesized by the low temperature (20K) photolysis of ozone in a solid CO_2 matrix. Five distinct frequencies were observed in the infrared spectrum. A structure with threefold symmetry would have only four distinct frequencies but any other structure for CO_3 would have six. Valence theory calculations[26-28] suggest that the proper structure is a planar C_{2v} form. In this shape the molecule is isostructural as well as isoelectronic with cyclopropanone; the addition of $O(^1D)$ to CO_2 to form CO_3 is analogous to the addition of CH_2 to CH_2CO to form cyclopropanone.

There are then experimental and theoretical arguments for the existence of CO_3. A number of observations in the gas phase can be nicely explained by the formation of a transient singlet state CO_3. For example, isotopic exchange of oxygen atoms in the ^1D state with oxygen atoms in CO_2 is facile but $O(^3P)$ atoms do not exchange[29,30].

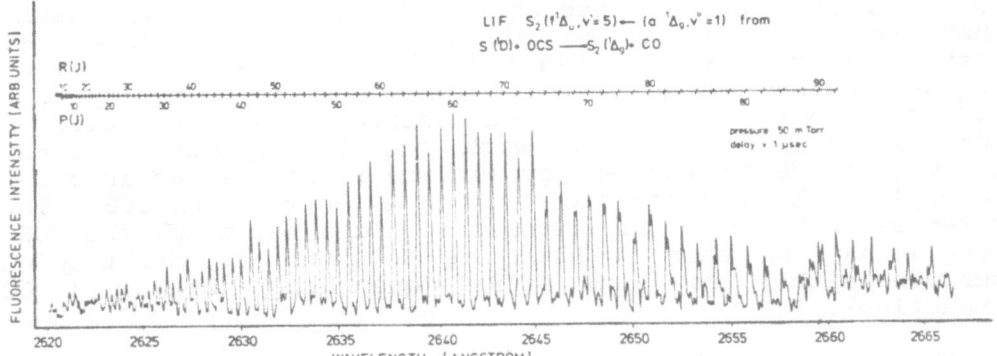

FIG. 5. LIF spectrum of the $(f^1\Delta_u, v' = 5) \leftarrow (a^1\Delta_g, v'' = 1)$ transition in S_2 obtained under single collision conditions ($P_{OCS} = 50$ nm Torr and a delay of 1μ ns.

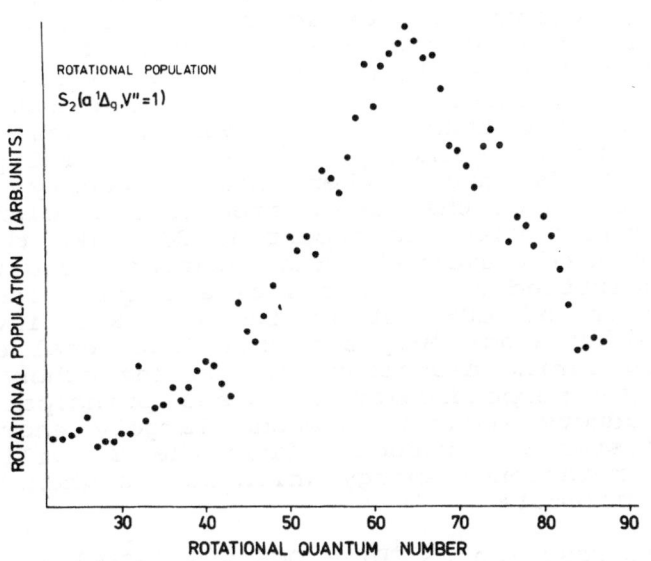

Fig.6 Rotational State Distribution of the $S_2(a^1\Sigma_g(v''=1))$ product inferred from data of Fig.5.

Let us construct a correlation diagram for the molecule COS_2 and its various fragments. First of all, we assume that there is a potential minimum for this molecule i.e. that it exists and is at least quasistable. In analogy with CO_3 we further assume that the minimum occurs for a planar C_{2v} structure. Fig.7 is a suggestive correlation diagram for the COS_2 system. As with the CO_3 system a slight maximum in the triplet energy surface is postulated in order to explain the low reactivity of $S(^3P)$ atoms with OCS. The potential barrier in the singlet exit channel traps the COS_2 long enough so that it has time to cross to the triplet surface either to form $S(^3P)$ (quenching) or S_2 + CO (reaction).

The experimental result on the diabatic reaction is that the S_2 product is vibrationally and rotationally cold. These facts have a simple relation to the C_{2v} COS_2 molecule. The intersection of the triplet and the singlet surfaces is postulated to be at an S-S distance which is close to the average bond distance in the $X^3\Sigma_g$ S_2 molecule, i.e. 1.89A. This provides a natural explanation of the lack of vibrational excitation. The lack of rotational excitation implies that the repulsive forces acting during separation of products are directed at the center of mass of S_2. This could happen, in principle, in a linear geometry but there are no acceptable bond structures that one could draw for a linear molecule nor are there isoelectronic analogs. Thus we are left with the C_{2v} geometry and the clear implication that both C-S bonds are broken simultaneously. The threefold coordination of the carbon atom implies planarity which in turn implies that the departing CO, like the S_2, will not be rotationally excited. The rotational cooling of the S_2 can be understood on the basis of a simple classical argument. Rotation of COS_2 about its C_2 axis involves motion of the sulfur atoms only and this rotational energy will be conserved during dissociation. On the other hand, rotation about the perpendicular in plane principal axis will, as the products separate, become largely energy of motion of the faster CO product. Thus the S_2 will lose roughly half the rotational energy which it had when it was part of the COS_2 molecule.

The adiabatic reaction $S(^1D)$ + OCS = $S_2(a^1\Delta)$ + CO

The $S_2(a^1\Delta_g)$ product of the adiabatic reaction has a vibrational distribution which is probably peaked at v=0 but has an appreciable value at v=1. In view of the large exothermicity, the small amount of vibrational excitation found in the newly formed S_2 bond is surprising. In many abstraction reactions with univalent atoms (e.g., $K+CH_3I$, $F+H_2$) the majority of the exoergicity is released into the newly formed bond. The prior expectation in the present case is that equal amounts of vibrational energy should appear in CO and S_2. The present reaction in which the newly formed S_2 has less than its a priori share strongly suggests an extensive redistribution of energy which is done by means

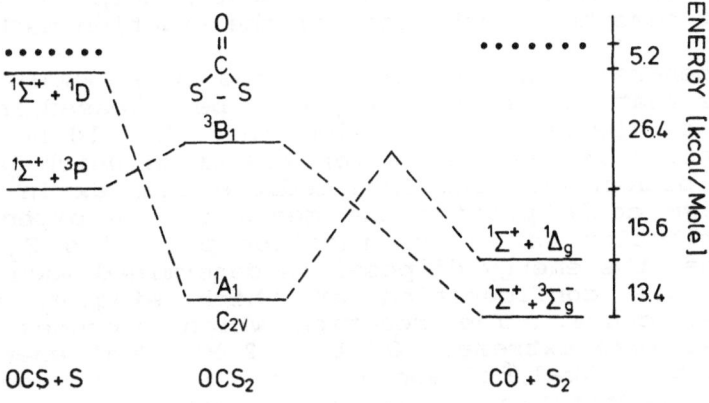

CORRELATION DIAGRAM FOR THE LOWEST ELECTRONIC STATES
IN THE REACTION OCS + S → CO + S$_2$

Fig.7 Correlation diagram for the lowest electronic states in the reaction of OCS + S = S$_2$ + CO. The dots indicate the relative kinetic energy of the reaction partners in the center of mass system.

of a reaction complex. Just as with the diabatic reaction we assume that the reactive collision begins with an addition of the divalent $S(^1D)$ atom to the C=S bond to form a COS_2 intermediate. We have previously postulated that this complex undergoes intersystem crossing in a symmetrical C_{2v} configuration and therefore both C-S bonds are broken in a concerted way. The complex is postulated to dissociate on its singlet surface by two consecutive C-S bond ruptures.

The narrow rotational distribution of the $S_2(a^1\Delta_g v=1)$ peaks at J=65. It does not have a Boltzmann shape and therefore cannot be parametrized by a rotational temperature. However J=65 for $S_2(a^1\Delta_g v=1)$ whose B value is $0.29cm^{-1}$ corresponds to 3.6 kcal/mol. The model of consecutive bond breakage which we propose would predict that the CO should also be rotationally excited. The rotational distribution of CO would help to clarify our present somewhat rough model of the reaction mechanism.

In general, for reactions of the type A+BCD = AB+CD, one expects that much more energy will be released into internal energy of the new bond AB than into the old bond CD. This expectation is reasonable for reactions which proceed by a direct mechanism. The intermediate complex in an indirect mechanism could provide the means for an extensive energy redistribution as in the reaction producing $S_2(X^3\Sigma)$. In this case the energy disposal is determined much more by the geometrical configuration at which singlet and triplet surfaces cross. The reaction which produces $S_2(a^1\Delta)$ is somewhat less extreme. Of the 42 kcal/mol energy release, typically ~3 kcal/mol appears as vibrational energy and 3-4 kcal/mol as rotational energy. In both cases, however, only a minority of the energy is released as internal energy of the new AB molecule.

$$F + RI = RF + I$$

In the two preceding sections applications of laser chemistry were given to three and then four center reactions. Now we consider a more complex reaction with at least six centers for which the main contribution of laser techniques is to detect a species at such an early time that it must be a primary reaction product. When fluorine atoms attack methyl iodide molecules, three reactions occur:

$$F + CH_3I = CH_3 + IF \qquad \Delta H_0 = -10.3 \text{ kcal/mol} \qquad 1)$$

$$F + CH_3I = CH_2I + HF \qquad \Delta H_0 = -36.5 \text{ kcal/mol} \qquad 2)$$

$$F + CH_3I = CH_3F + I \qquad \Delta H_0 = -70.9 \text{ kcal/mol} \qquad 3)$$

The three possible reactions provide a classic confrontation between statistical expectation and potential barriers. The most exothermic reaction has the greatest number of final states and is therefore statistically the most likely to occur. In fact the rate constant for reaction 3 is two orders of magnitude smaller than that for reaction 1 which is the dominant one. Reaction 1 has been studied by crossed molecular beams[31] and by laser induced fluorescence of the IF product[32]. Reaction 3 has been studied by experiments done with mixtures of SF_6 and CH_3I in two remarkably different experiments[33,34]. In one case the gas at 4 atm pressure was irradiated with high energy neutrons. The resulting (n, 2n) reaction produced ^{18}F atoms which were thermalized in the high presure gas in which CH_3I was a very minor component. The absolute rate constant for each of the three reactions was determined by radiochemical analysis of the products. In the second experiment the same gas mixture was irradiated but this time not with neutrons but with a CO_2 laser. Moreover the pressure was only 10^{-4} atm and the SF_6 and CH_3I pressures were comparable. The iodine atom product of reaction 3 was detected within one microsecond after its formation by a two photon laser induced fluorescence technique[35] In this method two photons are absorbed at 304.7 and 306.7 nm by the $^2P_{3/2}$(I) and the $^2P_{1/2}$(I*) atoms, respectively. The excited iodine atoms emit[2] an infrared photon followed by a vacuum ultraviolet photon, which is detected. The resulting fluorescence intensities for the I and I* atoms are shown in Fig.8 for methyl, ethyl, isopropyl and tert-butyl iodide. (Besides the iodine atoms produced in reactive collisions, iodine atoms are also produced during the probing pulse by a two photon absorption by the alkyl iodide.) From the relative intensities one sees that a tertiary iodide does not give iodine atoms, a secondary iodide does give some and a primary iodide gives much more. An exactly parallel behavior is shown by the Walden inversion in solution. Strong laser induced fluorescence of IF was seen for all the

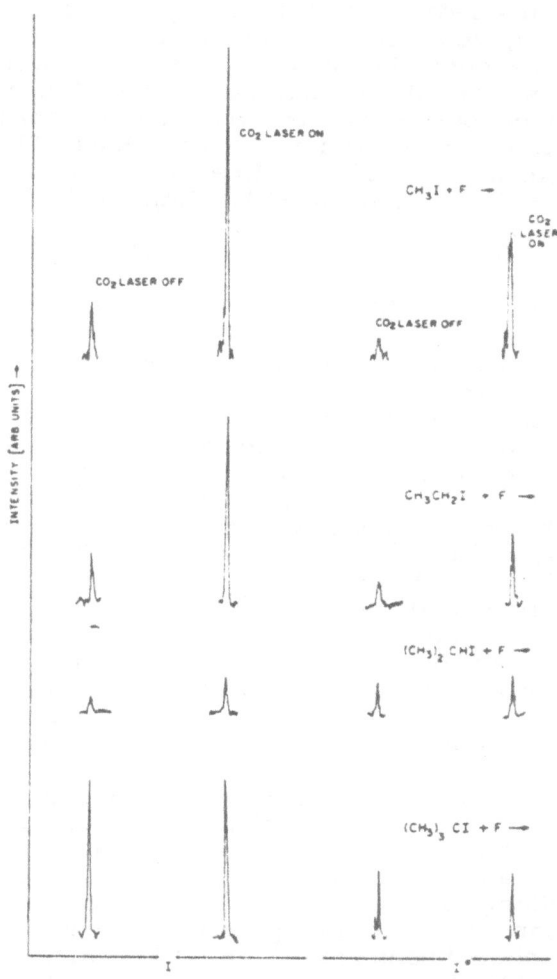

Fig.8 Fluorescence excitation signals of I and I* in the
presence and absence of fluorine atoms generated from SF_6 by
a CO_2 laser.

alkyl iodides discussed here and also CF_3I but iodine atom signals were not seen for t-BuI and CF_3I.

Reaction 3 is an extremely exothermic reaction. It is therefore reasonable to assume that some of the decline in potential energy takes place already in the entrance channel; in other words the fluorine atom is attracted to the back side of the CH_3I molecule just as it is to the front. Supporting this assumption is the fact that when excess argon was added to the reaction mixture to thermalize the fluorine atoms, no effect was seen on the I atom yield. (The F atoms photodissociated from SF_6 move relatively slowly but an appreciable fraction have kinetic energies of 2-4 kcal/mol[36].) However when sufficiently bulky groups are present, the fluorine atom can not approach closely enough to the carbon atom to feel its attraction and there is a barrier in the entrance channel.

How can one prove that the reaction producing iodine atoms is truly a Walden inversion i.e., that the F atom approaches from the back side of the carbon atom? A classical proof would begin with an optically active iodide and show that the product fluoride had the opposite configuration. A direct proof would be to carry out the reaction with beams of F atoms and oriented CH_3I molecules[37,38] and to show that the reaction cross section increased when the CH_3 group was oriented toward the F atom beam. Here we rely on a less rigorous argument. The reaction does not take place with an alkyl iodide such as tert-butyl iodide that is strongly hindered on its back side. Neither does it take place with CF_3I. On the other hand the IF-producing reaction appears to take place with all alkyl iodides. Thus we conclude that reactive attack of the F atom on the I atom side of the molecule always produces IF and does not result in the displacement of an I atom. Attack from the rear dislodges an I atom only when the steric hindrance is not too great.

Conclusion

Examples have been given of the measurement in the microsecond time domain by laser techniques of concentrations, isotopic product ratios, final vibrational and electronic distributions and above all of final rotational distributions. We not only know what the products really are but we are gradually learning just how the products are formed i.e. the structure of the transition state.

Thanks are due to the U.S.Department of Energy for support and to the Weizmann Institute, Rehovot,Israel where this review was written.

References

1. D.P.Gerrity and J.J.Valentini,
   J.Chem.Phys. 79, 5202 (1983)
2. E.E.Marinero, C.T.Rettner and R.N.Zare,
   J.Chem.Phys. 80, 6142 (1984)
3. P.Siegbahn and B.Liu, J.Chem.Phys. 68, 2457 (1978)
4. D.G.Truhlar and C.J.Horowitz
   J.Chem.Phys. 68, 2446 (1978); E71, 1514 (1979)
5. S.R.Ungemach, H.F.Schaefer and B.Lice, Disc.Farad.Soc.
   62, 330 (1976) and references therein
6. M.J.Frisch, B.Liu, J.S.Binkley, H.F.Schaefer and
   W.H.Miller Chem.Phys.Lett. 114, 1 (1985)
7. D.S.Perry and J.C.Polanyi, Chem.Phys. 12, 419 (1976)
   and references therein
8. E.Wurzberg and P.L.Houston, J.Chem.Phys. 72, 4811 (1980)
9. R.F.Heidner, J.F.Bott, C.E.Gardner and J.R.Melzer,
   J.Chem.Phys. 72, 4815 (1980)
10 A.Persky, J.Chem.Phys. 59, 3612, 5578 (1973)
11. D.M.Neumark, A.M.Wodtke, G.N.Robinson, C.C.Hayden and
    Y.T.Lee, J.Chem.Phys. 82, 3045 (1985)
12. J.T.Muckerman, Theor.Chem.Adv.Perspect. 6A, 1 (1981)
    and references therein
13. D.G.Truhlar, B.C.Garrett and N.C.Blais,
    J.Chem.Phys. 80, 232 (1984)
14. P.A.Whitlock, J.T.Muckerman and P.M.Kroger, in Potential
    Surfaces and Dynamics Calculations, D.G.Truhlar, ed.
    (Plenum, New York, 1981) p.551
15. R.J.Buss, P.Casavecchia, T.Hirooka, S.J.Sibener and
    Y.T.Lee, Chem..Phys.Lett. 82, 386, (1981)
16. R.Wallenstein, Opt.Commun. 3 3, 119 (1980)
17. K.S.Sorbie and J.N.Murrell,
    Mol.Phys. 29, 1387 (1975); 31, 905 (1976)
18. P.A.Whitlock, J.T.Muckerman and E.R.Fisher,
    J.Chem.Phys. 76, 4468 (1982)
19. R.E.Howard, A.E.Maclean and W.A.Lester,
    J.Chem.Phys. 71, 2412 (1979)
20. R.Schinke and W.A.Lester, J.Chem.Phys. 72, 3754 (1980)
21. S.W.Ransome and J.S.Wright, J.Chem.Phys. 77, 6346 (1982)
22. K.S.Sidhu, I.G.Csizmadia, O.P.Strausz and H.F.Gunning,
    J.Am.Chem.Soc. 88, 2412 (1966)
23. J.Tully, J.Chem.Phys. 61, 61 (1974)
24. N.G.Moll, D.H.Clutter and W.E.Thompson,
    J.Chem.Phys. 45, 4469 (1966)
25. E.Weissberger, W.H.Breckenridge and H.Taube,
    J.Chem.Phys. 47, 1764 (1967)
26. B.M.Gimarc and T.S.Chou, J.Chem.Phys. 49, 4043 (1968)
27. J.F.Olsen and L.Burnelle,
    J.Am.Chem.Soc. 91, 7286 (1969)
28. J.H.Sabin and H.Kim, Chem.Phys.Lett. 11, 593 (1971)
29. M.Yamazaki and R.J.Cvetanovic, J.Chem.Phys. 40, 482 (1964)
30. R.J.Collins and D.Husain,
    J.Chem.Soc. Farad. Trans.2 69, 145 (1973)

STUDIES OF RAPID INTRAMOLECULAR AND INTRAIONIC DYNAMIC PROCESSES WITH
TWO-COLOR PICOSECOND LASERS AND MASS SPECTROMETRY

D. A. Gobeli,[†] J. S. Simon,[*] Diane M. Szaflarski and
M. A. El-Sayed
Department of Chemistry and Biochemistry
University of California
Los Angeles, California 90024 U.S.A.

ABSTRACT. The availability of ultrashort laser pulses, with sufficient
energy per pulse to ionize and fragment a molecule, has permitted us to
develop a technique whereby the rapid dynamics involved in laser multiphoton ionization and dissociation processes may be followed. By using
two picosecond pulses of lasers of different colors (266 nm and 532 nm)
and variable delay between them as a source for a time-of-flight mass
spectrometer, it is possible to monitor the energy redistribution processes occurring in the excited electronic states of the parent molecule
and ion on the picosecond timescale. This is carried out by following
the changes in the observed mass spectrum as a function of the delay
between the two pulses.
     The technique is applied to understand the mechanism of formation
of some ionic fragments from 2,4-hexadiyne and 1,4-dichlorobenzene.

INTRODUCTION

One of the most important properties of pulsed lasers that is utilized
in chemistry today is its short pulse width. Pulses of duration in the
pico and femtosecond are now obtainable. This kind of width is on the
order or shorter than vibration periods in molecules. This makes
pulsed lasers useful in determining the rates of primary processes in
many photochemical and photobiological changes. (For a useful review
of this topic see: P. Rentzepis, Science (1982), 218, 1183; also see
ACS Symposium Series 1983, 236 (Multichannel Image Detectors, Vol. 2)
201-220.
     Another property of lasers not possessed by conventional light
sources is its high intensity. This makes lasers useful in depositing
more energy per unit time than any other conventional light (or any
other energy) source. A molecule like benzene could be degraded down

[†]Present address: Northrop Research & Technology Center, Palos Verdes
                     Peninsula, CA 90274

[*]Present address: Department of Chemistry, University of California,
                     San Diego, La Jolla, CA 92093

P. M. Rentzepis and C. Capellos (eds.), Advances in Chemical Reaction Dynamics, 41–55.
© 1986 by D. Reidel Publishing Company.

to C^+ or C^{++} by pulsed lasers available in many laboratories today with energies of mljouls/pulse. This is accomplished by the absorption of several photons within the pulse duration.

What is the mechanism of the ionization and dissociation of a molecule like benzene to give many fragments such as $(C_nH_m)^+$ where n = 1,2...6 and m = 0,1,...6? How long does it take an excited electronic state (formed by the absorption of light or by colliding with electrons or any other energy source) of a molecule or of an ion to redistribute its energy into vibrations prior to dissociation? In the present work, we have combined the two properties discussed above of pulsed lasers to develop a technique that could give results helpful in answering these questions.

We used two picosecond laser pulses, one at 266 nm and the other at 532 nm, as a source in a time-of-flight mass spectrometer. We studied the effect of delay between the two pulses as well as the intensity of the two pulses on the mass spectra produced. We will demonstrate that information can be obtained regarding the dynamics of some of the excited states reached by the absorption of different numbers of the UV photons that are responsible for the appearance of some daughter ions produced from the low energy dissociation channels. The molecule 2,4-hexadiyne ($H_3C-C\equiv C-C\equiv C-CH_3$) is used for this study, since a great deal of information is known about its energy level structure and that of its parent ion as well as the appearance potential of its daughter fragment ions. Some results on the photoionization photofragmentation dynamics of dichlorobenzene will also be presented.

The Quasi-equilibrium Theory (QET)[1]

Quasi-equilibrium theory (QET)[1] has been traditionally used to explain the ionization and unimolecular fragmentation behavior of molecules when they are subjected to an ionization source whose unit of energy is larger than their ionization potential. Such forms of ionization include 70-eV electron impact (e.g., in mass spectrometry) and single-photon photoionization using the 21.2-eV photons from a helium discharge (e.g., in photoelectron spectroscopy). The theory proposes that ionization occurs rapidly (10^{-15} s) and that because the ejected electron can have varying amounts of kinetic energy, a number of the excited electronic states of the molecular ion are formed. These excited electronic states of the molecular ion undergo rapid energy redistribution (internal conversion) into the vibrationally excited ground electronic manifold before fragmentation occurs. Thus energy redistribution from the excited electronic states occurs prior to dissociation. The latter then proceeds from a statistically equilibrated set of vibration-rotation levels of the parent ion ground electronic state. This theory has been found to be useful in accounting for the observed mass spectra of many large organic molecules.

Mechanisms of Multiphoton Ionization Dissociation

When the unit of energy of the ionization source is smaller than the ionization potential of the molecule, as in the case of laser multiphoton ionization dissociation, several mechanisms have been invoked[2] to explain the mass spectra produced. They all assume that the QET is valid. Very often, the system is excited past the ionization potential by the resonant absorption of several photons. The nature of the resonant intermediate states can have a large effect upon the ionization and fragmentation behavior displayed by the molecule and often causes a marked wavelength dependence.[3] In systems in which a rapid dissociative process effectively competes with the absorption of further laser photons, dissociation of the neutral molecule may occur followed by excitation and ionization of the resulting neutral fragments. This is the DI (dissociation then ionization) mechanism.

For systems in which the resonant intermediate state has no such dissociative channel, ionization to yield the molecular ion may occur through the sequential absorption of several photons. This is the ID mechanism. For systems which are not resonantly excited, ionization occurs by the simultaneous absorption of several laser photons.

Once the ionization potential has been reached, further absorption of photons may occur within the ionic manifold of the parent. In the ladder-climbing mechanism[4] of ionic fragmentation, the molecular ion is the direct precursor of all daughter fragments. Therefore, the absorption of several photons may take place past the appearance potential of low-energy daughter fragments in order to generate higher energy daughter fragments via this mechanism. In the ladder-switching mechanism[5] of ionic fragmentation, the molecular ion need not be the direct precursor of all other daughter fragments. Instead, it may fragment into low-energy daughter fragments which in turn absorb further photons and fragment into higher energy, smaller daughter ions.

Mass Spectrometry and Dynamics of Ionic Dissociation

Conventional mass spectrometry has been used to extensively study unimolecular decomposition processes of ions in the gas phase. The determination of specific rates of gas-phase ionic dissociation processes has, for the most part, been limited to cases that show metastable peaks in the mass spectrum.[6] These peaks, which usually appear as broadened, sometimes asymmetrically distorted features in the mass spectrum at nonintegral m/e values, arise when ionic fragmentation happens to occur on a time scale comparable to the extraction time. This is almost always on the microsecond time scale and is a severe limitation in that it restricts the study to those processes which occur within this narrow range of times. Thus, rapid dissociation processes as well as rapid energy redistribution processes occurring prior to dissociation in the 10^{-15}-10^{-7}-s range remain "out of reach" in conventional mass spectrometric investigations.

Attempts have been made to extend the range in which metastables may be detected for a variety of ionization methods including electron impact,[7] field ionization,[8] and multiphoton ionization-dissociation.[9] Some[9] have relied on altering the geometry of the mass spectrometer from conventional configurations in order to extend the range of detectability of metastable peaks. Others[8] have relied upon special properties of the ionization source to extend the range into

the picosecond region.[8] Some of these techniques[9] are still based
upon the analysis of metastables for the determination of ionic decomposition rates and are therefore limited by the extraction rate of
the instrument.

 We have developed a technique[10] for the determination of ionic
decomposition processes which is not based upon the analysis of metastable peaks and therefore is not limited by any of the constraints
which govern that method. In this technique, two picosecond lasers of
differing wavelengths serve as the ionization source of the mass
spectrometer. The relative delay between these two lasers is a variable parameter. The signals of the different fragment ions are followed as a function of the delay between the two laser pulses. The fast-
est process which this technique is capable of detecting is limited by
the pulse duration of the laser and therefore may be in the tens of
femtoseconds, since these are presently the shortest pulses available
today.[11]

The Technique and the Molecule Selected

 The two-color method is based upon the following idea. Consider
a molecule that ionizes and fragments via the ionic ladder-climbing
mechanism (which is expected to increase in probability as the primary
pump laser pulse width becomes shorter than the ionic dissociation
times). The first pump laser (e.g., the fourth harmonic of a Nd:YAG
laser at 266 nm) is used to excite and ionize the parent molecule and
to populate different electronic states of the molecular ion which are
separated in energy by the energy of at least one laser photon. Some
of these states lie above the appearance potential of daughter fragments whose intensity is constantly being monitored. According to
QET,[1] energy redistribution follows excitation. While this is occurring, a second pump laser (e.g., the second harmonic of the Nd:YAG laser
at 532 nm), delayed from 0 to 10000 ps with respect to the first pulse,
is introduced. It promotes some of the molecular ions to new and
higher electronic states. These new states could lie above the appearance potential of new and higher energy daughter fragments and hence
may result in their preferential formation at the expense of the
daughter ion produced from the UV excited states. If the absorption
cross-section for the secondary pump is different for ions before and
after energy redistribution, the rate of energy redistribution could
be measured from the change in the signals of the different daughter
fragments as a function of the relative delay between the two laser
pulses.

 The molecule chosen for study is 2,4-hexadiyne. The first excited
state of the molecular ion lies above the appearance potential of a
number of daughter fragments.[12] In addition, this is a system in
which radiative decay, and hence the rate of energy redistribution,
could be measured from the first excited state. The radiative lifetime of 20 ns[13] suggests that if quasi-equilibrium theory[1] is
applicable to this system, the energy redistribution rate from this
state should be approximately 20 ns. The photoion-photoelectron coincidence experiments of Danacher[14] suggest that the approximate

quantum yield of fluorescence of the molecular ion in its first excited electronic state is 40% and that the remainder of the system undergoes nonradiative decay. This value was obtained from the parent ion branching ratio. It was assumed that if the system did not undergo a radiative process, the excess internal energy would cause fragmentation.

When the fourth harmonic of the Nd:YAG is utilized as the primary pump laser, it should be possible to populate the first excited state of the molecular ion via the absorption of three laser photons by a one-photon resonant, two-photon ionization followed by the absorption of an additional photon within the ionic manifold. This three-photon state, which lies at 14.0 eV above the ground electronic state of the neutral molecule, is above the appearance potential of five daughter fragments (see Figure 1). Absorption of a 532-nm laser photon (the second harmonic of the Nd:YAG) would promote the system to the 16.33-eV energy level, which is above the appearance potential of several other daughter fragments (see Figure 1). By itself, the 532-nm laser output (0.5 mJ/pulse) is of insufficient intensity to cause ionization. The power of the fourth harmonic (between 1 and 10 µJ/pulse) is sufficient to cause ionization plus a moderate amount of fragmentation resulting mainly in the ions from the three and four-UV-photon states.

Experimental Section

A differentially pumped, time-of-flight mass spectrometer was constructed in our laboratory for the purpose of performing mass analysis following multiphoton ionization-dissociation of gaseous samples. The design of the ionization and extraction region was based roughly on the design of Wiley and McLaren.[15] It provided unit mass resolution well past m/e 120. Sample pressures in the ionization region were generally kept at less than 1×10^{-4} torr; the pressure in the field-free drift region of the mass spectrometer was maintained at about 2×10^{-7} torr. Variation of the pressure in the ionization region over several decades resulted in the detection of no ions with an m/e value greater than that of the molecular ion, indicating that no ion-molecule reactions were occurring.

Ions were generated with the doubled and quadrupled outputs of a Nd:YAG laser, passively mode-locked, producing 25-ps pulses (Quantel YG-471). The 532-nm pulse was generally kept at about 0.5 mJ/pulse; the 266-nm output ranged between 1 and 10 µJ/pulse. The pulse widths of the lasers were determined either by autocorrelation or by the MPI pulse width determination method.[16]

An adjustable delay ranging from 0 to 10000 ps was introduced between the 266-nm and 532-nm lasers by first separating them with a Pellin-Broca dispersion prism and then recombining them after the 532-nm pulse traversed an optical delay line with a dichroic mirror.

An electron multiplier (Galileo Electrooptics 4800 series channeltron) was used as the ion detector. The resulting signal was amplified by a X100 video amplifier (Pacific Precision Instruments), monitored on an oscilloscope, and/or sent to a boxcar integrator and signal analyzer (PAR Model 4420, 4422, and 4402). The boxcar integrator was set to average 10 laser shots per point. There were

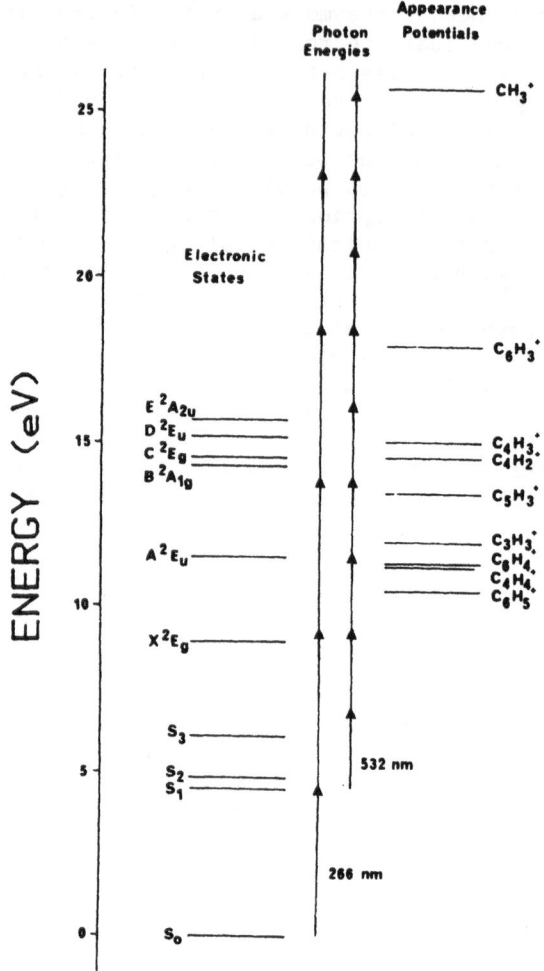

Figure 1. Energy level diagram of 2,4-hexadiyne

1024 points collected in a 10-ns delay range, meaning that approximately 10000 laser shots were fired per scan.

RESULTS

General Observation and the Breakdown Curve

The breakdown curves of hexadiyne[14] are shown in Figure 2. These curves give the relative probabilities of formation of the different daughter fragments as the internal energy imparted to the molecule increases. Since 2 photons of 266 nm laser ionize with very small

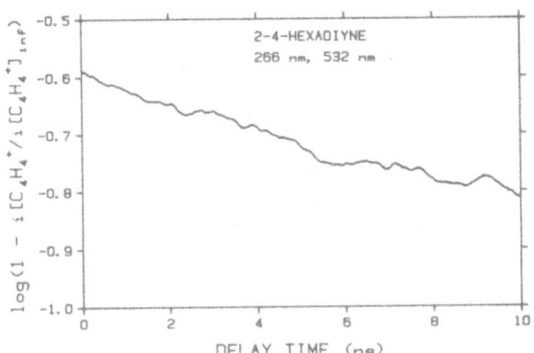

Figure 4. Plot of the decay of the precursor of the $C_4H_4^+$
daughter as a function of delay time between the two
laser pulses. The characteristic time is determined
to be about 20 ns, which is comparable to the observed
fluorescence lifetime of the X_2E_u state of the
parent ion of 2,4-hexadiyne.[13] This suggests that
this state is involved (directly or indirectly) in the
mechanism for the formation of $C_4H_4^+$.

The Effect of Delay on the $C_4H_4^+$ and $C_4H_3^+$ Currents:

As discussed earlier, the absorption of the 532 nm photons by molecules
in the 3-UV photon level should decrease the probability of formation
of $C_4H_4^+$ and increase that for the $C_4H_4^+$ fragment (see Figures 1, 2
and 3). As the delay between the green and UV laser pulses increases,
more ions in the 3-photon level can redistribute their electronic
energy and eventually dissociate into $C_4H_4^+$. Thus the mass peak of
$C_4H_4^+$ should increase and that of $C_4H_3^+$ decrease as the delay between
the two lasers increase, with a characteristic time corresponding to
the lifetime of the 3-photon level (t_1). Figure 5 shows these results.

The Temporal Behavior of the $C_6H_5^+$ Mass Peak, More Than One Isomer:

According to Figures 1 and 2, $C_6H_5^+$ should behave similar to $C_4H_4^+$,
i.e., it should originate from a level having a 20 ns lifetime.
Figure 6 (top) shows clearly that (at least) two components are
present. One has a long risetime of ~ 20 ns and the other has a
short risetime of ~ 200 ps. We propose that the short one arises from
another higher energy $C_6H_5^+$ isomer which is produced from the 4-photon
level. If this is the case, this would be the first time in which
kinetic results give indication that a mass peak represents more than
one isomer.

In support of this explanation is the fact that the component with

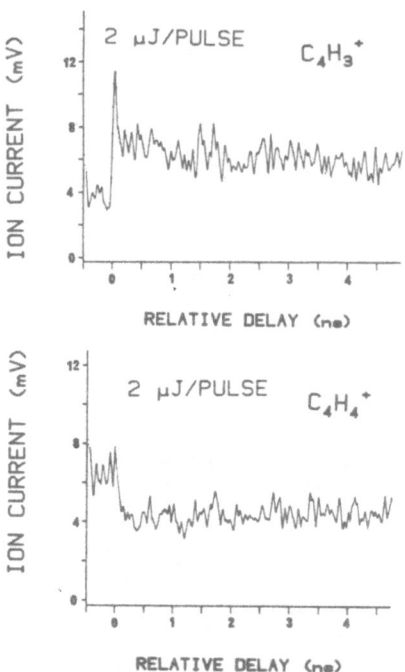

Figure 5.   The $C_4H_3^+$ and $C_4H_4^+$ ion current vs. the
relative delay time between the 266-nm and 532-nm laser
pulses. At zero delay, The $C_4H_3^+$ ion current
rises, followed by a slow decay of characteristic
time 20 ± 5 ns and the $C_4H_4^+$ ion current falls,
followed by a slow rise of characteristic time 20 ± 5 ns.
This suggests that the $C_4H_3^+$ fragment arises
from the absorption by the molecular ion of a single
green laser photon from the three UV photon state.
This conclusion is consistent with the known appearance
potential information on these ions which suggests that
the $C_4H_4^+$ ion (AP = 11.27 eV) may be formed after
the absorption of three UV laser photons, but the
$C_4H_3^+$ daughter fragment (AP = 14.05 eV) requires
the absorption of three UV plus one visible laser photon
for its formation.

the fast risetime seems to increase in relative amplitude as the UV
intensity increases (see Figure 6).

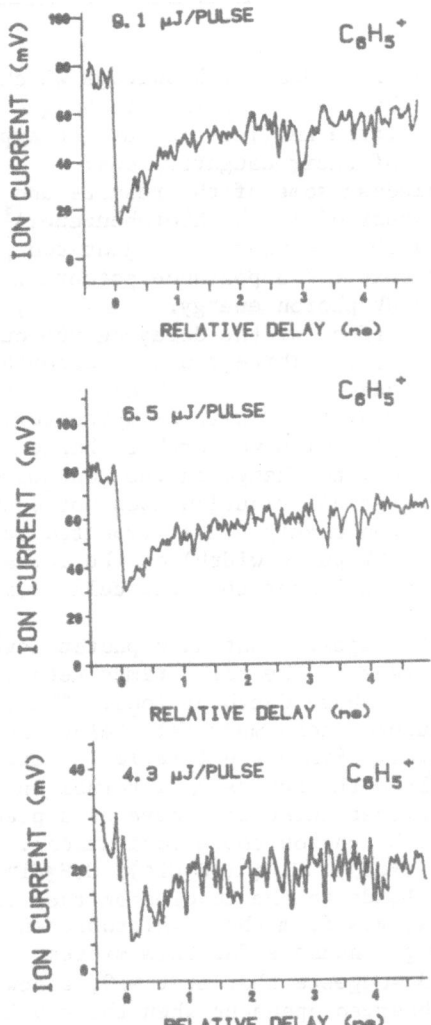

Figure 6. The ion current vs. relative delay time curve for
$C_6H_5^+$ taken at different UV laser intensities.
the graphs generated at the higher UV laser intensities
(top, middle) are clearly biexponential, with a fast
300 ± 100 ps component and a slower 20 ± 5 ns component.
Because the relative amplitudes of these two differing
decay rates are dependent upon the UV laser power, it
is suggested that the slow time arises from the three
UV photon state at 14.0 eV (the same as that for the
$C_4H_4^+$ ion) and that the faster component
arises from the four UV photon state at 18.66 eV above
the neutral ground state.

<u>General Applications of the Technique, Dynamics of Dissociation of</u>
<u>1,4-dichlorobenzene:</u> [17]

In hexadiyne, we have a great deal of knowledge which assisted us in
describing the results obtained. Can this technique be useful to apply
on molecules with very little knowledge about the appearance potential
or the breakdown curves of their daughter ions?

Below, we shall discuss some of the results and conclusions that
are obtained from the study of 1,4-dichlorobenzene. [17] Two UV photons
ionize this molecule by the one photon resonant-two photon ionization
process. We only know that the appearance potential of the $C_6H_4Cl^+$
daughter is below the 3 UV photon energy.

Figure 7 shows the effect of the delay on the current corresponding to the different masses.   Three types of behavior are observed:
1) an increase or a decrease at t = 0 with no further change at t > 0,
2) a decrease or increase at t = 0 with slow recovery at t > 0, and 3)
a spike at t = 0 for $C_6H_4Cl^+$ with no further change at t > 0.

Type 1 behavior (i.e., no change in the ion current signal for
t > 0) suggests that energy distribution does not occur during the
delay times used. This can result either from too rapid energy distribution (e.g., within the UV pulse width) or the formation from a precursor with a lifetime much longer than the delay time used (e.g., the
parent ion).

Behavior of type 2 suggests that an n-photon level with characteristic lifetime on the order of the delay times used is involved in the
mechanism of formation of these daughter ions. Thus, the appearance
potentials of these daughter ions must fall below the energy of this
particular n-photon level. Since the lifetime of excited electronic
states decreases rapidly with increasing internal energy, it is probably correct to assume that these ions have as a precursor the 3 UV
photon level (i.e., one UV photon above ionization). Thus the appearance potential of $C_6H_4^+(+Cl_2)$, $C_4H_4^+(+C_2Cl_2)$, $C_4H_3Cl^+(+C_2HCl)$, are
all below 14 eV. This leads to the conclusion that the elimination of
Cl_2 or acetylene derivatives from the dichlorobenzene parent ion constitute the lowest energy channels for this system.

Behavior of type 3 suggests that at t = 0, a new channel for the
formation of the ion observed opens up when the two lasers are overlapping in time (and of course in space). One such a mechanism in
dichlorobenzene is that when molecules in S_1 absorb one photon of green,
the rate of their dissociation could be calculated to be within the
pulse width. The radical produced $(C_6H_4Cl\cdot)$ could then absorb
additional UV and/or green photons to ionize. This should lead to a
decrease in the parent ion. This is also observed.

Figure 8 shows the temporal behavior of the signal corresponding
to the parent and $C_6H_4Cl^+$ ions. As one increases, the other decreases.
The increase in the signal of the daughter ion is observed tens of
picoseconds earlier than the decrease in the parent ion signal. An
understanding of this on a quantitative basis would require the calculation of the rates of the absorption of the different laser pulses
to yield the two ionic species as a function of the changing intensity
of the overlapping laser pulses as their delay changes around t = 0.

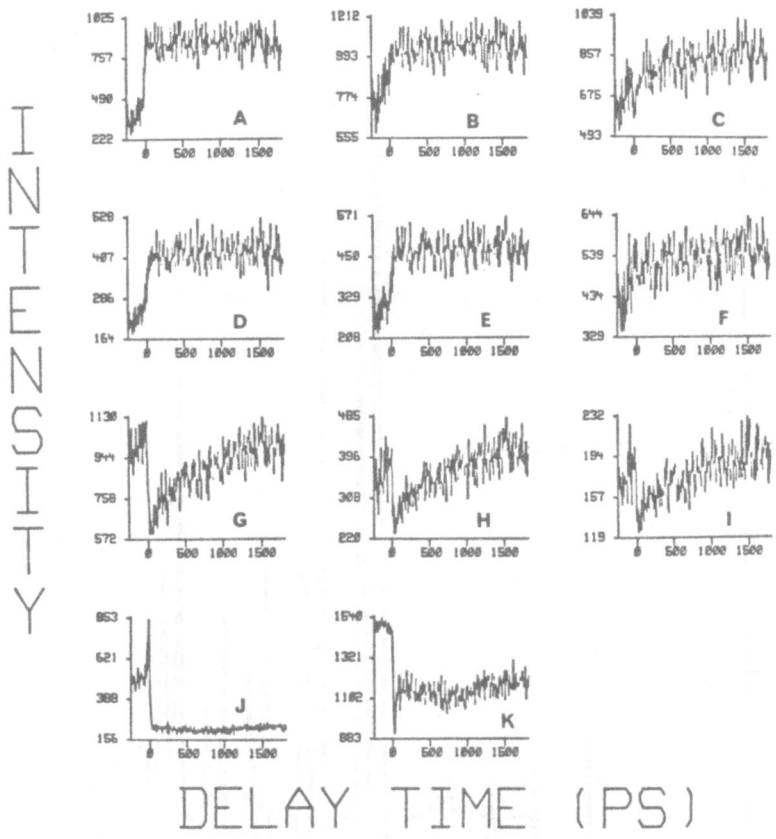

DELAY TIME (PS)

Figure 7. Ion current vs. delay time plots for several mass fragments. All plots were recorded simultaneously under the same experimental conditions. UV and green intensities were 14 μJ/pulse and .8 mJ/pulse, respectively. $A=C_4H_2^+$, $B=C_4H_3^+$, $C=C_4H_4^+$, $D=C_5H_2^+$, $E=C_6H_2^+$, $F=C_6H_3^+$, $G=C_6H_4^+$, $H=C_4H_3(35)Cl^+$, $I=C_4H_3(37)Cl^+$, $J=C_6H_4Cl^+$, $K=C_6H_4Cl_2^+$. Several different types of behavior are inhibited by these ions. This figure shows the advantages of being able to simultaneously monitor several of the mass fragments.

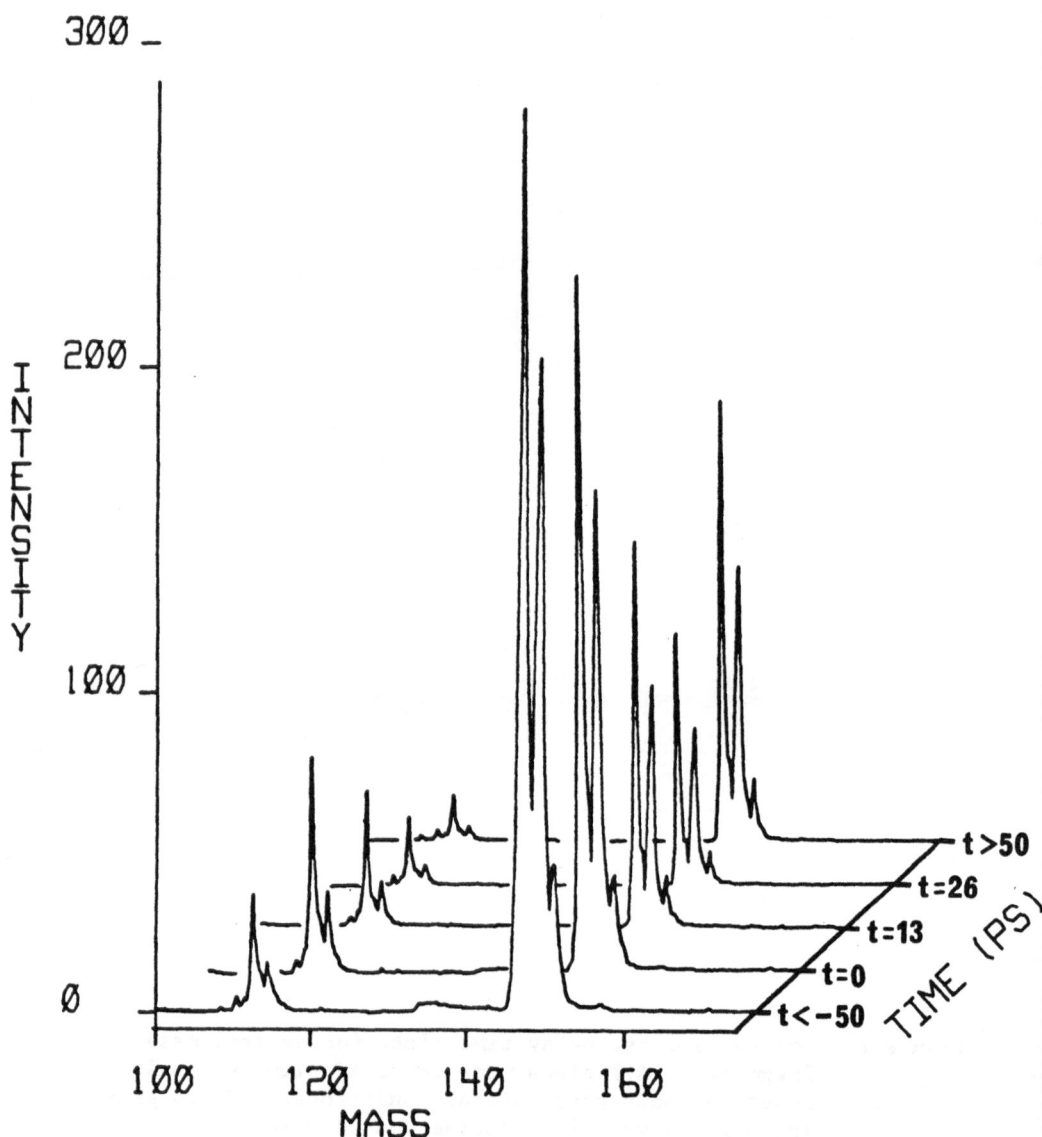

Figure 8.   The mass spectra in the range of m/e = 100 to m/e = 170 are
            shown, highlighting the region around t = 0.   These data
            show that the $C_6H_4Cl^+$ signal begins to increase prior to
            decrease in the parent signal.

ACKNOWLEDGMENT: The financial support of the National Science Foundation is gratefully acknowledged.

REFERENCES

1. H. M. Rosenstock, Adv. Mass Spectrom. 4, 523 (1968).
2. A. Gedanken, M. B. Robin, and N. A. Kuebler, J. Phys. Chem. 86, 4096 (1982).
3. J. J. Yang, D. A. Gobeli, R. S. Pandolfi, and M. A. El-Sayed, J. Phys. Chem. 87, 2255 (1983).
4. L. Zandee and R. B. Bernstein, J. Chem. Phys. 70, 2574 (1980).
5. U. Boesl, Neusser, and E. W. Schlag, J. Chem. Phys. 72, 4327 (1980).
6. J. A. Hipple, Phys. Rev. 71, 594 (1947).
7. O. Osberghaus and C. Ottinger, Phys. Lett. 16, 121 (1965).
8. H. D. Beckey, H. Hey, K. Levsen, and G. Tenschert, Int. J. Mass. Spectrom. Ion Phys. 2, 101 (1969).
9. U. Boesl, H. H. Neusser, R. Weinkauf, and E. W. Schlag, J. Phys. Chem. 86, 4857 (1982.
10. D. A. Gobeli, J. R. Morgan, R. J. St. Pierre, and M. A. El-Sayed, J. Phys. Chem. 88, 178 (1984).
11. B. I. Greene in "Ultrafast Phenomena", Springer-Verlag, West Berlin.
12. M. Momigny, M. L. Brakier, and L. D'Or, Bull. Cl. Sci., Acad. R. Belg. 48, 1002 (1962).
13. M. Allan, J. P. Maier, O. Marthaler, and E. Kloster-Jensen, Chem. Phys. 29, 311 (1978).
14. J. Danacher, Chem. Phys. 29, 339 (1978).
15. W. C. Wiley and I. H. McLaren, Rev. Sci. Instrum. 26, 1150 (1955).
16. (a) N. Morita and T. Yajima, Appl. Phys. B 28, 25 (1982);
    (b) D. M. Rayner, P. A. Hackett, and C. Willis, Rev. Sci. Instrum. 53, 537 (1982).
17. J. D. Simon, Diane M. Szaflarski and M. A. El-Sayed, J. Phys. Chem., in press.

DETECTION OF DIPOLAR CONTRIBUTION TO THE MECHANISM OF THE TRIPLET-TRIPLET ENERGY TRANSFER PROCESS IN MOLECULAR SOLIDS

Jack R. Morgan,[†] Hansjorg S. Niederwald[*] and M. A. El-Sayed
Department of Chemistry and Biochemistry
University of California
Los Angeles, California 90024 U.S.A.

ABSTRACT. It is commonly accepted that triplet-triplet excitation transfer between aromatic molecules takes place via an electron exchange mechanism. Since the probability of this mechanism falls much more rapidly (exponentially) with distance than the weak dipolar mechanism, it is possible that a distance can be found for which the dipolar mechanism begins to contribute to the observed transfer.

The technique of laser luminescence line narrowing is used at 4.2 K to study the spectral diffusion in 1-bromo-4-chloronaphthalene (BCN), a one-dimensional orientationally disordered solid. The 0,0 band of its singlet-triplet transition is inhomogeneously broadened. It is shown that this technique can be used to determine the donor decay, and thus the mechanism of triplet-triplet energy transfer, as the donor-acceptor distance is continuously changed by simply changing the wavelength of the exciting laser. At long wavelength, this technique is particularly convenient for use in disordered solids, i.e., solids with large inhomogeneous absorption line width.

At short distances, the donor phosphorescence decay is found to fit the expected one-dimensional electron exchange type mechanism. At distances longer than 10 Å, a deviation from one-dimensional exchange fit is observed, which could be accounted for by either a change in the dimensionality or a change in the transfer mechanism from exchange to dipolar. A microwave-phosphorescence double resonance experiment is used to determine the energy distribution of molecules in the different spin levels as a function of time. The results suggest that at long donor-acceptor separation, molecules in radiative spin levels (i.e., with transition dipole moments) transfer with higher probability than those in dark spin levels. This suggests that at long distances, the dipolar mechanism contributes to the triplet energy transfer in this solid.

[†] Present address: Department of Chemistry, University of Nevada, Reno

[*] Present address: Firma Carl Zeiss, 7082 Oberkochen/Wurtt, West Germany

P. M. Rentzepis and C. Capellos (eds.), Advances in Chemical Reaction Dynamics, 57–70.
© 1986 by D. Reidel Publishing Company.

INTRODUCTION

a. Distance Dependence of the Triplet-Triplet Energy Transfer Process

Let us use the excitation of the π-electrons in ethylene as an example
in the following discussion. In the ground state, ethylene has a $(\pi)^2$
configuration while the first excited state configuration is $(\pi)(\pi^*)$.
The transfer between an excited triplet molecule (the donor, D) and a
ground state molecule (the acceptor, A) can be visualized as:

Thus the mechanism expected at this level of approximating the wavefunctions of the electronic states is a double electron exchange. This
requires the molecular orbitals of the donor and those of the acceptor
to overlap. The transfer probability (P_R) thus decreases
exponentially[1,2] as the distance R increases, i.e.:

$$P_R \sim e^{-cR}. \quad\ldots\ldots\ldots\ldots \qquad (1)$$

where c is a constant. Using this relationship, the probability, $P_{(t)}$,
that the donor remains excited after time t is found[3] to be related
to time for transfer in one dimension as:

$$\ln P_{(t)} \sim -\ln t \quad\ldots\ldots\ldots\ldots \qquad (2)$$

Experimentally, $P_{(t)}$ is proportional to the donor intensity, $I_{D(t)}$,
where $I_{D(t)}$ is the donor phosphorescence intensity at time t after
excitation.

By spin orbit coupling, the triplet state could acquire transition
dipole moment. In this case, the energy transfer process can be
represented by:

In these processes, the de-excited electron of the donor and the excited
one of the acceptor change their spinning quantization directions during
the transition as a result of spin orbit interaction within each molecule. These interactions are stronger in systems with heavy nuclei,
e.g., bromine atoms. During these processes, the coupling between the
two molecules occurs as a result of the dipole moment induced during
the electronic transitions in the donor and acceptor molecules. The
probability of the energy transfer for this dipolar mechanism[1,2] has
the following distance dependence:

$$P_{(R)} \sim \frac{C'}{R^6} \qquad \ldots \ldots \ldots \qquad (3)$$

where C' is a constant that depends on the size of the transition dipole. The probability that the donor undergoing dipolar energy transfer remains excited at time t after a short excitation pulse is given by:[3,4]

$$\ln P_{(t)} \sim -C_t^{C/6} \qquad \ldots \ldots \ldots \qquad (4)$$

where D is the dimensionality of the system and 6 is for dipolar coupling. For 3-d dipolar interaction, ln P decreases with $t^{0.5}$.

STATEMENT OF THE PROBLEM

Let us select a system in which the π orbital overlap is maximum along one dimension, e.g., halogenated naphthalene.[5] As Fig. 1 shows, at short distance, the exchange mechanism dominates, since the size of the

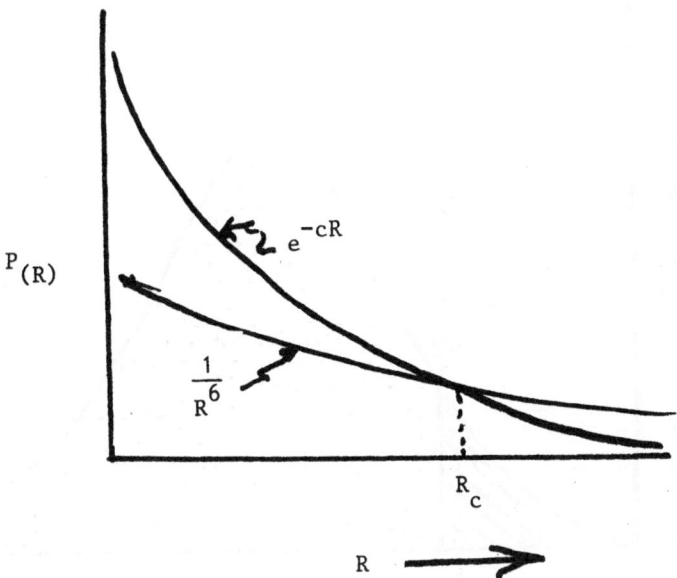

Fig. 1. Distance dependence of the 1-d exchange and
3-d dipolar mechanisms' transfer probability

transition dipole of singlet-triplet excitation is relatively small. Because of the difference in the form of the distance dependence (equations 1 and 3), the transfer probability of the one-dimensional exchange mechanism drops much more rapidly with distance than the 3-d dipolar mechanism. It is thus conceivable that at a certain distance

R_C, both would have equal probability. Beyond this distance the time dependence of the emission of the donor changes its form from that given by equation (2) to that given by equation (4).

A CONVENIENT METHOD FOR CONTINUOUSLY CHANGING DONOR-ACCEPTOR DISTANCE:

How can we change continuously the donor-acceptor distance? Solubility of one solid into the other is usually limited. Isotope mixed crystals are a possibility. However, in these crystals one is studying donor-donor as well as donor-acceptor energy transfer simultaneously. Lasers have both high intensity as well as sharp energy width to excite only a selected set of molecules having similar environment within a disordered solid, a solid with large distribution of sites (environments). These solids are characterized by a large inhomogeneous absorption linewidth. As shown in Fig. 2, excitation at ν_i with the laser makes molecules

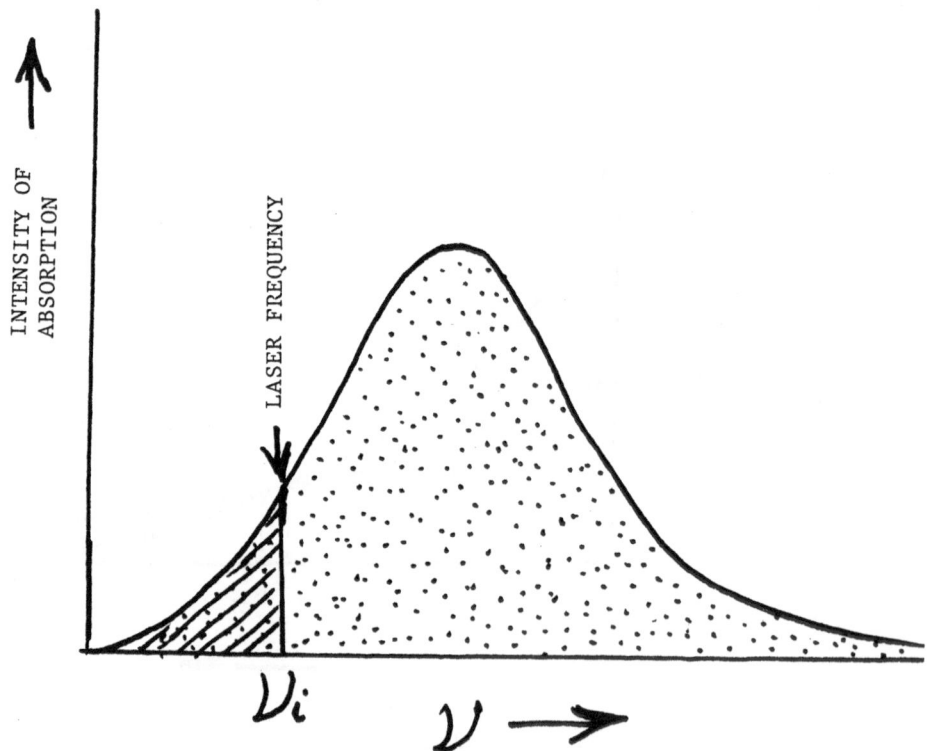

Fig. 2. The use of laser line narrowing techniques to excite donors in a selected environment absorbing at frequency ν_i. At low temperatures, energy transfer can only occur to the acceptor molecules absorbing at frequency $< \nu_i$. Molecules absorbing at frequency $> \nu_i$ do not accept the excitation energy and act as the solvent which separates the donors from acceptors. Thus by changing ν_i, the average distance (and the rate of the transfer) between donors and acceptors can be varied.

absorbing at ν_i energy donors. At low temperatures, e.g., 4.2 K, energy transport will only take place to molecules absorbing at equal or lower energy than $h\nu_i$. These are the acceptors. Molecules absorbing at higher frequency than ν_i are the solvent molecules separating the donors from the acceptors. Thus by increasing the exciting laser frequency of the laser, the density of the acceptor molecule increases and the average donor-acceptor distance decreases. Thus if one studies the decay characteristics of the donor emission as a function of the exciting laser frequency (i.e., the donor-acceptor distance), the distance at which the triplet-triplet energy transfer switches its mechanism might be found.

Absorption and Emission of 1-bromo-4-chloronaphthalene (BCN), an Ideal System:

The T_1-S_0 transition of BCN at low temperatures has an inhomogeneous width on the order of 100 cm^{-1}. This line width is about two orders of magnitude larger than the line width observed for the corresponding transition in 1,4-dichloronaphthalene (DCN) and 1,4-dibromonaphthalene (DBN), suggesting that the width in BCN is due to the static orientational disorder in the halogen positions in the crystal.[7] Comparative studies of the crystal structures[8] and Raman spectra[9] of this 1,4-dihalonaphthalene series show that the one-dimensional stacking feature and intermolecular interactions in BCN are similar to DBN for which one-dimensional exchange-type triplet excitons have been observed.[10]

The broad features observed in the T_1-S_0 absorption spectrum of BCN indicate that the singlet-triplet absorption in this system is inhomogeneously broadened. The nature of the inhomogeneous broadening most likely arises from the static orientational disorder in the bromine and chlorine positions in the crystal. The orientational disorder leads to an inhomogeneous distribution of the site energy due to the disorder in the static crystal shift from the gaseous excitation energy. The absorption profile of the 0,0 band of the singlet-triplet transition of neat BCN at 4.2 K is shown in Fig. 3. The general characteristics are in agreement with the spectrum reported in the literature.[7] Shown in Fig. 3 is a fit of the absorption profile to a Gaussian with a 32-cm^{-1} width (hwhm), suggesting an inhomogeneous type of broadening. A temperature-dependent study of the absorption line shape for BCN has shown that the homogeneous broadening due to interactions with phonons is much smaller than that due to the disorder.[7] The broad, structureless, and nearly Gaussian absorption profile lend support to the interpretation that the inhomogeneous profile is determined by the static structural disorder.

The phosphorescence profile of the 0,0 band of the T_1-S_0 transition of neat BCN is also shown in Fig. 3, which agrees with that previously reported.[7] The phosphorescence profile is observed to be narrower than the absorption profile and to originate from the low-energy sites observed in absorption. These results indicate that at 4.2 K a rapid phonon-assisted energy cascade from high to low energy sites occurs in neat BCN at 4.2 K. At temperatures where kT $\ll$

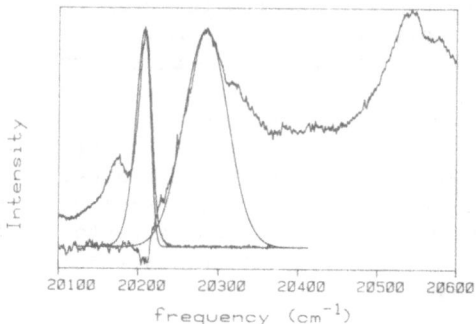

Fig. 3. The emission and absorption spectra of the T_1-S_0
transition of 1-bromo-4-chloronaphthalene (BCN) at
4.2 K. The observed absorption profile of the zero
phonon line of the 0,0 band is fitted to a Gaussian
centered at 20284 cm^{-1} with Γ = 32 cm^{-1} (solid

line). The observed narrowed emission profile is
centered at 20208 cm^{-1}.

inhomogeneous width, energy transfer from low to high energy sites will
be negligible since energy transfer from low to high energy sites
requires the population of phonons which will be small in this temperature regime.

EXPERIMENTAL

The BCN solid was synthesized and extensively zone refined. Crystals
were grown from the melt in a Bridgman furnace. The 4.2 K measurements were carried out with the sample immersed in liquid helium.
     Spectra were recorded with a 1-m Jarrell-Ash monochromator with
2-cm^{-1} resolution used throughout. Steady-state phosphorescence spectra
were obtained by excitation with the 3300-Å region of a 100-W mercury-
xenon lamp. Absorption spectra were obtained with a 80-W quartz-
halogen lamp. The signal was averaged with a PAR Model 162 boxcar
averager, recorded on a Tracor-Northern NS-570A multichannel digitizer,
and analyzed on a PDP-11/45 computer.
     For time-resolved measurements, a Quanta-Ray DCR-1 Nd:YAG pumped
pulsed dye laser at a repetition rate of 10 Hz with a spectral width
of 0.3 cm^{-1} and a pulse width of 6 ns was used as the T_1-S_0 excitation
source. Time-resolved spectra of the T_1-S_0 phosphorescence of the
0,0-321-cm^{-1} band in the wavelength domain were recorded with the PAR
boxcar averager. The temporal dependence of the donor phosphorescence
was performed by carefully tuning the monochromator to the donor
0,0-321-cm^{-1} band so as to follow the donor intensity. Spectra were
recorded with a Biomation 805 waveform digitizer, averaged with a
homebuilt signal-averaging computer, and analyzed on a PDP-11/45
computer. A special gated phototube was used in order to reject

scattered laser light. Due to interference from switching the focus
electrode, the first 10 μs of the signal following the laser pulse was
rejected. In order to estimate I_0, the intensity of time t = 0, I_0 was
determined by extrapolating a log I vs. log t fit to the first 10-50 μs
of signal to t = 0.   (6)
Microwave-induced delayed phosphorescence (MIDP) experiments
were done monitoring the 0,0 band (and sometimes the 0,0-234 cm^{-1}
vibronic band) of the phosphorescence emission. A Hewlett Packard 8690
sweep oscillator with 8699B and 8693A plug-in units was used as a
microwave source with an Alto Scientific microwave amplifier. The
microwave frequency was determined by the use of a Hewlett Packard 8557
spectrum analyzer. The microwave power was brought by a coaxial cable
to a helix surrounding the sample.

RESULTS

a. Spectral Diffusion:

The relationship between the absorption spectra and the emission spectra
of the triplet state in 1-bromo-4-chloronaphthalene has been reported
elsewhere[12] and shown in Fig. 3. It is to be noted that the emission in a steady-state experiment is observed from the low-energy sites.
Furthermore, the width of the emission is about 20 cm^{-1} which is much
narrower than the absorption line width but is still considerably
broader than that observed for the ordered 1,4-dichloronaphthalene
triplet emission.
Fig. 4 demonstrates that spectral diffusion[13] of the $T_1 \rightarrow S_0$
transition energy occurs in this orientationally disordered material.
The band shape of the first vibronic band was monitored as a function
of delay time. The spectra displayed were obtained by exciting a site
on the low-energy side of the $S_0 \rightarrow T_1$ absorption profile. The figure
shows that as the delay time increases, the resonant-type emission
decreases in intensity while emission from low energy acceptors (the
traps) increases in intensity. This figure also shows the same spectra
recorded at two different sample temperatures. The phosphorescence
spectra obtained under identical conditions, but at two different
sample temperatures (the top and bottom spectra in Fig. 4), clearly
indicate that the transfer rate is temperature dependent and, thus,
phonon assisted.[14]
The results of the effect of site-selective excitation on the rate
of spectral diffusion of the $T_1 - S_0$ energy is shown in Fig. 5. In
each case the spectrum is sampled after a 10-μs delay, using a 50-μs
sampling time. In order to minimize interference with stray light due
to the laser, we monitored the first vibronic band profile of the
emission. It can be seen that the observed emission profile is very
much dependent on the excitation energy. As the low-energy sites are
excited, a relatively narrow resonant emission is observed, indicating
a transfer rate slower than the sampling time. On the other hand, with
excitation of higher energy sites, the emission profile is broadened
indicating a faster transfer time to a broader distribution of the low-
energy sites. These observations are suggestive of a transfer rate

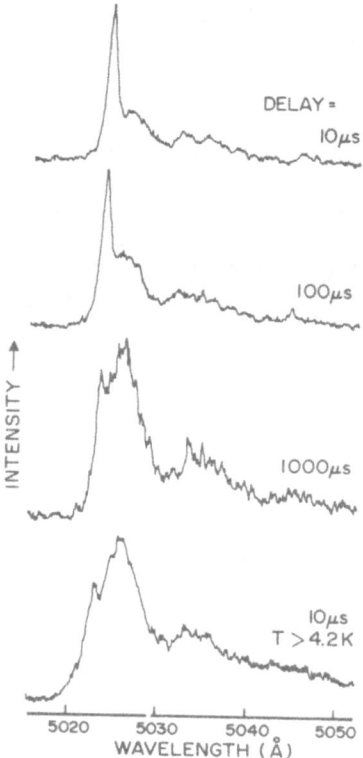

Fig. 4. Spectral diffusion of singlet-triplet excitation energy in an "amorphalline" solid at 4.2 K obtained with time-resolved phosphorescence line-narrowing techniques. The system is 1-bromo-4-chloronaphthalene excited at 4943 Å. The top three spectra are of the 0,0-321-cm^{-1} vibronic band of the phosphorescence emission recorded at different delay times after the laser pulsed excitation with a 50-μs sample time. The bottom spectrum illustrates the effect of temperature on the rate of spectral diffusion within the inhomogeneous profile of the 0,0-321-cm^{-1} band.

which at 4.2 K is dependent on the donor site energy (i.e., on the concentration of the acceptor's molecules).

b. Fit of Decay to Exchange Mechanism:

Fig. 6 shows the ln P vs. ln t fit of the 1-d exchange mechanism. An excellent fit for λ_{4940} Å (bottom decay) is observed. However as the wavelength increases, deviation at long times is seen. The fraction of the total decay that can be accounted for by the straight line exchange

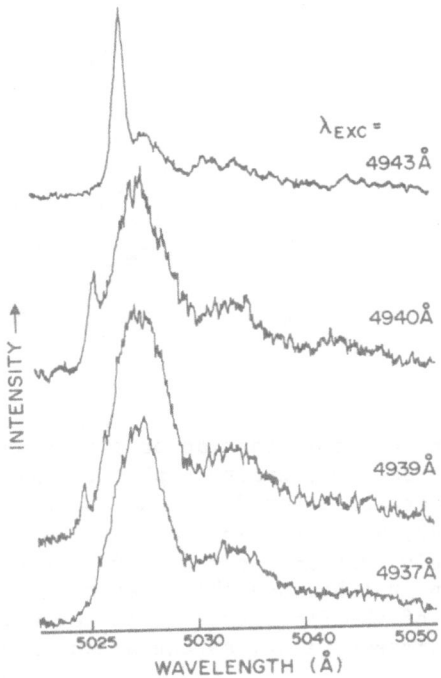

Fig. 5. Dependence of the spectral diffusion rate of the singlet-
         triplet excitation on the donor site energy in 1-bromo-4-
         chloronaphthalene at 4.2 K. The spectral intensity change
         of the 0,0-321-cm^{-1} band of the $T_1 \rightarrow S_0$ phosphorescence
         is monitored with a 10-μs delay by means of a 50-μs sampling
         time following pulsed laser excitation at different wavelength
         (i.e., different donor energies) within the inhomogeneously
         broadened 0,0 band of the $T_1 \leftarrow S_0$ absorption. The
         results show that the ratio of the line-narrowed (donor)
         phosphorescence to the lower energy broad emitting traps
         decreases as the donor (excitation) energy increases,
         suggesting an increase in the spectral diffusion rate.

fit decreases as λ increases.
     Fig. 7 shows the results of attempting to fit the portion that did
not fit to the exchange mechanism in Fig. 6 to the 3-d dipolar fit.
Fig. 7 suggests that at long time and wavelength (long donor-acceptor
distance), a 3-d mechanism could account for the data.
     While these results might[15] suggest a switching of the mecha-
nism at a few percent of acceptor concentration (> 10 Å for donor-
acceptor separation), the possibility of switching the dimensionality
from one to greater than one could equally be an explanation for the
observed deviation from the 1-d exchange equation. If one attempts to
make the fit of ln $P_{(t)}$ vs. ln^{D}t, where D is the dimensionality,

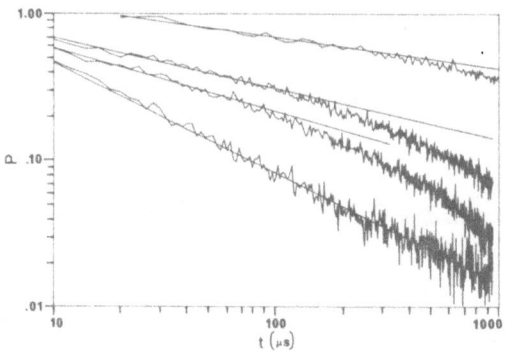

Fig. 6. The fit of the early portion of the decay of the triplet
         excitation due to triplet-triplet energy transfer to an
         exchange mechanism for different excitation wavelengths
         (4947, 4943, 4942, and 4940 Å from top to bottom,
         respectively) within the 0,0 band of the T_1-S_0
         transition in 1-bromo-4-chloronaphthalene at 4.2 K.
         The range of the fit increases as the excitation wave-
         length decreases, i.e., as the acceptor concentration
         increases.

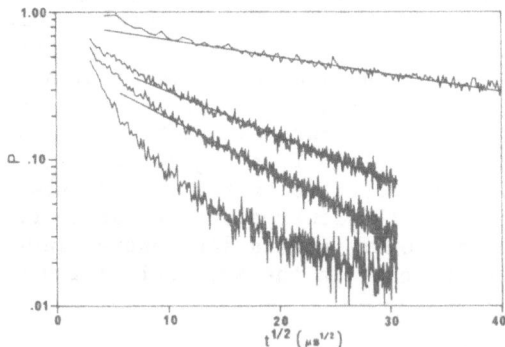

Fig. 7. The fit of the long time portion of the decay of the
         triplet excitation of the 0,0 band of the T_1-S_0
         transition of 1-bromo-4-chloronaphthalene at 4.2 K due
         to three-dimensional dipolar mechanism for the excitation
         wavelengths given in Fig. 6. The range of the fit is
         better at longer excitation wavelengths, i.e., at low
         acceptor concentrations.

straight lines are obtained[16] for D = 2.3 at λ = 4943 Å, 4945 Å and
4947 Å, the wavelengths for which the decay did not fit to the 1-d
exchange mechanism. Thus we cannot completely be sure that at very
long wavelength (i.e., large donor-acceptor distance) a contribution of
the dipolar mechanism is made to the triplet energy transfer in these
crystals. The experiments below are designed to test this contribution
directly.

c. Direct Experiment to Test for the Dipolar Contribution[17]

The triplet state of halogenated naphthalenes has three spin levels
separated by fractions of cm^{-1}, even in the absence of magnetic field.
This is the zerofield splitting which results from the anisotropic
spin-spin and spin orbit dipolar interactions.[18] Because of the
molecular plane of symmetry, only two levels of the π,π^* excited
triplet state of BCN are radiative (i.e., the electronic transitions
from these levels to the ground state carry a transition dipole moment) and one level is dark. It is thus expected that molecules in the
radiative levels will be the ones that can transfer their excitation
energy by the dipolar coupling mechanism (as well as exchange at short
distances). Molecules in the bottom dark level can transfer their
energy only by the exchange mechanism. It would thus be interesting
to test if there is any difference between the transfer rates of
molecules in the radiative and dark levels having singlet-triplet
energies at the lowest energy end of the absorption spectrum (i.e.,
separated from acceptors by large distances). We need to determine the
change in the site energy distribution of molecules having triplet
energy at the trap emission energies and occupying the radiative and
dark zerofield levels after 70 mls from turning off the excitation.
The determination of the change in the site energy distribution due to
energy transfer of the radiative levels in 70 mls is easy. We first
record the steady state trap emission. Then we simply turn off the
steady excitation source and record the spectrum again 70 mls later.
The results are shown in Fig. 8. The middle spectrum (b) (70 mls,
delayed) is weaker, sharper and shifted by 1-2 Å to the red from the
steady state spectrum (a) due to energy transfer. The question is how
much red shift (due to energy transfer) does the distribution of
molecules in the dark level undergo in 70 mls? It will be comparable
to the shift of the distribution in the radiative levels only if the
exchange mechanism is responsible for the triplet energy transfer in
BCN. However, if it is less, then the difference might be contributed
by the dipolar mechanism.

The study of the site energy distribution of molecules in the dark
level was made by phosphorescence microwave double resonance techniques.[6] If the steady state excitation is turned off for 70 mls,
the population of the radiative levels would decay mostly by radiative
decay (as seen from the much weaker middle spectrum as compared to the
top spectrum of Fig. 8) and to a small extent by energy transfer (as
seen from the slight spectral shift of the middle spectrum from that
on the top in Fig. 8). Molecules in the dark level can only decay in
70 mls by the less probable energy transfer process (occurring at the

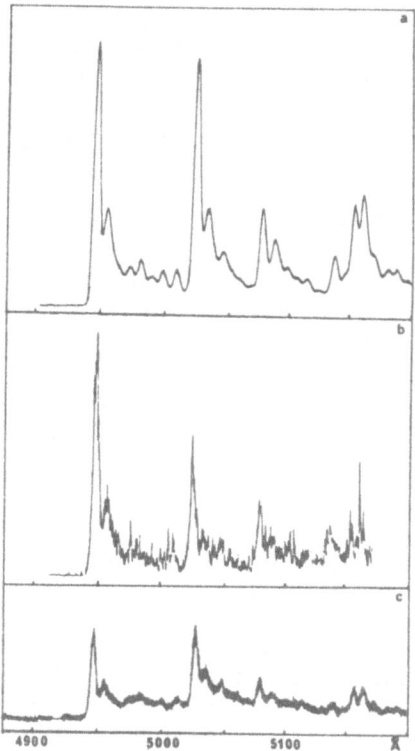

Fig. 8.   Phosphorescence spectra of a neat BCN crystal under cw
          excitation (top), of the same crystal, taken with a boxcar
          averager 70 ms after closing off the excitation light (middle)
          and of a 10^{-3} M solid solution of BCN in durene (bottom).
          The sample temperature was 1.8 K.   The excitation source was
          the 313-nm line of a 100-W Hg lamp.   The major difference
          between the cw spectrum and the delayed spectrum is in the
          relative intensities of the 0,0 band and the 0,0-321 cm^{-1}
          band.   The lines in the middle and bottom spectra are slightly
          narrower and red shifted by 1-2 Å.

large distances we are studying) and perhaps by spin lattice relaxation
processes. The latter processes are intramolecular in nature and do not
change the site energy distribution of excited molecules in the dark
levels in the crystal, it merely decreases the number of molecules in
the dark zerofield level at the different energy ranges by the same
fraction. If after 70 mls, a pulse of microwave of the appropriate
frequency is turned on, a certain fraction of the molecules in the dark
level is transferred to the radiative level and a pulse of light is

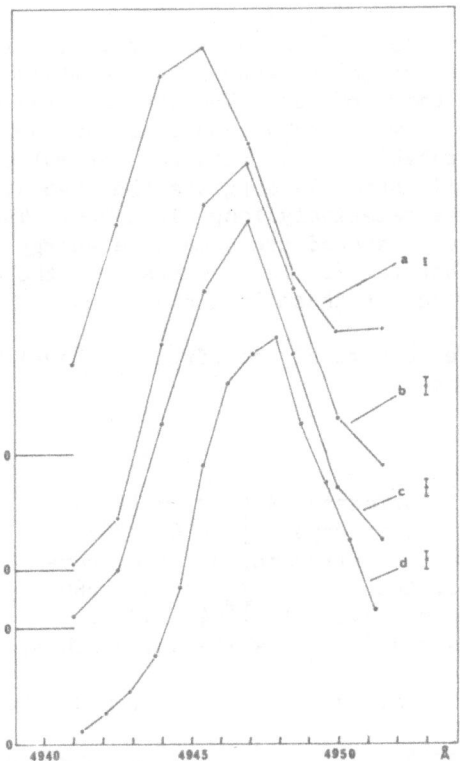

Fig 9. Inhomogeneous lineshape of the 0,0 phosphorescence transition
at 1.8 K, taken under four different conditions. Curve a
shows the cw intensity distribution; curves b and c show the
height of the leading edge of the MIDP signal at 1.05 GHz,
which is proportional to the population of the dark level,
after 70—ms (b) and 130—ms (c) delay time. Curve d is
derived from the delayed spectrum in the middle part of
Fig. 8 and represented the population of the radiative level
τ_L. With the red shift, curves b, c, and d also get slightly
narrower and asymmetric. As parts b and d show, the distribution of the radiative level τ_L (lifetime 50 ms) at 70 ms is
relaxed to lower energies than the distribution of the dark
level (lifetime 140 ms) at the same delay time.

observed. The intensity of this light pulse is proportional to the
population of the dark level having an energy equal to that of the
emitted radiation (since the zerofield splitting is very small as compared to the singlet-triplet energy separation). If this is done and
the intensity of the pulse of light emitted at the different trap
emission wavelength range is determined, the site energy distribution
of molecules in the dark zerofield level after 70 mls from turning off

the excitation can be determined.

Fig. 9 shows a comparison of the site energy distribution of molecules in the dark (b) and in the radiative (d) levels in the 0,0 region of the trap emission spectrum after 70 mls delay as compared to the steady state emission (a). The distribution of the radiative level shifts more to lower energy than that for the dark level, suggesting a more probable energy transfer for molecules in the radiative level. This strongly suggests the presence of a dipolar contribution at these relatively long distances. The fact that the 130 mls delay (curve c) showed the same site energy distribution of the dark level as that for 70 mls suggests that the exchange mechanism is probably frozen at these long distances.

ACKNOWLEDGMENT: The support of the Office of Naval Research is gratefully acknowledged.

REFERENCES

1. T. Forster, Z. Naturforsch. A $\underline{4}$, 321 (1949).
2. D. L. Dexter, J. Chem. Phys. $\underline{21}$, 836 (1953).
3. (a) M. Inokuti and F. Hirayama, J. Chem. Phys. $\underline{43}$, 1978 (1965).
    (b) A. Blumen, J. Chem. Phys. $\underline{72}$, 2632 (1980).
4. A. Blumen, J. Chem. Phys. $\underline{71}$, 4694 (1979).
5. R. M. Hochstrasser and J. D. Whiteman, J. Chem. Phys. $\underline{56}$, 5945 (1972).
6. J. Schmidt, D. A. Antheunis and J. M. van der Waals, Mol. Phys. $\underline{22}$, 1 (1971).
7. J. C. Bellows and P. N. Prasad, J. Phys. Chem. $\underline{86}$, 328 (1982).
8. J. C. Bellows, E. D. Stevens and P. N. Prasad, Acta Crystallogr., Sect. B $\underline{34}$, 3256 (1978).
9. J. C. Bellow and P. N. Prasad, J. Chem. Phys. $\underline{67}$, 5802 (1978).
10. R. M. Hochstrasser and J. D. Whiteman, J. Chem. Phys. $\underline{56}$, 5945 (1972).
11. P. N. Prasad, J. R. Morgan and M. A. El-Sayed, J. Phys. Chem. $\underline{85}$, 3569 (1981).
12. J. C. Bellows and P. N. Prasad, J. Phys. Chem. $\underline{67}$, 5802 (1979).
13. P. N. Prasad, J. R. Morgan and M. A. El-Sayed, J. Phys. Chem. $\underline{85}$, 3569 (1981).
14. T. Holstein, S. K. Lyo and R. Orbach, Phys. Rev. Lett. $\underline{36}$, 891 (1976).
15. J. R. Morgan and M. A. El-Sayed, J. Phys. Chem. $\underline{87}$, 2178 (1983).
16. C. L. Yang and M. A. El-Sayed, in preparation.
17. H. S. Niederwald and M. A. El-Sayed, J. Phys. Chem. $\underline{88}$, 5775, 1984.
18. See: S. P. McGlynn, T. Azumi and M. Kinoshita, "Molecular Spectroscopy of the Triplet State," Prentice Hall, Englewood Cliffs, N.J. (1969). Also, S. K. Lower and M. A. El-Sayed, Chem. Rev. $\underline{66}$, 199 (1966); M. A. El-Sayed, Accounts of Chemical Research $\underline{4}$, 23 (1971).

GAS PHASE CHAIN POLYMERIZATION

H. Reiss
Department of Chemistry
University of California Los Angeles
Los Angeles, California 90024

ABSTRACT.   Methods which allow, for the first time, the study of the
kinetics of true, homogeneous, gas phase, chain polymerization are
described. The detection mechanism involves nucleation and growth of
clusters. Reaction rates lower than <u>one product molecule</u> per cubic
centimeter per second are measured. The reaction is therefore noisy.
The scientific importance of such measurements is elaborated. Limitations of the present scheme of measurement are discussed and plans for
improving the level of precision are outlined.

I.   <u>Homogeneous Gas Phase Polymerization: Why Study it?</u>

This paper deals with a branch of reaction dynamics, namely gas
phase polymerization, which has only recently become amenable to
quantitative experimental study. As a result, only a few papers have
been published on the subject[1,2,3,4] and these have been concerned
chiefly with the demonstration of feasibility rather than the
quantitative measurement of reaction parameters.

One can find many papers in the literature on "gas phase
polymerization", but almost all are concerned with processes in which
the polymer grows on the wall of the reaction vessel while the growth
is fed by monomers in an ambient gas phase. The "gas phase
polymerization" with which we shall be concerned, is the true
<u>homogeneous</u> process in which the polymer is <u>also</u> in the gas phase. In
the past it has been almost impossible to study this process because
the polymer molecules are involatile and condense out of the gas. An
important attempt to measure the homogeneous process was made by
Melville and his coworkers[5,6,7] more than 40 years ago. A schematic
of the experimental arrangement used by these authors (reproduced from
one of their papers) is exhibited in the first slide. The reaction
vessel appears at the right and is denoted by the symbol R. In this
case it contained vinyl acetate vapor irradiated by uv light which
generated free radicals and induced the process of chain
polymerization. An immediate problem was the one already mentioned,
namely the condensation of the polymers to form a "smoke" or
"aerosol". The reaction then continued, primarily heterogeneously, by

71

P. M. Rentzepis and C. Capellos (eds.), Advances in Chemical Reaction Dynamics, 71–113.
© *1986 by D. Reidel Publishing Company.*

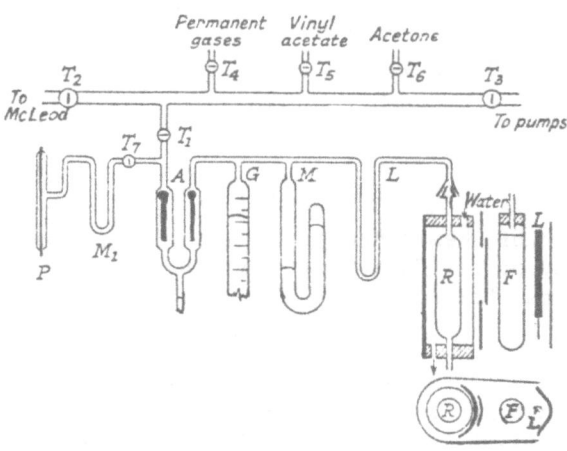

Slide 1:   A schematic of the experimental arrangement used by
Melville and coworkers.

growth on the aerosol particles. Nevertheless, an attempt was made to
study the rate of polymerization by monitoring the rate of decrease of
monomer gas pressure, and therefore, of the rate of disappearance of
monomer. The rate of polymerization (even in condensed systems) is
usually defined as the gross rate of disappearance of monomer. By a
tour de force Melville and coworkers were able to extract some useful
information from these experiments, but ultimately abandoned the
project, because of condensation. I have had some recent
correspondence with Professor Melville in which he explains the
reasons for the noncontinuance of the work.
     Clearly, the study of true gas phase polymerization requires a
technique in which the growing polymer molecules cannot encounter one
another and condense. During the past few years we have been able to
develop such a technique and, demonstrate its feasibility. The method
is so sensitive that it is possible to regularly study reaction rates
as low as one product molecule per cubic centimeter per second. This
rate is so slow that it is "noisy", much like radioactive decay.
Furthermore the method allows one to tune to the arrival of a polymer
product of a given degree of polymerization, so that rather than
measure the gross rate of disappearance of monomer we can, in
principle, measure the more informative rate of production of a
polymer of a <u>particular</u> size!
     Before proceeding to explain how this is done, I should mention
that our group is now the recipient of a generous grant from the
National Science Foundation which should allow the reduction of the
technique to a fully quantitative measurement. However, even though
the reviewers (at least those from the NSF polymer program—other NSF

divisions are also supporting this project) of our proposal were positive, almost all had the same question--"what is it good for?". For example, although taking the remarks out of context, in each case, the following quotations are worth reporting:

1. "The investigator made little effort to convince us of the practical aspects of the research."

2. "Most importantly, it is not clear why the work is being done. What, in specific, can one learn by this technique which is not available by other techniques?"

3. "When the program required to follow this particular principal investigator's amusement is as expensive as this one, one must ask questions about the relative importance of this work and other work."

4. "At its present stage, many questions could be raised regarding the precise value of such an approach to polymerization kinetics in general."

Since these questions have been asked by intelligent individuals, I feel it necessary to answer them in advance, in any presentation made to other intelligent individuals. Towards this end, I present slides 2, 3, 4, and 5. Slide 2 raises the question in three categories: (1) polymerization science, (2) polymer and other technologies, and (3) other aspects of polymer and nonpolymer science. Slide 3 elaborates the first of these categories into 11 subitems. I shall only discuss a few of these, leaving the discussion of the remaining ones to questions which might be asked during a discussion period following this paper. The term "polymerization science" refers to the process of polymerization itself, and, in particular, to the kinetic mechanisms involved.

The first item on the slide is of primary importance. In conventional studies involving free radical chain polymerization, in condensed media, the theory for the rate of disappearance of monomer (rate of polymerization) is developed, taking advantage of the fact that the instanteous concentration of free radicals is usually so low (10^{-8} cm^{-3}) that a steady state approximation may be made in which the free radical concentration can be eliminated from the mathematical expression containing the various rate constants. In contrast, in

WHY STUDY HOMOGENEOUS
GAS PHASE POLYMERIZATION?

(1) POLYMERIZATION SCIENCE
(2) POLYMER AND OTHER TECHNOLOGIES
(3) OTHER ASPECTS OF POLYMER AND NONPOLYMER SCIENCE.

Slide 2: The question in three categories.

POLYMERIZATION SCIENCE

(1)   DIRECT OBSERVATION OF GROWING FREE RADICALS OR IONS.

(2)   HIGH RESOLUTION - CAN TUNE TO RATE OF PRODUCTION OF
      POLYMER OF A PARTICULAR SIZE.

(3)   FAIRLY DIRECT MEASUREMENT OF DEPENDENCE OF PROPAGATION
      RATE CONSTANT ON SIZE.

(4)   SELF-THERMAL POLYMERIZATION WITHOUT IMPURITIES.

(5)   DIRECT MEASURE OF INITIATOR EFFICIENCY WITHOUT SOLVENT
      CAGE.

(6)   DIRECT MEASUREMENTS OF POLYMER SIZE DISTRIBUTIONS
      INDUCED BY CHAIN TRANSFER AGENTS.

(7)   DETERMINATION OF THE DISTRIBUTION OF POLYMER
      COMPOSITION IN COPOLYMERIZATION.

(8)   RATE STUDIES IN THE ABSENCE OF TERMINATION.

(9)   STUDIES OF IONIC POLYMERIZATION INDUCED BY WELL
      DEFINED FREE IONS.

(10) ACTUAL MEASUREMENTS OF THE "NOISE" IN THE
      POLYMERIZATION REACTION, AND USE OF THE INFORMATION
      WHICH IT CONTAINS.

(11) PARAMETERS AND PHENOMENA MAY BE UNIQUE (IN SOME CASES)
      TO HOMOGENEOUS GAS PHASE POLYMERIZATION - BUT CAN STILL
      SHED LIGHT ON CONDENSED PHASE PROCESSES.

Slide 3: Polymerization Science--11 subitems.

the method to be described, the free radicals are observed directly,
and the arriving polymers are, in fact, themselves free radicals. As
an addendum, it should be mentioned that, chain growth involving ions
can also be studied by the method. In this case the ions are directly
observed.

The steady state is established as a balance between the
processes of chain propagation and chain termination. In our method
we are able to work (in fact we must work) under circumstances such
that termination is absent. This is a unique situation.

Item 4 on the slide is another example of a determination unique
to the new method. It is well known that the free radicals which
initiate the chains can be generated photochemically or by the use of
"initiators" (activated either photochemically or thermally). Some
polymers are said to polymerize thermally by themselves. There is

however controversy over whether, in such cases, there may be a trace
initiating impurity which, in the end, is responsible for generating
the chain. As I describe later, our gas phase experiments are
conducted in a "cloud chamber" in which the monomer vapor is supplied
from a pool of liquid monomer. If an <u>involatile</u> radical scavenger is
added to this pool, it will ultimately consume any trace initiator,
and the vapor in which the polymerization is conducted will then not
contain initiator. Because the scavenger is involatile, it cannot
reach the vapor phase in order to inhibit polymerization, and if
polymerization occurs under such circumstances it can only be a
self-generated thermal process. Some of the controversy may therefore
be resolved.

Item 5 on the slide is also of interest. Polymer chemists
usually define "initiator efficiency" as the fraction of radicals,
produced by the initiator, not wasted by other processes not
contributing to chain growth. Ordinarily there is a rate constant
characterizing the rate of decomposition of the initiator to free
radicals and a quantity (fraction) F defined as the efficiency. It is
not easy to separate these two parameters by means of experiment. Our
gas phase method offers a possibility which will be elaborated further
in connection with slides 6, 7, and 8.

Item 7 dealing with copolymerization is also unique. Usually, in
copolymerization (non-block copolymerization) it is difficult to
measure the rates at which polymers of different copolymerized
compositions are produced. As it will be clear when I describe our
method, it is possible to measure the "noise" in the chemical rate.
When copolymerization is involved, the compositional variety
contributes to the noise, so that, in principle, it should be possible
to extract something about the dependence of the rate on composition
from this noise.

Item 11 on the slide is also to be noted. The phenomena and
parameters associated with gas phase polymerization may be unique to
the gas phase. For example the rate constants are probably not
identical with those observed in condensed systems. In our method the
polymer molecule is grown in a supersaturated vapor of its monomer.
As the polymer grows it may absorb monomer molecules from the vapor so
that it grows within a quasidrop of monomer liquid which becomes more
droplike as the polymer gets larger. Under these circumstances the
rate constant for propagation may depend on polymer size, out to quite
large sizes, unlike the situation in condensed phases. Although, in
these situations, the measurement of the rate constant may not be of
direct value to technologies based on condensed media methods, it
still has an intrinsic interest of its own. Furthermore, there are
features of the gas phase process which are of value to the
understanding of phenomena in condensed media. An example is item 4,
on the slide, dealing with self-thermal polymerization.

Slide 4 deals with the possible direct impact of gas phase
polymerization on technology. A process of some technological
interest involves the aerosol formed in the studies of Melville and
his coworkers. Although heterogeneous as well as homogeneous kinetics
are involved, the homogeneous processes are still part of the overall

POLYMER AND OTHER TECHNOLOGIES

(1) GAS PHASE AEROSOL POLYMERIZATION MAY HAVE SOME OF
     THE FEATURES OF EMULSION POLYMERIZATION, E.G.
     DECOUPLING OF RATE AND MOLECULAR WEIGHT. GAS PHASE
     STUDIES CAN DEFINE THE PARAMETERS INVOLVED AND SUGGEST
     THE THEORETICAL FRAMEWORK.

(2) THE TECHNIQUE FOR STUDYING THE GAS PHASE PROCESS
     INVOLVES, DIRECTLY, REACTION AND GAS TO PARTICLE
     CONVERSION. THIS IS OF DIRECT INTEREST TO
     ATMOSPHERIC ENVIRONMENTAL SCIENTISTS.

Slide 4: Possible direct impact of gas phase polymerization on
technology.

scheme, and any model developed for the description of the whole
process will benefit from a knowledge of the parameters associated
with its homogeneous parts. Such "aerosol polymerization" may have
some of the desired features of emulsion polymerization, especially
the feature of decoupling rate from molecular weight. As a
consequence the "aerosol process" could be of technological interest.
    Item 2 on slide 4 speaks for itself and could be of direct
interest to atmospheric scientists.
    Slide 5 deals with applications to phenomena involved either in
other aspects of polymer science besides polymerization itself, or in
nonpolymer science. I leave the six items on the slide to speak for

OTHER ASPECTS OF POLYMER AND
NONPOLYMER SCIENCE

(1) STUDY OF INTRAMOLECULAR PHASE TRANSITION. THE
     COIL-GLOBULE TRANSTION

(2) STATISTICAL MECHANICS OF A SINGLE POLYMER
     MOLECULE IN A SMALL CLUSTER OF ITS OWN MONOMER
     LIQUID. ABILITY TO CHECK THEORY BY MEASUREMENT.

(3) DIRECT MEASUREMENTS ON CONDENSATION NUCLEI.

(4) DIRECT MEASUREMENT OF NOISE IN CHEMICAL RATE
     PROCESSES AND RELATION TO IRREVERSIBILITY.

(5) MEASUREMENT OF PARTICLE TRANSPORT IN THE 10
     TO 100 ANGSTROM "BLIND" RANGE.

(6) DETECTION OF SINGLE FREE RADICALS.

Slide 5: Applications to nonpolymer science.

$$I \xrightarrow{k_d} 2R_1 \cdot$$

$$R_1 \cdot + M \xrightarrow{k_p} R_2 \cdot$$

$$R_2 \cdot + M \xrightarrow{k_p} R_3 \cdot$$

etc.

$$I = I_0 \, e^{-k_d t}$$

$$dP_1/dt = 2k_d FI - k_p MP_1, \qquad dP_2/dt = k_p MP_1 - k_p MP_2, \quad \text{etc.}$$

Where F is the initiator efficiency.

Slide 6: Initiation and propagation.

themselves, and refer any further comments to the discussion which follows this presentation.

As indicated earlier, the case of "initiator efficiency", item 5 on slide 3, will be elaborated further. Slides 6, 7, and 8 accomplish this. In slide 6 we show several reactions, the first being the disassociation of the Initiator to produce two free $R_1 \cdot$. The quantity k_d is the rate constant for this process. The next two reactions each involving a rate constant k_p describe propagation. In the first reaction, radical $R_1 \cdot$ adds to a monomer molecule denoted by M (usually a vinyl monomer with a double bond) to produce a larger radical $R_2 \cdot$ then $R_2 \cdot$ propagates to $R_3 \cdot$, etc. The decomposition of the initiator is a first order process and the concentration of the initiator, as a function of time t, is described by the fourth relation on the slide. Here we also use I to denote the concentration of initiator (whereas above it has been used to denote the initiator molecule), and I_0 is the initial initiator concentration. The last two equations on the slide are the rate equations corresponding to the first two propagation steps illustrated in the slide. In these equations M is the monomer concentration (in the above equations M is also used to indicate the monomer molecule) while P_1 and P_2 are the concentrations of radicals $R_1 \cdot$ and $R_2 \cdot$ while F represents the initiator efficiency.

Because all the differential equations are linear, the system is easily solved. Thus it is a simple matter to show that the total concentration of polymer radicals P_{tot} is given, as a function of time, by the first equation on slide 7. Furthermore, the concentration of polymer radicals of degree of polymerization n is given by the second equation on the slide. The last form on the slide represents the sum in the second line as the truncated exponential (incomplete gamma function) EXP_{n-1}. It will be noticed that all of these equations involve F as well as the efficiency k_d.

$$P_{TOT}(t) = \sum P_n(t) = 2FI_0(1 - e^{-k_dt})$$

$$P_n(t) = \frac{2k_dFI_0}{k_pM-k_d} \left(\frac{k_pM}{k_pM-k_d}\right)^{n-1} \left\{ e^{-k_dt} - e^{-k_pMt} \sum_{j=0}^{n-1} \frac{[(k_pM-k_d)t]^j}{j!} \right\}$$

$$= \frac{2k_dFI_0}{k_pM-k_d} \left(\frac{k_pM}{k_pM-k_d}\right)^{n-1} \left\{ e^{-k_dt} - e^{-k_pMT} \mathrm{EXP}_{n-1} [(k_pM-k_d)t] \right\}$$

Slide 7: Polymer radical concentrations.

From the differential equations on slide 6 it is also easy to derive an expression for the mean radical size $\langle n \rangle$ as a function of time. This appears as the first equation in slide 8. Furthermore it is also possible to derive an expression for the relative variance in radical size. This appears as the second equation in slide 8. It is important to notice that, in these equations, k_d appears, but not F. Thus, by measuring the moments implicit in the expressions on slide 8, we can measure k_d if we know k_p. Then by measuring the quantities expressed in slide 6 we can determine F, k_d having been already determined.

In condensed media it is also possible to perform measurements which separate k_d and F. A useful method, in this respect, is the so called "dead end" method of Tobolsky[8]. However, in this method it is also necessary to know the rate constant for termination k_t. Since our gas phase method will allow the measurement of the quantities appearing on the left of the equations in slides 7 and 8, it can accomplish the determination of initiator efficiency without requiring knowledge of k_t. However, that initiator efficiency may be peculiar to the gas phase, and not of much value in condensed media.

$$\langle n \rangle = 1 + k_pM \left\{ \frac{t}{1-e^{-k_dt}} - \frac{1}{k_d} \right\}$$

$$\frac{\Delta}{\langle n \rangle} = \frac{(\langle n^2 \rangle - \langle n \rangle^2)^{1/2}}{\langle n \rangle} =$$

$$= \frac{\left\{ \frac{k_pM}{k_d} \left(\frac{k_pM}{k_d} - 1\right) + \frac{k_pMt}{1-e^{-k_dt}} - \frac{k_p^2M^2t^2e^{-k_dt}}{(1-e^{-k_dt})^2} \right\}^{1/2}}{1 - k_pM/k_d \ (1 - (k_dt/(1-e^{-k_dt})))}$$

Slide 8: Moments of the polymer radical distribution.

II. Method of Study: Nucleation

After somewhat lengthy preamble, we finally come to a discussion of the method itself. In the method, growing polymers are detected by having them grow in supersaturated monomer vapor, nucleating drops of monomer liquid when they attain a certain size. The reaction is conducted in a cloud chamber where the monomer vapor can be maintained in a supersaturated state. Thus far, experiments have been performed in an upward thermal <u>diffusion</u> cloud chamber, but ultimately, as I will indicate later, the most quantitative results will require the use of an <u>expansion</u> cloud chamber. Sensitive detection of single polymer molecules is possible because of the extremely "critical" nature of the nucleation phenomenon.

In order to understand this phenomenon it is convenient to consider a supersaturated vapor, consisting of a single molecular

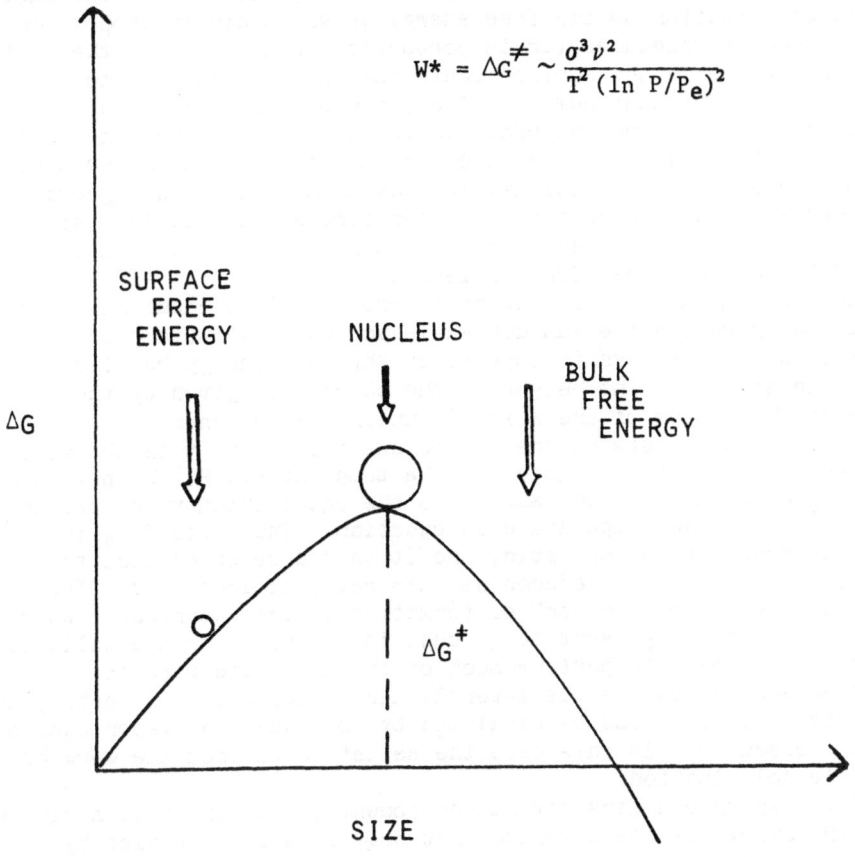

$$W* = \Delta G^{\neq} \sim \frac{\sigma^3 \nu^2}{T^2 (\ln P/P_e)^2}$$

Note that in a multicomponent system the peak is at a saddle point.

Slide 9: Origin of the nucleation free energy barrier.

species, in which a fragment or cluster of the stable liquid phase is
formed. Such clusters are often referred to as "embryos" of the new
phase. In the simplest theory the embryo is treated as a liquid drop,
having all the features of a macroscopic liquid, e.g. surface tension,
density, etc. One then investigates the reversible work (free energy)
which must be expended in forming a liquid drop of n molecules within
the original supersaturated vapor phase. Slide 9 addresses this
question. When the drop is very small reversible work must be
expended in the formation of the interface between the new drop and
the surrounding vapor. This involves an increase of free energy.
However, since the vapor is supersaturated, the formation of the bulk
of the drop involves a decrease in free energy. However, when the
drop is small the positive interface contribution dominates the whole,
so that increasing the size of the embryo (increasing the interface
area) requires an <u>increase</u> in free energy, and is therefore not
spontaneous. However, as the size of the embryo is further increased
the bulk contribution to the free energy grows in direct proportion
to the number of molecules (or in porportion to the cube of the radius
of the embryo) while the surface contribution goes only as the
two-thirds power of that number. Thus, the bulk contribution
eventually overtakes the interface contribution, and turns the curve
around so that a maximum is produced as in slide 9. An embryo having
the size corresponding to the maximum may undertake further growth
with a decrease in free energy, and therefore spontaneously. As
a result, the embryo corresponding to the size at the maximum is
referred to as a condensation "nucleus".

Thus the process of nucleation is essentially a chemical reaction
in which the products are van der Waals molecules (clusters or
embryos), and the process is impeded by the free energy barrier
sketched in slide 9. The height of the barrier is given by the
formula at the bottom of the slide in which σ represents the
interfacial tension between the liquid and its vapor, v is the volume
per molecule in the bulk liquid, T, the temperature, P, the pressure
in the supersaturated vapor, and P_e is the equilibrium vapor pressure
of the liquid at the temperature in question. The ratio P/P_e is
called the supersaturation ratio, and it should be noted that the
height of the barrier is reduced as this ratio is increased. The
barrier is created by the work of formation of the interface. When
surfaces are already present (e.g. dust in the vapor or the walls of a
container), the need to perform much of the interface work is
eliminated and the barrier is lowered; the condensation is "catalyzed"
by surfaces. It can also be catalyzed by ions when the vapor consists
of polar molecules. In this case the assist comes from the work of
dielectric polarization.

If the vapor contains several components, e.g. if it is a binary
vapor, the embryo is characterized not only by size, but also by
composition. In this case the simple free energy "hill" appearing in
slide 9 must be replaced by a free energy "surface" overlaying a base
plane in which a point denotes the numbers of molecules of <u>both</u>
species in the embryo. It is easy to show that this free energy
surface contains a saddle through which all the growth occurs. The

embryo corresponding to the nucleus now has the size and composition associated with the saddle point. In fact, this free energy surface is nothing more than the common potential energy surface familiar to ordinary chemical kinetics and to transition state theory. It is possible to show that the density fluctuations, to which the embryo sizes correspond, are canonical variables which can be canonically transformed to the intramolecular coordinates of the convential pontential energy surface. The nucleus corresponds to the transition state.

In order to derive the rate of nucleation, i.e. the number of droplets formed per cubic centimeter per second in the supersaturated vapor, it is necessary to calculate the rate at which fluctuations occur which carry embryos to the top of the free energy barrier. Thus in the case of a one component vapor we are confronted with the consecutive, reversible reactions illustrated in slide 10 where M_2

$$M + M \rightleftharpoons M_2, \quad M_2 + M \rightleftharpoons M_3, \text{ etc.}$$

$$J = KE^{-W^*/kT}$$

Slide 10: Consecutive, reversible reactions.

indicates an embryo containing two molecules, M_3 an embryo containing three molecules, etc. With the use of detailed balancing these consecutive reactions can be dealt with analytically to yield for the rate J (drops per cubic centimeter per second) the final equation in slide 10 in which, K is a prefactor which depends only slightly on the supersaturation ratio or "supersaturation", W^* is the height of the barrier prescribed in slide 9, k is the Boltzmann constant and, again, T is the temperature.

In order to see how "critical" the process of nucleation is we apply the equation for J, in slide 10, to the case of supersaturated water vapor at 300° K. The table in slide 11 lists values of J, drops

H_2O 300 K

$S = P/P_E$	Drops $cm^{-3} sec^{-1}$	$S = P/P_E$	Drops $cm^{-3} sec^{-1}$
1.2	3×10^{-996}	3.0	1.8×10^{-2}
1.4	8.1×10^{-275}	3.107	1.0 ← CRITICAL SUPER-
1.6	1.5×10^{-128}	3.2	22.3 SATURATION
1.8	4.9×10^{-73}	3.4	6.7×10^3
2.0	2.4×10^{-45}	3.6	7.1×10^5
2.2	3.2×10^{-29}	3.8	3.5×10^7
2.4	8.0×10^{-19}	4.0	9.5×10^8
2.6	1.2×10^{-11}		
2.8	2.0×10^{-6}		

Slide 11: Values of J as a function of supersaturation S.

$cm^{-3} \sec^{-1}$, in the second column, as a function of supersaturation S
(first column). A supersaturation at 1.2 corresponds to 120% relative
humidity. Under normal circumstances one would expect water vapor to
be condensing rapidly under this condition. In fact, the theory which
agrees well with experiment, indicates that the rate is of the order
of 10^{-996} drops $cm^3 \sec^{-1}$. Stated in another way, this means that on
the average it would require 10^{996} seconds for a single drop to appear
within one cubic centimeter. Since a year contains on the order of 2
X 10^7 seconds, condensation would, in effect, never occur at 120%
relative humidity. The fact that it does occur rapidly at 100%
relative humidity is due to the presence of dust particles, walls, and
other surfaces. If we increase the supersaturation to 2.8 (relative
humidity 280%) the table in slide 11 indicates that it would still
require on the order of a million seconds for a single drop to appear
within one cubic centimeter. At 300% relative humidity, we begin to
intersect the laboratory time frame, and only about 50 seconds is
required. Finally at 310.7% relative humidity, only 1 second is
required. Clearly the table in slide 11 leads to a plot of rate of
nucleation versus supersaturation having the character of the curve in
slide 12. There is a catastrophic collapse of the metastable state at
a particular value of supersaturation. This is not an equilibrium
phenomenon, but the point at which the metastable state collapses is
very reproducible, and the supersaturation, at this point, is referred
to as the critical supersaturation.

The process in which the condensation of supersaturated monomer
vapor is nucleated by a polymer molecule involves at least a binary,
nucleus containing both monomer molecules and the polymer. Processes
in which the nucleus contains more than one polymer molecule, and
possibly several polymer molecules having different degrees of
polymerization, are to be avoided, since it is almost impossible to
deconvolute and analyze the experimental data under such
circumstances. In fact, such multipolymer processes represent an
incipient return to the heterogenous aerosol process investigated by
Melville. It is necessary, therefore, to demonstrate that
experimental conditions can be achieved such that only one polymer
molecule is involved in the nucleus. The figure at the top of slide
13 suggests the configuration of such a nucleus. In the figure the
wandering line represents the polymer molecule while the M's are
monomer molecules. Thus the nucleus is a small drop of monomer liquid
in which a single polymer moleucle is dissolved. The free energy for
formation of this drop of solution characterizes the free energy
barrier of the nucleation process. A schematic of this barrier
appears at the bottom of slide 13. An "embryo" may be modelled as a
drop consisting of n monomer molecules and a single polymer molecule.
In the "reaction" exhibited at the middle of slide 13 an embryo
containing n monomer molecules acquires another monomer (reversibly)
to produce an embryo having n + 1 monomers. Because of this
reversibility, the growth of the embryo to the nucleus size may be
viewed as a random walk in size space, steps forward and backwards
corresponding, respectively, to the acquisition and loss of monomer
molecules.

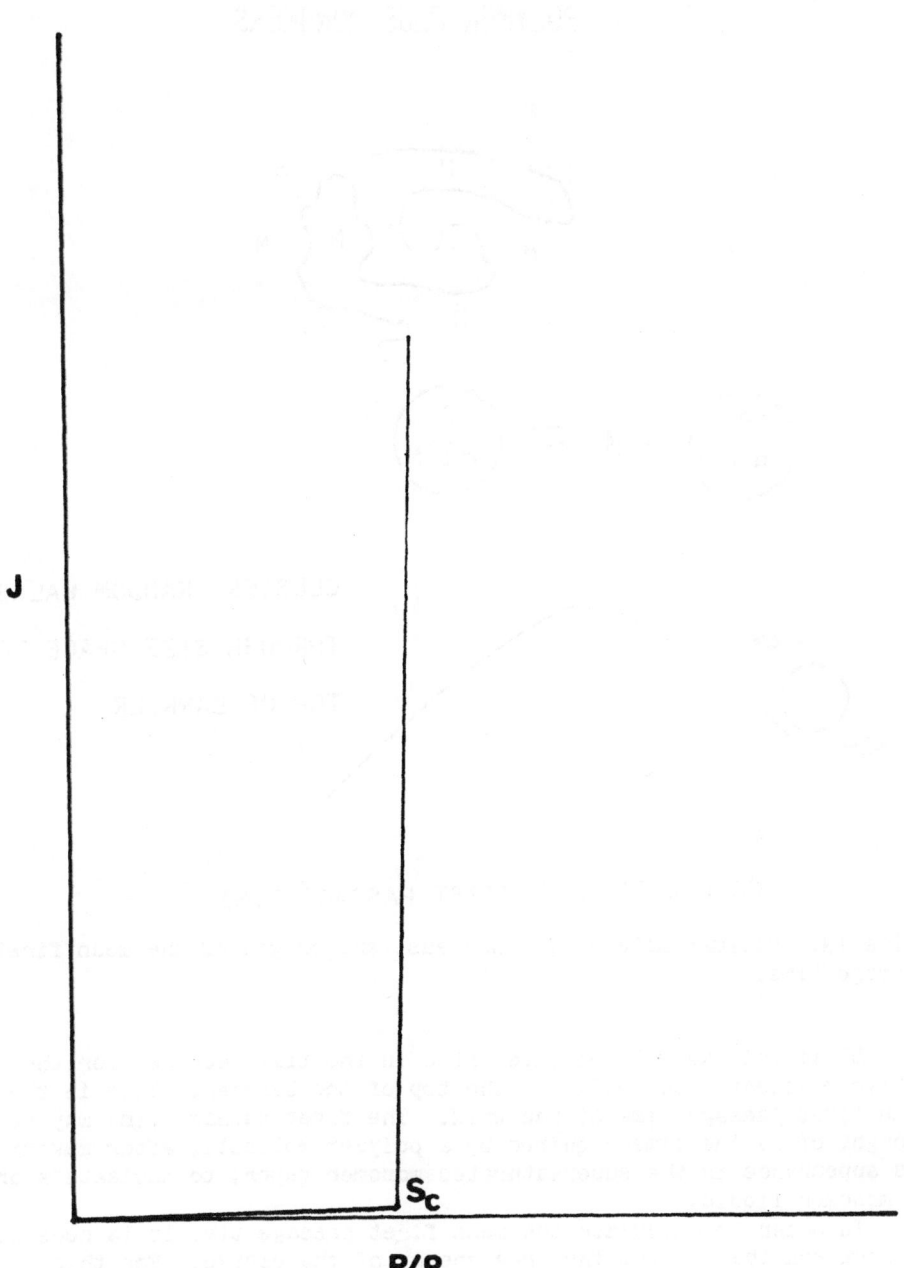

Slide 12: Nucleation versus Supersaturation.

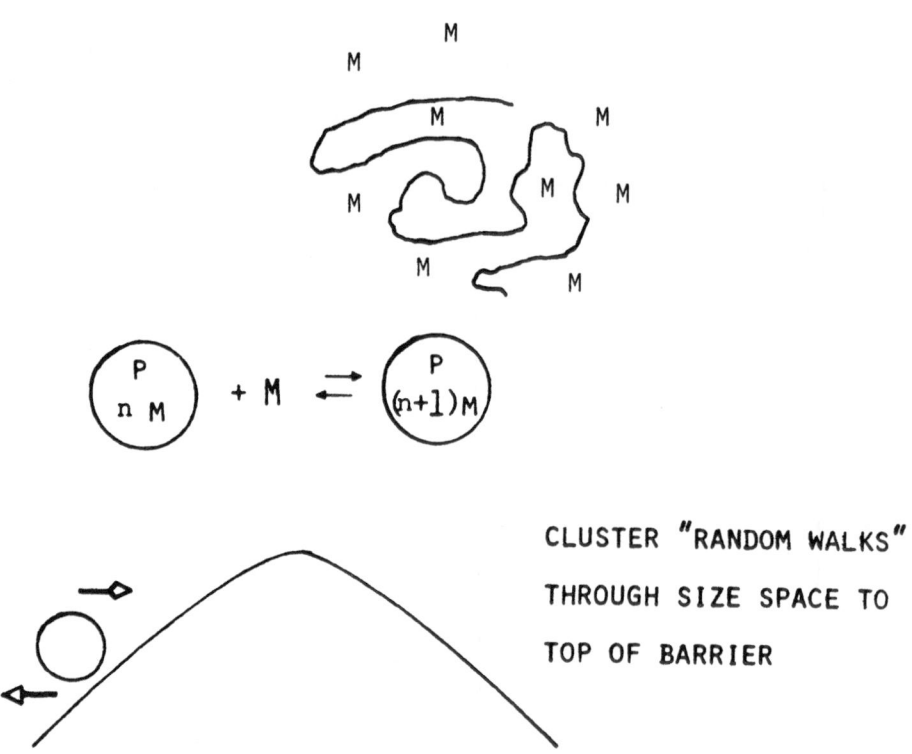

NUCLEUS CONSISTING OF SINGLE
POLYMER PLUS MONOMERS

CLUSTER "RANDOM WALKS"
THROUGH SIZE SPACE TO
TOP OF BARRIER

CALCULATE MEAN FIRST PASSAGE TIME

Slide 13: Polymer molecule, a nucleus, and origin of the mean first
passage time.

    Ultimately we will be interested in the time required for the
embryo or cluster to "walk" to the top of the barrier. This is the
mean first passage time of the walk. The first passage time may be
thought of as the time required by a polymer molecule, after making
its appearance in the supersaturated monomer vapor, to nucleate a drop
of monomer liquid.
    In order to calculate the mean first passage time it is necessary
to know something about the free energy of the embryo. For this
purpose a crude theory (reliable only for the purpose of demonstrating
trends) has been developed using the Flory-Huggins theory of polymer
solutions, modified to account for the interface between the drop of
solution (embryo) and the vapor. Slide 14 lists some results for the

VINYL ACETATE

280°K   CRITICAL SUPERSATURATION FOR HOMOGENEOUS
         NUCLEATION OF VINYL ACETATE VAPOR
         S_C = P/P_e = 3.81

NUCLEATION OF VINYL ACETATE BY POLYVINYL ACETATE

S_C = P/P_e	DEGREE OF POLYMERIZATION	MEAN FIRST PASSAGE TIME (SECS)
2.7	22	5.1
	26	2.0×10^{-4}
	28	5.3×10^{-6}
1.7	160	2.0×10^{6}
	183	1.1×10^{-6}
1.3	1500	7.9×10^{16}
	1610	7.0×10^{-7}
1.25	2000	1.0×10^{122}
	2700	4.0×10^{-6}

Slide 14: Mean first passage time vs. degree of polymerization.

mean passage time in the case where the monomers are vinyl acetate
molecules (a system which we have studied more thoroughly than any
other). The results are for a temperature of 280°K. Theory shows
that, for the case of homogeneous nucleation involving pure (no
polymers) vinyl acetate vapor, the critical supersaturation at which
the metastable, supersaturated state collapses catastrophically (the
critical supersaturation) is 3.81 (at 280°K). This critical
supersaturation is denoted by S_C and is listed, at the top of slide
14. In the table constituting the remainder of the slide, the first
column lists supersaturations, the second column, degrees of
polymerization, and the third column, the corresponding mean first
passage times in seconds.
    As an example, consider a monomer vapor at a supersaturation of
1.7. The table shows that at 280°K a polymer (polyvinyl acetate) with
a degree of polymerization of 160 will have a mean first passage time
of 2 X 10^6 seconds. On the other hand, increasing the size of the
polymer by about 12% to a degree of polymerization of 183, reduces the
first passage time to the order of a microsecond. A 12% increase in
the size of the polymer gives rise to a reduction of 12 orders of
magnitude in the first passage time. This illustrates again the
"critical" nature of the nucleation process. More important, however,
it illustrates the principle upon which our method of detection of
arriving polymer molecules is based. A supersaturation of 1.7 is
far less than 3.81, and no homogenous nucleation can be expected in
the absence of polymers. Furthermore, a polymer of size 160 will not

nucleate a drop. However, a modest increase in its size to 183, due
to propagation, allows it to nucleate a drop within a microsecond.
Since the propagation of the chain proceeds much more slowly than
this, nucleation is not rate controlling, and the appearance of the
drop signals the arrival of the molecule of the critical size. If the
propagation rate is much faster (as in the case of ionic chain
propagation) the critical polymer will have a larger size, and its

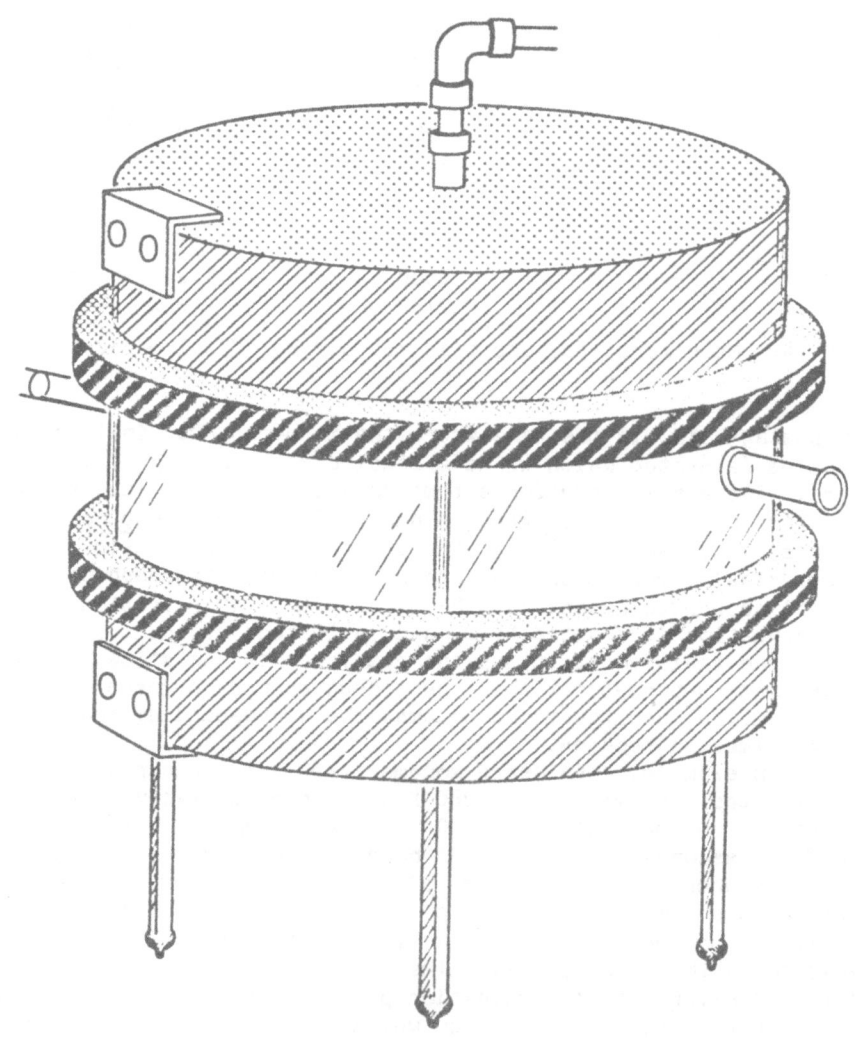

Slide 15:   Artist's drawing of cloud chamber.

mean first passage time will be even smaller, but there will always be a size for which the nucleation process will be so fast that it will not be rate controlling. It is obvious from the slide that one can "tune" to the arrival of polymer molecules of a specified size by controlling the supersaturation of the vapor.

In the case of "copolymerization" mentioned on slide 3, the cortical size will depend on the composition of the polymer, and will contribute to the "noise" observed in the cloud chamber experiments.

As I indicated earlier, the experiments are performed in a cloud chamber. At present, only an upward thermal diffusion cloud chamber has been used. (In the future it is desirable, as I explain later, to employ an expansion cloud chamber.) The diffusion cloud chamber is a simple device, useful for semiquantitative measurements on the kinetics of gas phase polymerization, and especially for the demonstration of feasibility.

Slide 15 is an artist's drawing of the device. It consists of two circular metal plates separated by a glass cylinder. The lower plate is heated while the upper one is cooled. The tubes attached to the glass ring, as shown in the slide, are graded seals which hold quartz windows so that uv light can be admitted to the chamber. The chamber is filled with helium gas, and, a shallow pool of liquid monomer (vinyl acetate) rests on the lower plate. Since the lower plate is heated monomer evaporates from the pool, and diffuses through the helium, to condense on the cooler upper plate where it forms a smooth film of liquid. This film drains to the side, and returns via the glass ring to the pool. Thus a constant state of reflux is established within the chamber. Convection within the chamber must be avoided, and this is accomplished by using helium as the carrier gas so that the hotter gas near the lower plate is nevertheless heavier by virtue of it being mass loaded with vinyl acetate.

Pressures and temperature vary with elevation in the chamber in the manner shown in slide 16. The curve labeled T, shows that the temperature drops, almost linearly with elevation, from the temperature of the lower plate to that of the top one. The partial pressure P of the diffusing vinyl acetate also drops almost linearly with elevation. However the equilibrium pressure P_e drops exponentially with temperature, according to the Clapeyron–Clausius relation, and has the form shown in the slide. The supersaturation S, i.e. the ratio of P to P_e therefore has the form exhibited on the right side of the slide. There is a maximum of supersaturation at an elevation corresponding to about three-quarters of the height of the chamber. In our experiments we attempt to position the uv beam, as shown in the slide, at the level of this maximum. The curves in slide 16 must be computed by simultaneously solving the equations for the transport of mass, energy, and momentum. Usually, the momentum equation can be neglected, and replaced by the requirement that the pressure be constant within the chamber. The chamber clears itself of unwanted condensation nuclei (dust particles), since they form drops which settle out of the vapor space.

Further detail of the experiment is shown in slide 17. Here we see the shallow pool (cross hatched) of vinyl acetate liquid, the

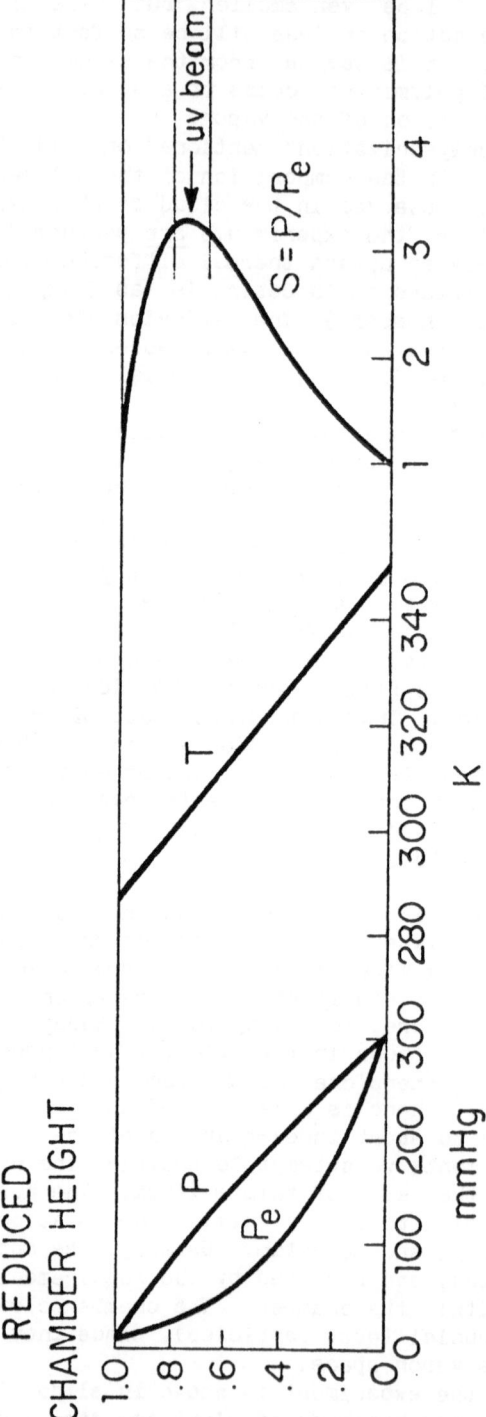

Slide 16: Variance of pressure and temperature with elevation.

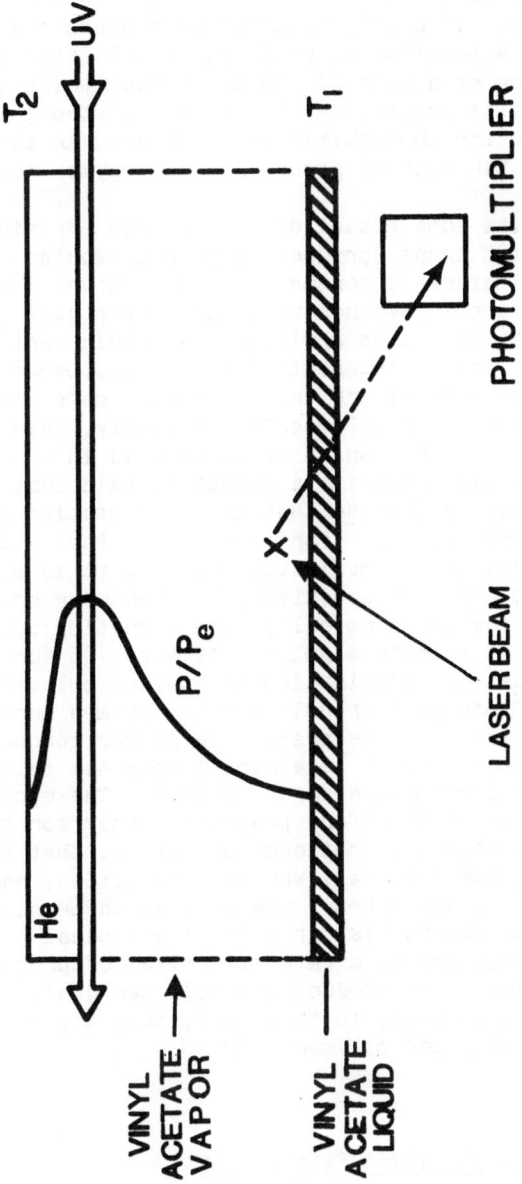

Slide 17: Further detail of the experiment.

supersaturation curve P/P_e, and the uv beam. The X marks the position
of a helium-neon laser positioned below the uv beam, and at right
angles to it. The uv photons produce free radicals which propagate
into polymer chains. These grow and diffuse. Eventually diffusing
polymer radicals reach the walls of the chamber where they are
adsorbed. This loss to the walls must be accounted for in the
analysis of the data. If a polymer molecule reaches the critical
size, when it is at a position directly above the laser beam, it
induces the formation of a drop of liquid monomer which then falls
through the beam and is counted by virtue of the strong light signal
which it scatters to the photomultiplier indicated in the slide. The
count rate measures the rate of production of polymer molecules of the
critical size.

Slide 18 exhibits some actual data from this experiment. Three
histograms, which plot drops (product polymer molecules) observed
during a given time interval, versus time, are shown. The bars in the
uppermost histogram are 10 seconds wide, and their heights indicate
the number of product molecules arriving (per cubic centimeter) in the
10 second interval. The uv intensity for this histogram is fairly
high, and the average rate of arrival of product molecules appears to
be in the neighborhood of 40 cm^{-3} sec^{-1}. Actually, this rate is
considerably higher than that which would be used in a careful
experiment. At this high rate it is almost certain that growing
polymers encounter one another so that the condensation nucleus
contains more than one polymer. For example, if the critical polymer
size involves a degree of polymerization of 100, it is possible for
polymers of sizes 30 and 70 respectively to encounter one another, and
even if they do not combine chemically, the resulting complex can, to
a first approximation, imitate a polymer of size 100 for the purposes
of nucleation. Almost any combination of smaller polymers adding to
size 100 will do. These smaller polymers are formed much more quickly
than the larger polymer, and are present in larger concentrations.

As I indicated earlier, it is almost impossible to deconvolute
the data when multipolymer nuclei are involved. The uppermost
histogram in the slide is therefore presented, only for the purposes
of illustration. It should be noticed, of course, that it is "noisy".
The two lower histograms involve lower uv intensities, and
consequently exhibit slower rates. The bars in these histograms are
20 seconds wide. The lowest histogram is more typical of a careful
experiment. It corresponds to a rate of arrival of product polymer
molecules equal to about one product molecule per cubic centimeter per
second. It is almost certain, in this case, that the condensation
nuclei each contain only one polymer.

III. Steady State Experiments: Feasibility

Slide 19 illustrates some reactions involved in the free radical
chain polymerization of vinyl acetate. The scheme includes initation,

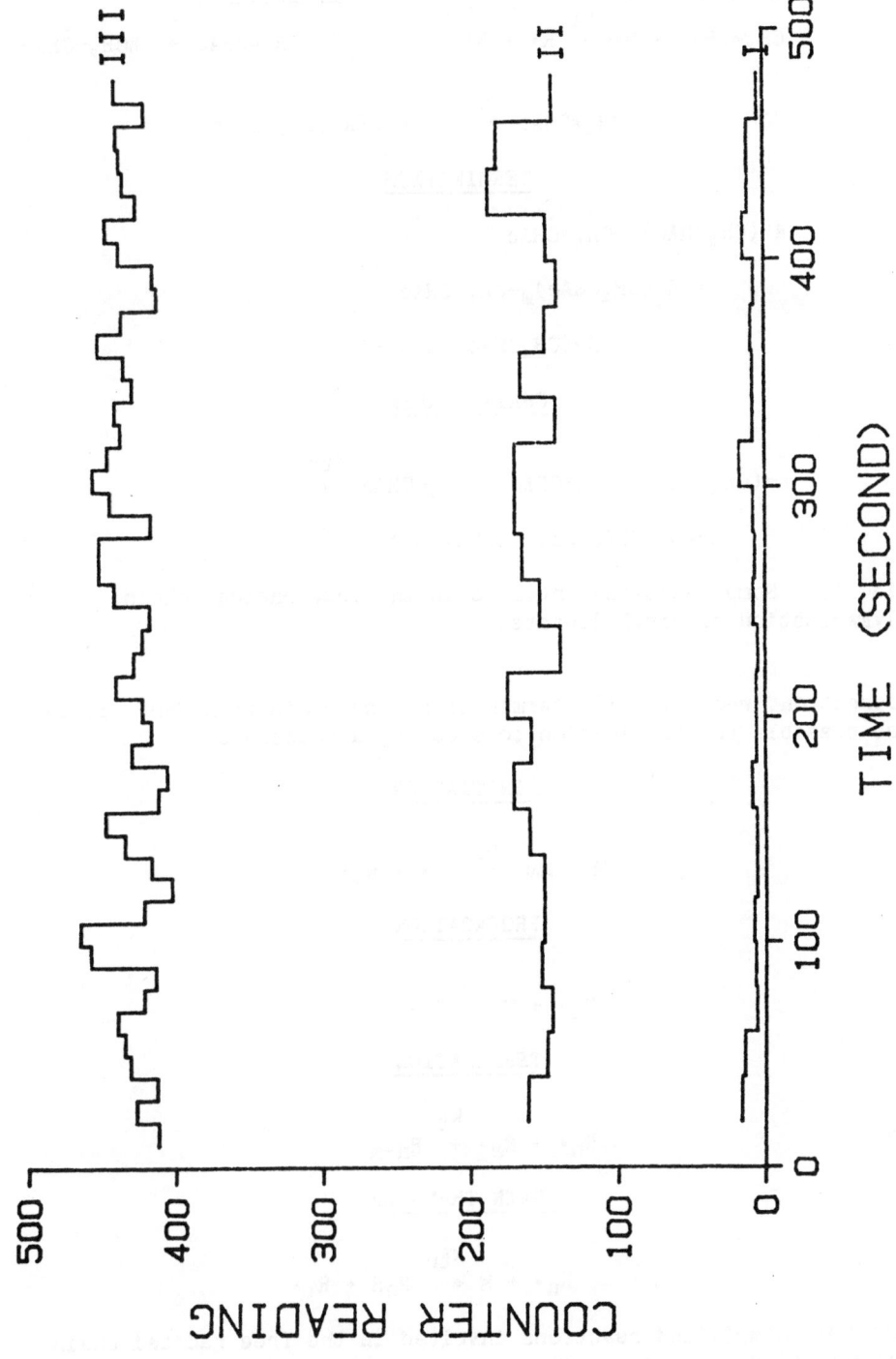

Slide 18: Data from experiment

<u>INITIATION</u>   <u>PROPAGATION</u>

$CH_2=CHAc + h\nu \xrightarrow{k_i} R\cdot + R'\cdot$ $R\cdot\ CH_2=CHAc \xrightarrow{k_p} RCH_2-\overset{\cdot}{C}HAc$

$RCH_2-\overset{\cdot}{C}HAc + CH_2=CHAc \xrightarrow{k_p} RCH_2-CH(Ac)CH_2-\overset{\cdot}{C}HAc$

<u>TERMINATION</u>

$R-(CH_2CHAc)_n-CH_2-\overset{\cdot}{C}HAc$

$\quad + R-(CH_2CHAc)_m-CH_2-\overset{\cdot}{C}HAc \xrightarrow{k_t}$

$\quad R-(CH_2CHAc)_{n+m+2}-R$

<u>CHAIN TRANSFER</u>

$R-(CH_2CHAc)_n-CH_2-CHAc + CH_2-CHAc \xrightarrow{k_{tr}}$

$\quad R-(CH_2CHAc)_{n+1} + CH_2=\overset{\cdot}{C}HAc$

Slide 19: Some reactions involved in the free radical chain
polymerization of vinyl acetate.

propagation, recombinative termination, and chain transfer. It is
also possible for termination to occur by a process of

<u>INITIATION</u>

$M + h\nu \xrightarrow{k_i} R_1\cdot + R_1'\cdot$

<u>PROPAGATION</u>

$R_1\cdot + M \xrightarrow{k_p} R_2\cdot$

<u>TERMINATION</u>

$R_n\cdot + R_m\cdot \xrightarrow{k_t} R_{n+m}$

<u>CHAIN TRANSFER</u>

$R_n\cdot + M \xrightarrow{k_{tr}} R_nH + R_1\cdot$

Slide 20: Simplified reactions involved in the free radical chain
polymerization of vinyl acetate.

disproportination, but I do not show this reaction in the slide.
Furthermore, if the reaction is performed in a cloud chamber we must
include a step in which the growing free radicals diffuse to the walls
where they are trapped. This step is not shown in the slide.

Slide 20 presents a simplified version of slide 19. Here M
represents a monomer molecule, and R_n. represents a free radical of
degree of polymerization n. H stands for a hydrogen atom. The
various rate constants k_i, k_p, k_t, k_{tr} corresponding to initation,
propagation, termination, and chain transfer are also shown.

If the uv source is steady, the reactions in slides 18 and 19
(augmented by the step involving loss to the walls) will achieve the
steady state. The concentrations of the various species will be
steady, and can be shown to be given by the formulas appearing in
slide 21. In this slide P_n represents the steady concentration of
free radicals of size n while $\overline{P}_n$ represents the steady concentration
of "dead" polymers resulting from termination. The quantities A and W
appearing in the expression for P_n are complicated functions of the
various rate constants, the monomer concentration M, effectively
constant during the reaction, and the light intensity I. In addition
to the four rate constants appearing in slide 20 there is an
additional one, k_D, which refers to "loss to the walls". One might
assume that dead polymers are effective as nucleating agents in the
same manner as "live" ones. However, under the conditions of our

IN THE STEADY STATE

$$P_n = AW^n$$

$$A = \left\{ \frac{k_i I}{k_p} + \frac{k_{tr}}{k_p} \left[-\frac{k_D}{2k_t} + \left(\left(\frac{k_D}{2k_t} \right)^2 + \frac{k_i IM}{k_t} \right)^{1/2} \right] \right\}$$

$$W = \left\{ \frac{k_p M}{k_p M + k_{tr} M + k_D + k_t \left[-\frac{k_D}{2k_t} + \left(\left(\frac{k_D}{2k_t} \right)^2 + \frac{k_i IM}{k_t} \right)^{1/2} \right]} \right\}$$

$$\overline{P}_n = \left\{ \frac{n-1}{2} \left(\frac{k_t}{k_p k_D} \right) \left[k_i I + k_{tr} - \frac{k_D}{2k_t} + \left(\frac{k_D}{2k_t} \right)^2 + \frac{k_i IM}{k_t} \right)^{1/2} \right. \right.$$
$$\left. \left. + \frac{k_{tr} M}{k_t} \right\} P_n$$

k_D REFLECTS LOSS TO THE WALLS

Slide 21: Steady concentration of polymer radicals.

experiments, the concentrations of dead polymers are neglible. An
important feature to be noted in the formulas of slide 21 is the
appearance of the light intensity in the denominator of the expression

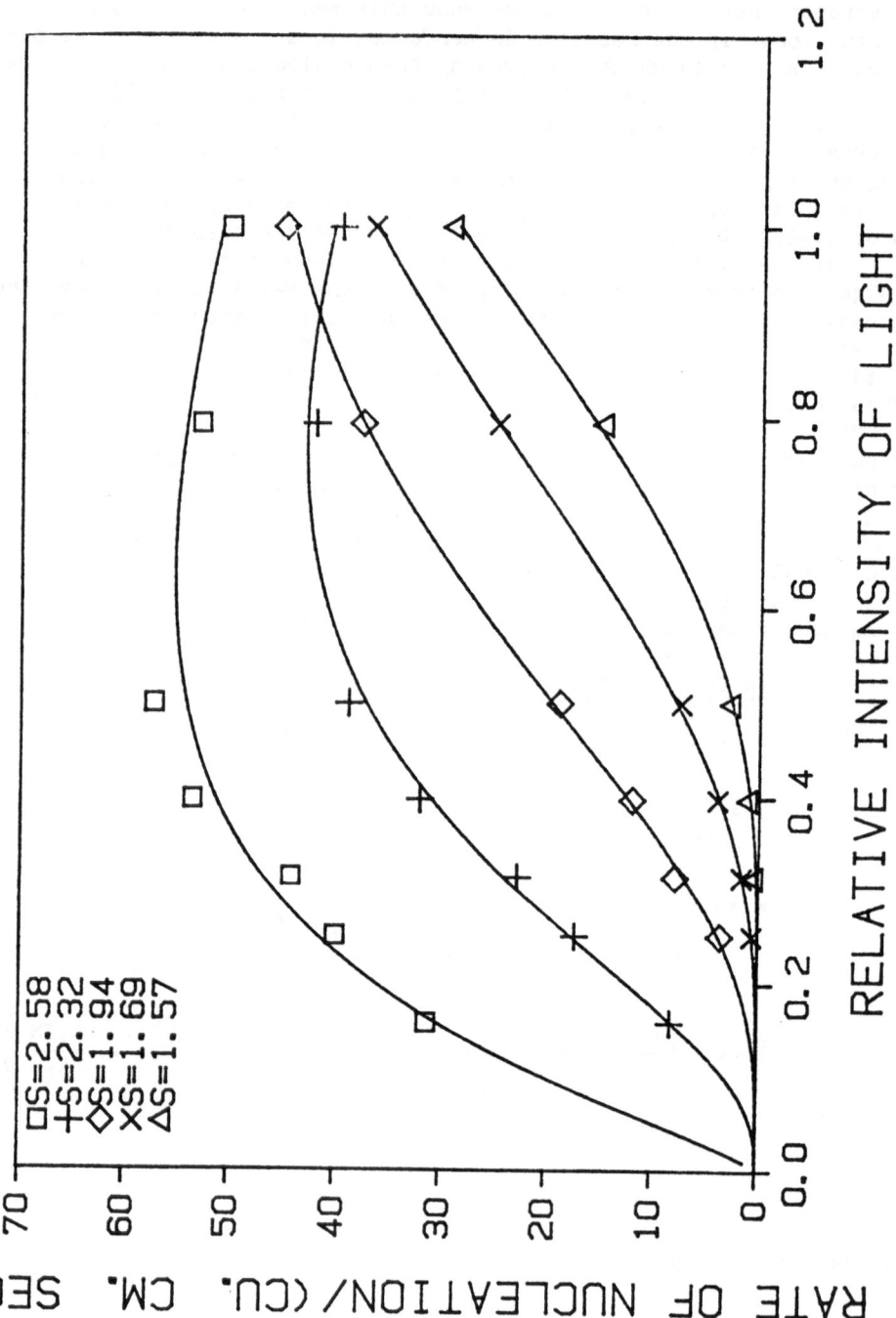

Slide 22: Plots of observed rates of nucleation in vinyl acetate vapor versus relative light intensity.

for W. Thus as the light intensity is increased W becomes smaller. Since W appears to the power of n in the formula for P_n, this implies that the concentrations of large enough polymers will be <u>reduced</u> as the light intensity is increased. Thus the steady state concentration of the polymer involved in the nucleation process will exhibit a maximum as a function of light intensity, and the same will be true for the corresponding rate of nucleation.

The experimental demonstration of this maximum is one of the features which confirms that the observed nucleation is due to the polymer process. Slide 22 contains plots of observed rates of nucleation (in vinyl acetate vapor) versus relative light intensity (270-280 nm). Five plots are shown, corresponding to different supersaturations ranging from 1.57 to 2.58. At the two highest supersaturations, 2.32 and 2.58, the plots do indeed exhibit maxima. It is probable that the three remaining plots would also exhibit maxima if measurements had been continued to higher light intensities. The maximum appears to move to higher light intensity as the supersaturation is reduced. This is consistent with theory. At a high enough supersaturation, nucleation should involve a nucleus containing only one polymer molecule. At lower supersaturations multipolymer nuclei are required. However the individual polymers in these nuclei should be smaller. According to the formulas in slide 21 the maximum occurs at smaller light intensities for larger molecules (larger n).

Actually, as I will soon show, the curve corresponding to a of 2.58 involves a single polymer nucleus, while that corresponding to a supersaturation of 2.32 involves nuclei containing two polymers.

If experiments are performed at low light intensities, i.e. in the regime on the left of slide 22, termination and chain transfer will be absent. Under these circumstances, the formulas of slide 21 indicate that the steady state concentration of polymers of the critical size will depend on the first power of the light intensity. To a high degree of approximation, it can be demonstrated that, if two polymers are involved in the nucleus, the nucleation rate will depend on the square of the light intensity. These claims are sketched in slide 23 from which it can be seen that, for the one polymer process,

FOR NUCLEATION INVOLVING A <u>ONE</u> POLYMER NUCLEUS THE
RATE SHOULD DEPEND ON THE <u>FIRST</u> POWER OF THE UV
INTENSITY

$$J_1 \sim I \qquad \ln J_1 \sim \ln I$$

FOR A <u>TWO</u> POLYMER NUCLEUS THE RELATION SHOULD BE
(IN <u>THIS</u> CASE APPROXIMATE)

$$J_2 \sim I^2 \qquad \ln J_2 \sim 2 \ln I, \text{ etc.}$$

ALL OF THIS IN <u>ABSENCE</u> OF CHAIN TERMINATION ONLY!!

Slide 23: One and two polymer processes.

a plot of the logarithm of the nucleation rate J_1 versus the logarithm
of light intensity will be a straight line of unit slope. Similarily,
for a two polymer process, a plot of the logarithm of the nucleation
rate J_2 versus the logarithm of light intensity will also be a
straight line, but with a slope of 2. Slide 24 exhibits such log-log
plots for experimental data obtained at the five supersaturations
corresponding to the curves in slide 22, but in the regime of low
light intensity. As predicted, they are all straight lines. It is
most important to note that the curve corresponding to the highest
supersaturation, 2.58, has a slope of 1.15, indicating, as I promised
earlier, that nucleation, at this supersatuation, is due essentially
to a one polymer process. The curve for the supersaturation of 2.32
has a slope of 2.03, indicating that two polymers are involved in the
nucleus. Thus it appears feasible to establish experimental
conditions such that the one polymer process is dominant. As I
indicated earlier, the establishment of such conditions is mandatory
if a quantitative interpretation of the data is to be made.
The straight line in slide 24 corresponding to the one polymer
process (supersaturation of 2.58), can be extrapolated to determine
the intercept on the rate axis. An analysis of the equations on slide
21 shows that if the size of the critical polymer in the nucleus is
known, and the values of k_i and k_p are available, then the activation
energy for propagation may be determined. Without entering into
detail, I can report that both k_i and k_p can be estimated in a
reasonably satisfactory manner using data in the literature and some
theory. Unfortunately, the same cannot be said about the critical
size. If we attempt this estimate by means of theory, then we have to
rely on the crude nucleation theory from which the table on slide 14
was derived. As I indicated, that theory is only useful for
demonstrating trends, and is by no means quantitatively accurate.
Accepting this limitation the best thing we can do is the following.
    We use the measured intercept to calculate a "what if?" curve.
That is, we suppose that we know the critical size. Choosing some
number for that size we then employ the measured intercept, the
estimated k_i and k_p, and the theoretical expressions on slide 21 to
calculate a value for the activation energy for propagation. The
curve in slide 25 is obtained in this manner. Thus for the measured
intercept it indicates what we would find the activation energy to be,
if the critical size has the value listed on the abscissa. Having
once plotted this "unbiased" curve we can then appeal to the available
crude nucleation theory for an estimate of that size. It happens
that, for the one polymer process in slide 24, the critical size
estimated by theory is 35. The curve in slide 25 indicates that if
the size was indeed 35, the activation energy predicted for
propagation would approximate 4,000 calories. Surprisingly, this
happens to be the value of the activation energy found for the
propagation of polyvinyl acetate in condensed media. However, because
of the crudeness of the nucleation theory, we cannot attach much
significance to this agreement. It is probably due to coincidence;
excellent agreement, usually been ascribed to compensating errors,
between theory and experiment is often found in studies of nucleation.

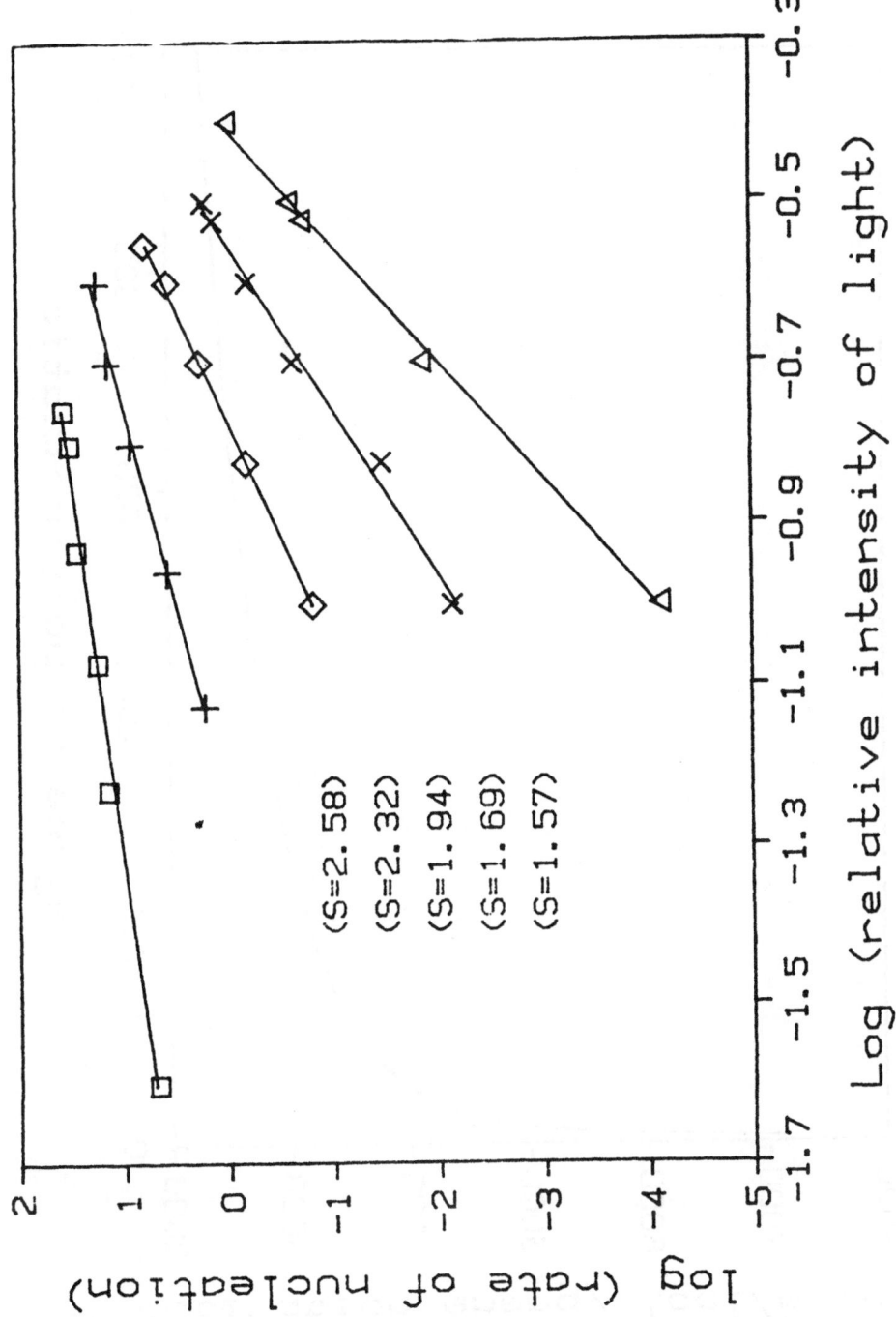

Slide 24: Log-log plots for experimental data obtained at five supersaturations in the regime of low light intensity.

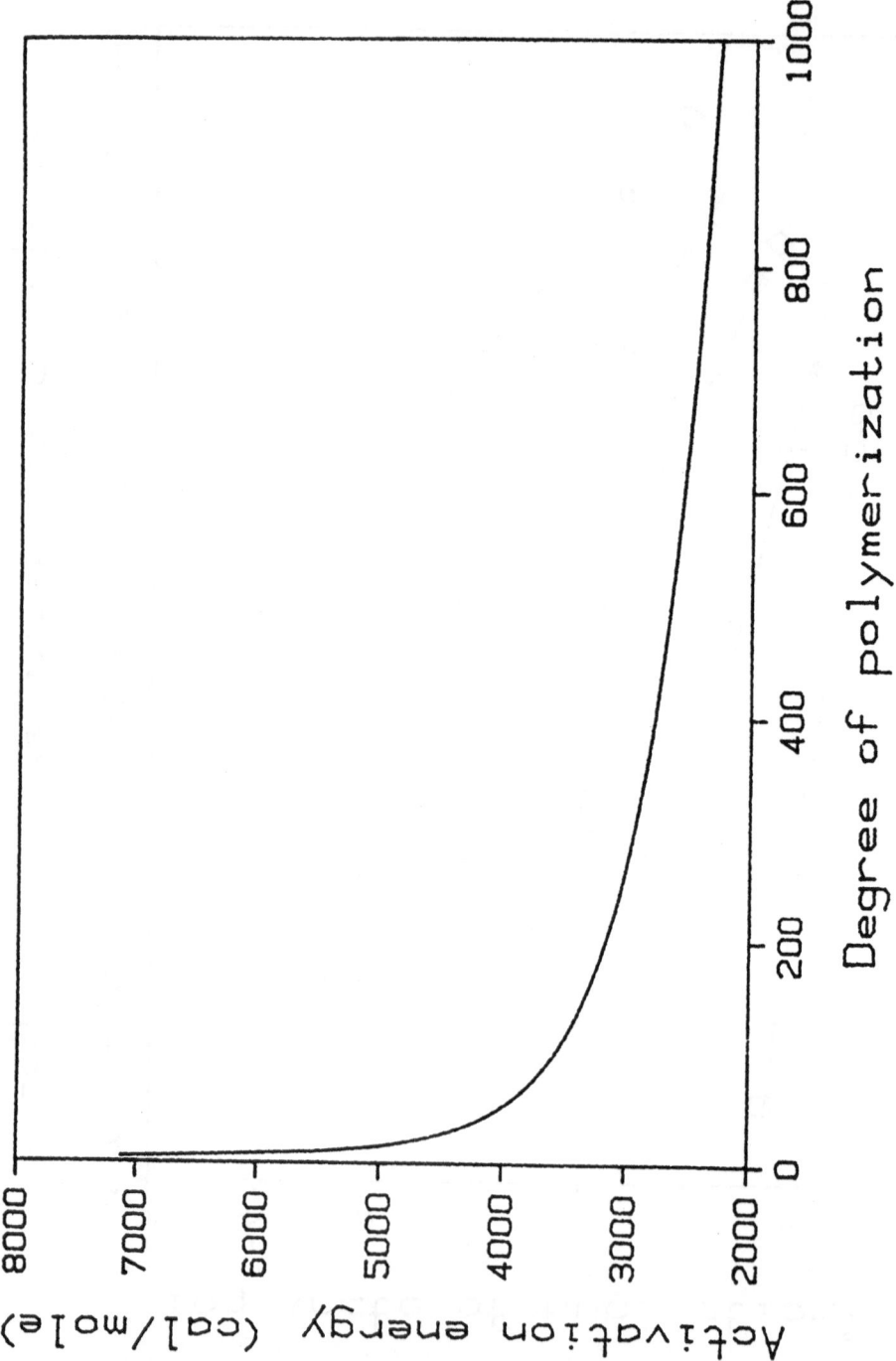

Slide 25: Plot of curve of estimated activation energy versus size of
critical polymer.

The best that can be said about slide 25 is that it reveals an
activation energy "in the ball park".

IV. Nonsteady Experiments: Avoiding Nucleation Theory

The discussion of slide 25 reveals a fundamental problem which
must be solved if the gas phase method, using nucleation for
detection, is to be a quantitative tool for the measurement of kinetic
parameters. It is absolutely necessary to know the size of the
critical polymer! Although there is the possibility that nucleation
theory will be improved (and many good theorists are working on the
problem), it is unrealistic to assume that it will be made precise
enough for our purposes. Thus it is desirable to find a means for
determining the size of the nucleating polymer, not dependent on
theory. There is a strong possibility of accomplishing this by
resorting to nonsteady rather than steady experiments.

Thus consider slide 26. The reaction indicated at the top of the
slide is a simple propagation step. Suppose that the only other

PROPAGATION AND LOSS TO WALLS ONLY

$$R_n\cdot + M \xrightarrow{k_p} R_{n+1}\cdot$$

INITIATION "PULSE" PRODUCES AN INITIAL CONCENTRATION
OF $R_1\cdot$ GIVEN PY P_0

$$P_n(t) = P_0((k_pMt)^{n-1}/(n-1)!)e^{-(k_pM + k_D)t}$$

Slide 26: The concentration of polymeric radicals of size n at time t.

processes, besides propagation, are initiation and loss to the walls.
Furthermore suppose that polymerization is initiated by a pulse of
light of negligible duration, such that a fixed initial concentration
of radicals of size 1 is available at the zero of time. No other
radicals are supplied to the system thereafter. Then it is an easy
matter to show that the concentration of polymeric radicals of size n,
at time t, is given by the last equation in slide 24 where P_0
represents the initial concentration of radicals in the pulse and M
is, again, the essentially constant monomer concentration. Suppose
that n in the equation refers to the size of the polymers responsible
for nucleation. Then, because the concentration $P_n(t)$ increases with
time, to a maximum, and eventually decays to zero, we should expect
the observed rate of nucleation to do the same. Furthermore by
fitting the observed nucleation rate to the expression for P_n in slide
26 it should be possible to determine n, the size of the critical
polymer.

We performed experiments aimed at implementing this program. Using the shutter mechanism of a Canon camera, we fashioned a one second uv pulse. The chamber was then "observed" by the photomultiplier for a 120 second period following the pulse. During this time drops occasionally appeared, and were counted. We employed a fairly high level of supersaturation in order to reduce the critical size so that the nucleation rate would be large enough to be conveniently measurable. Even so the reaction remained very "noisy" and the data obtained from a single run exhibited very little "structure". Therefore, the experiment was repeated until forty 120 second runs had been completed. The 120 second period was divided into 120 one second intervals and the number of drops accumulated in each interval over the 40 runs was recorded. At the outset, we did not have a suitable multichannel analyzer capable of performing the signal averaging electronically. Thus, signal averaging had to be done "by hand". This is the reason why the first experiment- involved only 40 runs. Slide 27 exhibits the results.

The curve shows the signal averaged rate of nucleation versus time (after the pulse). Each point on the slide has been signal averaged. Thus, for example, the fact that the point at 18 seconds indicates a rate of nucleation of about 1 drop per cubic centimeter per second means that about 40 drops were accumulated in the interval at 18 seconds during the 40 runs. As it may be seen from the slide the data, even though signal averaged, are still quite noisy. The solid curve in the slide is a least squares fit of the expression for P_n given on slide 26 (assuming the rate of nucleation to be porportional to P_n) to the points appearing in slide 27. The fit is reasonably good, and the critical polymer size derived from it is 6.8. The crude nucleation theory involved in the table on slide 12 predicts that the critical size should lie between 6 and 8. Again, this remarkable agreement has to be viewed as coincidence; all that we can be claimed is "ball park" agreement.

After these first experiments involving pulses were complete we acquired a suitable multichannel analyzer, and signal averaging could then be accomplished electronically. It was then easy to perform experiments involving as many as 400 runs. Slide 28 which is similar to slide 27 shows the results of one such set of runs. Because 400 runs, rather than only 40, are involved the data in slide 28 is less "noisy" than those of slide 27. However, there is evidence of a new phenomenon, at early times, i.e. at times less than 25 seconds. There appears to be an early "peak", followed by a second peak having its maximum at about 50 seconds. The early peak is probably due to multipolymer processes, i.e. to processes in which there is more than one polymer molecule in the nucleus. As I indicated earlier, such multipolymer nuclei can involve smaller polymers, the sum of whose molecular weights is equal to the critical size. Because these polymers are smaller, they are formed more rapidly, and so the nucleation process they induce is more prompt. However, why does it decay before the one polymer process is established?

The probable answer is indicated in slide 29. This slide shows the uv beam (schematically) and an absorbing plate (meant to represent

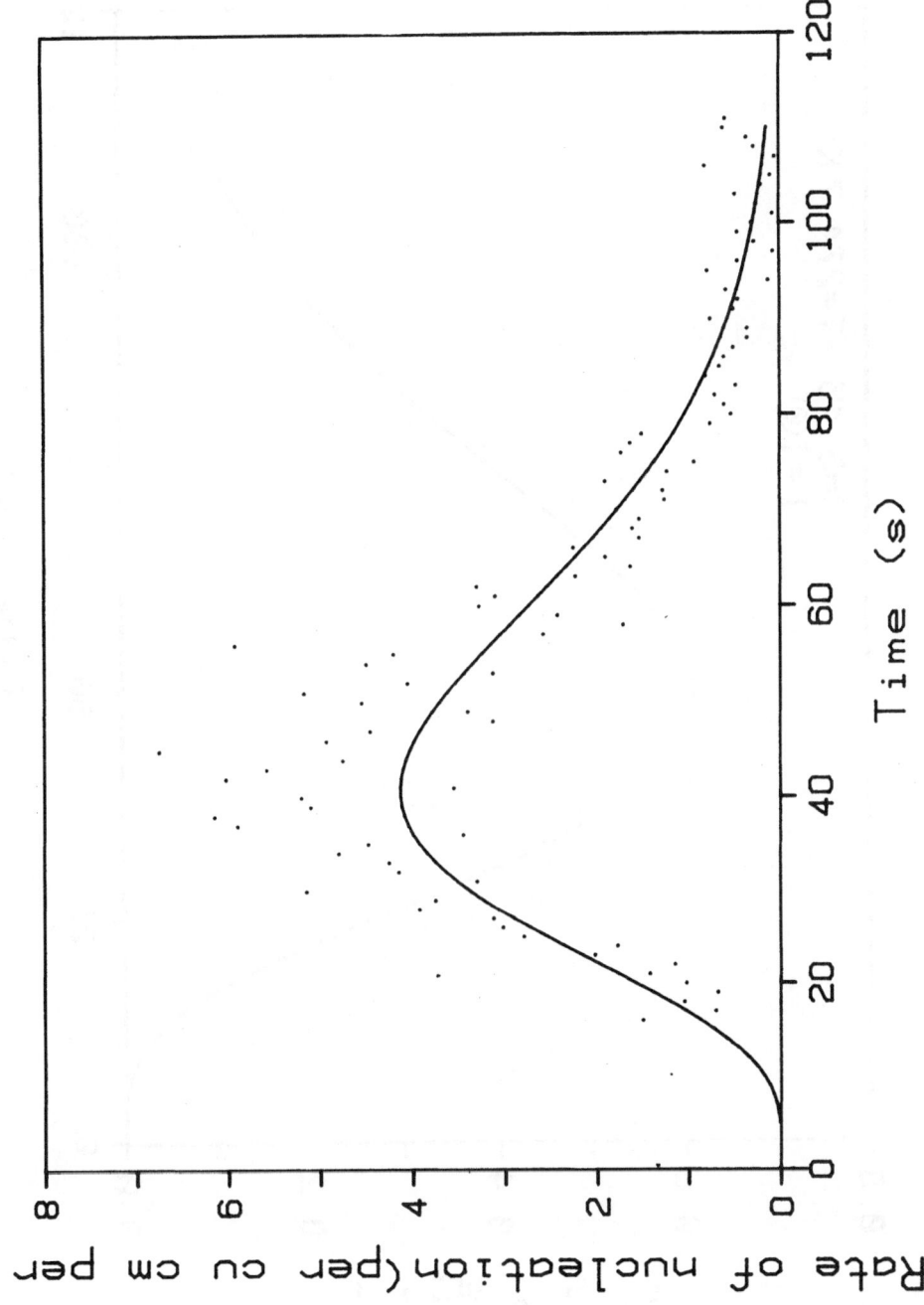

Slide 27: Nonsteady nucleation following uv pulse.

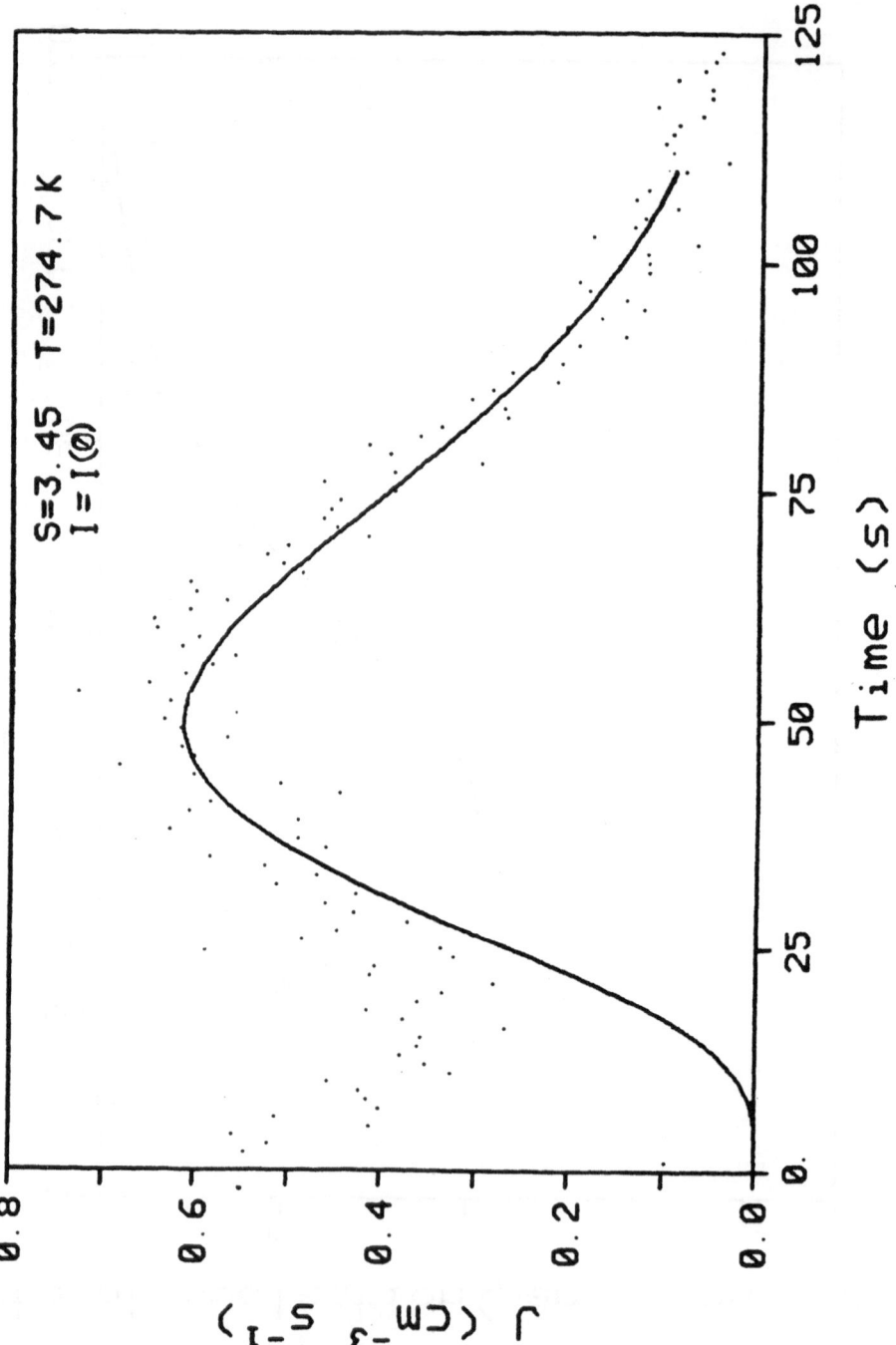

Slide 28: Nonsteady nucleation following uv pulse. Evidence of multipolymer nucleation.

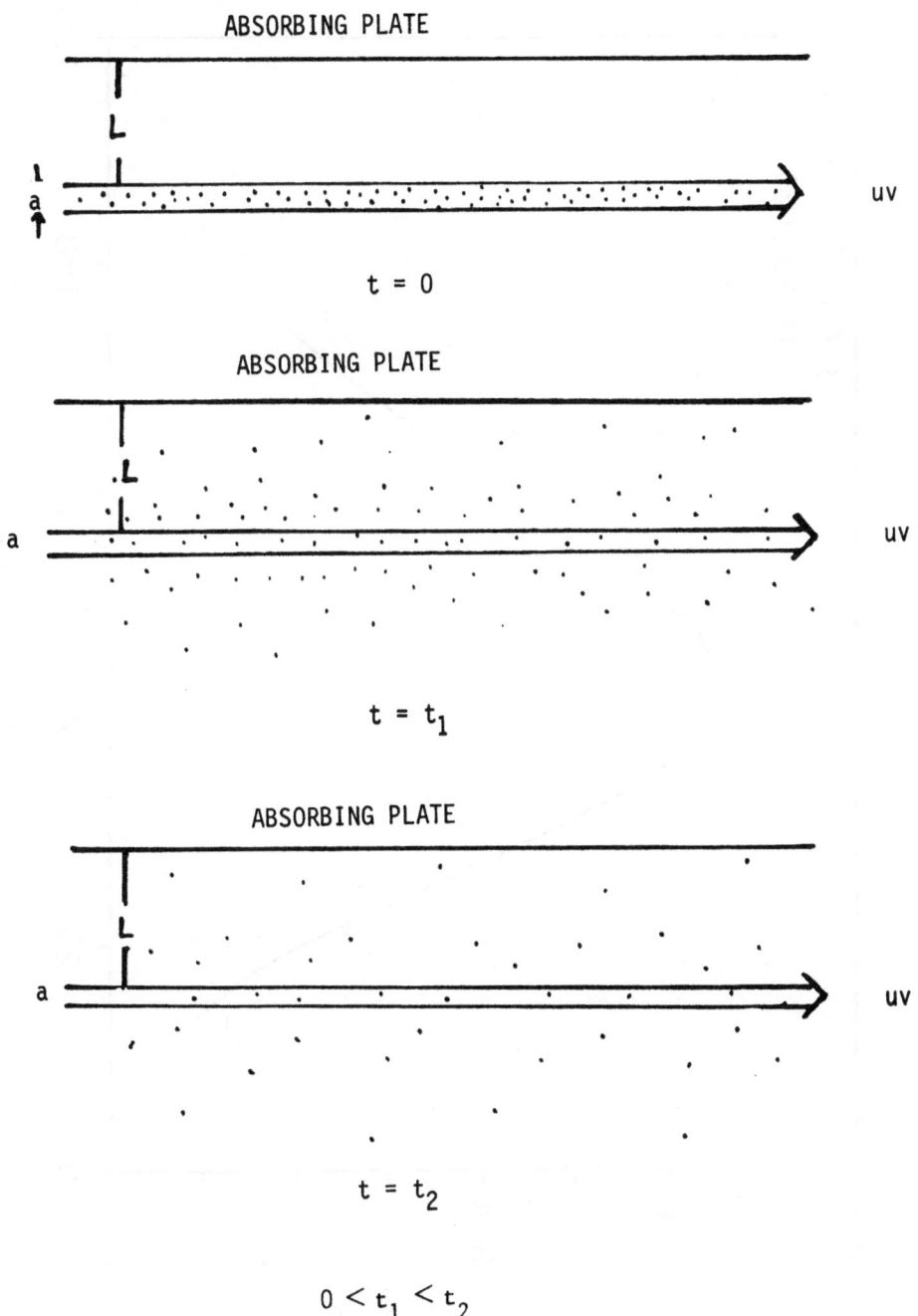

Slide 29: The uv beam and absorbing plate at a distance L from the beam.

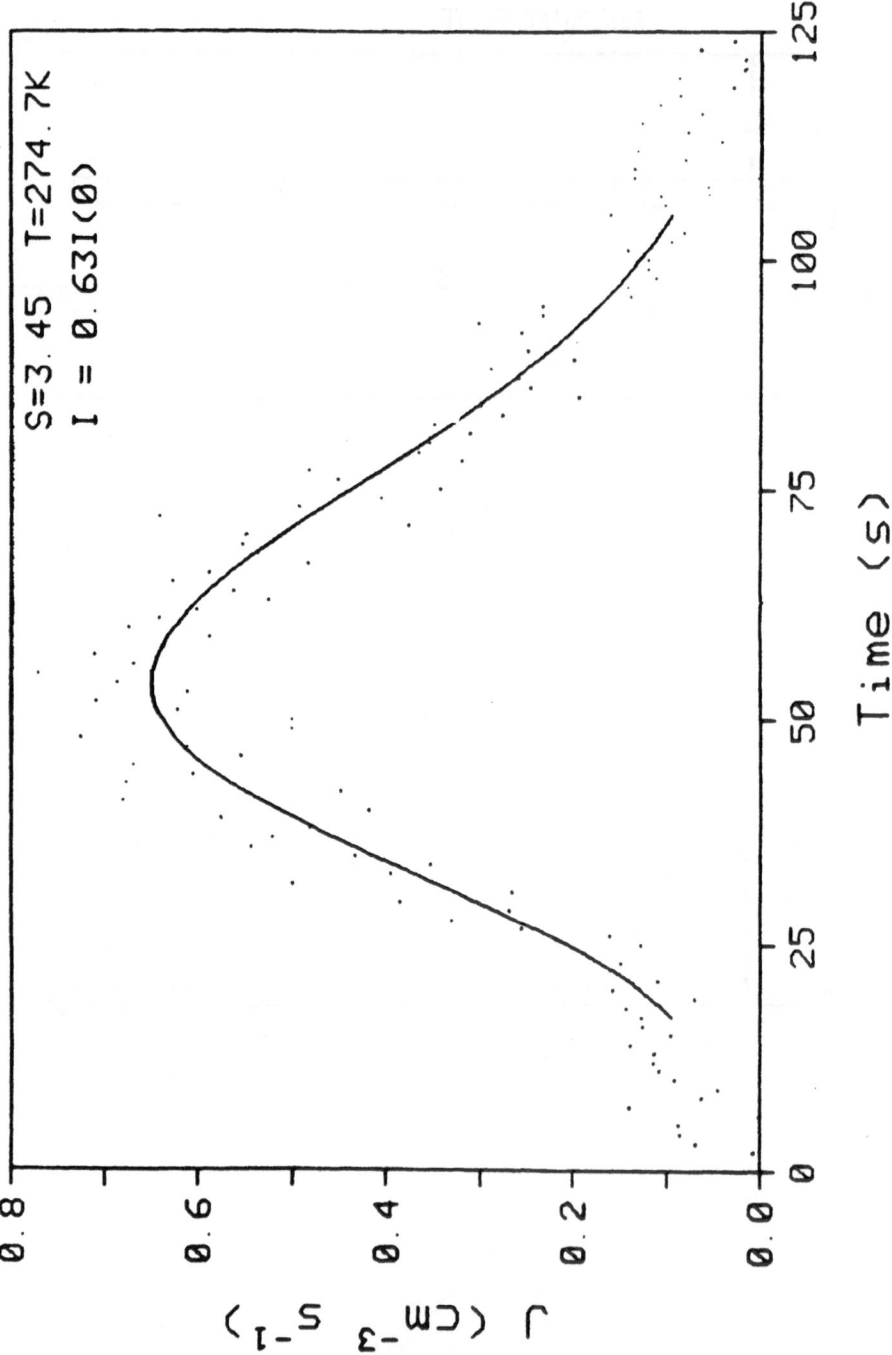

Slide 30: Nonsteady nucleation following uv pulse. No multipolymer nucleation.

the walls of the chamber) at a distance L from the beam. The
configuration at the top of the slide is intended to represent the
situation at zero time, just after the system has been pulsed. The
initial free radicals are indicated by the points within the uv beam.
These begin to propagate and, at the same ·time, diffuse. At early
times the propagating radicals are close enough to one another to
participate in a multipolymer nucleus. However at a later time say
t_1, as indicated by the middle picture on the slide, the concentration
of growing radicals has been diluted by outward diffusion. As a
result the multipolymer processes become disfavored, and eventually
disappears. This accounts for the decay of the early peak in slide
28. However, the single polymer nucleation process does not depend on
encounters between polymers so that it continues to develop through
later times (e.g. time t_2 in the slide), and gives rise to the main
peak in slide 28. The curve in slide 26 is a least squares fit of the
equation on slide 26 to the data of the peak and the tail.

    If this explanation is correct, it means that it is possible to
resolve multipolymer and single polymer processes, because they appear
as separate peaks. Furthemore the multipolymer processes should be
eliminated through a reduction of the concentration of radicals (by
reducing the light intensity) in the initial pulse. Thus if the
intensity is halved, and a three polymer process is involved, we might
expect the height of the three polymer process to be reduced to
one-eighth of its original value. At the same time the peak
corresponding to the one polymer process should be only halved.

    Slide 30 shows the result of reducing the light intensity to
about half of its value in slide 28. It is satisfying to see that the
early peak, assumed due to multipolymer processes, is essentially
gone. However the main peak, assumed to represent the one polymer
process, has retained its original height, and has not been haved as
expected. The most probable explanation of this is outlined in slide
31.

FIRST ORDER

$$\frac{dc}{dt} = -kc, \qquad c = c_0 e^{-kt}$$

SECOND ORDER

$$\frac{dc}{dt} = -kc^2, \qquad c = \frac{c_0}{1 + kc_0 t}$$

LARGE t $c \approx \dfrac{1}{kt}$

MEMORY OF INITIAL STATE IS LOST

Slide 31: First and second order reaction kinetics.

We learn, even in elementary physical chemistry, that in a first
order decay (such as that addressed at the top of the slide), the
concentration of reactant decreases exponentially with time in the
manner shown on the right of the relevant differential equation. In
this equation c_0 is the initial concentration of reactant while k is
the first order rate constant. Thus the concentration c always
"remembers" its initial state represented by c_0. In contrast, with
second order kinetics, as shown in the slide, the integration of the
rate equation leads to the formula on its right. Again c_0 is the
initial concentration and now k is the second order rate constant. If
c_0 is sufficiently large, then, at early times, one can ignore the
unity in the denominator so that, as shown on the slide, c_0 cancels
out of the expression for the concentration. Thus, a second order
process, eventually loses its memory of its initial state! The decay
of free radicals in the pulse is a second order one (bimolecular
recombination). Furthermore c_0 may be assumed to be roughly
porportional to light intensity. If that intensity is high enough,
a concentration of free radicals will eventually be established over
which the initial light intensity has no control; the system will have
lost its memory of the initial state. For the short times involved in
the multipolymer processes this memory has not been completely lost,
but for the longer time, corresponding to the peak of the single
polymer process, it may have been lost. The result is that the peak
of the single polymer process responds to an effective initial
concentration, independent of the initial intensity, while the
multipolymer processes respond to initial concentrations which do
depend on that intensity. As a result, the peak due to the
multipolymer processes is reduced while that due to the single polymer
one retains its original height.

V. Limitations of the Diffusion Cloud Chamber and Advantages of the Expansion Chamber

Up to this point the experiments described have all been
performed in the diffusion cloud chamber. Qualitatively, or at best
semiquantitatively, all of the observations are consistent with the
molecular level picture we have advanced, and on which our theory is
based. This is all part of the "bootstrap" approach. We are
satisfied that feasibility has been demonstrated, in the sense that we
can establish conditions under which a single polymer product molecule
is capable of inducing nucleation, and can therefore be counted.
However, although the diffusion cloud chamber represents an
inexpensive and relatively rapid means for accomplishing these ends,
it has many disadvantages when the reduction of the technique to a
quantitative tool for the study of polymerization kinetics is desired.
Slide 32 explains why.
It illustrates the diffusion cloud chamber, including both the
top and lower plates, and the uv beam employed for initiation as well
as the laser beam used for the counting of drops. The shaded vertical
column indicates the volume in which a nucleation event will be

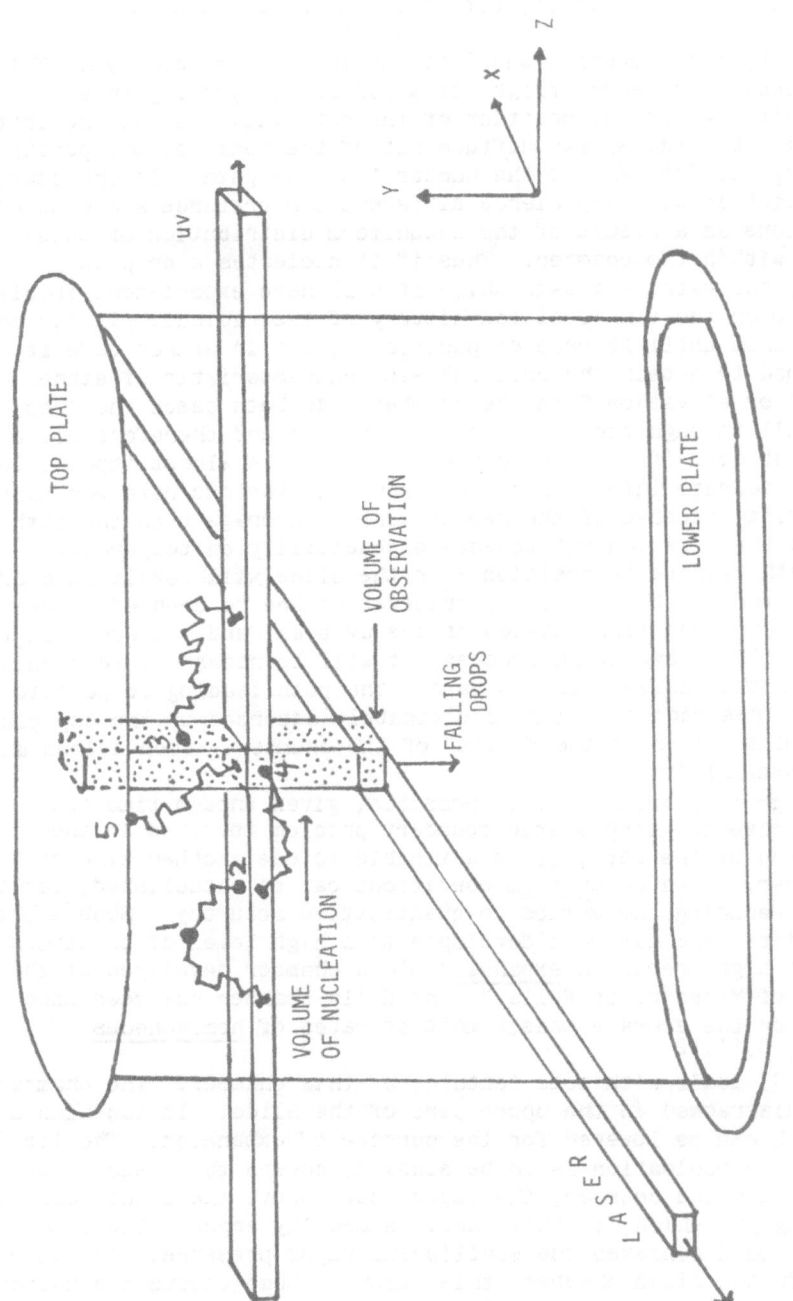

Slide 32: Limitations of the Diffusion Cloud Chamber.

counted, because the resulting drop will fall through the laser beam
and be observed in the optically defined volume of observation as a
result of the light which is scattered. The cloud chamber is a highly
nonuniform device. Within it, both the temperature and the
concentration of monomer varies with elevation. Thus consider the
path marked by the numbers 1 and 2 at the left of the diagram. This
path represents the random flight of a diffusing, growing free.
radical, initiated at the position of the retangular bar at the left.
The growing free radical may diffuse out of the beam to, say position
1, marked by the dot next to the number 1 on the path. In the course
of this flight it will experience different temperatures and monomer
concentrations as a result of the nonuniform distribution of these
quantities within the chamber. Thus if it nucleates a drop at
position 1, the rate of growth which it will have experienced involves
an average over the time-position history of the radical. It may not
nucleate a drop until it reaches position 2, but in either case it
will have had to attain the critical size characteristic of either
elevation 1 or elevation 2 in the chamber. In both cases the drop
will not fall through the volume of observation and therefore not be
counted. But even if it were counted, it would be almost impossible
to extract accurate information concerning the various rate constants
from the event, because of the requirement to average over the path.
For example the rate consant depends exponentially on temperature.

The path leading to position 3 in the slide will result in a drop
which falls through the volume of observation and be counted. However
the nucleation event lies outside of the uv beam, and, in addition to
the aforementioned averaging process, it will be necessary to consider
this fact in the analysis of the data. The path leading to position 5
indicates a free radical which is ultimately adsorbed on the top plate
and represents a loss to the "walls" of the chamber. Such a loss must
also be accounted for.

Although in principle it is possible, given enough time and
funds, to solve the complicated boundary problem involved in the
deconvolution of the data, it is advisable to use another kind of
cloud chamber, in which uniform conditions can be established, for the
purpose of reducing the method to quantitative accuracy. Such a cloud
chamber exists, and has been developed to a high level of precision.
This is the high precision expansion cloud chamber developed at the
University of Missouri at Rolla.[9] The Rolla chamber has been used
primarily for the acurate measurement of rates of homogeneous
nucleation.[10,11,12]

Slide 33 deals with some features of this chamber. The chamber
itself is diagrammed in the upper part of the slide. It contains a
piston which can be lowered for the purpose of expansion. The liquid,
in whose vapor nucleation is to be studied, covers the piston. As in
the diffusion cloud chamber, the vapor space above the liquid contains
a supporting gas which, in this case, is usually argon. The vapor
above the liquid achieves the equilibrium vapor pressure. In contrast
to the diffusion cloud chamber, this vapor is homogeneous and uniform.
When the piston is lowered rapidly the argon-vapor mixture is cooled
by adiabatic expansion, and becomes supersaturated to a degree

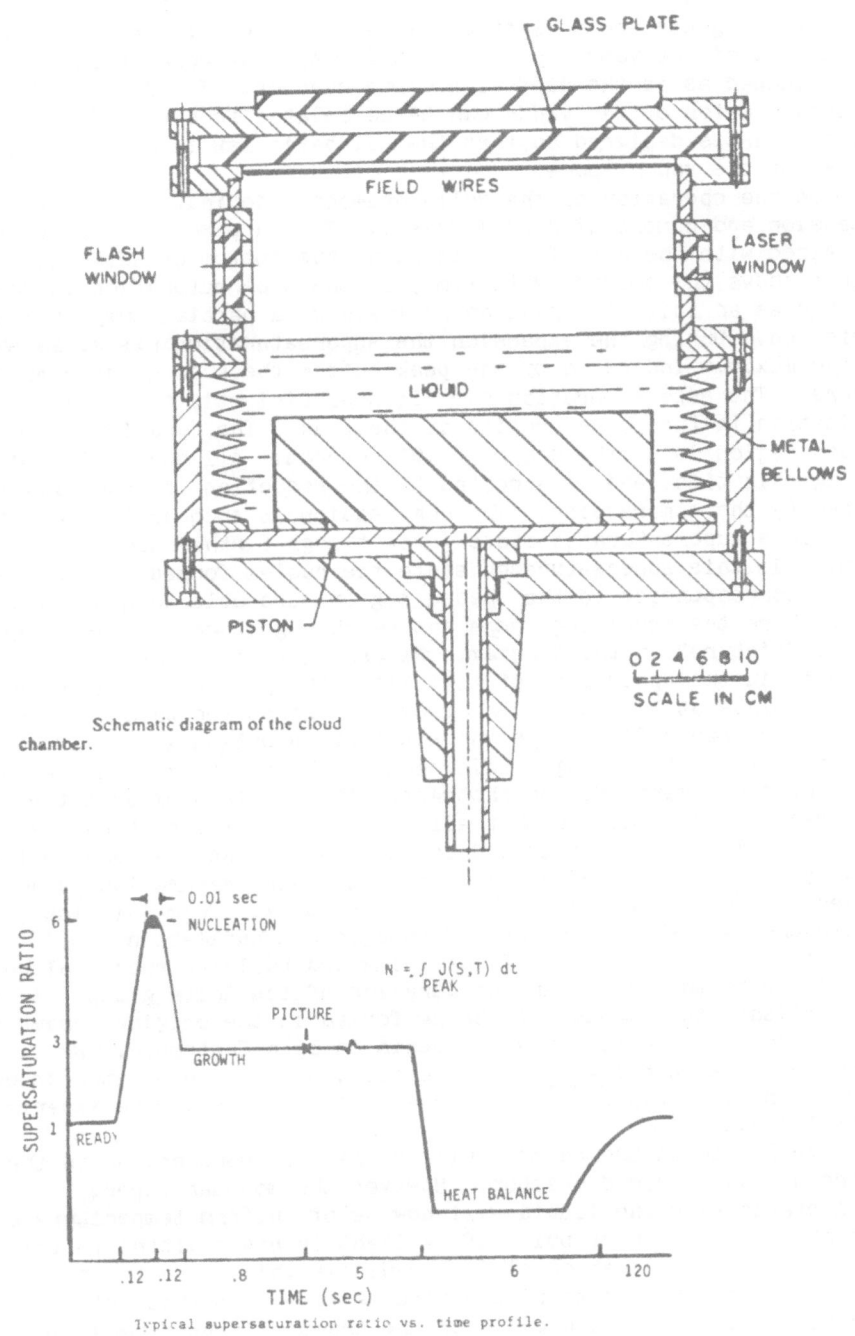

Schematic diagram of the cloud chamber.

Typical supersaturation ratio vs. time profile.

Slide 33:   Expansion Cloud Chamber.

depending on the extent of the expansion. If dust particles or other nucleating agents are present droplets form and they are allowed to settle out of the vapor space. In this way, the chamber is self-cleaned as is the diffusion cloud chamber. The degree of supersaturation of the vapor can be calculated from the adiabatic law. The chamber is designed so that the adiabatic law is accurate at the center of the vapor space.

In the operation of the Rolla chamber a definite cycle of expansion and compression is involved. The purpose of this can be explained with the aid of the figure at the bottom of slide 33. This figure shows the course, with time, of the supersaturation in the chamber as an initial expansion followed by a partial compression is performed. During the expansion the supersaturation rises, as shown, to the maximum indicated by the peak before the subsequent compression occurs. The supersaturation has risen sufficiently to cause nucleation in the shaded region at the peak. The duration of the shaded region is of the order of .01 seconds. All nucleation occurs during this time, and is arrested by the reduction of supersaturation caused by the compression. The compression is such as to leave the vapor in a supersaturated state (even though further nuclei do not form). In this supersaturated state the nuclei formed during the pulse, corresponding to the shaded region, are able to grow to the point where the resulting drops can be photographed, at the point marked "picture" in the diagram. Referring to the sketch of the cloud chamber, in the upper part of the slide, the picture is taken through a glass plate at the top of the chamber. Light for this picture is supplied by xenon flashlamps whose output is optically configured into a "slab" of light, having a thickness of about 0.5 cm. The depth of focus of the camera used to photograph the droplets exceeds the thickness of the slab, so that all the drops in the slab can be recorded. The number of such drops is counted, and represents the integrated production of nuclei during the supersaturation pulse corresponding to the shaded region at the peak. From this integrated number one can derive the rate of homogeneous nucleation.

We are in the process of designing and building such a cloud chamber, with the advice and cooperation of the Rolla group. In fact, our earliest experiments will be performed in the original chamber at Rolla, and they are scheduled to begin in late September. We will not be concerned with <u>homogeneous</u> nucleation, but in the polymer process which forms the subject of this paper. The scheme of the experiments is as follows.

The liquid in the chamber will consist of monomer, as in the case of the diffusion cloud chamber. However the monomer vapor, equilibrated with the liquid will now be of uniform temperature and concentration. A short pulse of uv light is now admitted to this vapor, and the propagation of free radicals which results is allowed to continue for some specified period of time. At this point the expansion and compression cycle of the chamber is performed, just as in the case for the measurement of the rate of homogeneous nucleation. However, now the depth of the expansion is reduced as to avoid homogeneous nucleation, and the drops which form are those nucleated by polymer molecules <u>larger than or equal</u> to the critical size for nucleation at the attained supersaturation. These are photographed and counted, just as in the case of homogeneous nucleation. The experiment is repeated using different "waiting times" between the light pulse and the subsequent expansion. Thus the polymer distribution is allowed to develop through different times.

All of the development occurs at a single temperature and monomer concentration; conditions in the cloud chamber are uniform. The expansion and compression cycle, to the time of the photograph, must be short on the scale of the rate of polymer propagation, so that additional polymers which forms after the waiting time are not counted. Such an arrangement of times is feasible.

The analysis of the data is explained in slide 34. At the top of

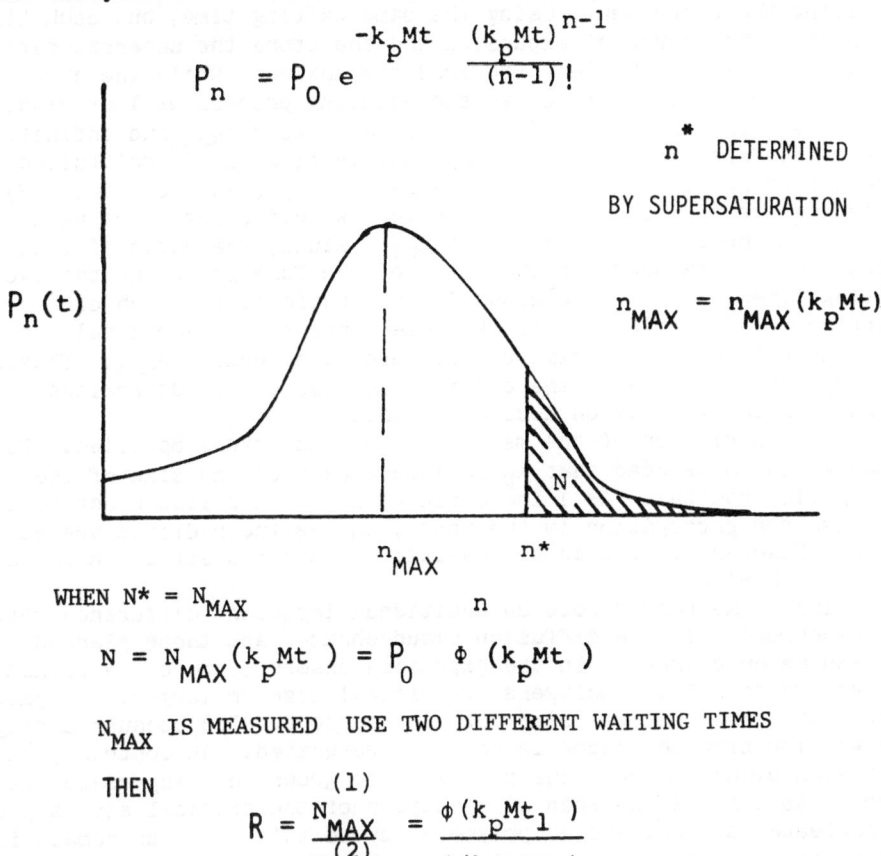

$$P_n = P_0 e^{-k_p Mt} \frac{(k_p Mt)^{n-1}}{(n-1)!}$$

n^* DETERMINED

BY SUPERSATURATION

$$n_{MAX} = n_{MAX}(k_p Mt)$$

WHEN $N^* = N_{MAX}$

$$N = N_{MAX}(k_p Mt) = P_0 \, \phi \, (k_p Mt)$$

N_{MAX} IS MEASURED   USE TWO DIFFERENT WAITING TIMES

THEN

$$R = \frac{N_{MAX}^{(1)}}{N_{MAX}^{(2)}} = \frac{\phi(k_p Mt_1)}{\phi(k_p nt_2)}$$

R, M, t_1, t_2 ARE MEASURED AND k_p

CAN BE DETERMINED

Slide 34: Analysis of expansion cloud chamber data.

the slide we show the same formula for the concentration of polymers of size n that appeared on slide 26, except for the fact that k_D, associated with the rate of loss to the walls, is not included. This is because the expansion cloud chamber is large enough so that within reasonable "waiting time" there will be no effect of the walls. In the formula the waiting time is denoted by t. Again P_0 represents the concentration of free radicals in the initial pulse, and M denotes the now uniform monomer concentration. A plot of P_n, after a particular waiting time t, resembles the curve shown in the slide. The size corresponding to the maximum is denoted by n_{max}. The starred value of n represents the critical size for the supersaturation achieved during the expansion. The shaded region represents the integrated number of drops N produced by the expansion and counted in the photograph. By repeating the experiment, using the same waiting time, but each time increasing the degree of expansion and therefore the supersaturation, n^* can be moved to the left (towards the maximum) while the curve remains invariant. In this way the counting process will eventually locate n_{max} and the area under the curve between n_{max} and infinite size. This area is denoted by N_{max} and is given by P^0 multiplied by a function of $k_p Mt$. This function is denoted by ϕ in the slide. By repeating the series of experiments for two different fixed waiting times, and therefore two different N_{max} values, the ratio of these two values can be measured, as the ratio of the function ϕ for the two waiting times. The only unknown in this ratio is k_p which can therefore be determined. With k_p determined the measured value of N_{max} for either waiting time can be used to determine n_{max}. Thus not only k_p, but the size of the critical polymer can be determined without having to rely on nucleation theory.

The description of the method is somewhat oversimplified. For example, it is assumed that k_p is independent of the size of the propagating radical. As I indicated earlier, this liable not to be the case for propagation in the vapor, unless the radicals are very large. However, it should be possible to handle a situation where k_p varies with size.

Finally we should note an additional important difference between the experiments in the diffusion cloud chamber and those planned for the expansion chamber. In the expansion chamber, the drops formed correspond to all the polymers of critical size or larger. Polymers larger than the critical size are allowed to develop because during the waiting time the vapor is not supersaturated. In contrast, in the diffusion cloud chamber, the polymers are grown in a supersaturated vapor. As a result as soon as a polymer of the critical size appears it nucleates a drop, and no polymers beyond this size can remain in the vapor. In the diffusion cloud chamber the rate of nucleation is a direct measure of the rate of production of polymers of the critical size. Things are less direct in the expansion cloud chamber, but there is still the overwhelming advantage of having the polymers grow in a vapor, uniform in both temperature and concentration. Expansion cloud chamber experiments (and the expansion cloud chamber), are far more expensive than experiments in the diffusion cloud chamber, but they represent the proper approach for converting the gas phase method into a quantitative tool for the study of polymerization kinetics.

Since experiments involving molecular beams represent an important component of the discussion at this conference, it worth remarking that the expansion cloud chamber method may only be a step on the way to utilizing molecular beams for the study of homogeneous gas phase chain polymerization. With a nozzle beam the polymers could be allowed to develop during a "waiting time" just as in the case of the expansion cloud chamber. Expansion through the nozzle of the beam could then be performed so that clusters or drops could be nucleated on polymers which have attained the critical size. Methods for counting these clusters could certainly be developed. Finally, we should note that the techniques described in connection with free radical chains could be developed for chains propagated by ions. The opportunity to study such ionic chain propagation with well defined free initiating ions would constitute a valuable application of the method.

References

1. H. Reiss and M. A. Chowdhury, J. Phys. Chem. 87, 4599 (1983).
2. M. A. Chowdhury, H. Reiss, D. R. Squire, and V. Stannett, Macromolecules 17, 1436 (1984).
3. H. Reiss and M. A. Chowdhury, J. Phys. Chem. 88, 6667 (1984).
4. Y. Rabin and H. Reiss, Macromolecules 17, 2450 (1984).
5. H. W. Melville, Proc. R. Soc. London, Ser. A. 163, 511 (1937).
6. H. W. Melville, Proc. R. Soc. London, Ser. A. 167, 99 (1938).
7. H. W. Melville and R. F. Tuckett, J. Chem. Soc. Part 2, 1201, 1211 (1947).
8. R. D. Bohme and A. V. Tobolsky, in "Encyclopedia of Polymer Science and Technology" vol. 4, p. 599, John Wiley and Sons Inc., New York, N. Y. 1966.
9. J. L. Schmitt, Rev. Sci,. Instrum. 52 (11), 1749 (1981).
10. J. J. Schmitt, G. W. Adams, and R. A. Zalabsky, J. Chem. Phys. 77 2089 (1982).
11. J. L. Kassner, Jr., J. C. Carstans and L. B. Allen, J. Atmos. Sci. 25, 919 (1968).
12. L. B. Allen and J. L. Kassner, Jr., J. Colloid and Interface Sci. 30, 81 (1969).

...

References

COMMENTS ON CLUSTERS IN BEAMS, SPECTROSCOPY, AND PHASE TRANSITIONS

H. Reiss
Department of Chemistry
University of California Los Angeles
Los Angeles, California 90024

ABSTRACT. This paper contains a discussion of "clusters" found
typically in 1) molecular beams, and 2) in saturated and supersaturated
vapors. Important differences as well as similarities between these two
kinds of clusters are discussed. The point is made that most clusters
in beams are large "molecules" (and are treated by theory as such)
rather than fragments of matter intermediate between vapor molecules and
bulk matter, and that careful consideration is necessary in order to
avoid the confusion and "paradoxes" which have plagued the field of
nucleation.

I. Introduction: Clusters or Molecules?

Recently, considerable attention has been focused on "clusters" in
molecular beams, especially nozzle beams[1]. The clusters which appear in
molecular beams range from van der Waals molecules, through molecules
bound by polar forces, to small condensed fragments of metal. They may
consist of a single component or be multicomponent. Frequently, the
study of such clusters (and the concomitant expense) is justified by
pointing out that they are an important separate form of matter, lying
between molecules and bulk substance. As such, they are supposed to
form a "bridge" between the two extremes.

Although the clusters found in beams may occasionally have this
character, more often they do not, and misunderstanding can be
generated by insisting that they fulfill this intermediate role. This
point will be elaborated later.

However, the fact that the clusters in beams may not provide the
above mentioned "bridge" does not mean they are not important or useful.
They can be (and have been) employed as specialized tools for other
purposes (see reference 1), and comprise a distinct field by themselves.
Applications have involved (1) the structure of van der Waals molecules,
(2) the effect of rare gas ligands on spectral lines, (3) the study of
surface states, (4) the dependence of melting point on crystal size, (5)
the promotion of concerted reactions, (6) band theory, surface states,
and especially study of the threshold (with increase of size) to

115

P. M. Rentzepis and C. Capellos (eds.), Advances in Chemical Reaction Dynamics, 115–133.
© *1986 by D. Reidel Publishing Company.*

macroscopic properties, and (7) the measurement of fragmentation energies.

To a certain extent, the development of this field and its various applications has been "technique" and "instrument" driven. Nozzle beams were "available" and a variety of spectroscopic tools (including mass spectrometry) were "in place" for the interrogation of such clusters. An approach along the route of molecular beams, involves a path on which attention moves upward from the microscopic to the macroscopic domain. "Clusters" have also been important in the study of phase transitions (both equilibrium and dynamic). Especially in the field of "nucleation", the path has led downward from the macroscopic towards the microscopic. Along this path, molecular ideas were grafted somewhat uncritically onto a macroscopic theory. As we shall see below, this process led to misunderstanding and paradox. There is the same danger if workers are not critical about the role of clusters, along the opposite path (from the microscopic to the macroscopic). The present paper is offered as a contribution toward the avoidance of this problem.

At the outset, we should understand which "clusters" do provide a bridge between the microscopic and the macroscopic. This question can be raised in the two contexts which exhaust all possibilities, namely, (1) equilibrium between phases, and (2) the dynamics of phase transitions.

The clusters which appear in beams, e.g. van der Waals molecules, are physical clusters. That is to say, they are defined by some physical convention, e.g. they may be limited to a certain energy[2] or confined to a certain region of configuration space[3,4]. Although there have been theories of imperfect gases and phase transitions (e.g. condensation) in which physical clusters have been used[5,6], such theories have generally been unsuccessful. Usually they treat an imperfect gas as an ideal gas mixture of clusters, each cluster of a given size being regarded as a molecule of a distinct species. It is extremely difficult to advance theories of imperfect gases or condensation (except for gases, only slightly imperfect) using physical clusters.[2,7] The problem arises in the interaction between the clusters themselves. This, in turn, makes it difficult to define what is meant by a "physical" cluster.

In order to make progress it is necessary to use mathematical clusters, e.g., as in the Mayer theory[8], in which the clusters (based on diagrams) are in reality "bookeeping" devices. Although it turns out that the imperfect gas may be treated formally as an ideal mixture of mathematical clusters (i.e., Dalton's law holds for mathematical clusters) the concentrations of some mathematical clusters may actually be negative! This strange behavior results from the fact that mathematical clusters account for those density fluctuations in excess of fluctuations expected on a purely random basis. If, for example, the molecules of the gas interact only repulsively, one expects density fluctuations to be smaller than random. To account for this it is necessary to have negative concentrations of mathematical clusters. In contrast to equilibrium, theories dealing with the dynamics of condensation (nucleation) have involved "physical" clusters. However, this is because nucleation experiments have usually been confined to

conditions under which the supersatured vapor is only <u>slightly</u>
imperfect.[9] When nucleation in <u>highly</u> imperfect systems is
involved[2,7,10,11,12], physical clusters soon run into trouble.

In cases where physical clusters are useful (i.e. in the dynamics
of phase transitions, in slightly imperfect gases), they are, for the
most part, different from the clusters studied in molecular beams. For
example, they are always defined (even if heuristically) in a way which
attempts to account for their interaction with the surrounding gas[13].
In the so called <u>capillarity</u> <u>approximation</u> this interaction is introduced, in the free energy of the cluster, by the appearance of the interfacial tension between the liquid and the vapor (see below). However
crude this approximation, it is still made in such a way that the cluster
is indeed treated as a "bridge" between the molecular and bulk states.

Under what conditions can the clusters found in beams be related to
those involved in nucleation theory, and therefore regarded as
intermediate between the molecular and bulk states? The answer appears
to be--"only under conditions in which the binding energy is so large
compared to the temperature, and the vapor pressure so low, that the
vapor phase is essentially absent." Also, under these conditions, even a
<u>small</u> pressure of vapor represents a very high supersaturation, and, as
will be seen below, the nucleus for condensation would then be very
small. Metal clusters or those involving polar bonds (e.g. clusters of
NaCl) may fall into this class. The same may be true of the clusters of
carbon atoms involved in soot formation. Clusters held together by
dispersion forces (van der Waals molecules) could fall into this class
at the very low temperatures achieved by beams. But in all these cases,
the vapor representing the free molecular state is essentially absent
and, therefore, the clusters are things unto themselves.

As a result, they can be studied without ever accounting for their
contact with molecules of the same species in the molecular state.
Thus, in a classic study, Hoare, Briant and their co-workers[14,15,16,17],
using a computer, were able to identify the global potential minimum of
a cluster of thirteen atoms interacting by means of Lennard-Jones
forces, and to study the vibration and rotation of such clusters. The
possibility that the cluster might lose an atom by evaporation was not
considered, reflecting the fact that the relaxation time for internal
degrees of freedom was exceedingly short compared to the time for
evaporation. This is typically, a low temperature calculation.
Calculations concerning the structures of smaller van der Waals
molecules or metal clusters (or for that matter of any <u>molecule</u>) assume
similar conditions.[18,19,20,21,22,23] Thus the clusters are treated as
large <u>molecules</u>, rather than as intermediate entities capable of
illuminating the transition from the molecular to the bulk state. These
large "molecules" are certainly interesting in themselves, and are worth
studying for what they can reveal. However, it is the cluster involved
in nucleation theory which is more appropriately the "bridge" between
molecular and bulk states.

The clusters in molecular beams are often formed by a series of
irreversible additions of single molecules, and involve a three body
recombination law. For example heteromolecular clusters, involving the
rare gas atoms Ar, Kr, or Xe and a tetracene molecule exhibit a Poisson

distribution with respect to size.[1] This is indicative of a sequence of irreversible steps. In such cases the conventional free energy barrier to nucleation does not exist, and the clusters represent final products rather then "embryos" of the new phase. In the more conventional nucleation process the barrier opposing nucleation is usually high, and the nucleus forms by a series of highly reversible additions of molecules. As a result, the nucleus is in quasi-equilibrium with the molecules of the original metastable phase. Furthermore at any given instant, the concentration of the cluster of the size of the nucleus is vanishingly small, and the kinetic process falls into a quasi-steady state.[24] There are, however, situations in which the steady state is established only slowly, and the relaxation time associated with this process must be accounted for by the theory.[25,26,27,28,29]

Molecular beam experiments, together with theory, can shed light on the difficult phenomena involved in nucleation. However, the theory should view the clusters in proper perspective, and be careful to avoid the pitfalls which trapped investigators who developed theories by extrapolating downward from the macroscopic level. For this reason, we outline, in the next section, some features of the existing theories of nucleation, and clarify the misconceptions which led to some of the traps.

II. Nucleation Theory: Formulation, Misconceptions, and Paradoxes

The most popular theory of nucleation has been developed by many workers. Among the pioneers are Farkas[30], Becker and Döring[31], Volmer[32], and Frenkel and Zeldovich[9]. Detailed summaries of the theory can be found in a number of places.[9,24,33,34,35] In this section we merely outline the approach, focusing on the most important points. We shall concentrate on the nucleation of liquid drops from a supersaturated vapor. All of these theories rely on the so called "capillarity approximation". This amounts to assuming that the clusters of the new phase are liquid drops, having all of the features of macroscopic drops, including a well defined interfacial tension.

The quantitative expression of the capillarity approximation appears in the specification of the free energy of a liquid cluster containing n molecules. Thus for the free energy F_n of such a cluster we write

$$F_n = \mu_\ell n + \alpha n^{2/3} \tag{1}$$

where

$$\alpha = 4\pi\sigma(3v/4\pi)^{2/3} \tag{2}$$

and μ_ℓ is the chemical potential of the bulk liquid at the ambient temperature T and at the pressure p outside of the cluster (because of the interfacial tension the pressure inside the drop-cluster is higher than the pressure outside). The quantity σ is the interfacial tension

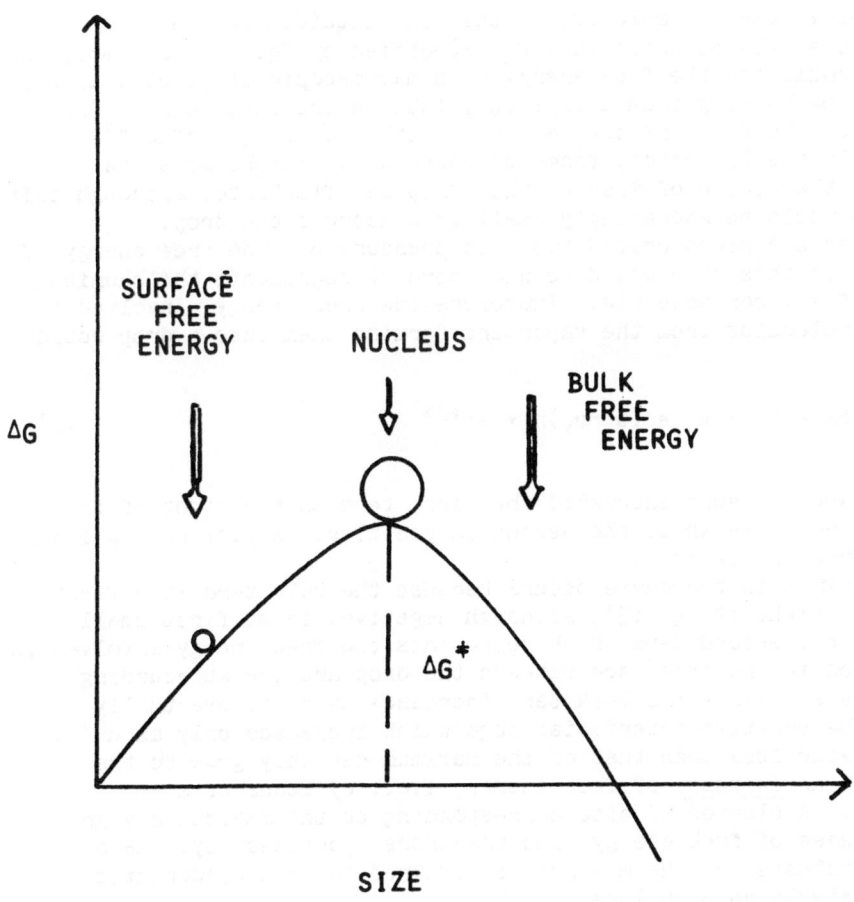

Figure 1 – Illustration (schematic) of the free energy
           barrier for nucleation for a one component
           system. ΔG* is the barrier height.

and v is the volume per molecule in the bulk liquid. For later purposes, it should be noted that F_n, specified by Eq. (1) is simply the standard formula for the free energy of a macroscopic drop, e.g. a drop which might be hanging from a capillary tube in the laboratory. The boundaries of the drop are defined by a "Gibbs dividing surface"[36] which is at rest in the laboratory frame of reference. Furthermore the position of the center of mass of this drop may fluctuate, although this fluctuation would be exceedingly small in a macroscopic drop.

Consider a supersaturated vapor at pressure p. The free energy of n molecules in this drop would be $n\mu_v$ where μ_v represents the chemical potential of a vapor molecule. Therefore the free energy involved in removing n molecules from the vapor and forming them into a drop would be

$$\Delta F_n = F_n - \mu_\ell = (\mu_\ell - \mu_v)n + \alpha n^{2/3} \tag{3}$$

Since the vapor is supersaturated the first term on the right of this equation is negative while the second is positive. A plot of ΔF_n looks like the curve in Figure 1.

The maximum in the curve occurs because the bulk term (the first term) on the right of Eq. (3), although negative, is at first small compared to the second term which represents the free energy involved in the formation in the interface between the drop and the surrounding vapor. However, since the bulk term increases as n, it eventually overtakes the positive interfacial term which increases only as $n^{2/3}$. A cluster of size less than that of the maximum can only grow to the maximum with an increase of free energy, i.e., by means of a fluctuation. A cluster of size corresponding to the maximum may grow with a decrease of free energy, and therefore spontaneously. As a result, the cluster at the maximum is referred to as a condensation nucleus or simply as a nucleus.

The rate of nucleation is conveniently defined as the number of condensation nuclei produced by fluctuations, per cubic centimeter per second. It is assumed that all such nuclei go on to become macroscopic drops. To compute this rate, which we denote by J, we assume the following series of reactions

$$\begin{align} A + A &= A_2 \\ A_2 + A &= A_3 \\ \text{etc.} \end{align} \tag{4}$$

In Eq. (4) the symbol A represents a molecule of gas, and A_2, A_3, etc. represent clusters which are dimers, trimers, etc (van der Waals molecules if dispersion forces are involved). Growth is assumed to occur only by encounters between clusters and single molecules, because under most experimental conditions the concentrations of clusters are very small. Furthermore experiments are usually performed in the presence of a supporting gas (He, Ar, N_2, etc) and the individual molecular steps indicated in Eq. (4) involve three body collisions in which a molecule of the supporting gas carries away the excess energy.

The individual steps are shown as reversible.

Although many chemical kineticists are not familar with it, the phenomenon of nucleation is really a branch of chemical dynamics, and fairly conventional methods are used to evaluate the rate determined by the set of reactions in Eq. (4). In particular, the forward rate constants for the indicated reactions are easily evaluated, but the reverse rate constants depend sensitively on cluster size, and are usually determined by invoking the principle of detailed balance. The application of this principle requires that we apply a formal constraint which disallows the existence of clusters beyond a certain size (say the size of a nucleus). Then the system is forced into an equilibrium in which the equilibrium distribution of clusters, according to size, is prescribed, by a simple statistical mechanical argument to be

$$N_n = N\exp(\mu_v n/kT) \tag{5}$$

where N_n is the concentration of clusters of size n, N is the concentration of vapor molecules in the system, q_n is the canonical ensemble partition function for the cluster, and k is the Boltzmann constant. It is a standard result of statistical mechanics that

$$q_n = \exp(-F_n/kT) \tag{6}$$

Substituting Eq. (6) into Eq. (5), and noting Eq. (3), leads to the result

$$N_n = N\exp(-\Delta F_n/kT) \tag{7}$$

Using this relation it is possible to determine the reverse rate constants from the forward ones, and the kinetics contained in Eq. (4) can easily be summarized by the following expression for the rate of nucleation.

$$J = K\exp(-\Delta F_{n*}/kT) \tag{8}$$

in which the superscript on the n indicates that it represents the size of the nucleus, and that the free energy increment in the exponent is therefore the height of the barrier exhibited in Fig. (1). The preexponential factor K in Eq (8) depends slightly on the degree of supersaturation in the system, but ΔF_{n*} depends sensitively on the supersaturation as follows,

$$\Delta F_{n*} \sim \sigma^3 v/[T\ln(p/p_e)]^2 \tag{9}$$

In this equation p_e represents the saturation or equilibrium vapor pressure at the temperature T, and p/p_e is the supersaturation.

The theory, outlined thus far, is actually a transition state theory, and the free energy barrier indicated in Fig. (1) is nothing more than a free energy of activation. If the system is a multicompoment one, and the clusters are therefore heteromolecular, a cluster must be identified by specifying the molecular numbers of the

Table 1

H_2O 300 K

$S = P / P_E$	DROPS CM^{-3} SEC^{-1}
1.2	3×10^{-996}
1.4	8.1×10^{-275}
1.6	1.5×10^{-128}
1.8	4.9×10^{-73}
2.0	2.4×10^{-45}
2.2	3.2×10^{-29}
2.4	8.0×10^{-19}
2.6	1.2×10^{-11}
2.8	2.0×10^{-6}
3.0	1.8×10^{-2}
3.107	1.0 ←————CRITICAL SUPERSATURATION
3.2	22.3
3.4	6.7×10^{3}
3.6	7.1×10^{5}
3.8	3.5×10^{7}
4.0	9.5×10^{8}

Dependence of Nucleation Rate in H_2O
on Supersaturation

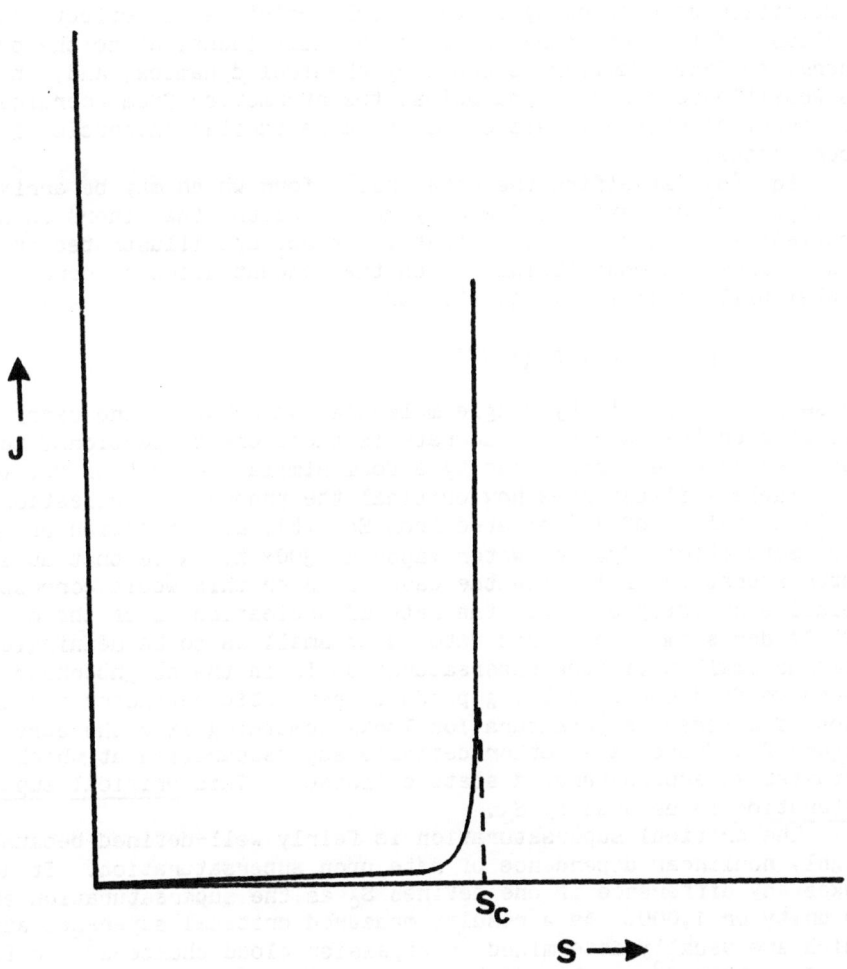

Figure 2   Schematic of the dependence of nucleation rate J on supersaturation S (as in Table 1). S_c is the critical supersaturation.

various species within it. Thus, in the case of a binary system, the free energy "hill," appearing in Fig. (1), becomes a free energy "surface" covering a base plane in which the coordinate axes denote the molecular numbers of each species. It is easy to show that this surface posesses a saddle point, and that the majority of drops are formed by clusters which pass through this saddle point.[37] The cluster having the composition corresponding to the saddle point is, in effect, the nucleus. This free energy surface is fully identical to the potential energy surfaces familar to ordinary chemical dynamics, and, in fact, it is possible to perform a canonical transformation from coordinates which represent cluster composition to the more familar intramolecular coordinates.

Eq. (8), specifing the rate, has a form which may be arrived at by a simple approximation. One only has to assume that there is a bottleneck, or slow step, so that the reactions illustrated in equation 4 are almost in equilibrium. Then the concentration of condensation nuclei will be given by Eq. (7), or

$$N_{n*} = N\exp(\Delta F_{n*}/kT) \tag{10}$$

These nuclei are hit by single molecules which stick and carry the nuclei over the barrier. The rate is therefore proportional to N_{n*} and may therefore be represented by a form similar to that of Eq. (8).

Table 1 illustrates how critical the process of nucleation can be. It lists values of J (computed from Eq. (8)) as a function of the supersaturation p/p_e for water vapor at 300° K. Note that at a supersaturation of 1.2 (in the case of water this would correspond to a relative humidity of 120%) the rate of nucleation is of the order of 10^{-996} drops/cm³ sec. This rate is so small as to be meaningless. It remains small until the supersaturation is in the neighborhood of 3.1 where we find one drop being produced per cubic cetimeter per second. A plot of J versus supersaturation looks something like the curve in Figure 2. There is a rather definite supersaturation at which the metastable, supersaturated state collapses. This <u>critical</u> <u>super</u> <u>saturation</u> is denoted by S_c.

The critical supersaturation is fairly well-defined because of the highly nonlinear dependence of rate upon supersaturation. It hardly makes any difference if one defines S_c as the supersaturation at which J is unity or 1,000. As a result, measured critical supersaturations which are usually determined in expansion cloud chambers[38], diffusion cloud chambers[39], or by adiabatic expansion through nozzles[40], show excellent agreement with the theory outlined above, for a large number of substances.[41,42,43,44] However, most of these experiments have not been concerned with the measurement of the actual <u>rate</u> of nucleation but only with the measurement of S_c. The actual measurement of the <u>rate</u> is difficult.[38] Because of the rapid increase of J with supersaturation, illustrated in Figure 2, a large error in the measurement of rate can still correspond to a small error in the value of S_c.

We now turn to the subtleties (mentioned in the last section) of the seemingly simple theory. The development of an accurate theory and a clear understanding of the phenomenon of nucleation (either by workers who have been concerned with it in the past, or by individuals entering the field because of an interest in clusters in molecular beams) is desirable in view of the broad importance of nucleation to science and technology. There is hardly a field in which it does not play a role. For example, it is important in chemical processing, air pollution, atmospheric and cloud physics, photography, metallurgy, solid state electronics, aerodynamics, turbine design, polymer morphology, boiling and cavitation, planetary formation, vulcanism, superconductivity, and molecular biology. Furthermore nucleation has been increasingly utilized as a means of detection and amplification in high energy physics and chemical physics.[45,46,47,48,49,50,51,52] There is ample reason to develop good theories and better methods of measurement.

The conventional theory, outlined above, furnishes an example of "extrapolation downward." The cluster is modeled as a macroscopic liquid drop even though it may, at times, contain less than 10 molecules. In attempts to take account of the molecular nature of nucleation phenomena, the simple theory has usually been modified by an uncritical grafting of molecular ideas onto its initial macroscopic form. Most workers realized that the agreement between theory and experiment (even in regard to the prediction of critical supersaturations as opposed to actual rates of nucleation) was better than it should have been, and it was widely believed that the unexpected agreement was do to a fortuitious cancelation of errors. One of the earliest attempts to improve the theory involved making the drop "less macroscopic." For example, it is well known that the interface between a liquid and its vapor is not discontinuous but involves a transition zone having a thickness of several molecular diameters within which the density of the liquid decreases continuously to the density of the vapor. When the radius of the drop is comparable with the thickness of this transition zone,. the effects of "curvature" on surface tension must be taken into account.[53] Modifications aimed at accounting for curvature were undertaken at an early date.[54] However, these modifications, themselves, were necessarily uncritical, because among all the clusters of different sizes and compositions, only the drop corresponding to the nucleus (saddle point) could be truly in equilibrium (albeit unstable equilibrium) with the supersaturated vapor. Thus, strictly speaking, the equilibrium concept of interfacial tension had no meaning for clusters other than the nucleus.

Another modification which led to years of controversy involved the so called "replacement free energy." This modification was first attempted by Lothe and Pound[55]. The basis of the modification is most easily understood by reference to Eq. (6). In that equation, the free energy F_n which appears in the exponent is the free energy of a "drop" at rest in the laboratory coordinate system, and is specified by Eq. (1). Lothe and Pound therefore reasoned that the partition function q_n appearing in Eq. (6) belonged to an entity, at rest, having no translational or rotational degrees of freedom, and, therefore, could only account for the internal degrees of freedom of the cluster. They therefore mutliplied q_n by appropriate translational and rotational partition functions to compensate for this deficiency. When the new partition function was inserted into the theory, the value of N_n, for the nucleus, given formerly by eq. (7), was increased by a factor of about 10^{18}. The rate J, given by Eq. (8) was therefore increased by a corresponding factor. This increase was enormous enough to affect, in spite of the hypernonlinearity exhibited in Figure 2, the value of the critical supersaturation predicted by the theory. The Lothe-Pound theory predicts critical supersaturations 30-50% lower than those of the original theory.

In the majority of cases, however, measurement supports the original theory. Ultimately the controversy was resolved for most workers (some are still not convinced) when it was noticed, as noted in the previous section, that the drop at rest in the laboratory coordinate system (i.e. whose Gibbs dividing surface was at rest) had a center of mass which could still fluctuate.[56,57,58] For a macrascopic drop, this fluctuation is small, but for a cluster of the size of a nucleus it is considerable. Since internal degrees of freedom refer to motion relative to the center of mass, the free energy specified by Eq. (1) therefore accounts for some translational motion. Furthermore there is no restriction on rotation. The fluid drop simply does not rotate rigidly, but on a fine grained basis it does possess angular momentum. When all of this is carefully taken into account the correction to q_n in Eq. (6) amounts to a factor of about 10^3 or 10^4. [56,57,58] Such a relatively small correction has almost no effect on the value predicted for the critical supersaturation, but it does lead to a significant change in the actual rate of nucleation. Recent experiments[59] on the actual rate seem to confirm this correction.

The liquid drop, (capillarity approximation) model has led to still another inconsistency. In the thermodynamic treatment of a small drop, it is recognized that the pressure within the drop will exceed the ambient pressure by a factor $2\sigma/r$ where σ is the interfacial tension, and r, the radius of the drop. A molecule located sufficiently deep within the drop (if the drop is large enough so that its interior is uniform) should only be aware of the drop because of this increase in pressure. In other words, its chemical potential should be modified, only because of the pressure. The condition of equilibrium, is then the usual one, i.e. the chemical potential of the species in the vapor phase (or the external phase if the outside phase is not a vapor) should be equal to the chemical potential within the drop. However, if one attempts to use a model based on the capillarity approximation, in say a binary system in which the nucleus corresponds to the saddle point, taking due account of the dependence of interfacial tension on composition, the condition of equilibrium (unstable equilibrium) which identifies the saddle point[60,61], is not the one quoted above. Instead, additional terms arise involving the derivatives of interfacial tension with respect to composition. This inconsistency has been repaired, recently[62], by an analysis which shows that the composition at the interface must enter the picture, and that a proper application of the Gibbs adsorption equation leads to the standard condition involving the equality of chemical potentials. The conclusion is that, even though the capillarity approximation defines a model, that model must be dealt with consistently within the requirements of thermodynamics.

There has even been some attempt to formulate a theory of

nucleation, based on "mathematical clusters."[63] However, as indicated earlier, mathematical clusters are simply bookkeeping tools, and, in the end, the theory proves to have no fundamental basis.

Measurements of critical supersaturations or critical supercoolings (in the case of liquid-liquid transitions) near critical temperatures seem to indicate that the metastable state is anomalously stable near the critical point.[10,64,65] Recent theoretical work[66,67,68], and some very recent experimental studies[69], have shown that the "stabilizing" of the metastable state is only apparent. The effect is due to the slow rate of growth (due to "critical slowing down") of the drops to the size where they can be observed by the methods of light scattering. This anomaly emphasizes the fact that nuclei themselves (because of small size and exceedingly small concentration) are never observed; only the entities which develop from them can be interrogated. New methods which might allow the direct observation of nuclei would be valuable.

The specification of a metastable state on a purely microscopic basis is, itself, a difficult matter.[70] Ultimately, the state is constrained into existence by a kinetic constraint (bottleneck) rather than an equilibrium restriction. This gives rise to a series of thermodynamic questions dealing with constraints, kinds of work, and independent thermodynamic variables.[71]

The nucleus in the simple theory, is identical with the drop which can remain in unstable equilibrium with the surrounding vapor. In other words, the pressure of the surrounding supersaturated vapor merely reflects the increased vapor pressure of the drop due to its curvature. From this, it should be clear that increasing the degree of supersaturation decreases the size of the nucleus. At very high supersaturations the nucleus may be so small that the liquid drop model is badly in error. It should be mentioned that the capillarity approximation can be used when the condensed phase is a crystalline solid rather than a liquid. In this case it may be necessary to determine the equilibrium form of the crystalline nucleus by means of the so called Gibbs-Wulff construction.[72,73,74] No matter which case is involved (liquid or crystal), at high supersaturations it is unreasonable to model the nucleus as a macroscopic entity having an interface. Indeed, the nucleus will then be more like the clusters found in molecular beams, i.e, it should be regarded as a molecule. In the nucleation of ionic solids, either from solution or the vapor phase, there is evidence that the nucleus is a unit cell of the crystal. For example, this seems to be the case in the formation of crystalline ammonium chloride from NH_3 and HCl vapors.[75] As indicated in the last section, high supersaturations are usually involved when the nucleated clusters are strongly bound, because then the vapor pressure of the condensed species is very low. At the very low temperatures achieved in beams even van der Waals clusters may be strongly bound (on a relative basis). However the van der Waals molecules which are observed may be smaller than or exceed the relevant nucleus in size, since the expansion limits the amount of condensate available for growth.

At such high supersaturations where the nucleus is very small, steady state nucleation kinetics (as in the simple theory) are not likely to be achieved. Among other things, the concept of the quasi equilibrium of the nucleus with single vapor molecules will no longer be valid. In any event, the theory must deal with nonsteady kinetics, the limit of which is the process leading to the Poisson distribution of sizes referred to in reference 1. In the very rapid expansion of Ar vapors, achieveable in nozzle devices[17,40], the nucleus could be an Ar dimer or trimer. In this case the most effective way to advance the theory would be to treat the process as a strict chemical reaction (under the influence of dispersion forces) and to perform trajectory analyses in the manner which chemical kineticists are accustomed to. If the nucleus is a van der Waals molecule consisting of only a few atoms (or molecules), there might be several isomeric modifications which need to be considered for their relevence to condensation. These would correspond to the fluctuation of shape of the conventional liquid drop nucleus (considered only through an inversion of the order of averaging in the conventional theory--the liquid drop itself constituting the average shape). Such fluctuations or isomerizations are associated with the entropy of activation when considered from the chemical kinetic point of view.

Still another problem which has been investigated concerns the removal (by a third or additional bodies) of the excess energy of the two combining species in Eq. (4). In the quasi macroscopic theory this has involved studies of accomodation.[76,77,78]

It is important to mention that if the supersaturation is sufficiently high, no nucleation may be necessary for the process of condensation. This is because the barrier, expressed by Eq. (9), may vanish.

Their are cases, involving conventional nuclei, in which non-steady kinetics may be involved. An example, is the binary condensation of sulfuric acid-water vapor mixtures.[27,28] In these mixtures the concentration of sulfuric acid molecules maybe so low (less than 10^6 molecules/cm^3) that very long times are required for collisions between such molecules. The long induction time demands nonsteady kinetics. A similar example involves a mixture of NH_3 and HCl vapors condensing to form solid ammonium chloride.[75] If either vapor component is present in only small amounts, long induction times and nonsteady kinetics will result. Still another example involves nucleation in the neighborhood of a critical temperature.[29] Here the large size of the condensation nuclei, and the effect of "critical slowing down" on the dynamics, leads to important nonsteady problems.

It should also be pointed out that in place of the conventional capillarity approximation, in which the nucleus is viewed as a macroscopic liquid drop, separated discontinuously from the vapor by an interface, more realistic theories have been advanced in which the liquid density within the nucleus is not regarded as uniform.[79,80] Such nonunifom nuclei have been analysized theoretically in a number of ways, prominent among which is an application of the so called "square gradient theory" first introduced by Cahn.[81] Also, it should be mentioned that direct monte carlo methods have been applied to the study of nucleation. For example several authors have studied the appearance of crystals from a Lennard-Jones liquid using monte carlo methods and periodic boundary conditions.[82,83]

As a final note, we remark that nucleation and the formation of clusters in vapors in which the molecular species can "associate"[84] (even in the <u>saturated</u> vapor) presents a special theoretical problem.[85] In a sense, the molecules can be "stored" in the vapor by the process of association, and critical supersaturations are actually, apparently, higher for associated vapors. This goes <u>counter</u> to intuition, since one feels, at first, that association is a step on the way to nucleus formation. Association can also involve multicomponent vapors, e.g. the formation of hydrates in sulfuric acid-water vapor mixtures[85] (not unlike the association of rare gas atoms with tetracene). Thus, paradoxically, here is a case in which the "cluster-molecules" formed in molecular beams could actually be a step in a direction <u>away</u> from condensation!

III. Conclusion

It should be clear from the discussion in the preceeding section that, in the development of conventional theories of nucleation, considerable confusion has been generated by the rather cavalier extrapolation of macroscopic ideas downward. In the analysis of clusters in the molecular beams, especially in the context of the kinetics of condensation (a context which is often but not exclusively mentioned) the same sort of cavalier extrapolation <u>upward</u> is taking place. Furthermore, there is almost no reference to the extensive existing literature of which the 86 references in the current paper form but a small sample. Such a noncritical approach will simply lead to another set of time consuming confusions. Since the molecular beam approach does hold promise as a means for studying the important phenomenon of nucleation it is well worth the effort to take every precaution to avoid uncritical thinking. Hopefully, the superficial discussion of this paper will alert workers to the pitfalls which are possible.

IV. References

1. U. Even, A. Amirav, S. Leitweiler, M. J. Ondrechen, Z. Berkovitch-Yellin, and J. Jortner, <u>Faraday Discuss Chem. Soc.</u> **73**, 153 (1982).

2. T. L. Hill, <u>J. Chem. Phys.</u> **23**, 617 (1955).

3. F. H. Stillinger, <u>J. Chem. Phys.</u> **38**, 1486 (1963).

4. J. K. Lee, J. A. Barker and F. F. Abraham, <u>J. Chem. Phys.</u> **58**, 3166 (1973).

5. W. Band, <u>J. Chem. Phys.</u> **7**, 324, 927 (1939).

6. R. Ginell, *J. Chem. Phys.* **23**, 2395 (1955); **34**, 992, 1249, 2174 (1961).

7. H. P. Gillis, D. C. Marvin, and H. Reiss, *J. Chem. Phys.* **66**, 214 (1977).

8. J. E. Mayer, *J. Chem. Phys.* **5**, 67 (1937); J. E. Mayer and P. C. Ackerman, ibid., **5**, 74 (1937); J. E. Mayer and S. F. Harrison, ibid., **6**, 87, 101 (1938).

9. J. Frenkel, 'Kinetic Theory of Liquids,' ch. VII, Oxford University Press, New York, 1946.

10. R. B. Heady and J. W. Cahn, *J. Chem. Phys.* **58**, 896 (1973).

11. R. McGraw and H. Reiss, *J. Stat. Phys.* **20**, 385 (1979).

12. R. H. Heist and M. Heeks (A study of nucleation in vapors near the critical temperature.) Private communciation to the author)--to be published.

13. D. J. McGinty, *J. Chem. Phys.* **55**, 580 (1971).

14. M. R. Hoare and P. Pal, *Advances in Phys.* **24**, 645 (1975).

15. M. R. Hoare and J. McInnes, *Faraday Discuss. Chem. Soc.* **61**, 13 (1976).

16. C. L. Briant, *Faraday Discuss. Chem. Soc.* **61**, 25 (1976).

17. M. R. Hoare, P. Pal, and P. P. Wegener, *J. Colloid and Interface Sci.* **75**, 126 (1980).

18. J. J. Burton, *J. Chem. Phys.* **52**, 345 (1970); *Nature* **229**, 335 (1971).

19. C. L. Briant and J. J. Burton, *J. Chem. Phys.* **63**, 2045 (1975).

20. J. J. Burton, *Cat. Rev.-Sci. Eng.* **9**, 209 (1974).

21. J. J. Burton and C. L. Briant, *Advances in Colloid and Interface Science* **7**, 131 (1977).

22. F. F. Abraham, M. R. Muzik, and kG. M. Pound, *Faraday Discuss. Chem. Soc.* **61**, 34 (1976).

23. M. J. Ondrechen, Z. Bekovitch-Yellin, and J. Jortner, *J. Am. Chem. Soc.* **103**, 6586 (1981).

24. H. Reiss, *Ind. Eng. Chem.* **44**, 1284 (1952).

25. W. G. Courtney, J. Chem. Phys. **36**, 2009 (1962).

26. F. F. Abraham, J. Chem. Phys. **51**, 1632 (1969).

27. W. J. Shugard and H. Reiss, J. Chem. Phys. **65**, 2827 (1976).

28. F. J. Schelling and H. Reiss, J. Chem. Phys. **74**, 527 (1981).

29. Y. Rabin and M. Gitterman, Phys. Rev. **A 29**, 1496 (1984).

30. L. Farkas, Z. Phys. Chem. (Leipzig) **A125**, 236 (1927).

31. R. Becker and W. Döring, Ann. Phys. (Leipzig) **24**, 719 (1935).

32. M. Volmer, J. Phys. Chem. **25**, 555 (1929).

33. M. Volmer, 'Kinetic der Phasenbildung,' Theodor Steinkopff, Dresden and Leipzig, 1939. (Available as a Translation by the Intelligence Department, AMC under number ATI No. 81935.)

34. F. F. Abraham, 'Homogeneous Nucleation theory,' Academic Press, New York and London, 1974.

35. A. C. Zettlemoyer, 'Nucleation,' Marcel Dekker, New York, 1969.

36. H. Reiss, 'Methods of Thermodynamics,' p. 168, Blaisdell Publishing Co., New York, 1965.

37. H. Reiss, J. Chem. Phys. **18**, 840 (1950).

38. J. L. Schmitt, Rev. Sci. Instrum. **52**, 1749 (1981).

39. J. P. Franck and H. G. Hertz, Z. Phys. **143**, 559 (1956).

40. P. P. Wegener and B. J. C. Wu, in Nucleation Phenomena, edited by A. C. Zettlemoyer, pp. 325-417, Elsevier, New York, 1977.

41. M. Volmer and H. Flood, J. Phys Chem. **170**, 293 (1934).

42. J. L. Katz and B. J. Ostermeir, J. Chem. Phys. **47**, 478 (1967).

43. J. L. Katz, J. Chem. Phys. **52**, 4733 (1970).

44. J. L. Katz, J. Chem. Phys. **62**, 448 (1974).

45. D. C. Marvin and H. Reiss, J. Chem. Phys. **69**, 1897 (1978).

46. P. Mirabel and J. L. Clavelin, J. Chem. Phys. **70**, 5767 (1979).

47. J. L. Katz, F. C. Wen, T. McLaughlin, R. J. Reusch, and R. Patch, Science **196**, 1203 (1977).

48. F. C. Wen, T. McLaughlin, and J. L. Katz, <u>Phys. Rev.</u> **A26**, 2235 (1982).

49. H. Reiss and M. A. Chowdhury, <u>J. Phys. Chem.</u> **87**, 4599 (1983).

50. A. W. Gertler, J. O. Berg, and M. A. El-Sayed, <u>Chem. Phys. Lett.</u> **578**, 343 (1978).

51. M. A. Chowdhury, H. Reiss, D. R. Squire, and V. Stannett, <u>Macromolecules</u> **17**, 1436 (1984).

52. H. Reiss and M. A. Chowdhury, <u>J. Phys. Chem.</u> **88**, 6667 (1984).

53. S. Ono and S. Kondo, 'Handbuch der Physik,' **10**, Springer-Verlag, Berlin and New York, 1960.

54. G. M. Pound and V. K. La Mer, <u>J. Chem. Phys.</u> **19**, 506 (1951).

55. J. Lothe and G. M. Pound, <u>J. Chem. Phys.</u> **36**, 2082 (1962).

56. H. Reiss, J. L. Katz, and E. R. Cohen, <u>J. Chem. Phys.</u> **48**, 5553 (1968).

57. H. Reiss, <u>Advances in Colloid and Interface Science</u> **7**, 1 (1977).

58. R. Kikuchi, <u>Advances in Colloid and Interface Science</u> **7**, 67 (1977).

59. J. L. Schmitt, R. A. Zalabsky and G. W. Adams, <u>J. Chem. Phys.ee</u> **79**, 4416 (1983).

60. G. J. Doyle, <u>J. Chem. Phys.</u> **35**, 795 (1961).

61. H. Reiss and M. Shugard, <u>J. Chem. Phys.</u> **65**, 5280 (1976).

62. G. Wilemski, <u>J. Chem. Phys.</u> **80**, 1370 (1984).

63. W. J. Dunning, in <u>Nucleation</u> edited by A. C. Zettlemoyer, Marcel Dekker, New York, 1969.

64. J. S. Huang, W. I. Goldburg, and M. R. Moldover, <u>Phys. Rev. Lett</u> **34**, 639 (1975).

65. R. G. Howland, N. C. Wong, and C. M. Knobler, <u>J. Chem. Phys.</u> **73**, 522 (1980).

66. K. Binder and D. Stauffer, <u>Adv. Phys.</u> **25**, 343 (1976).

67. J. S. Langer and L. A. Turski, <u>Phys. Rev.</u> A **8**, 3230 (1973).

68. J. S. Langer and kA. J. Schwartz, <u>Phys. Rev.</u> A **21**, 948 (1980).

69. E. D. Siebert and C. M. Knobler, <u>Phys. Rev. Lett.</u> **32**, 1133 (1984).

70. O. Penrose and J. L. Lebowitz, J. Stat. Phys. **3**, 211 (1971).

71. H. Reiss, Be. Bunsen Gesellschaft **79**, 943 (1975).

72. G. Wulff, Z. Kristallog **34**, 449 (1901).

73. C. Herring in Structure and Properties of Solid Surfaces, edited by R. Gomer and C. S. Smith, Cambridge University Press, London, 1953.

74. W. J. Dunning in Nucleation, edited by A. C. Zettlemoyer, pp. 20-23, Marcel Dekker, Inc., New York, 1969.

75. T. T. Kodas, S. E. Pratsinis, and S. K. Friedlander, in press J. C. I. S. (1985).

76. R. Probstein, J. Chem. Phys. **19**, 619 (1951).

77. A. Kantrowitz, J. Chem. Phys. **19**, 1097 (1951).

78. C. Y. Mou and R. Lovett, J. Chem. Phys. **70**, 3488 (1979).

79. J. W. Cahn and J. E. Hilliard, J. Chem. Phys. **31**, 688 1959).

80. Y. Rabin and H. Reiss, Macromolecules **17**, 2450 (1984).

81. J. W. Cahn, Acta Met **9**, 795 (1961); **10**, 179 (1962).

82. A. Rahman, M. J. Mandell, and J. P. McTague, J. Chem. Phys. **64**, 1564 (1976).

83. M. J. Mandell, J. P. McTague and A. Rahman, J. Chem. Phys., (1976).

84. R. H. Heist, K. M. Colling, and C. S. Dupuis, J. Chem. Phys. **65**, 5147 (1976).

85. J. L. Katz, H. Saltsburg, and H. Reiss, J. Colloid and Interface Sci. **21**, 560 (1966).

86. R. H. Heist and H. Reiss, J. Chem. Phys. **59**, 665 (1973).

ENERGY TRANSFER IN COLLISIONS OF POSITIVE HALOGEN IONS WITH RARE GAS ATOMS

Matti Hotokka, Department of Physical Chemistry,
Abo Akademi, SF-20500 Abo, Finland
B. Roos, Department of Physical Chemistry 2,
Chemical Center, P. O. Box 740, 522007 Lund 7, Sweden
K. Balasubramanian, Department of Chemistry,
Arizona State University, Tempe, Arizona 85287
R. B. Sharma, N. M. Semo and W. S. Koski, Department of
Chemistry, The Johns Hopkins University, Baltimore, MD 21218

ABSTRACT. The collisions of F^+ with Ne and Br^+ with Kr have been
studied by measuring the energy spectra of the positive halogen ions
scattered at 0° to the beam direction. Transfer of electronic energy
from the halogen ion to translation energy of the rare gas and vice
versa is observed. The transfer is taking place by a Landau-Zener type
process. The results are interpreted by means of potential energy
curves calculated by ab-initio quantum chemical methods.

1. INTRODUCTION

There has been considerable investigation of formation and de-exitation
of neutral rare gas halide molecules.[1-6] Similarly, there have been
extensive experimental and theoretical studies carried out in the rare
gas oxides because of the potential application of these systems to
lasers.[7-8] On the other hand, there has been a very limited number of
studies on the rare gas halide diatomic ions. These rare gas-halogen
positive ion systems are isoelectronic with the rare gas oxides and as
such may have similar laser applications. In addition, in hot atom
studies of halogen systems, rare gases have been used as moderators and
the assumption is generally made that they are inert. Recently, it has
been demonstrated that this assumption is not applicable to the halogen
hot-atom systems since collisional studies between Br^+ and Kr show that
Kr is not inert but participates in an active chemical way in the hot
atom process.[9]

In the 1960's, with the discovery of stable xenon and krypton
fluorides, it was shown by mass spectroscopic techniques that XeF^+ and
KrF^+ are stable ions.[10] The noble gas halide ions, $KrCl^+$, KrF^+, $ArCl^+$,
and ArI^+ have been reported as stable ions.[11, 12] The ion $KrBr^+$ has
been prepared and its dissociation energy inferred from a threshold

P. M. Rentzepis and C. Capellos (eds.), Advances in Chemical Reaction Dynamics, 135–143.
© 1986 by D. Reidel Publishing Company.

measurement of a collision dissociation reaction.[13] ArF[+] has been reported to have a dissociation energy of $\geq$ 1.655 eV.[14] On the other hand, attempts to make NeF[+] and HeF[+] have been unsuccessful.[14] In order to get a better insight into the nature of the diatomic rare gas halides ions, we are investigating the collisions between halogen positive ions and rare gas atoms and at this time, we report our preliminary results on the Ne-F[+] and Kr-Br[+] systems.

2. EXPERIMENTAL RESULTS

The collisions of F[+] with Ne and Br[+] with Kr were carried out using a tandem mass spectrometer which has been described previously.[15] It consisted of an ion source, a 180° electrostatic analyzer and a quadrupole mass spectrometer as an input section. The beam of halogen positive ions from this section was passed through a shallow reaction chamber containing the target gas which was Ne in the case of F[+] and Kr in the case of Br[+]. The ions scattered at 0° to the beam direction were then passed through a second quadrupole mass spectrometer which was followed by a second 180° electrostatic analyzer and an electron multiplier. The beam width was 0.1 eV (FWHM). The ions were prepared by 100 eV electron bombardment of CH_3F in the F[+] case and of CH_3Br in the Br[+] case. The beam composition in both cases was 30% X[+] ($^1D_2$) and 70% X[+] ($^3P_{0,1,2}$) where X[+] is the halogen positive ion.

3. RESULTS AND DISCUSSION

Using this beam as the projectile and the rare gas as the target, a kinetic energy spectrum of the halogen positive ion was obtained as is illustrated in Figure 1 for the F[+]-Ne case. The central peak is the unperturbed F[+]. The satellite peaks on each side of the central peak are separated by 2.5 eV from the main peak.

The spacing between the $^3P_2$ ground state of F[+] and the first excited state $^1D_2$ is 2.587 eV.

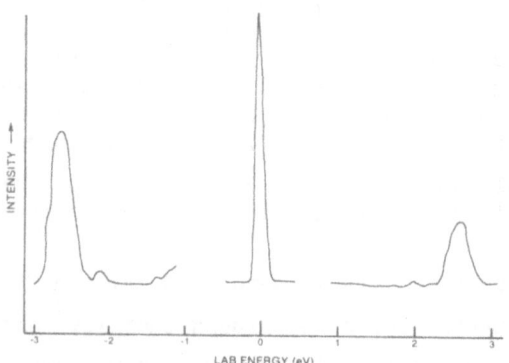

Figure 1. Kinetic energy spectrum of F[+] for F[+]-Ne collisions at 20 eV lab.

Apparently, on collision of F^+ with Ne there is a transfer of electronic energy into translational energy and vice versa. The peak on the right corresponds to the transition $F^+(^1D_2) \rightarrow F^+(^3P_{0,1,2})$. This peak is not present if the primary ion beam is completely in the ground state. The more intense satellite on the left is due to the transition $F^+(^3P_{0,1,2}) \rightarrow F^+(^1D_2)$. The spacings corresponding to the various j values are too small to be resolved in the spectrum because of the limitations of our instrumental resolution. At kinetic energies below 5.0 eV (lab) the F^+ peak in the left side of Figure 1 is not observed.

In view of the fact that the rare gas halide ions such as $ArCl^+$, $KrBr^+$, and XeI^+ are easily prepared by electron bombardment of a mixture of the appropriate rare gas and halogen molecule, we carried out similar experiments on $Ne-F_2$ mixtures. Mass spectrometric analysis of the emitted ions clearly showed a mass 39 peak which may correspond to NeF^+, however, the peak intensity was not high enough to carry out corroborative experiments such as collision induced dissociation to support the assignment. In this connection it is interesting to note that Berkowitz and Chupka[14] were not successful in producing NeF^+ by the reaction of $F_2^+(Ne,F)NeF^+$.

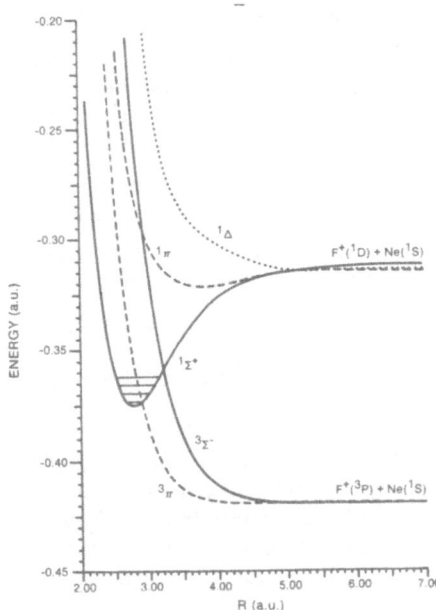

Figure 2. Potential energy curves for F^+-Ne.

In order to interpret these results, it is necessary to have some knowledge of the $Ne-F^+$ potential energy curves. The pertinent curves were calculated for this system using the CASSCF method[16] and the results for the five lowest states are summarized in Figure 2. As has been noted previously[17], the ground state is repulsive and the only bound curve is the $^1\Sigma^+$ curve correlating with $F^+(^1D) + Ne(^1S)$. The binding energy D_e for this state is 1.644 eV compared to 1.69 eV for the isoelectronic F_2. Of particular interest to this study are the curve crossings between $^3\Sigma^-$ and $^1\Sigma^+$ and $^1\Pi$ and $^3\Pi$ and $^1\Sigma^+$. We believe that the observed energy transfer for the

F^+-Ne collisions take place at these crossings by means of a Landau-Zener type of process. Work on the dynamics and transition probability is currently in progress.

A second closely related series of observations involve the Kr-Br^+ system. The experimental procedure is the same as in the Ne-F^+ case. Figure 3 gives a typical Br^+ energy spectrum at 0° to the beam direction.

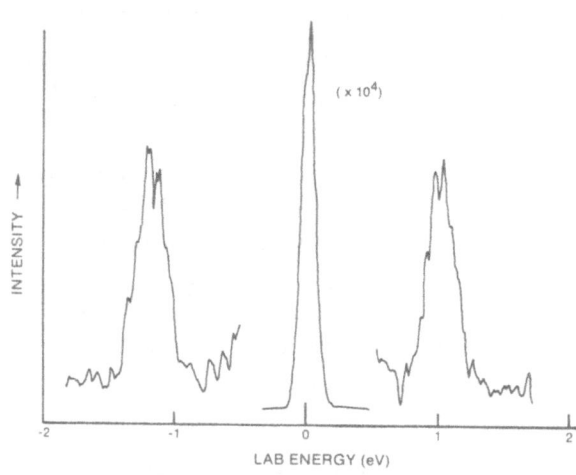

Figure 3. Kinetic energy spectrum for Br^+-Kr collisions at 20 eV (lab) and at 0° to the beam direction.

The curve is similar to the F^+ case but with some interesting differences. First there is a doubling of the satellite peaks indicating that the energy transfer process involves the $j = 0$ and $j = 1$ states of the ^3P state. These states are separated by 87 meV. We are apparently observing the transitions $^3P_{0,1} \longleftrightarrow {}^1D_2$ but we are not seeing the $^3P_2 \longrightarrow {}^1D_2$ transition even though the $j = 1$ and $j = 2$ states are separated by 389 meV which should be readily observable.

The $KrBr^+$ ion can be easily prepared[13] by electron bombardment of a Kr-Br_2 mixture and it is readily observed mass spectroscopically indicating that it has a lifetime longer than 20 μs which is the ion transit time in our instrument. Collision induced dissociation measurements by Watkins et. al.[13] indicate that $KrBr^+$ has a dissociation energy of approximately 1.5 eV.

In order to shed additional light on these observations the potential energy curves for the $KrBr^+$ system were calculated using the relativistic codes developed by Christiansen, Balasubramanian and Pitzer[18]. The results for the pertinent potential energy curves are given in Figure 4 and in Table 1. Nine low lying ω-ω states have been calculated. All nine are included in Table 1 and for the sake of clarity only seven these are included in Figure 4. It is clear that the ground state of $KrBr^+$,(2(1)), is repulsive. The only state that is appreciably bound is the $0^+(111)$ state correlating with $Br^+(^1D_2) + Kr(^1S_0)$.

In the Landau-Zener model the energy transfer between several states induced by collisions, takes place at points where the potential energy curves of the appropriate electronic states of the rare gas-halogen ion complex cross. From Table 1 and Figure 4 one can see that the states 2, 1, and $0^+(1)$ of $KrBr^+$ correspond to $Kr(^1S_0)$ + $Br^+(^3P_2)$. Similarly, the 1(11) and (0^-) states correlate with $Kr(^1S_0)$ + $Br^+(^3P_1)$ and 2(11), 1(111) and $0^+(111)$ correspond to $Kr(^1S_0)$ + $Br^+(^1D_2)$. In the Landau-Zener model the transition between $Br(^1D_2)$ to $Br^+(^3P_1)$ is allowed if one of the curves which dissociates into $Kr(^1S_0)$ + $Br^+(^1D_2)[2(11), 1(111), 0^+(111)]$ crosses with one of the curves which dissociates into $Kr(^1S_0)$ + $Br^+(^3P_1)[1(11), 0^-(1)]$. There

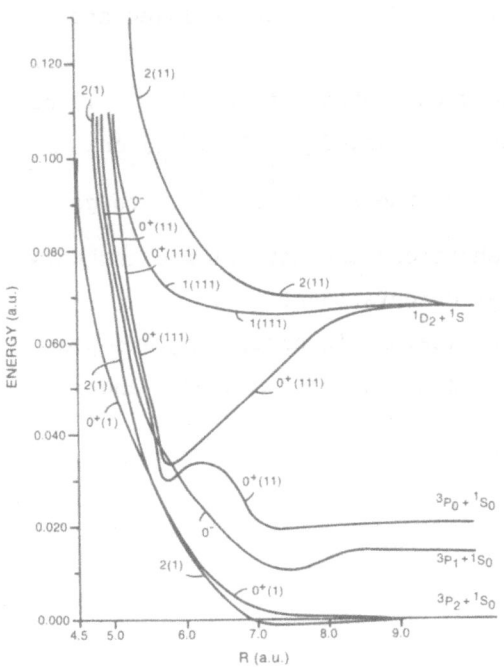

Figure 4. Potential energy curves for the Br^+-Kr system.

are a few restrictions imposed by symmetry. For example, two curves of the same $(\omega-\omega)$ symmetry cannot cross. It also appears that the selection rule for predissociation should be applicable for this case to determine among the several channels, the one that is allowed by symmetry restrictions. This selection rule for predissociation[19] is $\Delta\Omega = 0, \pm 1, +\nleftrightarrow-$. Thus, for example, if $0^+(111)$ curve crosses with 1(11) then the transition $^1D_2 \longleftrightarrow {}^3P_1$ for Br^+ is allowed. From Table 1 it can been seen that the $0^+(111)$ and 1(11) curves cross at 5.75 Bohr which is consistent with the observed $^1D_2 \longrightarrow {}^3P_1$ transition.

Although the $0^+(11)$ curve does not directly cross with $0^+(111)$, 1(111) or 2(11) curves, it crosses with the 1(11) curve between 5.75 – 6.0 Bohrs. Since the 1(11) curve is a channel for the $^1D_2 \longleftrightarrow {}^3P_1$ transition, a complex formed by $Br^+(^1D_2)$ which goes through this

channel could meet the $0^+(11)$ state. Thus by this process a Br^+ which
starts as $^1D_2$ could go to $0^+(11)KrBr^+$ state which dissociates into
$Kr(^1S_0)$ and $Br^+(^3P_0)$. So the transitions $^1D_2 \quad ^3P_0$ are allowed and
are observed experimentally.

The curves which dissociate into ground state atoms $(2, 0^+, 1)$ do
not cross with curves which dissociate into $Kr(^1S_0) + Br^+(^3P_0)$ or
$Kr(^1S_0) + Br^+(^3P_1)$. Thus there are no channels for $^1Br^+(^1D_2)$ to go
into $^3 Br^+(^3P_2)$ and the $^1D_2 \quad ^3P_2$ transitions are not allowed and this
is in agreement with the experimental observations.

Work on the dynamics of Br^+-Kr collisions is continuing and the
collisions of other rare gas-halogen ion pairs such as Ar-Cl^+ and
Xe-I^+ are being investigated.

Table 1.

Potential Energy Curves of KrBr$^+$. (Energies in hartrees).

R	O$^+$(I)	2(I)	1(I)	1(II)	O$^-$	O$^+$(II)	O$^+$(III)	1(III)	2(II)
4.5	.1044	.1599	.1639	.1987	.1705	.1720	.2016	.1999	.3186
5.0	.0492	.0684	.0073	.0990	.0801	.0814	.1110	.1090	.1899
5.25	.0395	.0431	.0478	.0725	.0556	.0585	.0677	.1038	.1285
5.5	.0289	.0289	.0334	.0502	.0419	.0454	.0468	.0771	.1036
5.75	.0227	.0213	.0242	.0344	.0350	.0301	.0341	.0713	.0963
6.0	.0163	.0169	.0146	.0224	.0287	.0331	.0381	.0696	.0846
7.0	.0021	-.0010	-.0001	.0119	.0129	.0205	.0497	.0662	.0711
9.0	.0014	-.0008	-.0007	.0135	.0137	.0206	.0635	.0671	.0690
11.0	0.0	0.0	0.0	.0145	.0145	.0213	.0667	.0667	.0667

References

1. L. G. Piper, J. E. Velazco and D. W. Setser, J. Chem. Phys., 59, 3323 (1973).

2. J. E. Velazco, J. H. Kolts and D. W. Setser, J. Chem Phys., 65, 3468 (1973).

3. J. Tellinghuisen, A. K. Hays, J. M. Hoffmann and G. C. Tisome, J. Chem. Phys., 65, 4473 (1976).

4. D. L. King, L. G. Piper and D. W. Setser, J. Chem. Soc. Faraday Trans. III, 73, 177 (1977).

5. M. Rokni, J. H. Jacobs and J. A. Mangano, Phys. Rev., A16, 2216 (1977).

6. R. E. Olson and B. Liu, Phys. Rev., A17, 1568 (1978).

7. T. H. Dunning, Jr. and P. J. Hay, J. Chem. Phys., 66, 3767 (1977) and references therein.

8. Cohen, Wadt and P. J. Hay, J. Chem Phys., 71, 2955 (1979) and references therein.

9. H. P. Watkins, N. E. Sondergaard and W. S. Koski, Radiochim. Acta, 29, 87 (1981).

10. J. H. Holloway, Noble Gas Chemistry (Methuen, London (1968)).

11. A. Henglein and G. A. Muccini, Angew. Chem., 72, 630 (1960).

12. I. Kuen and F. Howorka, J. Chem. Phys., 70, 595 (1979).

13. H. P. Watkins and W. S. Koski, Chem. Phys. Lett., 77, 470 (1981).

14. J. Berkowitz and W. A. Chupka, Chem. Phys Lett., 7, 447 (1970).

15. K. Wendell, C. A. Jones, Joyce J. Kaufman and W. S. Koski, J. Chem. Phys., 63, 750 (1975).

16a. Per M. Siegbahn, A. Heiberg, B. Roos and B. Levy, Physica Scripta, 21, 323 (1980).

16b. B. O. Roos, P. R. Taylor and Per M. Siegbahn, Chem. Phys., 48, 157 (1980).

17. J. F. Liebman and L. C. Allen, J. Am. Chem. Soc., 92, 3539 (1970).

18. P. A. Christiansen, K. Balasubramanian and K. S. Pitzer, J. Chem. Phys., **76**, 5087 (1982).

19. G. Herzberg, Spectra of Diatomic Molecules, Van Nostrand Rheinhold Co., New York.

CORRELATIONS IN PICOSECOND PULSE SCATTERING: APPLICATIONS FOR FAST REACTION DYNAMICS

B. Van Wonterghem° and A. Persoons

Laboratory for Chemical and Biological Dynamics
University of Leuven
Celestijnenlaan 200 D
3030 Leuven
Belgium

ABSTRACT. A new intensity fluctuation correlation technique for (depolarized) scattered light pulses will be described. This method is based on the short time autocorrelation of the scattering power of a system, probed by two consecutive picosecond laser pulses separated by a variable delay time. The time correlation function measured this way contains information on the reaction dynamics of an equilibrium reaction mixture, if the polarizability tensor of the reacting molecules is affected by the reaction process. Changes in the absorption dipole moment (magnitude or direction) can be detected by resonance scattering. This new technique will be applied to the study of intramolecular proton transfer kinetics in solution.

1. INTRODUCTION

As known already for a long time, light scattering from solutions is caused by spatial fluctuations of the polarizability[1]. These fluctuations occur spontaneously, even in a system in thermodynamic equilibrium. A method able to probe the time dependence of these fluctuations gives information about the kinetics of the various modes of fluctuation dissipation. Each scattering center or molecule can be described by a position vector and a (complex) polarizability tensor, the elements of which describe the polarizability in a certain direction. Translational diffusion causes fluctuations by changing the interference between the waves scattered by different molecules. A second cause is the fluctuations of the scattering power of the molecules, determined by the polarizability tensor. This can be changed by reorientation of the molecules (rotational diffusion) or by changes in the tensor elements because of a chemical transformation of the molecule[2]. This includes conformational changes and bond breaking and forming processes (e.g. intramolecular proton transfer). The relaxation time of these fluctuations can be determined by measuring the autocorrelation function of the instantaneous scattering power $f(t)$ of the scattering volume. The autocorrelation function $g_2(\tau)$ gives a measure of the correlation in scattered intensities at two moments a time τ apart and is

P. M. Rentzepis and C. Capellos (eds.), Advances in Chemical Reaction Dynamics, 145–150.
© *1986 by D. Reidel Publishing Company.*

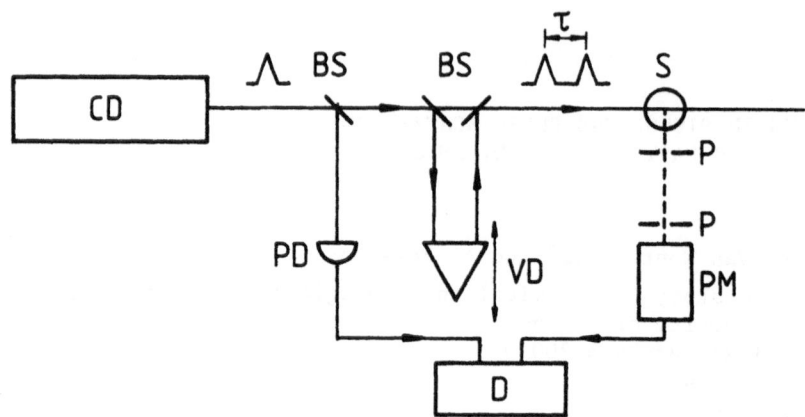

Figure 1. Principle of the apparatus. (CD: cavity dumper, PD: photodiode, VD: variable delay line, BS: beamsplitter, S: scattering cell,
P: pinhole, PM: photomultilplier, D: electronic data acquisition)

mathematically defined as:

$$g_2(\tau) = \frac{<I_s(t)I_s(t+\tau)>}{<I_s(t)>^2} \tag{1}$$

$$= \frac{<f(t)f(t+\tau)>}{<f(t)>^2} \tag{2}$$

where the scattered intensity $I_s(t)$ is defined as $I_0(t)f(t)$, I_0 being
the intensity of the incoming laser light.

2. EXPERIMENTAL DETERMINATION OF THE AUTOCORRELATION FUNCTION

In the case of slow relaxation times (> 100 nanoseconds), one can use
a digital correlator to calculate the correlation function of the
fluctuations in the photon count rate of a photomultiplier (PM) that
looks at the scattering of a sample, illuminated by a CW laser beam.
In order to measure much faster correlation decays, we searched for a
technique based on picosecond laser pulses, rather than CW laser light.
The principle of our apparatus is schematically depicted in figure 1.
The source of the pulses is a synchronously pumped cavity dumped dye
laser, pumped by a CW modelocked frequency doubled Nd-YAG laser. This
system delivers pulses with a duration of about 2 picoseconds at a repetition rate variable up to 4 MHz. These pulses are divided by a beamsplitter in two equal parts and recombined, one part being delayed by
a variable delay line. The pulse pairs generated this way are sent on

a sample cell that contains the solution under study. The scattered pulses are detected by a fast photomultiplier tube, with a response time of about 3 nanoseconds. Because the delay time between two pulses of a pair is less than the response time of the PM, the light intensity scattered by the two pulses is integrated into one anode output pulse by the PM. We shall mention two methods to extract the autocorrelation function of the scattering power from these anode pulses.

2.1. Anode pulse height analysis

If the intensity of an incoming pulse is denoted by I_p, the total intensity scattered by a pulse pair I_t equals

$$I_t(t) = f(t)^2 I_p + f(t+\tau)^2 I_p \qquad (3)$$

The autocorrelation can then be related to the second moment $<I_t(t)^2>$ of $I_t(t)$, assuming stationarity of $f(t)$:

$$<I_t(t)^2> = 2<I_p^2>\{<f(t)^2 f(t+\tau)^2> + <f(t)^4>\} \qquad (4)$$

which per definition (1) equals

$$<I_t(t)^2> = 2<I_p^2>\{g_2(\tau) + g_2(0)\} . \qquad (5)$$

This second moment can be calculated from the experimental anode pulse height distribution accumulated by a multichannel pulse height analyzer[3]. Repeating this while varying the delay time τ gives us the complete autocorrelation $g_2(\tau)$.

2.2. Pulse counting

A second method to determine $g_2(\tau)$ makes use of the joint photon count probability function $P(n,m,\tau)$[4] i.e. the joint probability that n photons are detected at time t and m photons at time $t+\tau$. In our experiment it is very easy to determine $P(0,0,\tau)$. In our approach this is the probability that a pulse pair with internal delay τ did not produce a photoelectron on the PM cathode surface. This is accomplished by counting the number of anticoincidences between an exciting pulse pair, monitored by a fast photodiode, and the anode output of the photomultiplier. This effect can be explained as follows. The probability that a scattered pulse does not produce ay photoelectron on the PM is $P(0)$. The joint probability that two pulses a time τ apart don't produce any photoelectron can be written as

$$P(0,0,\tau) = P(0)P(0,t+\tau|0,t) \qquad (6)$$

where $P(0,t+\tau|0,t)$ is the conditional probability that a pulse at time $t+\tau$ did not produce any photoelectron if a pulse a time τ earlier did neither. If τ is much smaller than the correlation time of the scattering process, $P(0,t+\tau|0,t)$ goes to unity. If τ on the contrary goes to infinity and the scattering of the two pulses is no more correlated,

$P(0,t+\tau|0,t)$ becomes equal to $P(0)$. So we conclude that

$$P(0,0,\tau = 0) < P(0,0,\tau = \infty) \tag{7}$$

Once $P(0,0,\tau)$ is determined for different delay times τ, $g_2(\tau)$ can be calculated or fitted according to the following relationship[5]:

$$P(0,0,\tau) = \{1 + 2<n> + <n>^2(1 - g_2(\tau))\}^{-1} , \tag{8}$$

where $<n>$ is the mean number of photoelectrons produced by a scattered pulse pair. The advantages of this method are the simplicity of detection (one only needs a discriminator and a fast photon counter) and the high processing speed, which allows a pulse repetition rate of about 100 MHz. The only difficulty is that the scattering cross-section of the solution must be fairly high to detect a variation in $P(0,0,\tau)$ when varying the delay time τ.

These two methods were verified in two extreme cases: static scattering on a glass plate and totally uncorrelated scattering (in our time range at least) from CCl_4. Scattering from solutions is still a· problem because of the low scattering intensity. Resonance scattering[6] at the absorption band of a molecule could be a solution because the scattering intensity is enhanced by a factor 100 to 1000.

3. SCATTERING FROM CHEMICALLY REACTING SYSTEMS

As we have mentioned in the introduction, light scattering from a reactive mixture consists of two parts: a diffusive and a reactive one. In order to give an idea how it is possible to extract kinetic information from the autocorrelation function of the scattering power of such a mixture, we shall consider a simple example. This consists of a mixture of two types of molecules A and B in a dynamic equilibrium:

$$A \underset{k_b}{\overset{k_a}{\rightleftarrows}} B . \tag{9}$$

If the reaction rate $k=k_a+k_b$ is much greater than the diffusion rate K^2D_t (note[7]), the fluctuations due to the chemical reaction can be separated from the diffusive ones by their different time scale. In this case the autocorrelation of the scattering power, probed by depolarized scattering is proportional to[8]

$$g_2(\tau) \propto \{Pe^{-(\overline{D}_t + 6\overline{D}_r)/\tau} + Qe^{-k/\tau}\} , \tag{10}$$

where $\overline{D}_t$ and $\overline{D}_r$ are the mean translational and rotational diffusion coefficients of A and B. A crucial aspect of eq. 10 is the amplitudes of the two exponentials, expressed as P and Q. P depends upon the square of the sum of the polarizability anisotropies of A and B, while Q depends upon the square of their difference. As a consequence the diffusive scattering will always be more intense than the reactive scatter-

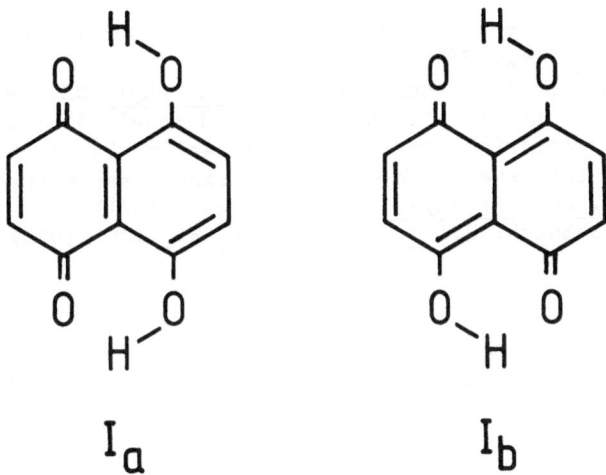

I_a I_b

Figure 2. Two possible structures of the Naphtazarin molecule

ing. Note also that the diffusive scattering is dependent on the scattering angle while the reactive one is not.
        An example of the system described above is the intramolecular
proton tranfer in internally hydrogen bonded molecules, like Naphta-
zarine (fig. 2). IR studies indicated the existence of a fast proton
tunneling from structure I_a to I_b [9]. Our technique will be applied to
this type of reactions.

4. CONCLUSION

A new method is tested in order to explore the possibilities of light
scattering as a probe of fast reaction kinetics in solution. The time
resolution of this method depends only upon the width of the laser
pulses used, i.e. about one picosecond. Our goal is the study of intramolecular proton transfer across hydrogen bonds and to make a contribution to the study of proton tunneling in such systems.

5. NOTES AND REFERENCES

° B. Van Wonterghem is a research assistant of the National Fund for
  Scientific Research (Belgium).

1. Einstein, A.: Ann. Phys. (Leipzig) 33, 1275 (1910)
2. Berne, B.J., Pecora, R.: Dynamic Light Scattering,
   Wiley-Interscience, New York . London (1976)
3. Arecchi, F.T.: IEEE J. Quant. Electr. QE-2, 341 (1966)
4. Saleh, B.: Photoelectron Statistics, Springer Series in Optical
   Sciences, volume 6, Springer Verlag, Berlin.Heidelberg.New York
   (1978)

5. Furcinitti, P., Kuppenheimer, J., Narducci, L.M., Tuft, R.A.:
   J. Opt. Soc. Amer. **62**, 792 (1972)
6. Bauer, D.R., Hudson, B., Pecora, R.: J. Chem. Phys. **63**, 588 (1975)
7. K is the length of the scattering vector and equals $4\pi n\sin(\theta/2)/\lambda_0$,
   where θ is the scattering angle, n the refractive index of the test
   solution and λ_0 the wavelength of the exciting laser pulses.
8. Pecora, R.: Ann. Rev. Biophys. Bioeng. **1**, 257 (1971)
9. Bratan, S., Strohbusch, F.: J. Mol. Struct. **61**, 409 (1980)

PICOSECOND TRANSIENT RAMAN STUDIES: TECHNIQUE AND APPLICATIONS

John B. Hopkins
Department of Chemistry
Louisiana State University
Baton Rouge, LA 70803 USA

and

P. M. Rentzepis
AT&T Bell Laboratories
Murray Hill, NJ 07974 USA

ABSTRACT. Picosecond spectroscopy[1] is a well established field; however picosecond Raman is a rather new technique which has recently emerged from the rapid advances in laser technology. This manuscript will discuss two examples of vibrational spectroscopy which illustrate the detailed information which may be obtained to elucidate the dynamics and structural changes occurring during the course of ultrafast chemical reactions.

1. DIRECT PICOSECOND TIME RESOLVED STUDIES OF VIBRATIONAL RELAXATION IN THE CONDENSED PHASE

There are several phenomena currently being studied which may be listed under the general area of vibrational relaxation. First there is the process which occurs in an isolated molecule when the density of vibrational levels becomes very large. In this case, excitation which is initially localized in a single well defined vibrational mode decays into other vibrational degrees of freedom which constitute a "heat bath" within the isolated molecule. This form of vibrational relaxation is truely speaking a vibrational randomization process since there is no net loss of energy between the molecule and the surroundings. A second form of vibrational relaxation currently under investigation occurs in the gas phase with an inelastic molecular collision. Here, vibrational energy is distributed into different vibrational, rotational, and translational quantum states of the colliding partners as a result of the collision. Thirdly, the subject which will be discussed in this paper is the dynamic behavior of a molecule imbeded in a low temperature matrix. The molecule under study is initially excited to vibrational level which is in excess of the thermal vibrational population given by the normal Boltsmann distribution. This "non thermal" vibrational energy decays into the surroundings until equilibrium is achieved.

Condensed phase vibrational relaxation is an extremely important photophysical process because the vibrational energy content of a molecule is known to effect both the reaction rate and branching ratios in a photochemical reaction. In the condensed phase these parameters will exhibit dynamic behavior due to the fact that the vibrational energy content

151

P. M. Rentzepis and C. Capellos (eds.), Advances in Chemical Reaction Dynamics, 151–163.
© *1986 by D. Reidel Publishing Company.*

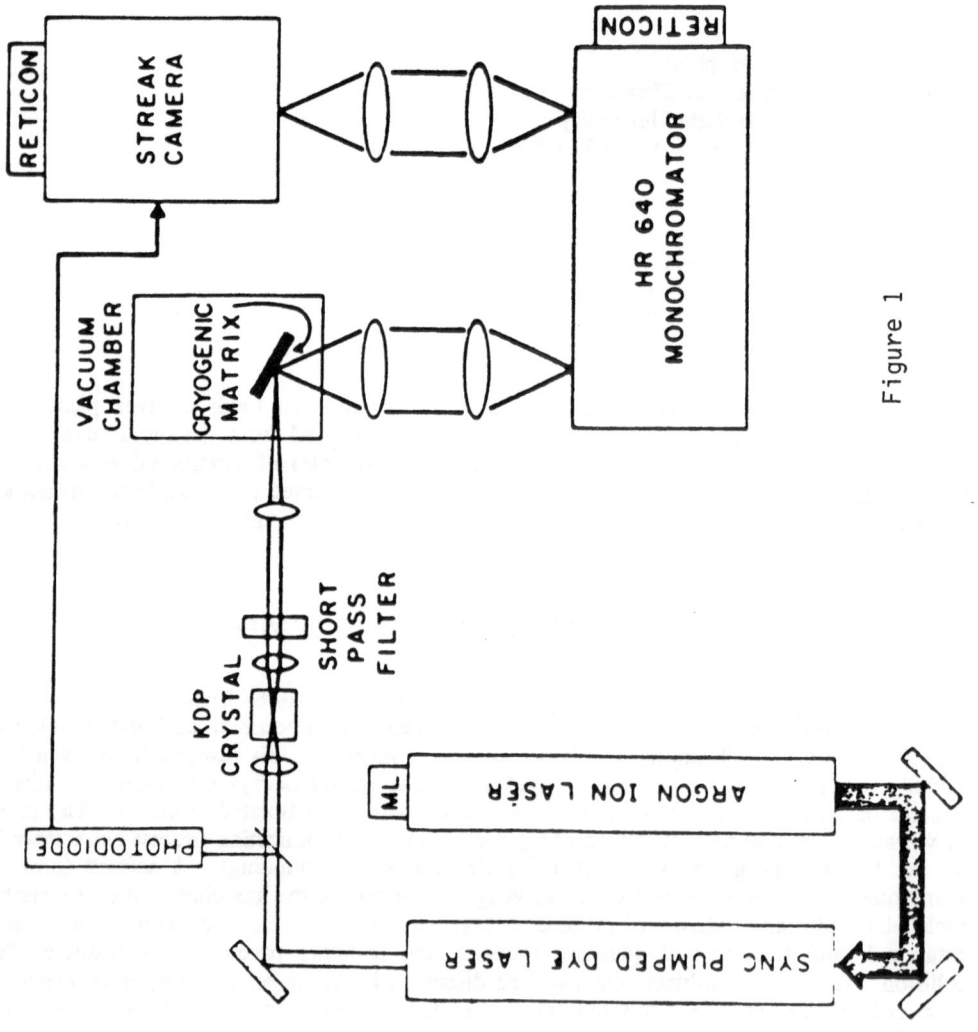

Figure 1

of the molecule is rapidly changing as the thermalization process proceeds. In addition to the photochemical aspects of vibrational energy, detailed information regarding energy flow between the solute molecule and the solvent can yield valuable information concerning the molecule-medium coupling interactions.

Electronic transitions of molecules at room-temperature even in condense phase molecules are generally very broad and void of vibrational structure due to band sequence congestion and inhomogeneous broadening processes. However, low temperature rare gas matrix isolation, a well known technique,[2] elliminates most of these problems and provides well resolved vibronic spectra for even very large molecules. This technique has been employed in conjunction with studies in the picosecond time domain to measure directly the kinetics and mechanism of vibrational energy dissipation in large aromatic molecules.

It should be pointed out that time-resolved spectroscopy has been previously utilized to investigate vibrational relaxation in diatomic and triatomic molecules which are isolated in low temperature rare gas matrixes.[3,4] In the case of small molecules with large vibrational spacings it was found that the vibrational relaxation time scales were in the microsecond to millisecond range. Additionally it was found that the mechanism of the relaxation process was dominated by the rotational motion of the molecule in the matrix. In other words vibrational to rotational energy relaxation is the predominant decay channel. This is followed by fast relaxation to the matrix phonon modes due to the strong coupling of the molecular rotational levels with the low frequency matrix phonon modes.

We shall present in this paper a discussion on the direct observation of picosecond kinetics and spectra of large polyatomic molecules is rare gas matrixes. Several consequences of the molecular size and structure will be noted with regard to their effect on the rate and mechanism of the vibrational relaxation process. First the large molecules under study here are rigid and incapable of significant rotation, therefore we shall only consider the vibrational degrees of freedom. Second the density of vibrational states reduces the vibrational spacing and therefore the amount of energy which much be exchanged, at a selected period of time, with the matrix is reduced. This has the effect of increasing the overall decay rate from $\sim 10^6$ sec^{-1} to $\sim 10^{11}$ sec^{-1}.

To collect the experimental data; lifetime (ps) and spectra, we used the experimental apparatus shown in figure 1 which has been described in detail in references.[5,6] Very briefly, the output of a dye laser, synchrously pumped by a mode locked argon ion laser, is frequency doubled by means of a KDP crystal. This second harmonic is utilized for excitation of the rare gas matrix isolated molecule. The emitted fluorescence, after passing through appropriate filters to remove spurious light, is dispersed by a 0.6 meter monochromator and its output imaged onto the slit of a Hamamatsu synchroscan streak camera. The azulene sample was used without purification as obtained from Aldrich Chem. Co.

The emission spectrum of azulene in an argon matrix at 4K obtained a by picosecond excitation to the vibrationless level of S_2 is shown in figure 2. The dynamics of vibrational energy relaxation were studied by exciting specific upper vibrational levels of S_2 and monitoring directly with the streak camera the emission lifetime of the individual vibronic level. However, it was found that the spectrum shown in figure 2 is not effected by the vibrational level excitited which indicates that the decay lifetime of the upper vibronic levels is very fast compared to the fluorescence lifetime. Our data therefore suggests that the vast majority if not all of the vibrational population has decayed to $S_2(v = 0)$. The vibrational decay rate can in fact be measured, however by the emission risetime of the $S_2(v = 0)$ state subsequent to excitation at higher vibrational levels. We find that, the decay rate can be

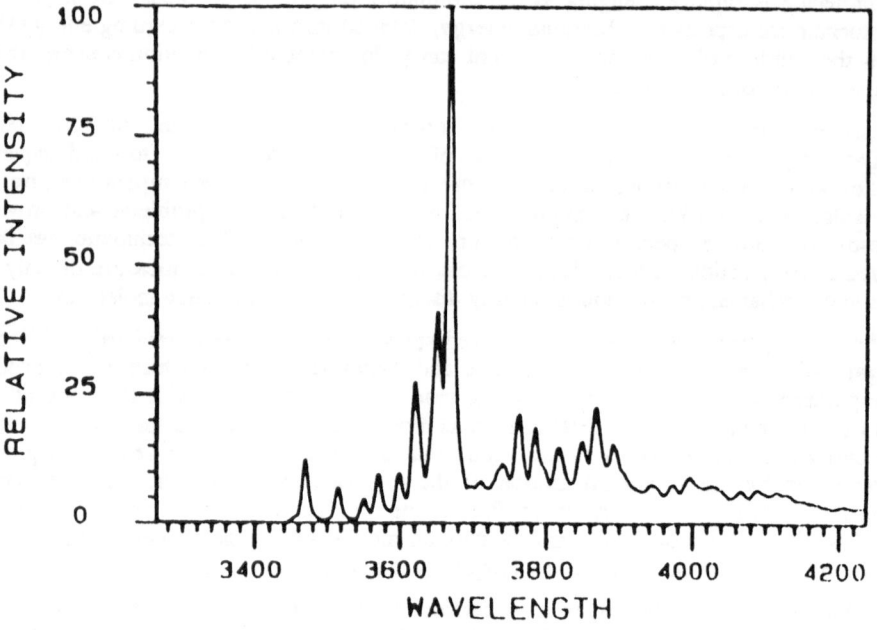

Figure 2

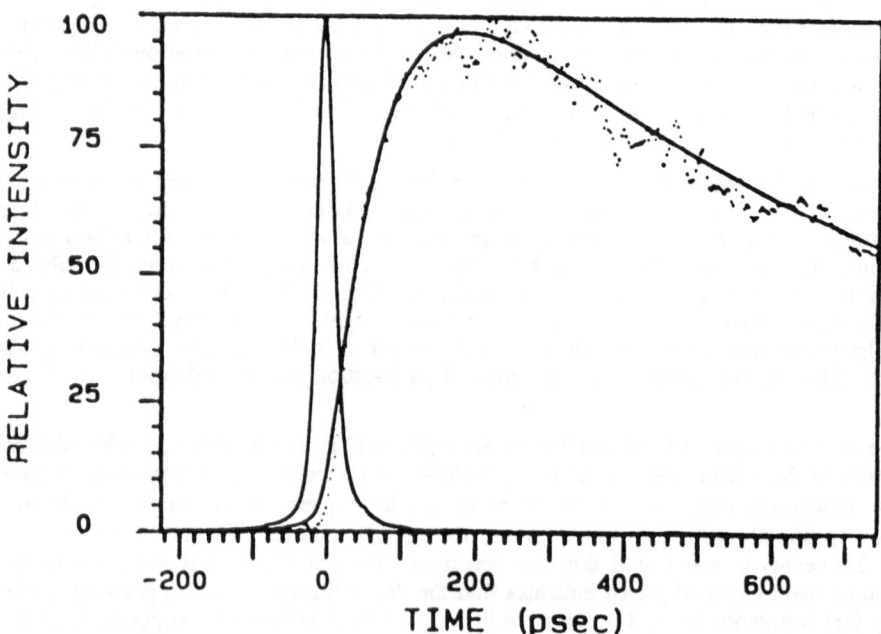

Figure 3

fitted to a single process which represents the summation of all decay pathways which populate the $S_2(v = 0)$ state from the nth excited vibronic level. In principle if the decay process has several channels with varying rates the sum of all these processes will be a complicated multiexponential. It is found experimentally however, as shown in figure 3, that in this case the risetime of the relaxed fluorescence can fit a rather smooth single exponential.

This experimental method and treatment of results allowed for the vibrational decay rates to be determined as a function of the initial vibrational excitation. The data are listed in Table 1. The most likely model based on the invariance of the measured decay rates for vibrational energy relaxation is a very fast, 2-5 ps, decay to low lying bottleneck states followed by a slower decay ~ 14 ps to the S_2 vibrationless level. It is interesting to note that the bottleneck states as shown in table 1 are not the lowest lying vibrational levels of the S_2 manifold. This indicates that the decay rate of an individual level depends on the symmetry and normal mode character of the vibrational mode in addition to the vibrational state density rather than simply the location of excess energy above $v = 0$. It is observed however that above ~ 1500 cm^{-1} excess energy the lifetime of these upper levels is faster than the low lying ones.

Table 2 shows the rates of the vibrational decay process which were measured at constant values of vibrational excitation in different matrix hosts. It is found experimentally that the decay rates decrease with increasing atomic weight of the matrix. This agrees with the theoretical descriptions[7,8] which considers an isolated intramolecular vibration weakly coupled to a dense manifold of matrix phonon states. This rather simplified temperature independent rate model agrees with the Fermi Golden rule expression.[7] The vibrational and phonon mode dependence to the decay rate is depicted as a vibrational wavefunction overlap which predicts the energy gap to be as observed and also depict that as the atomic weight of the matrix increases there is a decrease in phonon frequency. This effect causes an increase in the energy gap between the phonon states and molecular states and results in turn to a decrease in the coupling and consequently slower decay rates from the bottleneck states.

One of the consequences of such statistical models is that molecular matrixes will have much higher intermolecular coupling terms for exchange of vibrational energy compared to atomic matrices. In the molecular matrix, vibrational modes exist with large vibrational frequencies which greatly reduce the energy gap between the vibrational levels of the guest and the host molecules of the matrix. It is therefore expected that under these conditions intermolecular energy exchange should be exceedingly fast.

Experimentally, this prediction can be investigated by measuring the rates of intermolecular energy exchange between a quest molecule such as Azulene in a N_2 matrix at 4K. In the N_2 matrix the Nitrogen intra-molecular vibrational frequency is ~ 2200 cm^{-1}. The vibrational decay within the S_2 electronic manifold of azulene was measured as a function of excitation energy above the S_2 vibronic level. It was found that the rate of repopulation of $S_2(v = 0)$ was essentially constant over the region of 1500-2600 cm^{-1} of initial excitation in S_2. The implication of this result is the opposite of what was predicted theoretically in that the decay pathway is the same for Azulene in a N_2 matrix as well as atomic matrixes.[5] The energy decays very quickly within the Azulene molecule populating low lying bottleneck states which then decay into $v = 0$ on a much slower time scale. Evidently, the channel for intermolecular energy exchange which is open at the initial level of excitation cannot compete with the rate of intramolecular energy relaxation within the azulene molecule.

In figure 3 the experimental results are depicted schematically. Large molecules such as

TABLE 1

DEPENDENCE OF VIBRATIONAL RELAXATION RATES ON EXCESS ENERGY FOR S_2 AZULENE IN NEON AT 4°K.

Vibrational Energy Content (cm⁻¹)	Decay Lifetime (ps)
3980	20
2642	23
2407	14
1519	13
1108	15
800	14
300	≤3

TABLE 2

VIBRATIONAL ENERGY RELAXATION RATES FOR MATRIX ISOLATED AZULENE AT CONSTANT INITIAL LEVEL OF 2400 cm⁻¹ OF INITIAL VIBRATIONAL EXCITATION.

Matrix	Temperature (K)	Decay Lifetime (ps)
Neon	4	14
Argon	4	18
Xenon	4	50
Xenon	25	49
Xenon	50	47

Azulene posesses very high density of vibrational states in the vacinity of 2200 cm^{-1} (the N_2 molecular frequency). The decay pathway which involves the azulene vibrational manifold and the phonons of the matrix, shown schematically in figure 4, dominates the overall decay process due to the small amount of energy which must be exchanged with the matrix from each step down the strongly coupled vibrational ladder in the S_2 electronic manifold. For smaller molecules especially in diatomics where the vibrational level density is low the decay characteristics may be altered and the overall decay process may be dominated by an intermolecular energy exchange to the N_2 stretching vibration.

In the case of fluid media at room temperature the procedure described above would be obviously incapable of resulting in useful information because the vibrational structure will be broaden due to thermal effects and collisional inhomogeneous broadening. To circumvent the inherent weakness of the emission and absorption techniques we have need for a method which exhibit resolved vibrational structure even in the room temperature solution. Techniques such as Raman, and CARS, are known to meet these criteria. To a large extent however, they have not been exploited on the picosecond time scale largely because of experimental difficulties in detecting the weak signals generated by Raman scattering and infrared absorption processes. Picosecond Transient Raman spectroscopy has been recently demonstrated.[9,10] In the next section we describe a high repetition rate experimental apparatus which was recently designed by us to provide transient Raman spectra with picosecond time resolution.

2. PICOSECOND RAMAN SPECTROSCOPY

The photochemistry of diphenylethylene in liquid and gas phase has been extensively studied and recently reviewed[11,12] in the current literature. Upon photoexcitation to the S_1 electronic state the photochemistry is dominated by a cis-trans isomerization which occurs with a barrier of about 1200 cm^{-1} from the zero point in the trans excited state potential well. The geometry change associated with this isomerization is a large scale molecular motion which can be used as a classic case to test the theoretical foundation which currently describes photochemical reactions, nonradiative transition theory, and activated barrier crossing processes in solution.[13-15]

Detailed information regarding the role of vibrational motions in the isomerization process is difficult to obtain due to the lack of vibrational structure frequently observed in the conventional picosecond spectroscopic techniques such as absorption and fluorescence spectroscopy. To obtain vibrational information of solvated molecules we have utilized transient Raman spectroscopy which is well documented in its capability to provide resolved vibrational structure even in condensed phases. Raman data from short lived transients are extremely difficult to detect due to the inherently low cross sections of Raman scattering. Recent technological advances however have made it possible to generate observable picosecond transient Raman spectra for some species.[10,17] In these, picosecond Raman experiments a high repetition rate (~800 KHz) amplified sync pumped dye laser was used.

In our case we utilized a recently developed high sensitivity picosecond Raman apparatus[(18)] which consists of a dye laser synchronously pumped by a cw Nd/YAG mode locked laser and amplified by a high repetition rate (5 KHz) high energy laser source. The 5 KHz picosecond Raman system described here is ideal because it provides a rather high energy per pulse which is essential for the low cross section Raman process. The average energy is two orders of magnitude higher than that used in the previous picosecond Raman experiments and therefore results in the generation of large concentrations of the transient intermediates which consequently allow for their detection. It is important also to note that

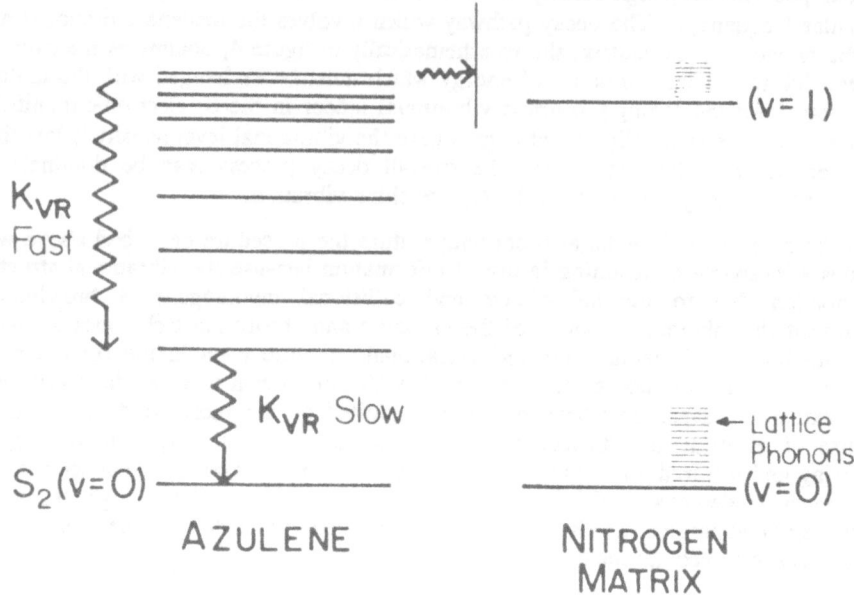

Figure 4

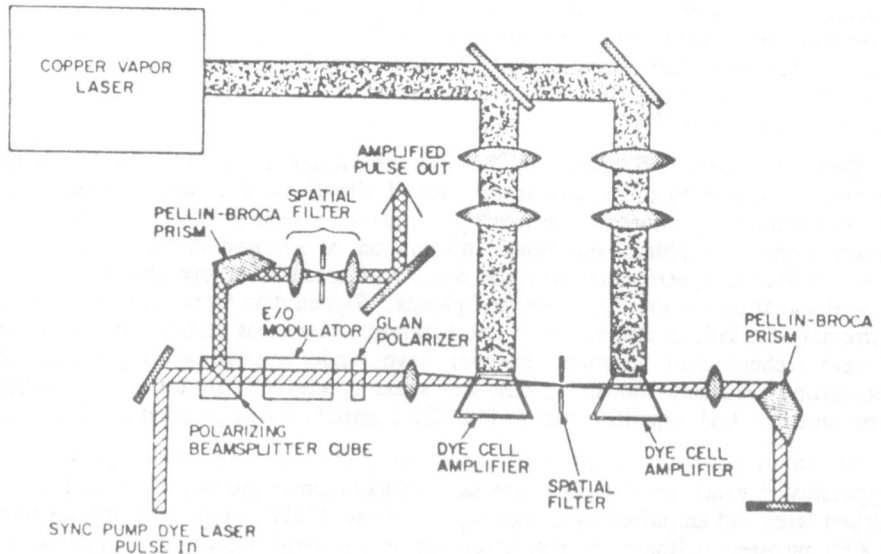

Figure 5

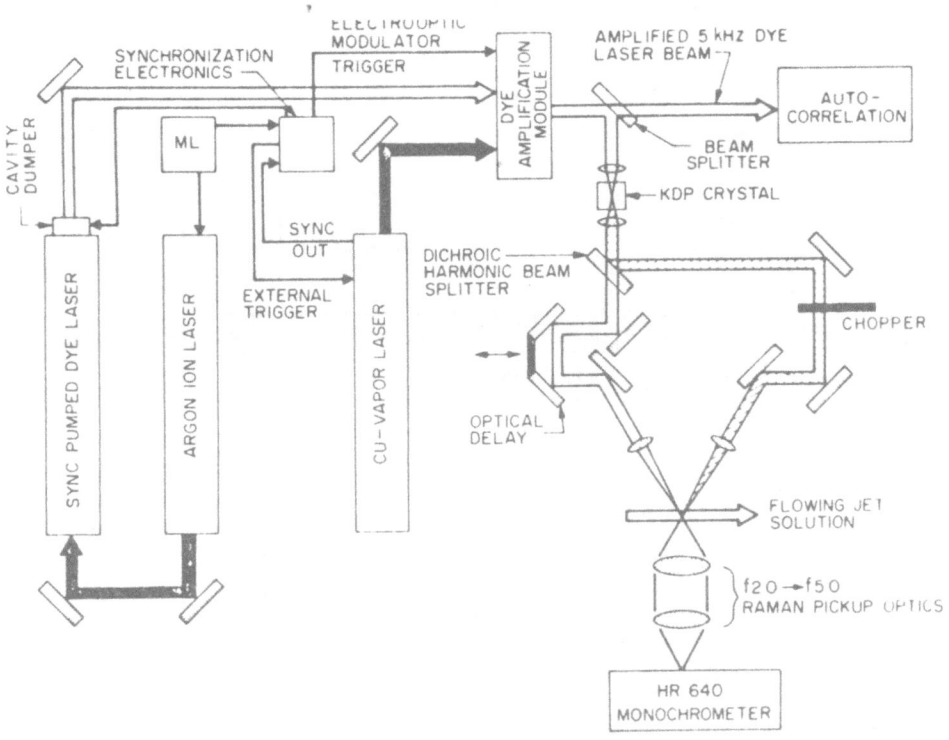

Figure 6

the peak powers are not sufficiently high to induce multiphoton absorption and other nonlinear processes such as stimulated Raman scattering. The application of this new experimental technique is discussed here with respect to the study of excited state processes in substituted diphenylethylenes.

The 5 KHz amplification system which was used to generate the tunable picosecond pulses is shown in Fig. 5 and is described in detail in Reference 16. The picosecond pulses are formed in a cavity dumped syncrouously pumped dye laser pumped by a mode locked CW-Nd:YAG laser. The dye laser output is a train of ~5 ps, 30 nJ pulses at an 800 KHz repetition rate. An electro-optic modulator is used to select pulses and inject them into a dye amplifier cavity pumped by a copper vapor laser (CVL) which operates at 5 KHz. A multipass optical arrangement is necessary to achieve high gain amplification because the CVL pulse is long (~45 ns) with respect to the dye lifetime (~4 ns). Due to the long pulse width of the CVL laser, the energy in the pulse cannot be efficiently stored in the dye gain media during a single pass extraction. The apparatus configuration shown in Fig. 6 uses two passes of the picosecond dye pulse through the gain media. Although additional passes could be introduced with the insertion of more optical elements, we find the current arrangement to be sufficiently effective to achieve the desired amplification. Upon exiting the electro-optic modulator with zero retardation voltage on the crystal, the outgoing pulse has a polarization which is rotated 90° with respect to the input laser beam. A polarization

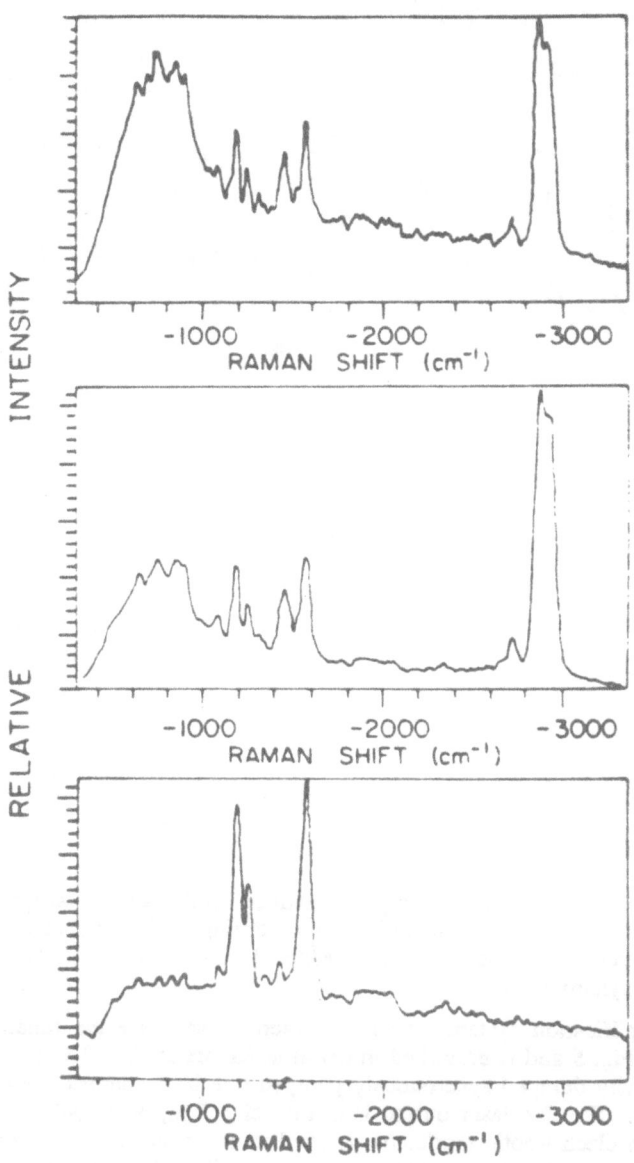

Figure 7

beam splitting cube separates the two beams of opposite polarization, and amplified spontaneous emission is removed from the beam with spatial filtering. The resulting pulses have an integrated pulse energy of 20-50 μJ and ~5 ps pulse width.

For Raman experiments the visible dye laser beam is frequency doubled in KDP and separated into two beams using a dichroic harmonic separator. After an appropriate optical delay, the beams are imaged onto the same focal waist of 0.6 mm on the sample, as shown in Fig. 6. In the current experiment the second harmonic, at 300 nm, is used to excite the first excited state singlet state while the visible, fundamental, used to induce the transient resonance Raman spectra.

In experiments, such as this, where weak signal levels are expected, great care must be taken to eliminate background components from the total observed signal. It is typically found that large fluorescence and Raman signals originate from materials commonly used to construct sample cells. In order to reduce this background, a free flowing solution was used in place of a traditional sample cell. The geometry of the jet solution provided an optical path length of ~ 300 μm, and its flow rate was such that the 200 μm illuminated sample area was replaced with fresh material between every laser pulse.

Raman scattering from the laser irradiated sample is collected by at f2.0 optics and expanded into f5.0 to match the input optics of a Jobin Yvon HR640 single monochromator. A Schott RG2 filter is used to attenuate the raleigh scattered light and eliminate the fluorescence.

The two color pump and probe Raman experiment described above will in general have a large background signal due to fluorescence from the model compound. In addition, there are strong Raman bands from the solvent which are present in the two color spectrum due to the normal Raman scattering from the visible laser beam beam. In order to obtain the transient Raman spectrum with the highest signal to noise ratio possible the pump and probe laser beams are mechanically chopped at different frequencies of ~500 Hz and 700 Hz respectively. Photon counting detection electronics are employed for extremely high sensitivity detection. After discrimination, the electronic signal representing a single photon event is routed to three separate photon counters. (Aston model 721). These counters are gated with the reference output of the choppers such that: Counter A is active only when the probe laser alone is imaged onto the sample, Counter B is active only when the pump laser alone is imaged on the sample and counter C when both laser beams, excitation and probe, impinge on the sample. After a suitable integration period, a very accurate measurement of the signal and background spectral components is achieved. The transient Raman spectrum is then simply the result of subtracting the two single laser only spectra from the two color spectrum. This fast chopping photon counting method is also found to be quite effective in reducing noise signals generated by fluctuations in the laser intensity and optical changes in the flowing sample solution. In addition because of the noise reduction method employed in separating the two laser signals from the background components, the transient spectra can be easily isolated even when the transient and precursor species have similar vibrational frequencies.

The picosecond transient Raman data obtained after S_1 state excitation of trans Stilbene at $t = 0$, i.e. both beams impinge upon the sample simultaneously, is shown in figure 7. The two color spectrum is shown in the top trace. Subtraction of the spectrum taken simultaneously with the ultraviolet laser only is given in the middle trace. The lower spectrum, Fig. 7, is the resulting transient spectrum obtained by further subtraction of the spectrum generated by the visible laser alone. It is readily seen that this data recovery method is quite powerful in isolating the transient spectrum from the background components and the Raman bands generated by the ground state of the procursor and solvent species. Detailed studies of the ground and excited states of isotopically substituted trans stilbenes have been published previously using Raman spectroscopy. The 1100 cm^{-1} to 1600 cm^{-1} region of the spectrum is dominated by three bands which are believed to be the symmetric C-phenyl stretch (1148 cm^{-1}), the phenyl-CH in plane bend and the 1566 cm^{-1} band due to either a C-C stretch or phenyl ring stretch v8a. These assignments based on extrapolation from similar frequencies observed in the ground state where the assignments are more accurately known.

The true normal mode character of the 1566 cm^{-1} band has been investigated in the experiment described here in order to develop an in depth knowledge of the S_1 electronic potential surface which is pivotal to the isomerization process. Substitution of heavy groups in one of the rings of t-Stilbene should provide information which is crucial for the determination for the assignment frequency of the 1566 cm^{-1} vibrational mode. If this vibration is a C-C stretch, the 1566 cm^{-1} vibration would be expected to shift to lower frequencies with substitution. However, the spectrum of the phenyl ring stretch v8a would be expected to split into two vibrations for the substituted and unsubstituted ring. In addition the splitting should be large due to the fact that the substitution occurs at a ring carbon which involves a large amplitude of the v8a vibration.

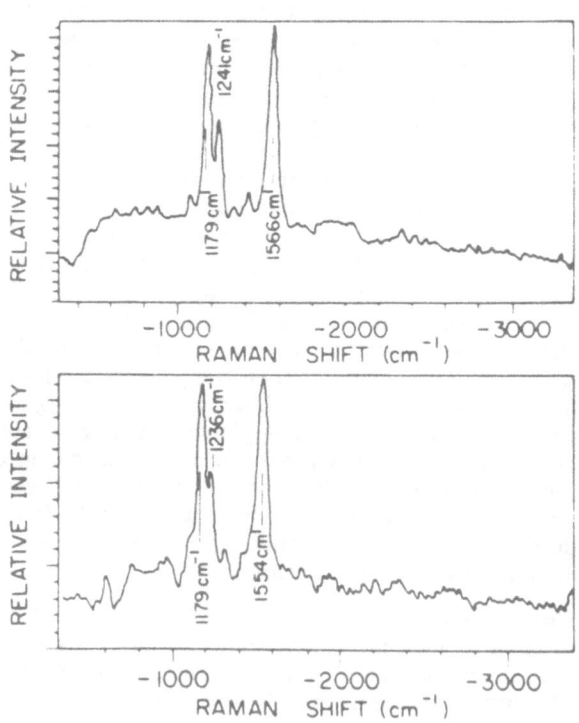

Figure 8

The transient Raman data for two compounds t-Stilbene and 4-methoxy stilbene is shown in figure 8. Consider the 1566 cm^{-1} vibration of t-Stilbene; The substituted compound spectrum shown in figure 8a depicts only a single vibration which is slightly shifted to lower frequencies by ~ 25 cm^{-1} with no apparent broadening. This data strongly suggest that this vibration belong to the C-C stretch, and agrees with the previous tentative assignment of this band.[11]

The dynamics of the S_1 excited state of t-Stilbene were followed by changing the optical delay between the pump and probe laser. No dynamic evolution of the vibrational spectrum was observed other than changes in the overall intensity which decayed with a lifetime of ~ 300 ps. Unfortunately, as a result of the extremely selective nature of the resonance Raman enhancement, to only a single frequency, we cannot observe transients which undergo isomerization leading away from the S_1 potential surface. At least in principle, however, the method as described here may be a powerful tool for detecting the transition state and final product species. At the present time we are constructing an experimental picosecond Raman system capable of tuning the probe laser frequency into resonance with the changing absorption of the transient molecular species. We expect that this system would make it possible to detect the changing spectrum and time-resolve the transient Raman spectra. This method will provide knowledge of transient events and species with high accuracy and establish a data base for excited state Raman spectroscopy which to a very large extent is unknown today.

REFERENCES

[1] P. M. Rentzepis, Chem. Phys. Letts. 2, 117 (1968).
[2] E. Whittle, D. A. Dows and G. C. Pimentel, J. Chem. Phys. 22 1943 (1954).
[3] V. E. Bondybey and L. E. Brus, Advan. Chem. Phys. 41, 269 (1980). L. Young and C. B. Moore, J. Chem. Phys. 81, 3137 (1984); V. E. Bondybey, T. A. Miller, J. H. English, Phys. Rev. Lett., 44, 1349 (1980); P. M. Rentzepis and B. E. Bondybey, J. Chem. Phys. 80, 4727 (1984); V. E. Bondybey, S. v. Milton, J. H. English, and P. M. Rentzepis, Chem. Phys. Lett., 97, 130 (1983).
[4] L. Abouf-Marquin, H. Dubost, and F. Legay, Chem. Phys. Lett. 22, 603 (1973).
[5] J. B. Hopkins and P. M. Rentzepis, Chem. Phys. Lett., 117, 414 (1985).
[6] D. Huppert and P. M. Rentzepis, J. Phys. Chem. in press.
[7] D. J. Diestler, J. Chem. Phys. 60, 2692 (1974).
[8] A. Nitzan, S. Mukamel, and J. Jortner, J. Chem. Phys. 60, 3929 (1974).
[9] T. L. Gustafson, D. M. Roberts, and D. A. Chernoff, J. Chem. Phys. 79 1559 (1983).
[10] T. L. Gustafson, D. M. Roberts, and D. A. Chernoff, J. Chem. Phys. 81, 3438 (1984).
[11] H. Hamaguchi, C. Kato and M. Tasumi, Chem. Phys. Lett., 100, 3 (1983).
[12] H. Humaguchi, C. Kato and M. Tasumi, Chem. Phys. Lett. 106, 153, 1984.
[13] J. A. Syage, W. R. lambert, P. M. Lambert, P. M. Felker, A. H. Zewail, and R. M. Hochstrasser, Chem. Phys. Lett., 88, 266 (1982).
[14] Photoselective Chemistry, J. Jortner, R. D. Levine, S. A. Rice ed. Interscience, 1981.
[15] W. M. Gelbart, K. F. Freed, and S. A. Rice, J. Chem. Phys. 52, 2460 (1970).
[16] J. B. Hopkins and P. M. Rentzepis, App. Phys. Lett., 47, 776 (1985).
[17] J. B. Hopkins and P. M. Rentzepis, Chem. Phys. Lett. in press.

PICOSECOND SPECTROSCOPY OF IRON PORPHYRINS AND HEMOPROTEINS

K.D. Straub
John L. McClellan Memorial Veterans Hospital and
University of Arkansas for Medical Sciences
Little Rock, Arkansas 72205

P.M. Rentzepis
AT&T Bell Laboratories, Murray Hill, New Jersey 07974

ABSTRACT. The mechanisms of energy dissipation in Fe(II) and Fe(III) protoporphyrin IX complexes have been studied by picosecond spectroscopy. The existence of different oxidation states, different electronic spin states, and different numbers of axial ligands in these compounds affords a way to study effects of electronic configuration on the decay of the excited state. Thus Fe(II) Protoporphyrin IX dimethyl ester in benzene with no axial ligands and spin state S=2 has an absorption of the excited state which disappears in 20 psec while the ground state reappears in 45 psec. Similar kinetics occur with reduced cytochrome C and deoxymyoglobin. This kinetic behavior indicates the existence of an intervening d orbital from the iron which serves as an intermediate energy level in the decay process. The pyridine complex of Fe(II) Protoporphyrin IX dimethyl ester has a biphasic relaxation with a long-lived (> 250 psec) state which is due to photoejection of the pyridine ligand from a d state. All Fe(III) porphyrin complexes studied have extremely fast relaxation (<< 25 psec) to the ground state, presumably due to multiple low-lying charge transfer states.

Of central importance in the study of many fast reactions is the understanding of mechanisms of energy dissipation in the excited electronic state. While the majority of excited states of interest are generated during the course of chemical reactions, much can be learned from the easier-to-study optically generated electronic excited states. Because of the large number of different pathways of energy dissipation of the electronic excited states, metallo-porphyrins have provided a large amount of information about the mechanisms of non-radiative decay from the optically excited state (1-4). Thus, while the radiative lifetime of free-base porphyrins is approximately 20 nsec allowing them to be strongly luminescent, metalloporphyrins have lifetimes which range from the sub-picosecond range up to 10 nsec (5,6). The electronic interaction between the d

P. M. Rentzepis and C. Capellos (eds.), Advances in Chemical Reaction Dynamics, 165–170.
© *1986 by D. Reidel Publishing Company.*

electrons of transition metals and the π electrons of the porphyrin give rise to several different effects which cause shortening of the natural lifetime of the parent porphyrin. Because iron porphyrins have several stable electronic configurations in the ground state, there are a number of possible pathways for non-radiative decay of the excited state. The present experiments are designed to study the effect of oxidation state, spin state, and axial ligands on the picosecond relaxation of the excited state of iron porphyrins and hemoproteins.

MATERIALS AND METHODS

The picosecond spectra of porphyrins and hemoproteins were taken with an experimental set-up as described before (7,8). All spectra were obtained using a 2 mm quartz cell adjusted to an optical density of approximately 0.6 at 530 nm. Pumping was done at 530 nm with 25 psec excitation pulse. Relaxation kinetics were constant over the power range of 0.2 to 2.0 m joules/pulse. All solutions were sealed under N_2. Preparation of the Iron(II) Protoporphyrin IX dimethyl ester was performed in a 95 percent N_2 - 5 percent H_2 atmosphere over a Pd catalyst to remove traces of oxygen (9). Decay times were calculated from the slope of the first order kinetic curves at the given wavelengths. All spectra were performed at room temperature.

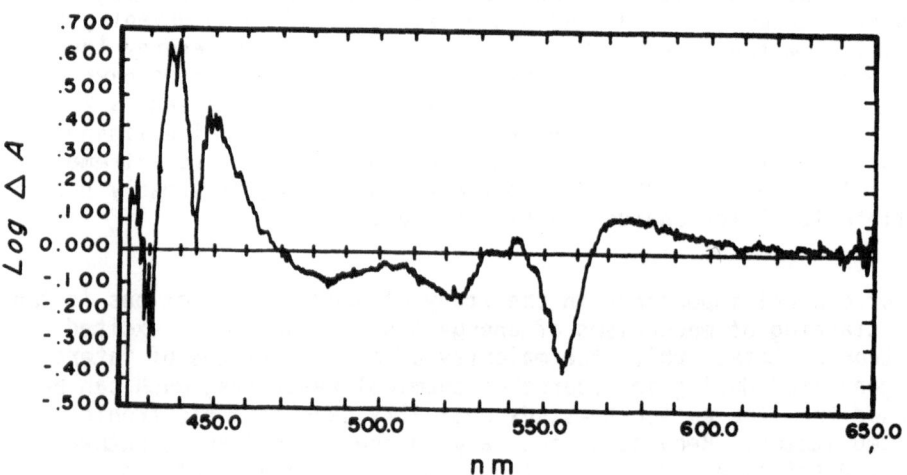

Figure 1. Difference spectrum of Fe(II) Protoporphyrin IX dimethyl ester in pyridine. The spectrum is the difference between the ground state and the excited state after excitation at 530 nm with a 25 psec pulse. This spectrum was taken 8 psec after arrival of the excitation pulse peak.

RESULTS

Iron (II) Protoporphyrin IX Dimethyl Ester in Pyridine:

The excitation difference spectrum of Fe(II) Protoporphyrin IX dimethyl ester in pyridine 8 psec after the arrival of the excitation pulse is shown in Figure 1.

Excitation of this compound in pyridine (2 axial ligands; spin state, S=0) at 530 nm gave a biphasic decay curve. With the probe. wavelength at 570 nm where a new absorbtion is seen, there is a 90 psec relaxation with a longer recovery of approximately 260 psec. However, the absorbance of the excited state at 450 nm decays with a 41 psec time constant followed by the longer time constant between 230 and 300 psec. Bleaching at 555 nm follows the same time course as absorbtion at 450 nm.

Iron(II) Protoporphyrin IX Dimethyl Ester in Benzene:

The excited state of iron(II) Protoporphyrin IX dimethyl ester (no axial ligands; spin state, S=2) in benzene relaxes with a rapid decay of the excited state absorption at 460 nm with a time constant of about 20 psec. At the bleaching wavelength of 570 nm and in the absorbance band at 600 relaxation takes about 45 psec. No long-lived decay similar to pyridine is seen.

Iron (II) Cytochrome C:

Iron Protoporphyrin IX is covalently bound into this small protein (MW - 10,000) and is known to have two axial ligands in the Fe(II) state with low spin (S=0). However, its excited state relaxation kinetics are more similar to iron (II) Protoporphyrin IX dimethyl ester in benzene (no axial ligands) than to pyridine with two axial ligands. The excited state absorption at 450 nm decays with a time constant of 20 psec while the absorbtion band at 570 nm decays with a time constant of 33 psec.

Iron (II) Deoxymyoglobin FeII:

Deoxymyoglobin has a non-covalently bound Fe(II) Protoporphyrin IX which is in the high spin state (S=2) with only one axial ligand. In this case the excited state absorption relaxes with a time constant of 34 psec while a broad excited state absorption centered at 600 nm relaxes with a time constant of approximately 60 psec. No ground state bleaching can be observed. No long-lived state is evident.

Fe(III) Myoblogin:

With the iron in oxidation state +3, a second axial ligand from an anion is present. When CN- is the axial ligand a low spin S=1/2 electronic state is produced while F- allows the high spin electronic state (S=5/2) to form. In neither case could bleaching of ground state or absorption of excited state be observed outside the 25 psec exciting pulse. From our earlier work, we believe these decay times to be shorter than 0.5 psec (10).

DISCUSSION

It is evident that for all the Fe(II) Protoporphyrin IX complexes studied an excited state absorption appears with absorption between 400 - 500 nm. This state lasts between 20 and 40 psec. A second absorbtion increase at 570-600 nm in the α band recovers with a time constant between 33 and 91 psec for these same complexes. These observed decay processes may be explained by at least two different possibilities. A likely explanation would be that the initially excited multiplet state undergoes rapid intersystem crossing (too fast to observe), and the resulting state with absorbtion at 450 nm decays one of the d levels of the iron (20-40 psec) with absorbtion at 570 - 600 nm which in turn decays to ground (33-91 psec). An alternate explanation would have the intersystem crossing occur with a time constant of 20-40 psec and a subsequent decay to ground through intermediate d states with an overall time constant of 33-91 psec. We view the second pathway as less likely because the time constants are independent of spin-state which is not expected if intersystem crossing is one of the observed steps. Low lying (~ 1 eV) intervening d levels in the decay process have been postulated for Ni(II) Protoporphyrin dimethyl ester in benzene in which the return to ground state takes 260 psec, while decay of the first observed excited state occurs in 10 psec (11).

   The biphasic decay process seen in pyridine solvent is a special case. The initial evolution of the excited state is followed by a long-lived species with residual absorption at 450 nm and at 570 nm. Dixon et al (12) have observed similar kinetics for iron porphyrins with other basic ligands and have found that the long-lived decay time is a function of the concentration of axial ligand in the solution. They conclude that this long-lived decay is really a ground-state 5 ligand species which is generated by ejection of the 6th ligand from the excited state. These authors think that the most likely photodissociative state is a low lying (d_π, d_{z^2}) state. This is compatable with our proposed decay sequence with the dissociation coming from the second observed state (d level).

   Comparison with the Cu and Ag complexes of protoporphyrin can give some indication of why the intersystem crossing might be so rapid (Table 1). The difference in lifetime between these two metalloporphyrins in similar non-ligating solvents is likely due to the large increases in spin-orbit coupling when going from Cu(II) to Ag(II). Fe(II) Protoporphyrin IX will also have large amounts of spin-orbit coupling no matter what its electronic spin state happens to be and this magnetic perturbation allows much more rapid intersystem crossing.

   Finally, the series of d orbitals belonging to iron will have their energy levels shifted according to the oxidation state of the metal and the strength of the axial ligands (13). In the Fe(II) compounds of the present study, the decay of the d state appears to be only moderately affected by the axial ligands present (33-91 psec), therefore we postulate that the intermediate has mostly $d_{(x^2-y^2)}$ character.

TABLE I

Proto-porphyrin Complex	Ref.	Step 1	Decay Pathway Step 2	Step 3
Cu(II)	15	$^2S_1 \rightarrow {}^2T_1$ (<<8)	$^2T_1 \rightarrow {}^4T_1$ (480)	
Ag(II)	15	$^2S_1 \rightarrow {}^2T_1$ (<<8)	$^2T_1 \rightarrow {}^4T_1$ (11)	
Ni(II)	11	$S_1 \rightarrow S_d$ (10)	$S_d \rightarrow S_0$ (260)	
Fe(II)	This study	$S_1 \rightarrow T_1$ (<<25)	$T_1 \rightarrow T_d$ (21)	$T_d \rightarrow T_0$ (46)

Proposed decay pathways for some metalloporphyrins. Relaxation times in picoseconds are shown in ( ). Benzene is solvent in all cases.

The Fe(III) porphyrin complexes have very short lifetime regardless of spin state or axial ligands. The rapid decay may be due to the low-lying d-π charge transfer states although other mechanisms cannot be ruled out (14).

REFERENCES

1. Becker RS, Kasha M: J Am Chem Soc 77: 3669, 1955.
2. Allison JB, Becker RS: J Chem Phys 32: 1410, 1960.
3. Eastwood D, Gouterman M: J Mol Spectry 30: 437, 1969.
4. Gouterman M: In: The Porphyrins, Vol. 3 (Dolphin D, ed), Academic Press, NY, 1978, ch. 1.
5. Magde D, Windsor MW, Holten D, Gouterman M: Chem Phys Letters 29: 183, 1974.
6. Straub KD, Rentzepis PM, Huppert D: J Photochem 17: 419, 1981.
7. Reynolds AH, Rentzepis PM: Biophys J 38: 15, 1982.
8. Hilinski EF, Rentzepis PM: Anal Chem 55A: 1121, 1983.
9. Brault D, Rougee M: Nature New Biol 241: 19, 1973.
10. Huppert D, Straub KD, Rentzepis PM: Proc Natl Acad Sci USA 74: 4139, 1977.
11. Kobayashi T, Straub KD, Rentzepis PM: Photochem Photobiol 29: 925, 1979.
12. Dixon DW, Kirmaier C, Holten D: J Am Chem Soc 107: 808, 1985.
13. Adar F, Gouterman M, Aranowitz S: J Phys Chem 80: 2184, 1976.

14. Adar F: In: The Porphyrins, Vol. 3 (Dolphin D, ed), Academic
    Press, New York, 1978, p 198.
15. Kobayashi T, Huppert D, Straub KD, Rentzepis PM: J Chem Phys
    70: 1720, 1979.

INTERMOLECULAR PROTON TRANSFER

D. Huppert and E. Pines
Raymond and Beverly Sackler Faculty of Exact Sciences
School of Chemistry, Tel Aviv University, Ramat Aviv
69978 Israel

ABSTRACT.   In this review we describe some recent experimental development in the field of intermolecular proton transfer. It was observed,
several decades ago, that upon excitation, aromatic amines and alcohols
become more acidic in aqueous solutions. Excitation of aqueous solution of these compounds by picosecond laser pulses results in proton
transfer from the excited compound to the solvent. The proton transfer
rates of some of these compounds was determined by time resolved
fluorescence techniques. The measured time constants vary from a few
tenths of a picosecond to several tenths of a nanosecond. The proton
transfer process creates an ion pair which reversibly recombines on a
very short time scale. Picosecond time resolved fluorescence measurements were able to reveal this phenomenon. Detailed kinetic analysis
showed that geminate recombination is very efficient and most of the ion
pairs recombine within a few hundred picosecond. The geminate recombination process is time dependent and is governed by diffusion and coulombic
attraction. Finally we describe a method for light induced proton
concentration jump. Intense short laser pulses can generate enough
protons as to cause an observable reduction in the bulk pH. A relaxation
kinetics analysis is carried out in order to follow the time evolution
of the proton concentration.

INTRODUCTION

The rates of proton transfer reactions have been shown to span at least
15 orders of magnitude.(1-5). Since the stopped flow method is limited
to relatively slow reactions, measurements of rapid reactions has been
obtained by NMR or perturbation techniques.(6-7). In addition to
temperature ("T-jump") and pressure ("P-jump"), light may also be used
since the acidities and basicities of excited molecules are different
from those in the ground state .(8). Given that the initial excited
state is formed in ca 1 fsec, the equilibrium constant for proton
dissociation or association can be changed on a very short timescale.
Diffusion is not now the limit for rate of proton transfer reactions,

P. M. Rentzepis and C. Capellos (eds.), Advances in Chemical Reaction Dynamics, 171–178.
© *1986 by D. Reidel Publishing Company.*

and the motions of the proton within or near the hydration sphere of the
donor/acceptor may be examined.

EXCITED STATE PROTON TRANSFER

Förster interpreted the pH-dependence of the fluorscence maximum of 1-
aminonaphthalene-4-sulfonate as a change in the excited state pK_a. (9).
Aromatic amines and phenols are more acidic in the S_1 state, while
aromatic acids, ketones and certain nitrogen heterocyclics are more
basic. The most straightforward method for determination of the excited
state pK_a is the Förster cycle.(10). On the assumption that ΔS is the
same for excited and ground state dissociations, one may estimate the
pK_a difference from Eq.1. The average of absorption and emission is
taken to account for Franck-Condon effects.

$$pK_a^* - pK_a = (h\nu_A - h\nu_{AH})/2.3 \ RT \tag{1}$$

$h\nu_A$, $h\nu_{AH}$ = frequencies of electronic transition between the ground and
excited states for A and AH, respectively.
　　　Alternatively, direct measurements on the picosecond time scale of
the forward and reverse reactions can now be made.(11-12). The parameters
are shown in Eq.2, in which $\overrightarrow{k}$ is the first order rate constant for proton
transfer to the solvent, $\overleftarrow{k}$ is the bimolecular rate constant for the
reaction between the anion and the protonated solvent, k_r and k'_r are
the radiative rate constants for $AH(S_1)$ and $A(S_1)$. k_{nr} and $k_{nr'}$ are the
nonradiative rate constants for $AH(S_1)$ and $A^-(S_1)$.

$$AH^*(S_1) + H_2O \underset{\overleftarrow{k}}{\overset{\overrightarrow{k}}{\rightleftharpoons}} A^{-*}(S_1) + H_3O^+ \tag{2}$$

$$AH(S_0) + h\nu_{AH} \quad \overset{k_r \quad k_{nr}}{\swarrow \quad \searrow} \quad AH(S_0) \qquad A^-(S_0) + h\nu_A \quad \overset{k'_r \quad k'_{nr}}{\swarrow \quad \searrow} \quad A(S_0)$$

The rate equations for the reaction scheme in Eq.2 are shown in Eqs.3
and 4.

$$-d[AH \ (S_1)]/dt = (\overrightarrow{k}+k_r+k_{nr})[AH \ (S_1)]-\overleftarrow{k}[H_3O^+][A^-(S_1)] \tag{3}$$

$$-d[A^-(S_1)]/dt=(k_{r'}+k_{nr'})[A^-(S_1)]+\overleftarrow{k}[H_3O^+][A^-(S_1)]-(\overrightarrow{k}+k_r+k_{nr})[AH(S_1)] \tag{4}$$

The integrated equations show a complex relationship between the
concentrations of the excited state acid and anion on the rate constants.
(4). At high pH and high concentrations of AH(S), the concentrations
are given by Eqs.5 and 6.

$$[AH \ (S_1)] = [AH \ (S_1)]_0 \ \exp\text{-}t/\tau \tag{5}$$

$$[A^-(S_1)] = [A^-(S_1)]_0/(\Sigma k) \; [\exp{-t/\tau'} - \exp{-t/\tau}] \qquad (6)$$

in which $(\Sigma k) = (\overrightarrow{k} - \overleftarrow{k}[H_3O^+] + k_r - k_{r'} + k_{nr} - k_{nr'})$,
τ is the lifetime of AH (S_1) and τ' is the lifetime of $A^-(S_1)$ at
$pK^* < pH < pK$.

$$\tau = 1/(\overrightarrow{k} + k_r + k_{nr}) \; \text{and} \; \tau' = 1/(k_{r'} + k_{nr'})$$

The steady state assumption can be used to derive two equations
based on relative quantum yields. (13,14).

HYDROXYARENES

Derivatives of aromatic hydrocarbons (phenols, amines) are well adapted
to the study of excited state proton transfer reactions. The properties
of the ground state compounds are easily measured, and fluorescence is a
common property of these classes of compounds. The proton transfer
rates of 2-naphthol have been thoroughly studied, using phase fluorimetry
(15), nanosecond (16) and picosecond emission spectroscopies.(11-12).
An electron-withdrawing group, as in 2-naphthol-6-sulfonate, decreases
the pK^*_a by one unit, mostly by a large increase in deprotonation rate.
An additional sulfonate group (2-naphthol-3,6-disulfonate) decreases
the pK^*_a by another unit, with a three-fold increase in deprotonation
rate.   A linear relationship between the log of the deprotonation rate
constant and the excited state pK_a exists over more than 2 orders of
magnitude.
        The deprotonation rate of HPTS in H_2O increases linearly with
pressure (17), rising from $8 \times 10^9 sec^{-1}$ at 294K and 1 bar to $25 \times 10^9 sec^{-1}$
at 294K and 9kbar (the liquid-ice VI transition point). Dedeuteration
rates (294K) in D_2O rise from $2.7 \times 10^9 sec^{-1}$ to $10 \times 10^9 sec^{-1}$ at 8kbar,
indicating that the isotope effect is constant over the pressure range
examined. The activation volume (ΔV^*) derived from the rate constants
is -6 cc/mole, compared to a ΔV_0 for acid base dissociation of -12 ± 3
cc/mole. Since ΔV seems insensitive to the structure of the acid, the
fact that $\Delta V^*/ \Delta V_0 = 0.5$ suggests that the transition state for proton
dissociation is halfway between the acid and the dissociation products.

GEMINATE RECOMBINATION

HPTS is known to have a $pK^*=0.5$ (12), a 7 units change from its ground
state equilibrium constant. Thus, in the excited state the HPTS molecules are strong acids. In aqueous solution the dissocation process
occurs within 100 psec with a very high $(\Omega = 70\%)$ yield of separation,
which was measured long time (20 nsec) after the excitation (18).
Proton concentrations up to $10^{-4}M$ were observed in the so called "pH-
jump" experiments (19) in $10^{-3}M$ aqueous HPTS.
        Steady state fluorescensce measurements in very acidic aqueous
solutions revealed that the reactivity of the excited state remained

very high, and the back homogeneous protonation reaction $RO^{*-} + H^{+}$
-----> ROH^{*} proceeded with a diffusion-controlled rate constant of
5×10^{10} M^{-1} s^{-1}. (20)

The very high yield of separation of the ionic pairs and the very
high diffusion-controlled reactivity of the excited state seem to
contradict each other. The well known escape probability of Onsager
(21) which measures the very long time yield of separation of the ion
pair is given by Eq. (1).

$$\Omega_{(\infty)} = \exp(-R_D/r_0) \tag{7}$$

where R_D is the Debye radius, $R_D = |Z_1 Z_2| e^2/\epsilon k_B T$ and r_0 is the initial
separation between the ionic pair. Here Z_1 and Z_2 are the charge numbers
of the reactants, e the electron charge, ϵ the static dielectric
constant of the medium, k_B the Boltzmann constant and T the absolute
temperature. From Eq.(7), putting $R_D=28\text{Å}$, and $\Omega=70\%$, r_0 is 80 Å. This
distance is far greater than the distance where the proton is expected
to thermalize after dissociation. Haar et al. (18) were able to show
that 70% is the expected yield of separation if only ground state
recombination was important. Thus, in their model the separation yield
of the ion pair is determined by the life time of the excited molecule.
However, their model treats the excited state molecule as a totally
nonreactive species in a marked contrast to the experimental findings.

We propose a very simple model which is based on our recent time
resolved results (Fig.1).

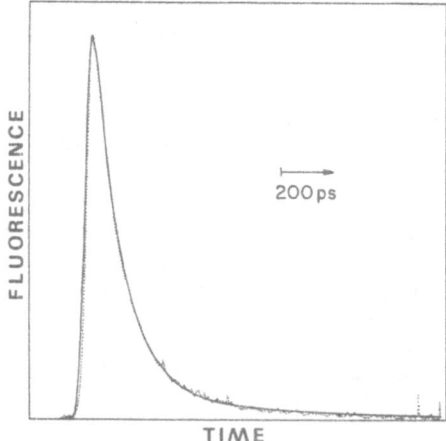

FLUORESCENCE

200 ps

TIME

Fig. 1. Time resolved fluorescence of HPTS in aqueous
solution monitored at 450nm. The dotted line
is the experimental data and the full line is
a computer fit using the expression of Eq. 11.

The model explains all the previous observation in terms of
partial equilibrium which exists between the dissociating parent molecule
and the newly formed, closely packed, ion pair. This quasi equilibrium
state is disturbed by the diffusion which gradually causes the ion pair
to separate. It is in the form of Eq.(8).

$$R*OH \underset{k_2(t)}{\overset{k_1}{\rightleftharpoons}} R*O^- \ldots H^+ \tag{8}$$

where k_1 is the rate constant for generating the ion pair and $k_2(t)$ is a time dependent rate constant for the back geminate recombination process.

In order to describe $k_2(t)$, the rate of the geminate recombination, we used the expression given by Hong and Noolandi (22) for the long time rate of ion pair recombination after it was generated by a δ function distribution at a mutual separation distance of r_0.

For the diffusion-controlled limit and with $R_D \gg a$, the contact radius, their expression takes the form:

$$k_2(t) = k_2 t^{-3/2} \tag{9}$$

k_2 is given by:

$$k_2 = R_D \exp(-R_D/r_0)/2(\pi D_{AB})^{\frac{1}{2}} \tag{10}$$

D_{AB} is the relative diffusion coefficient between the ionic pair.

The solution of the kinetic system shown in Eq.8 is given by

$$\frac{[ROH]_t}{[ROH]_0} = 1 - \frac{k_1 \int_0^t \exp(k_1 t' - 2k_2 t'^{-\frac{1}{2}}) dt'}{\exp(k_1 t - 2k_2 t^{-\frac{1}{2}})} \tag{11}$$

Eq.(11) has the same characteristics as the observed decay, namely a fast and almost exponential decay which is followed by a long and non-exponential "tail". Numerical solution of Eq.(11) is compared with the experimental decay of Fig.1. The best fit was achieved with $k_1 = 1.5 \times 10^{10} s^{-1}$ and $k_2 = 3.5 \times 10^{-6} s^{\frac{1}{2}}$. With $R_D = 28$ Å and $D_{AB} = 10^{-4} cm^2 s^{-1}$, r_0 is calculated according to Eq.(10) to be 35 Å. Although these results are only indicative some general conclusions could be drawn from them.

The large magnitude of r_0 shows that it represents some average distance of a broad distribution function rather than a δ function value. Since in our model dissociation and recombination occur repeatedly, this conclusion seems natural.

Furthermore, according to the Hong and Noolandi's model, the initial distribution function for protons in aqueous solution, characterized by a diffusion coefficient $D \simeq 10^{-4} cm^2 sec^{-1}$, is diffusing within 10 psec to form a distribution function with a half width of roughly $R_D/2$, or 14 Å for the given reaction parameters. This period of time is shorter by a factor of 2 than our experimental time resolution. It follows that the real physical situation as well as our time resolution allow us to detect only a diffused distribution of ion pairs distances rather than a sharp distribution.

PHOTON-PROMOTED pH JUMPS

Based on the rate constants for the intermolecular proton transfer and
pK* already cited, we may infer that compounds with a pK*≃0 will transfer
protons to water in a few picoseconds. Geminate recombination is
completed shortly after proton ejection, and within a nanosecond or two,
a homogeneous transient proton population is obtained.

Several intense picosecond pulsed lasers(mode locked Nd/YAG,ruby)
or nanosecond lasers(Nd/YAG, nitrogen, excimer) can deposit more than
10mj ultraviolet energy ($\sim 10^{16}$ photons/pulse) and produce a photon
"concentration" of greater than 10^{-3} einsteins in a target. If the
absorption coefficient of the ground state ROH is large enough to absorb
10% of the incident radiation, a final pH of 4 may be attained. Since
most of the acids (ROH) used have a pK_a of 7 or more, a pH jump of at
least three pH units can be obtained. A pH jump of this magnitude would
be useful for the study of ATP synthesis and acid-catalyzed tautomeriza-
tion of cis-trans isomerization.

The light induced concentration jumps of H^+ and RO^- are limited in
duration since the excited RO^- relaxes back to the ground state.
(τ for HPTS 6 nsec). The equilibrium constant for dissociation of ROH
is seven orders of magnitude smaller than that for ROH*. The return of
the pre-pulse solution pH occurs within 100 μsec. The ground state
anion can be used as the indicator for the pH jump. In the case of
HPTS, both acid (400nm) and anion (450nm) have large absorption
coefficients (30000 $M^{-1} cm^{-1}$). The maxima differ in position enough
(50nm) for transient absorption techniques to be useful for monitoring
the relaxation of the pH jump. The analysis introduced by Eigen for
small perturbations (7) can be used to extract the pH jump time dependence.

The operation of the pH jump approach was demonstrated with a
proton-emitter (ROH)-indicator(In) system in water. (Eq.21)

$$ ROH \underset{k_1}{\overset{k_2}{\rightleftharpoons}} RO^- + H^+; \quad In^- + H^+ \underset{k_4}{\overset{k_3}{\rightleftharpoons}} InH \qquad (11) $$

The time dependence of the concentration involved is derived from
two coupled differential equations, Eqs. 12 and 13, leading to the
solution shown in Eq. 14.

$$ dx/dt = -\{k_1([RO^-] + [H^+] + k_2)\}x + (k_2[RO^-])y \qquad (12) $$

$$ dy/dt = k_3[In]x - [k_3([In] + [H^+]) + k_4]y \qquad (13) $$

in which x is the induced change in the equilibrium concentration of
the proton emitter, [ROH], and y is the induced change in the indicator
concentration, [In]. (23).

$$ y = [k_3[In]x_o/\gamma_1 - \gamma_2][exp(-\gamma_2)t - exp(-\gamma_1)t] \qquad (14) $$

in which x_0 is the initial $[H^+]$ after the pH jump and γ_1 and γ_2 are constants related in a complex way to the rate constants and concentrations of the species.

The pH jump method has been used to study (a) fast proton transfer reactions of a number of pH indictors in homogeneous aqueous solution (23-25), (b) proton diffusion in ice (26) (c) proton diffusion and proton transfers in micellar dispersions (27-29) and (d) proton diffusion in model biological systems (30).

References

1. Bell, R.P. 1973. "The Proton in Chemistry", 2nd edn., Chapman and Hall, London, 310pp.
2. Caldin, E.F. and Gold, V. 1975. "Proton-transfer Reactions", Chapman and Hall, London, 448pp.
3. Schulman, E.F. 1977. "Fluorescence and Phosphorescence Spectroscopy", Pergamon Press, New York.
4. Ireland, J.F. and Wyatt, P.A.H. 1976. Adv. Phys. Org. Chem. 12:131-221.
5. Gutman, M. 1984. Methods of Biochemical Analysis 30:1-103.
6. Grunwald, E., Ralph, E.K. 1971. Acct. Chem. Res, 4:107.
7. Eigen, M. 1964. Angew. Chem. Int. Ed. Eng. 3:1.
8. Weller, A. 1961. Prog. Reaction Kinetics 1:189.
9. Forster, T. 1949. Naturwissenschaften 36:186.
10. Forster, T. 1950. Z. Elektrochem. 54:42.
11. Campillo, A. J., Clark, J.H., Shapiro, S.L., Winn, K.R. 1978. In Proceedings of the First International Conference on Picosecond Phenomena, eds. Shank, C.V., Ippen, E.P. and Shapiro, S.L. Berlin: Springer-Verlag, Berlin.
12. Smith, K.K., Huppert, D., Gutman, M., Kaufmann, K.J. 1979. Chem. Phys. Lett. 64:522-27.
13. Weller, A. 1958. Z. Physik. Chem. N.F. 15:438.
14. Weller, A. 1952. Z. Elektrochem. 256:662.
15. Demjaschkewitch, A.B., Zaitsev, N.K., Kuzmin, M.G. 1978. Chem. Phys. Lett. 55:80.
16. Gafni, A., Modlin, R.L., Brand, L. 1976. 80:898.
17. Huppert, D., Jayaraman, A., Maines, R.G., Steyert, D.W., Rentzepis, P.M. 1984. J. Chem. Phys. 81:5596-5600.
18. Haar, H.P., Klein, U.K.A. and Hauser, M. 1978. Chem. Phys. Lett. 58:525.
19. Gutman, M. and Huppert, D. 1979. J. Biochem. Biophys. Methods 1:9.
20. Weller, A. 1958. Z. Physics Chem. 17:14.
21. Onsager L. 1934. J. Chem. Phys. 2:599.
22. Hong, K.M. and Noolandi, J. 1978. J. Chem. Phys. 68:5162.
23. Pines, E. and Huppert, D. 1983. J. Phys. Chem. 87:4471-78.
24. Gutman, M., Nachliel, E., Gershon, E., Giniger, R., and Pines, E. 1983. J. Am. Chem. Soc. 105:2210-2216.
25. Gutman, M. and Huppert, D. 1981. J. Am. Chem. Soc. 103:3709-13.

26. Pines, E., Huppert, D. 1985. Chem. Phys. Lett. 116:295.
27. Gutman, M., Huppert, D., Pines, E., Nachliel, E. 1981. Biochimica et Biophysica Acta 642:15-26.
28. Politi, M.J., Fendler, J.H. 1984. J. Am. Chem. Soc. 106:265.
29. Bardez, E., Goguillon, B.T., Keh, E., Valeav, B. 1984. 88:1909.
30. Gutman, M., Nachliel, E., Gershon, E., Giniger, R. 1983. Eur. J. Biochem. 134:63-69.

Molecular Dynamics of Liquid Phase Reactions by Time-Resolved
Resonance Raman Spectroscopy

George H. Atkinson
Department of Chemistry and
Optical Sciences Center
University of Arizona
Tucson, Arizona 85721

The molecular dynamics of liquid phase reactions at room temperature
can be elucidated with respect to structural and conformational changes
by time-resolved resonance Raman (TR^3) spectroscopy. The vibrational
structures and conformations of transient species as well as their
kinetic properties can be determined directly from TR^3 spectra. Time
resolution as short as a few picoseconds has been used to record TR^3
spectra. The information available from these measurements is illustrated by results on the ionic intermediates in the photolysis of
stilbene ions and substituted anthraquinones, excited electronic states
of chrysene and phenanthrene, excited triplet state formation and
reactivity of retinal isomers, and the photochemical reactions of
bacteriorhodopsin.

Introduction

Reactions occurring in the gas phase have been examined extensively
with respect to molecular dynamics largely because of the excellent
experimental opportunities to isolate reactants from collisional perturbations. Even in cases where studies of a completely isolated
species are not feasible, the collisional perturbations can be quantitatively controlled to the extend that "isolated molecule properties"
can be often separated from those caused by collisions. The contributions of collisionally-initiated dynamics are, of course, also of fundamental importance.
     High resolution data in both the frequency and time domain have
provided the information needed to construct detailed views of the
molecular structure and kinetic properties of even large polyatomic
molecules in the gas phase. Mechanisms of vibronically excited-state
photophysics and energy transfer reactions also have been elucidated
for these molecular systems. The dynamics associated with photochemistry (e.g., dissociation) and molecular rearrangements (e.g., isomerization) have been studied more commonly for small molecules although the
extensive studies of the isomerization of stilbene stand as exceptions.
     This same level of detailed dynamical information on changes in

179

P. M. Rentzepis and C. Capellos (eds.), Advances in Chemical Reaction Dynamics, 179–205.
© *1986 by D. Reidel Publishing Company.*

both molecular structure and kinetic properties has not been routinely
available for liquid phase reactions primarily because of the experimental limitations associated with isolating environmental perturbations. The structure-sensitive, time-resolved spectroscopies used to
study gas phase reactions have not been as successfully utilized in the
characterization of liquid phase dynamics due to the presence of highly
perturbative collisions. As a result, the amount and type of dynamical
information available in the two phases is significantly different.
The enormous importance of chemical reactivity in the liquid phase
requires that a more detailed and complete description of the molecular
dynamics become available.

As one considers methods for experimentally examining liquid phase
dynamics, vibrational spectroscopy stands as a clearly attractive op-
tion. The structural sensitivity of infrared spectroscopy has long
been evident while the rapid development of vibrational Raman spectroscopy stimulated by advances in laser technology has earned it an important role in structure and conformation determinations. To construct a
model for dynamics in a liquid phase reaction, however, it is necessary
to obtain time dependent information and usually for transient species
at low, instantaneous concentrations. It is widely recognized that
conventional infrared absorption methods are instrumentally difficult
to adapt to time-resolved measurements and that infrared absorption is
not particularly sensitive. Recent developments in Fourier-transform
infrared (FTIR) and infrared absorption from diode lasers have begun to
alter this situation and attention should be directed to work in this
area [1]. Time-resolved FTIR and diode laser measurements applied to
dynamical studies will not be treated here although these types of data
are often complementary to the time-resolved Raman experiments which
will be discussed in detail.

The material presented in this paper is largely taken from recently published work or from results currently submitted for publication
elsewhere and is intended to provide a tutorial review of the current
research in time-resolved resonance Raman (TR^3) spectroscopy as it
applies to liquid phase chemistry. This paper is not an exhaustive
review of the field nor does it select examples of experimental results
from a wide range of research groups actively using TR^3 spectroscopy.
Although the illustrative results are taken primarily from work done in
my own laboratory, they are representative of the type of data and
information currently obtainable.

Time-Resolved Resonance Raman Spectroscopy

Vibrational spectroscopy derived from laser excitation is especially
useful for time-resolved measurements because the temporal resolution
is largely determined by the properties of the excitation laser. This
characteristic pertains to vibrational Raman spectroscopy which is now
routinely obtained with laser excitation. Since it is the inelastic
scattering of a sample resulting from its interaction with a radiation
field, the degree to which the radiation field can be effectively
coupled to molecular vibrations determines the strength of the vibra-

tional Raman signal. The capability of laser sources to provide strong radiation fields in resonance with molecular transitions is the basis for high detection sensitivities. In general, the versatility with which Raman spectroscopy has been applied to molecular systems has been primarily due to the availability of versatile lasers. This is certainly true of the application of vibrational resonance Raman in time-resolved experiments. Several specific advantages of TR^3 spectroscopy should be noted in detail:

a. The vibrational degrees of freedom are sensitive not only to changes in molecular structure, but also to conformational differences. This point has been documented extensively within the context of the normal mode approximation by a vast literature on stable molecules [2]. Conformational changes are thought to be particularly critical in controlling liquid phase reactions, especially those involved in biochemical phenomena such as membrane transport.

b. By generating Raman scattering through the direct excitation into electronic transitions, resonantly-enhanced Raman signal can be readily obtained [3]. Resonance excitation effectively coupling the exciting radiation field to the electronic transition moment and as a consequence, the resonance Raman (RR) scattering is commonly 10^6 times larger than that observed from normal Raman scattering. RR scattering thereby provides exceptionally high detection sensitivity. A general benchmark is that RR signals with signal-to-noise ratios of 10 can be recorded for molecules in the 10^{-6} M concentration range when electronic transitions having absorption coefficients of approximately 10^{4} mole 1^{-1} cm^{-1} are pumped. Such detection sensitivities are sufficient to permit many chemical and biochemical transients formed under normal conditions of reactivity to be monitored by RR scattering.

c. Resonantly-enhanced Raman scattering provides excellent selectivity with respect to detecting either one chromophoric part of a complex molecule or one component within a complicated reaction mixture. When the electronic transition used to obtain resonantly-enhaned Raman is contained within one chromophoric group of a large molecule (e.g., porphyrin moieties within proteins), RR scattering will be observed primarily from the vibrational modes of that chromophore alone. The same type of selectivity obtains for one chromophoric component within a reaction mixture. By tuning the probe laser wavelength into resonance with a particular electronic transition, RR scattering can be used to selectively examine a particular chromophore or only one component of a reacting system.

d. The temporal properties of Raman scattering are determined by the radiation field primarily and therefore, will follow the temporal behavior of the probing laser excitation. Although not rigorously correct for the resonance case where the distinctions between Raman scattering and resonance fluorescence merge, the lifetime properties of the excited electronic states of the molecule are not a major contributor to the temporal properties measured by RR scattering for time

scales as short as a few picoseconds. As a consequence, kinetic information into the picosecond time regime can be readily extracted from RR
data.

   These characteristics of TR^3 spectroscopy have been utilized widely to record vibrational spectra of transient, intermediate species in
chemical and biochemical reations [4-8]. There are several other types
of measurements that can be made particularly well with TR^3 spectroscopy and which are especially useful for elucidating liquid phase dynamics. Two of these are: (a) time-resolved excitation profiles (TREP)
and (b) real-time kinetic measurements derived from pump-probe experiments.

e. The resonantly-enhanced nature of Raman scattering not only provides increased detection sensitivity, but also is the basis for
assigning vibrational Raman signals to particular vibronic transitions.
Since the intensity of RR scattering derives from the interaction of
the exciting radiation field with a specific vibronic transition
moment, all of the resulting RR bands will maintain the same relative
intensities as the wavelength of the exciting radiation changes. Excitation profiles (EP) are obtained by measuring the intensity of a RR
band as a function of the wavelength of the exciting radiation.
Following the principal stated above, the EP for bands originating from
the same resonantly-enhanced vibronic transition would be the same.
   When a transient species is monitored by RR scattering, the kinetic parameters characterizing the lifetime of the intermediate must be
considered. In this case, the existence of a particular electronic
transition now occurs whenever the transient species is present during
the reaction. For a fixed period of time after the formation of the
transient (i.e., fixed delay time after the reaction begins), an EP can
be measured. Such a time-resolved EP (TREP) has the same characteristics as the EP of a stable species, namely that the TREP with be the
same for all vibrational Raman bands originating with a particular
transient species at a fixed time delay. A TREP is especially useful
for unraveling complex Raman spectra containing contributions from a
variety of stable and transient species and from several transient
species that are either photophysically or photochemically formed.
   This same type of EP information also has been shown recently to
be of importance in calculating absorption spectra. Excitation profiles can be used via theoretical methods based on either transformation [9] or Kramers-Kronig [10] relationships to calculate the absorption spectra from which resonantly-enhanced Raman signals are derived.
In stable species, electronic absorption spectra are often readily
available and these methods are used primarily to calculate excitation
profiles. Both absorption and RR spectra are more difficult to obtain
for transient species and therefore, the information contained in TREP
may be viewed as potentially more interesting since it provides the
opportunity to calculate unknown spectroscopic properties. An
intriguing application of TREP data encompasses cases in which the
energetic separation of excited electronic states changes as a reaction
procedes making both the absorption and RR spectra a function of reac-

tion time.

f. The use of TR^3 spectroscopy to measure the kinetic properties of
transients has yet to be widely exploited. Transient absorption spectroscopy is generally regarded as an easier experimental method for
obtaining this information and in cases where there is no difficulty in
assigning absorption spectra to a specific transient species, this
conclusion is valid. It is becoming increasingly well-recognized,
however, that liquid phase transients are not uniquely identified by
their absorption spectra, especially when conformational or structural
changes are involved. In these cases, TR^3 spectroscopy can provide the
detailed structural and conformational identification required to
uniquely monitor only one transient species and as a result, the capability to extract kinetic information directly from TR^3 data is critical to characterizing the reaction dynamics. The use of TR^3 data to
measure kinetic properties should increase as these techniques are more
widely used.

Experimental Techniques

A variety of methods have been employed to initiate reactions for study
by TR^3 spectroscopy including rapid flow mixing, temperature jump by
rapid heating, and photolytic excitation [4-8]. TR^3 spectroscopy can be
utilized to monitor any of these reactions although the time required
to initiate the reaction will be a major factor in determining the time
resolution of the measurement. Photolytically-initiated reactions are
of prime interest for this discussion and therefore, will be the only
experimental techniques discussed in detail.
    Two general approaches have been used to obtain TR^3 data following
photolysis: (1) single laser excitation and (2) two laser excitation
using a pump-probe configuration. Both types of experiments normally
introduce the sample rapidly into the laser beam either to provide
time-resolution or to exchange the sample volume exposed to laser beam
rapidly enough to avoid multiple excitations.

1. Single laser experiments

A single laser operating at one wavelength can be used both to photo-
lytically-initiate a reaction and to generate RR scattering from the
reaction mixture. If the single laser is continuous (i.e., cw), time
dependence can be introduced into the measurement by rapidly moving the
sample through the laser beam. The laser beam is usually focused at
the sample in order to increase the time resolution. Rapidly flowing
jets of liquid sample or spinning cells containing the liquid sample
are commonly used. These TR^3 experiments have obtained time resolutions as fast as 10^{-8} seconds, but are more readily applicable for
events occurring in 10^{-7} seconds or longer.
    The major disadvantages of this approach derive from restricted
number of reactions that can be studied due to the limited wavelength
range of cw lasers, the requirement that photolytic initiation and RR
detection must occur at the same wavelength, the absence of direct

kinetic information such as rate coefficients, and the fact that the RR scattering observed comes from a mixture of the starting reactants plus all intermediates present during the period over which the sample intercepts the laser beam. The first and last disadvantage can be minimized when a single pulsed laser is substituted for cw radiation. Not only are the wavelength ranges available from pulsed lasers themselves wider than their cw counterparts, but nonlinear optics can be used efficiently with the higher peak powers available from pulsed lasers to extend these ranges into both the infrared and ultraviolet regions. The complexity of the reaction mixture from which RR scattering is collected can also be simplified by the faster time resolution available from pulsed laser excitation. The duration of the laser pulse replaces the rate of liquid flow as the factor determining the time resolution of the TR^3 measurement. An example of this experimental approach using picosecond laser pulses from a mode-locked dye laser and Nd:YAG pumping is presented in Figure 1. Within the context of these limitations, single laser experiments have been exceedingly useful methods for opening chemical and biochemical systems to study by TR^3 spectroscopy.

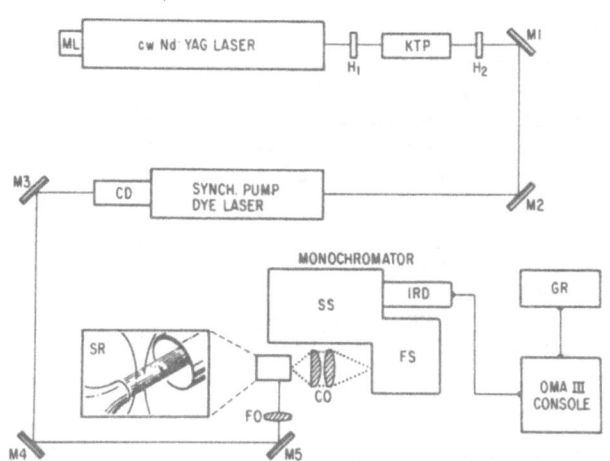

Figure 1: Single laser PTR^3 instrumentation. ML: mode locker; H_1: halfwave plate (1064 nm); H_2: halfwave plate (532 nm); M1-M5: front surface dielectrically coated mirrors; CD: cavity dumper; FO: focusing optics, microscope objective; SR: sample region, laser passes vertically through flowing sample; CO: collecting optics; FS: filter stage; SS: spectrograph stage; IRD: intensified reticon detection; GR: graphics.

2. Two laser experiments

Pump-probe configurations utilizing two, independent pulsed lasers provide the most versatile approach to TR^3 spectroscopy of any of the

methods reported to date. When used with rapidly flowing or spinning
samples, high repetition rate (e.g., megahertz), pulsed lasers can be
used to obtain high signal-to-noise ratio RR spectra without exposing
the reaction mixture to multiple excitations. The tunable laser wavelengths available extend from 220 nm to 1.1 μ with a variety of fixed
wavelength lines over an even wider spectral range. Of major
importance is the flexibility to photolytically initiate the reaction
at one wavelength and to monitor intermediates by RR scattering at a
second, independently selected wavelength. This experimental
configuration also permits the intensities of the pump and probe radiation to be independently chosen, thereby opening intensity-dependent
phenomena to study. An additional advantage is the opportunity to
measure rate data that can be used to calculate rate coefficients.
This last capability, in combination with the structural and conformational sensitivity of TR^3 spectroscopy in liquid phase reactions, makes
it feasible to examine the kinetic properties of conformational changes
in great detail. Finally, the complexity of reaction mixtures can be
unraveled to a large degree by the presence of a short duration probe
pulse optimized to initiate RR scattering only from those intermediates
present at a fixed time after photolysis. The duration of the laser
pulse used to generate RR scattering thereby determines which part of
the time dependent reaction contributes to the RR spectrum. The
combination of flexibility with respect to each of these experimental
parameters gives two pulsed laser, pump-probe technique its overall
versatility.

Two examples of this experimental approach are given in Figures 2
and 3. TR^3 spectra can be readily obtained with 10^{-8}s time resolution

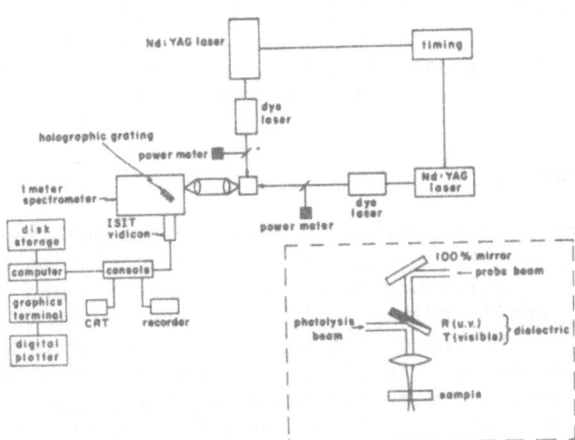

Figure 2: Instrumentation for nanosecond TR^3 spectroscopy. Excitation
geometry for sample regions is shown as an insert.

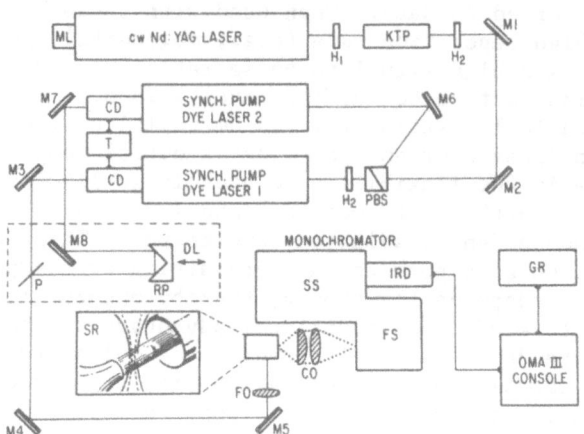

Figure 3: Two laser PTR^3 instrumentation. ML: mode locker; H$_1$: halfwave plate (1064 nm); H$_2$: halfwave plate (532 nm); M1-M8: front surface dielectrically coated mirrors; PBS: polarizing beam splitter; CD: cavity dumper; T: timing synchronization; DL: delay line; RP: retroprism; P: dielectrically coated pellicle beam splitter; FO: focusing optics; SR: sample region showing colinear laser beams; CO: collecting optics; FS: filter stage; SS: spectrometer stage; IRD: intensified reticon detector; GR: graphics.

using two Q-switched Nd:YAG pumped dye lasers together with a rapidly flowing or spinning sample (Figure 2). Picosecond ($\sim 2 \times 10^{-12}$s) time resolution can be obtained by using two synchronously-pumped, mode-locked dye lasers operating with an optical delay line. Pumping sources can be obtained from either mode-locked ion lasers or mode-locked cw Nd:YAG lasers. The repetition rates of the lasers are controlled by cavity dumping each dye laser. The velocity of the sample through the laser beams relative to the laser repetition rates must be carefully considered in order to avoid multiple excitation of the reactants.

3. Spectrometers

The requirements for spectral dispersion are determined by the competing needs of spectral throughput versus resolution and straylight rejection. A single, one meter monochromator using a Czerny-Turner mounting is readily adaptable for TR^3 experiments (Figure 2). RR bands within 100 cm^{-1} of the excitation line can be recorded for transients in the 10^{-6} M regime. Triple monochromators incorporating two stages of subtractive dispersion for increased straylight rejection, together with a concave grating section, are effective when multichannel detection is used since a flat focal plane of several centimeters is obtained. The versatility of changing dispersion in the final stage of these monochromators is provided by having more than one concave grating that can be readily introduced into the light path. This is especially advantageous in TR^3 spectroscopy where both high resolution spectra focusing on small wavelength shifts and low resolution survey spectra are often required for a complete analysis.

4. Detectors

Multichannel detectors based on either vidicon or reticon technology
are commonly part of the detection system. The multiplex advantage of
detection over a wide spectral range has been critical to the development of TR^3 spectroscopy since a significant part of the vibrational
Raman spectrum can be recorded simultaneously during each pump-probe
cycle. The spectral response of these detectors is centered in the
visible although recent improvements have extended this sensitivity
range into the ultraviolet (via scintillators) and infrared (via the
detector material in reticons). To obtain the detection sensitivity
routinely required for TR^3 spectroscopy, intensifiers are used in
conjunction with both vidicons and reticons. Multichannel detectors
rely heavily on computer storage of the large quantities of spectral
data and on versatile software to accommodate the analysis routines
needed to correct data for nonlinearities in wavelength and sensitivity
response. Software capabilities are also important in facilitating the
comparison of data from different TR^3 experiments. The usefulness of
multichannel detection can be severely limited without the proper
computer support.

Results

TR^3 spectroscopy is sufficiently versatile to be used in monitoring
liquid phase, molecular intermediates for essentially any type of room
temperature reaction involving changes in the vibrational degrees of
freedom. The statement is inaccurate only when the experimental limitations of the instrumentation are reached such as in the cases of time
resolution, laser wavelengths, and radiation flux. Data on both photochemical and photophysical intermediates formed in the nanosecond and
picosecond time regimes are described here to illustrate this versatility.

1. Ionic intermediates

The existence of ionic intermediates in electron transfer chemistry has
been proposed from transient absorption spectroscopy for an enormously
large group of reactions. Detection of the ions formed during electron
transport often can be observed by transient absorption spectroscopy
due to the large changes in the population of electronic energy levels,
but the structural characterization of the ion via its electronic
absorption spectrum is more difficult. The broad spectral profiles of
these transient absorption spectra also makes it difficult to confiden-
tially conclude that the same ionic species is formed under different
chemical conditions. TR^3 spectra are especially useful in addressing
these questions.

Photolytically-induced electron detachment in stilbene ions illustrates the value of TR^3 spectra for firmly identifying an ionic

species. Transient absorption measurements had suggested that when
electron photodetachment was used to form the radical anion of trans-
stilbene (T⁻·) from the dianion of trans-stilbene (T²⁻), an intermediate
ionic species was formed, specifically the radical anion of cis-
stilbene (C⁻·) [11]. TR³ spectra of the $T^{2-} + h\nu \rightarrow T^{-} + e^{-}$ reaction
were recorded with 532.0 nm (17 ns pulsewidth) excitation of a solution
of T²⁻ in tetrahydrofuran at room temperature [12]. TR³ scattering was
generated by 506.0 nm (1 μs pulsewidth) laser radiation occurring 2 μs
after photolysis. The TR³ spectrum obtained is presented in Figure 4B
together with the RR spectra of the T²⁻ solution (4A) and a chemically-
stabilized sample of T⁻ only (4C). It is apparent from a comparison of

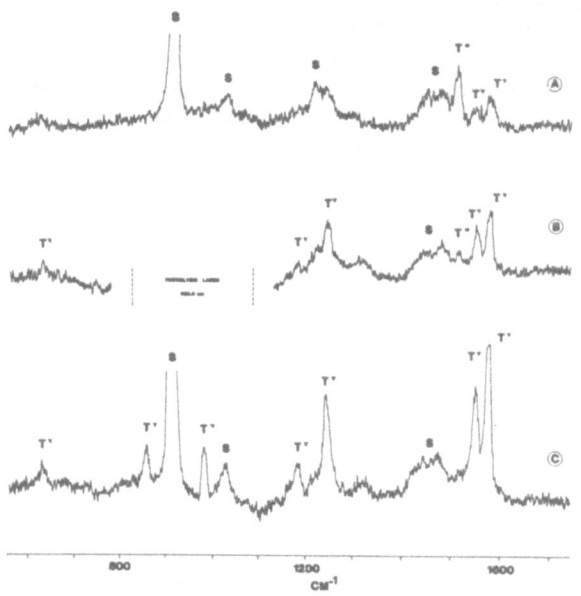

Figure 4: (A) RR spectrum of T²⁻ with 506.0 nm excitation, (B) TR³
spectrum of T²⁻ sample with 506.0 nm excitation 2 μs after 532.0 nm
excitation, (C) RR spectrum of a chemical stabilized T⁻ sample with
506.0 nm excitation.

these spectra that the ion formed within 2 μs of electron detachment
from T²⁻ is T⁻. No vibrational Raman bands that are unassignable to T⁻·
were observed in the TR³ spectrum and therefore, there is no evidence
to support the formation of another ionic intermediate. Subsequent TR³
studies have shown that the vibrational Raman spectrum of the radical
anion of a cis-stilbene analogue, 5H-dibenzo(a,d) cyclohepten-t-
one(DBCH⁻·), was significantly different than that of T⁻ [13]. If the
C⁻· species were formed, its RR spectrum would be easily recognizable.
      The ionic intermediates formed during electron transfer in anthra-
quinone derivatives are of fundamental interest in the photosensitiza-
tion of a wide range of photochemical processes such as those underlying solar energy conversion [14]. The role of each ion and its

photophysical precursor (e.g., excited triplet-states) remains to be
firmly established, but it is clear that the effectiveness of anthra-
quinone derivatives as photochemical sensitizers is directly related to
the presence of specific ions. The strong influence of pH, temperature, and solvent on the formation of different ions has been documented via transient absorption spectroscopy. The identification of which
ionic species are actually present and their molecular structure has
not been resolved, however, and it is in this area where TR[3] spectroscopy has begun to make significant contributions.

     Water-soluble anthraquinone obtained by sulfonation was examined
by TR[3] spectroscopy using nanosecond time resolution [15,16]. The
instrumental approach was similar to that presented in Figure 2 except
that excimer laser pumped dye lasers were used. The central issue to
be described here focuses on the identification of the ionic intermediate formed in pure aqueous solutions of anthraquinone-2,6-disulphonate
(AQ26DS). In the presence of a strong electron donor, the radical ion
(AD26DS$^-$) should rapidly appear and under acid conditions subsequently
protonate to produce the semiquinone radical. The characterization of
this reaction and the chemical conditions which facilitate it is an
important step in characterizing the overall photocatalytic properties
of anthraquinones.

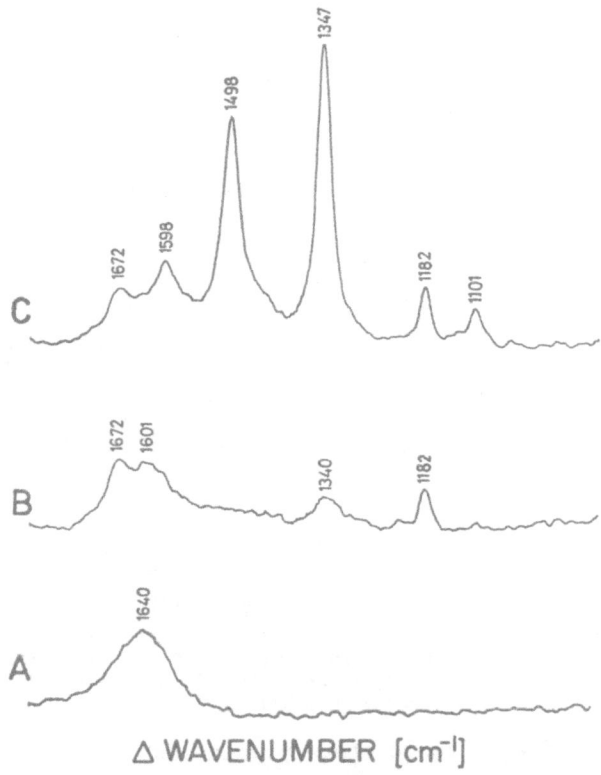

Figure 5: Raman spectra obtained using a probe laser wavelength of 480
nm of (A) water only, (B) AQ26DS and NaNO$_2$, and (C) TR^3 spectrum
observed at a time delay of 100 ns after 337 nm excitation.

The TR^3 spectra of AQ26DS in the presence of NaNO$_2$ is shown in Figure 5. The RR spectra of the water solvent (A) and the AQ26DS sample with probe laser only (B) are also shown for comparison. With 337 nm pumping and RR scattering generated by 308 nm radiation 100 ns later, several RR bands assignable to AQ26DS$^-$ are clearly seen. These same RR bands can be seen in Figure 6 to change intensity as a function of reaction over the 50 ns to 5 µs interval. These RR spectra have firmly established the existence of AQ26DS$^-$ during the reaction and determined its kinetic behavior.

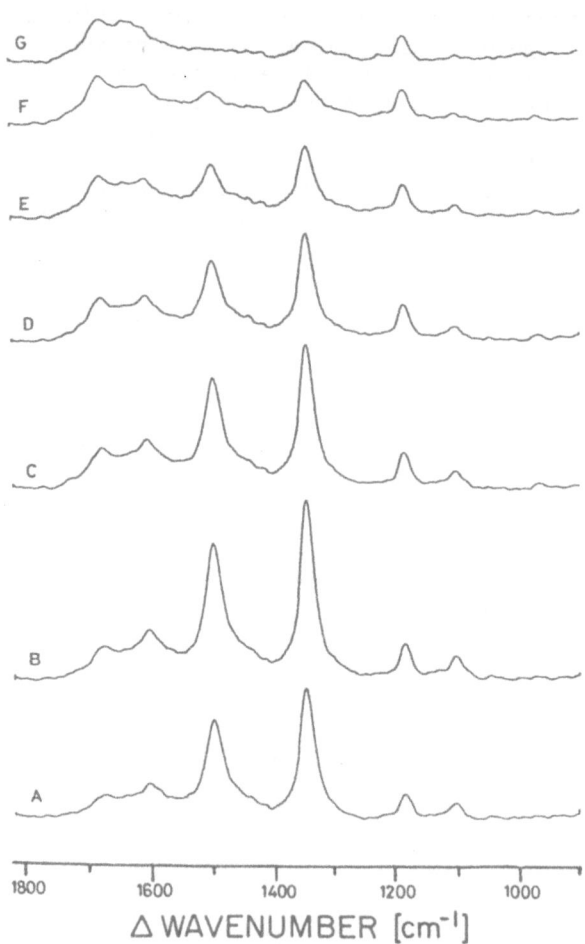

Figure 6: Time dependence of the transient AQ26DS RR spectrum of time delays of (A) 50 ns, (B) 100 ns, (C) 500 ns, (D) 1 µs, (E) 2 µs, (F) 5 µs, and (G) probe only.

2. Excited electronic state populations

The initial intermediates formed in photolysis are the excited electronic states populated during and subsequent to excitation. These photophysical intermediates act as precursors to the dissociations, electron and proton transfers, and molecular rearrangements that are generally described as photochemical. The population of different potential energy surfaces is usually manifested by significant changes in the vibrational degrees of freedom which are evident in the vibrational Raman spectra. These TR^3 spectra of excited electronic states can be analyzed with respect to vibrational normal modes to the same extent as those of the analogous spectra of ground-state species. The polarization properties and time dependent intensities can also be measured directly in order to establish transition moments and transient populations.

Excellent examples of these TR^3 measurements are provided by cyclic hydrocarbons such as phenanthrene and chrysene. The TR^3 spectra of the lowest-energy excited triplet state of phenanthrene-h_{10} and -d_{10} are shown in Figure 7 [17]. Each RR band given a wavenumber label originates in the excited $^3B_2$ state of phenanthrene. The deuterium shifts and the polarization properties measured 55 ns after 330 nm excitation into the $S_0 \rightarrow S_1$ transition of phenanthrene are presented in Table I. These data are in complete analogy with those obtained for

TABLE I   Frequency positions (cm^{-1}) for $^3B_2$ phenanthrene-h_{10} and -d_{10} observed by time-resolved resonance Raman (TR^3) spectroscopy.[a]

Phenanthrene-h_{10}				Phenanthrene-d_{10}			
Freq. (cm^{-1})	Rel. I	P[b]	$\rho_\perp$[c]	Freq. (cm^{-1})	Rel. I	P[b]	$\rho_\perp$[c]
394	(s)	P	0.30	381	(s)	P	0.30
				650[d]	(w)	-	-f
				756[d]	(vw)	-	-f
				810	(vs)	P	0.40
823	(m)	P	0.22	836	(vw)	-	-f
930	(w)	-	-f	883	(s)	P	0.24
981	(vs)	P	0.33	975	(vs)	P	0.41
1028[d]	(w)	P	-e				
1040	(w)	P	-e	1047	(m)	P	0.37
1166	(w)	P	-e	1183[d]	(vw)	-	-f
				1239	(m)	P	0.44
1335	(vs)	P	0.33	1325	(s)	P	0.38
1395	(w)	P	-e	1381	(m)	P	0.13
1433	(m)	P	0.27	1427	(m)	P	0.37
1448	(m)	P	0.25	1466	(s)	P	0.36
1473	(s)	P	0.31				
1504	(w)	-	-f				
				1565[d]	(vw)	-	-f
				1696[d]	(vw)	-	-f

[a] frequency position accurate ± 1 cm^{-1} except where noted.

[b] P = polarization.

[c] $\rho_\perp \equiv$ depolarization ratio ($= I_\perp / I_\parallel$); $\pm 10\%$.

[d] frequency positions accurate to ± 2 cm^{-1}.

[e] band intensities permit measurement of polarization properties but not $\rho_\perp$.

[f] band intensities too weak to permit measurement of polarization properties.

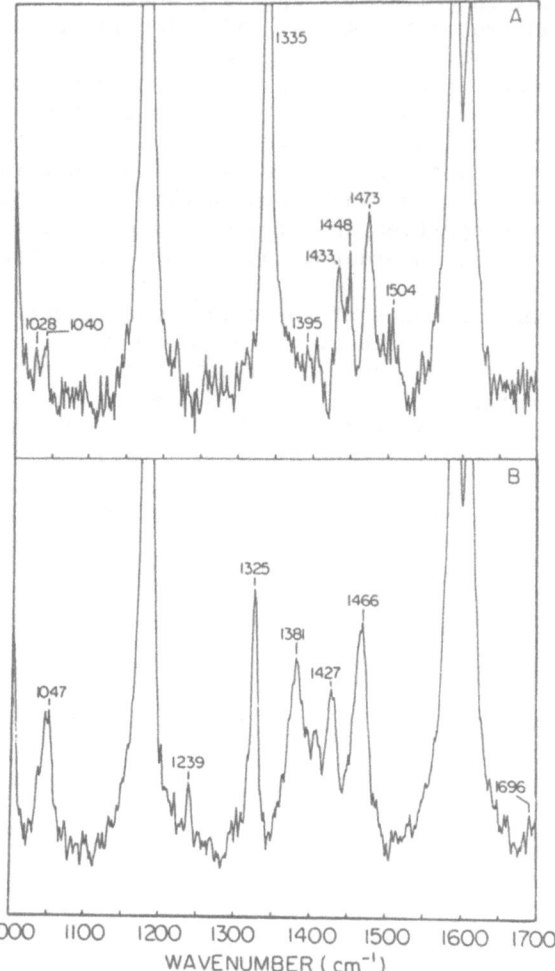

Figure 7: TR^3 spectra of excited $^3B_2$ phenanthrene-h$_{10}$ (A) and -d$_{10}$ (B). The $^3B_2$ bands are labelled with wavenumber positions while solvent bands are unlabelled.

stable species. The TR^3 spectra of ^3B_u state of chrysene-h$_{12}$ and -d$_{12}$ are presented in Figure 8 [18]. These spectra were used to monitor the excited-state population of ^3B_u chrysene during $S_1 \rightarrow T_1$ and $T_1 \rightarrow S_0$ intersystem crossing. The general kinetic behavior is shown in Figure 9 for the time-dependent intensity of the 982 cm^{-1} band from ^3B_u chrysene-h$_{12}$ [19]. The kinetic analysis of these data requires that the transient absorption properties of the sample attributed to the $T_1 \rightarrow T_n$ transition be separated from the TR^3 signal. These absorption data can be obtained directly from the intensities of the Raman bands of the solvent and used to correct the intensity dependence observed in TR^3 bands from the excited state. This procedure provides kinetic information in the form of rate coefficients for intersystem crossing

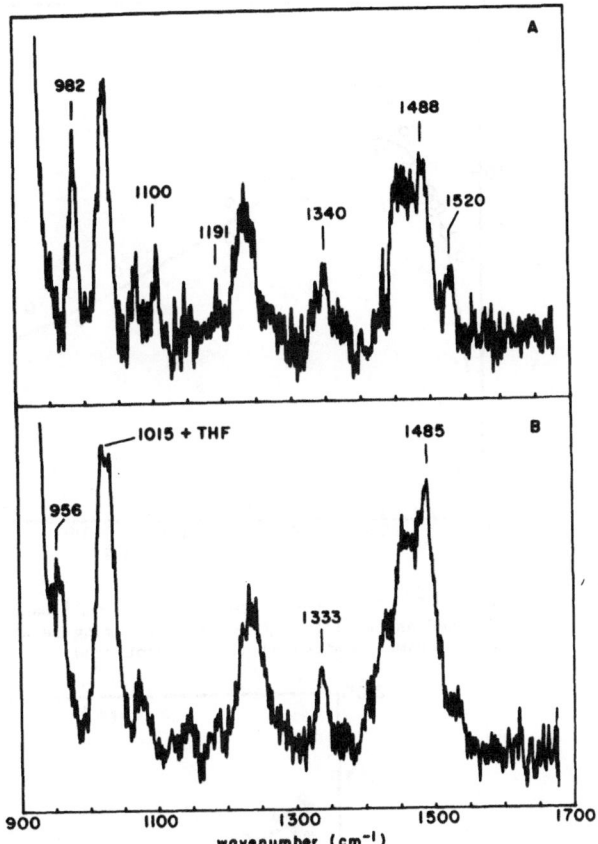

Figure 8: TR^3 spectra of excited ^3B_u chrysene-h$_{12}$ (A) and -d$_{12}$ (B). The ^3B_u bands are labelled with wavenumber positions while the solvent bands are unlabelled.

that exhibit <5% uncertainties. An example of these data are shown in Figure 10 for the first order decay of excited ^3B_u chrysene via $T_1 \rightarrow S_0$ intersystem crossing. The deuterium effect seen in this rate experimentally confirms a prediction made from radiationless transition theory [20,21].

A more demanding application of TR^3 spectroscopy to the study of excited electronic states encompasses cases where a significant competition between photophysical and photochemical decay processes exists. The experimental challenge then centers on the capability to record the TR^3 spectrum of the excited electronic state of only one, well-defined component of the photochemical reaction.

Perhaps the most complex system yet studied is the excited triplet state of all-trans retinal (ATR). ATR readily isomerizes to other retinal isomers following optical excitation into its excited electronic state manifolds. Photophysical decay such as intersystem crossing occurs with comparable yields. The measurement of the TR^3 spectrum of an excited electronic state of one retinal isomer, there-

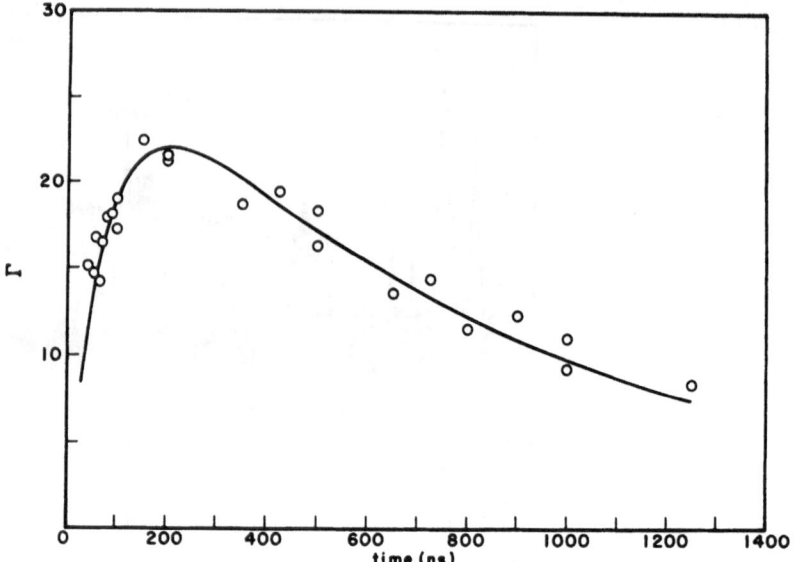

Figure 9: Time dependence of the 982 cm⁻¹ band of ³B_u chrysene Γ is
defined as the intensity of RR scattering relative to the intensity of
normal Raman scattering of a solvent band.

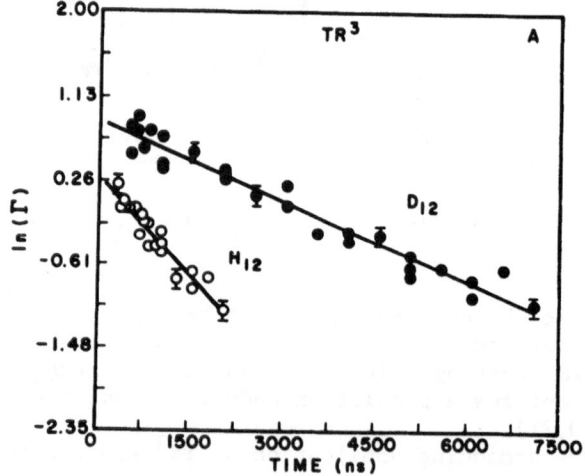

Figure 10: First-order decay of excited ³B_u chrysene-h₁₂ and -d₁₂ as
measured by TR³ bands at 982 cm⁻¹. Γ is defined in Figure 9.

fore, requires conditions which minimize the competitive isomerizations. It also requires experimental confirmation that significant
concentrations of the other isomers are not formed in the sample under
study.

The TR³ spectra obtained from ATR following 354.7 nm excitation
into the $S_0 \rightarrow S_1$ transition are shown in Figure 11 [22]. The spectrum

obtained with probe laser radiation only (470 nm) contains bands from the S_0 state of ATR and thereby confirms that the original sample is in the all-trans configuration. The TR^3 spectrum recorded with a 40 ns time delay shows substantial changes, including a new band at 1555 cm^{-1} with major intensity. The TR^3 spectrum recorded with a much longer delay time (20 μs) is the same as the probe only spectrum. This last spectrum demonstrates that retinal has relaxed photophysically from the excited electronic states populated by 354.7 nm excitation to reform primarily ATR. The presence of significant concentrations of other isomers would be evident from the 20 μs TR^3 spectrum. The transient species detected by TR^3 scattering at 40 ns is therefore identifiable with ATR and based on several types of data [22] can be assigned to the lowest-energy triplet state of ATR. It should be emphasized that TR^3 spectra are the basis for both detecting and characterizing the transient intermediate and confirming that no significant amount of permanent photochemistry (i.e., isomerization) has occurred during the measurement.

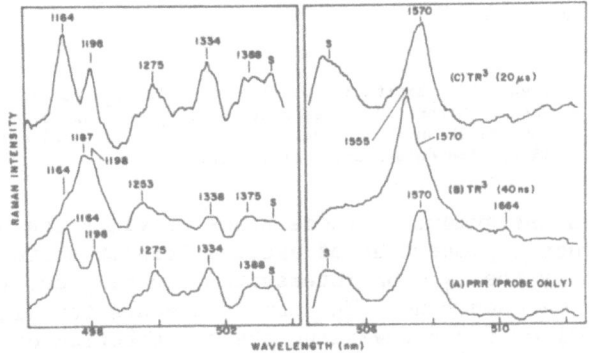

Figure 11: (A) RR spectrum of all-trans retinal (ATR) in methanol with 470 nm excitation only, (B) TR^3 spectrum of ATR 40 ns after 354.7 nm excitation and with 470 nm probing and (C) TR^3 spectrum of ATR 2 μs after 354.7 nm excitation and with 470 nm probing.

The availability of these data on the excited triplet state of one retinal isomer makes it feasible to examine the excited-state mechanism for isomerization in retinal. To firmly assign the 1555 cm^{-1} TR^3 band to triplet state ATR, the TREP for both the 1555 cm^{-1} and 1570 cm^{-1} bands were recorded [23]. Those data, together with schematic representation of the $S_0 \rightarrow S_1$ and $T_1 \rightarrow T_n$ absorption spectra of ATR are presented in Figure 12 [23]. The overlapping region of the TR^3 spectrum (40 ns delay) is shown as an insert. The EP of the 1570 cm^{-1} band follows closely the $S_0 \rightarrow S_1$ absorption spectrum as would be anticipated for RR scattering in preresonance with an electronic transition. The TREP (40 ns) for the 1555 cm^{-1} is dramatically different in that it follows closely the $T_1 \rightarrow T_n$ absorption spectrum. The large differences between these two types of excitation profiles provides solid evidence for their respective assignments. This example also provides an excellent example of the value of TREP data in unraveling a complex TR^3 spectrum.

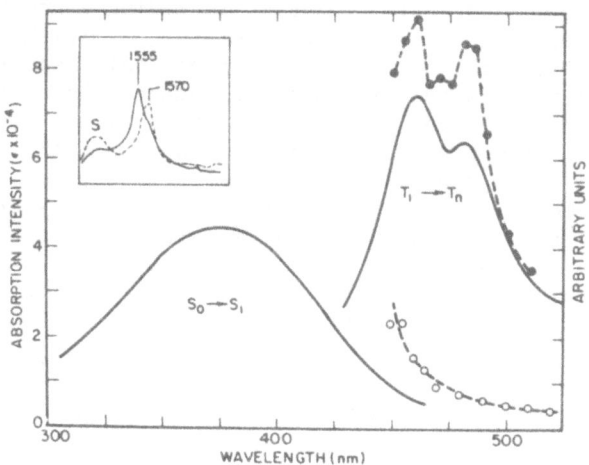

Figure 12: Excitation profiles for ATR (T_1) (● data) and ATR (S_0) (○ data). Schematic representation of $S_0 \rightarrow S_1$ and $T_1 \rightarrow T_n$ absorption spectra shown. TR^3 spectra at 40 ns (solid line) and 20 μs (dashed line) time delays shown as insert.

The firm assignment of these bands provides the basis for examining the kinetic properties of excited triplet state ATR. Of particular interest is the $T_1 \rightarrow S_0$ intersystem crossing decay channel and the influence of O_2 quenching. The same procedure for kinetic analysis described earlier for normalizing the contribution of transient absorption to TR^3 data [19] was used to obtain the decay rate of excited triplet state ATR in the presence and absence of O_2 [23]. The O_2 concentration was held sufficiently high to insure pseudo-first order kinetics. The rates observed, shown in Figure 13, exhibit a strong dependence on O_2 and are the first measured values for the O_2 quenching of one excited state in a single isomer. Although the O_2 quenching rate is an important value to compare with the decay rate of excited triplet state ATR by itself, it is clear that this method provides the opportunity to examine the quenching mechanism separately, especially with respect to its dependence on the conformation of the retinal. Such detailed studies of the influence of conformation on reactivity in the liquid have not been previously feasible.

3. Picosecond dynamics of biochemical intermediates

The fundamental value of TR^3 spectroscopy for the study of dynamics in liquid phase reactions can be best seen in its applications to biochemical systems where the molecular changes often encompass conformational transformations in molecules that are structurally complex. In these cases, the unique advantages of TR^3 spectroscopy described above can be utilized to elucidate dynamical and structural

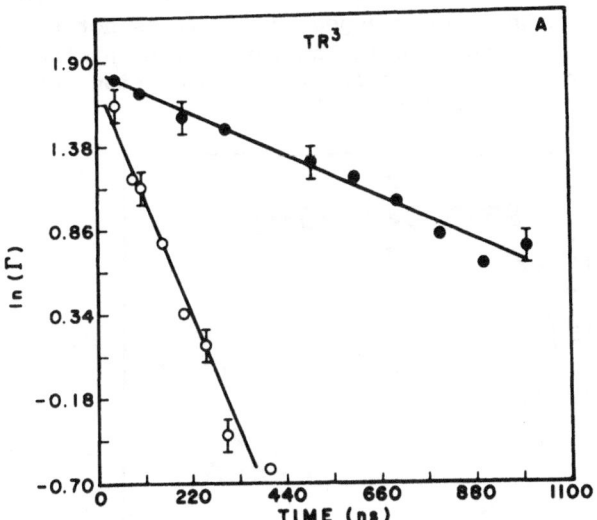

Figure 13: Decay kinetics of ATR (T_1) in the absence (● data) and presence (0 data) of O_2 as measured directly from TR^3 bands. Γ is defined in Figure 9.

properties not previously accessible to study. TR^3 spectroscopy has been used by numerous research groups to examine several biochemical systems including extensive work on heme proteins and on visual pigments [4-8]. An extremely wide range of time scales has also been studied for these systems. In this paper, our attention will focus on recent results concerning intermediates formed on the picosecond time scale in the bacteriorhodopsin (BR) system.

BR is a well-characterized protein which facilitates proton and ion transport across the membrane of the Halobacterium halobium [24-26]. BR contains the same chromophore as found in the visual pigment rhodopsin, namely retinal. The BR system differs from rhodopsin in that it undergoes a series of reactions which form a cycle to reproduce the starting material, BR-570. (The numberical part of the notation gives the maximum of the absorption spectrum for each species.) During the cycle, major changes in the isomeric structure and conformation of retinal occur which are thought to be directly correlated to the proton and ion transport across the BR membrane. Two of these changes which have been characterized are the isomerization around the C_{13}-C_{14} bond from all-trans retinal (BR-570) to 13-cis retinal (M-412) and the deprotonation of the Schiff base linkage between the retinal chromophore and the lysine of the protein backbone (i.e., BR-570 protonated to M-412 deprotonated). The schematic representation of the BR photocycle is presented in Figure 14.

The initial molecular transformations occurring within the first 100 ps, involving BR-570 and K-590, are of interest for this discussion. Instrumentation designed around two picosecond lasers in a pump-probe configuration (Figure 3) was used to record TR^3 spectra of the K-590 intermediate. The maxima of the absorption spectra of the initial BR intermediates shift slightly through the visible spectral region, but remain strongly overlapped [27]. It is not until M-412 is formed on the 40 μs time scale that the absorption spectra of BR intermediates become well separated. These absorption spectra for the reaction of 1°C are shown in Figure 15.

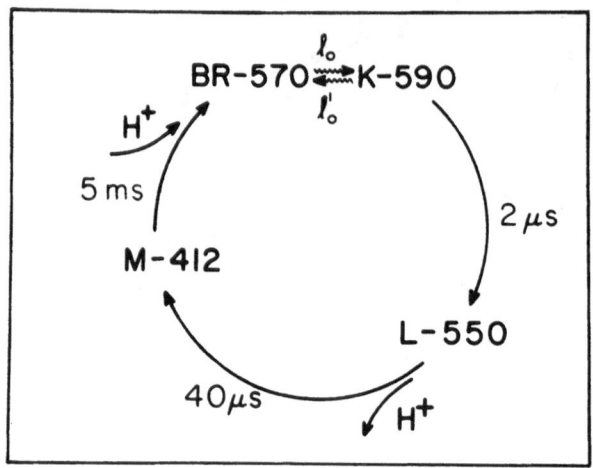

Figure 14: Photocycle of bacteriorhodopsin (BR) [notation and ℓ_0 defined in text].

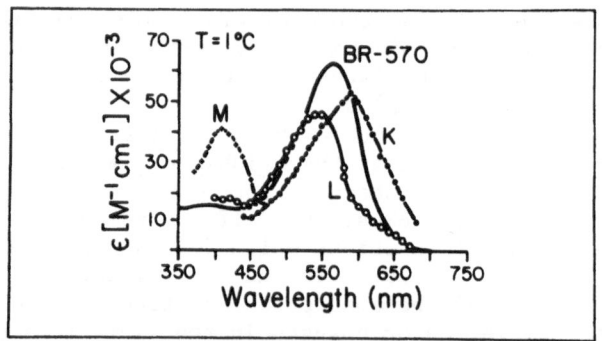

Figure 15: Absorption spectra of the intermediates of the BR photocycle at 1°C [adapted from Lozier, R.H., Bogomolni, R.A., and Stoekenius, W. (1975) Biophys. J. 15, 955].

To monitor the K-590 intermediate, the probe laser wavelength must be selected to be in resonance with the absorption band having a maximum at 590 nm. It is clear from Figure 15, however, that 590 nm radiation also remains partially in resonance with the starting compound, BR-570. Thus, the probe laser can produce RR scattering from the K-590 intermediate and can photolytically initiate the photocycle from BR-570. Analogously, the photocycle needs to be initiated by laser radiation occurring within the absorption band having its maximum at 570 nm. This same radiation, however, generates RR scattering

from BR-570 which will overlap the RR scattering from K-590 generated
by 590 nm excitation.

It is also thought that the K-590 species can be photolytically
converted back to BR-570. As a consequence, pumping radiation at
either 570 nm or 590 nm can be expected to create the photolytic
equilibrium between BR-570 and K-590. The BR case illustrates the
often encountered complexity of a biochemical reaction which must be
addressed in a TR^3 experiment. The emphasis in this case is on the
quantitative control of the experimental parameters of laser wavelength, laser power per pulse duration (i.e., peak power), and time
delay between pumping and probing.

Several questions must be addressed in order to establish that the
vibrational Raman spectrum of the K-590 intermediate can be obtained
using 570 nm excitation to initiate the cycle and 590 nm to generate RR
scattering from K-590. What is the RR spectrum of BR-570 using picosecond (8 ps autocorrelation times) laser pulses at 590 nm? How does
the intensity of the picosecond pulses at 590 nm affect the RR spectrum
observed? Can the RR spectrum of K-590 be obtained as a function of
reaction time on the picosecond time scale? Data addressing these
three questions are presented here.

The spectrum of BR-570 has been reported by several workers, but
all of this work utilized cw laser excitation with a single laser operating at substantially different excitation wavelengths. The photostationary equilibrium between BR-570 and K-590 means that the relative
concentration of the two species is a sensitive function of the peak
power of the radiation source, especially when only one laser acts both
to initiate the photocycle and to generate RR scattering. Clearly, a
RR spectrum of BR-570 alone is best obtained with the lowest laser
power feasible. A comparison between the cw laser RR spectra and that
obtained by picosecond pulsed laser excitation is an important part of
the analysis since the respective peak powers are substantially
different. Such a comparison is made in Figure 16 [28]. There is very
good agreement in general between these two spectra, but in detail there
are significant differences which can be attributed to changes either in
excitation wavelengths and therefore, changes in the resonance effect,
or in the peak power and therefore, to changes in the relative concentrations of BR-570 versus K-590. This latter point is substantiated in
the data presented here on intensity dependence (vide infra) [29].
The spectra in Figure 16 demonstrate that the RR spectrum of BR-570
alone can be readily obtained using picosecond laser excitation as long
as the peak power remain below a specific threshold for K-590 formation.

As the peak power of a single picosecond laser operating at 590 nm
increases, significant changes in the RR spectrum occur which can be
assigned to the formation of the K-590 intermediate. Examples of these
changes are presented in Figure 17 for spectral regions which characterize the $C_{13}-C_{14}$ isomerization (i.e., fingerprint region at 1130-
1240 cm^{-1}) or hydrogen-out-of-plane vibrations (i.e., HOOP region at
910-1050 cm^{-1}), respectively. The peak power used in these experiments was quantitatively characterized by the parameters $\ell_0 t$ which
incorporated the size of the sample overlapped by the laser radiation
and the approximate quantum yields for the photo-induced reactions [30].

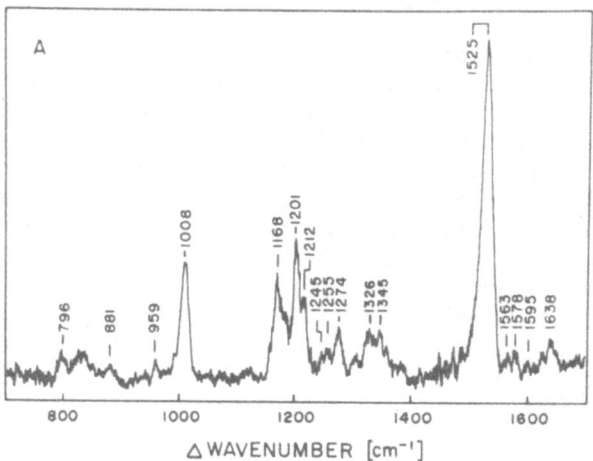

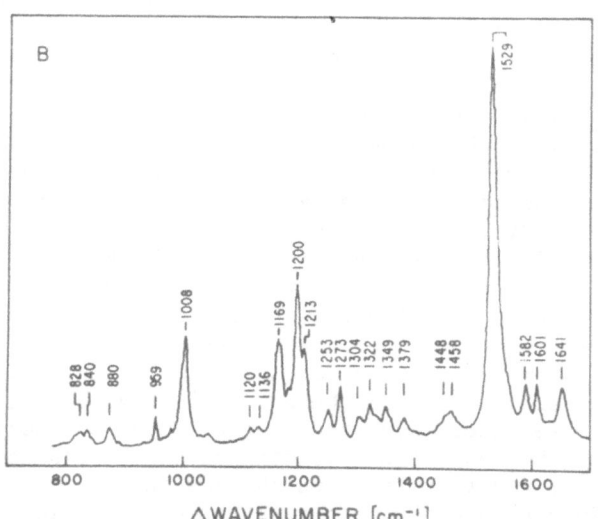

Figure 16: (A) PTR³ spectrum of BR-570 in H₂O suspension (8 ps pulsewidth, 590 nm, average power = 1.7 mW, repetition rate = 1 MHz, $\ell_0 t \sim 0.12$), (B) cw spectrum of BR-570 in H₂O suspension (514.5 nm, $\ell_0 t$ < 0.1). ℓ_0 defined in text.

The larger the value of $\ell_0 t$, the larger the number of photons delivered to the sample volume per unit time. The new RR bands observed in the bottom spectra of each region (i.e., larger $\ell_0 t$), therefore, are due to photolytically-induced phenomena, namely the establishment of the photostationary equilibrium between BR-570 and K-590. As the $\ell_0 t$ value increases, more K-590 is formed during the pulsewidth of the laser. These experiments at 590 nm establish the intensity threshold below which the photostationary equilibrium has not become significant. This information is critical to a TR³ experiment designed to detect K-590.

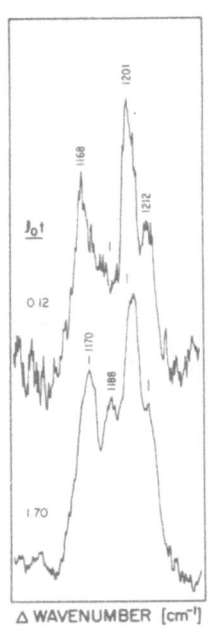
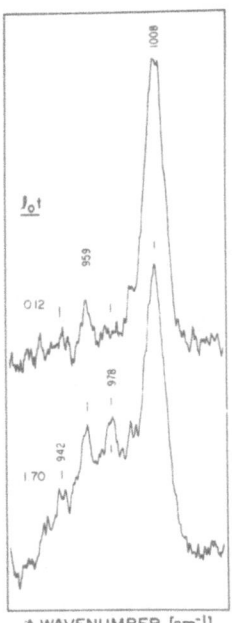

Figure 17: Single laser (590 nm) PTR^3 spectra of BR suspension in H$_2$O
with low excitation power (upper traces) and high excitation power
(lower traces). Spectra at left are from the fingerprint region and at
right from the HOOP region.

Finally, on the basis of these previous results, a pump-probe TR^3
experiment designed to detect K-590 as a function of the picosecond
reaction time can be performed [31]. The photocycle was initiated by 8
ps laser pulses at 570 nm operating with 21 mw and RR scattering from K-
590 was generated by 8 ps laser pulses at 590 nm operating with 5 mw
(both lasers have 1 MHz repetition rates) [32]. These strong pump-weak
probe parameters were chosen to maximize K-590 concentrations. The time
delay between the pump and probe pulses was controlled by an optical
delay line.

Picosecond TR^3 (PTR^3) spectra in the fingerprint and HOOP regions
are presented in Figure 18 for three delay times: 0 ps, 40 ps, and
probe only [33]. The new RR bands which increase in intensity as a
function of reaction time are the same as those observed to increase as
a function of laser intensity (Figure 17). The assignment of these new
spectral features to K-590 or to a K-like intermediate is a consistent
conclusion from both types of experiments. Similarily, the general
decreasing intensities of several RR bands also appear to be the same in
the two types of experiments leading to the conclusion that the BR
intermediate decreases in concentration as a function of both reaction
time and laser intensity.

The PTR^3 experiments provide a direct spectroscopic monitor of
conformational and structural changes that occur in the retinal
chromophore during the first 100 ps of the BR photocycle.
Specifically, the all-trans to 13-cis isomerization around the C$_{13}$-C$_{14}$

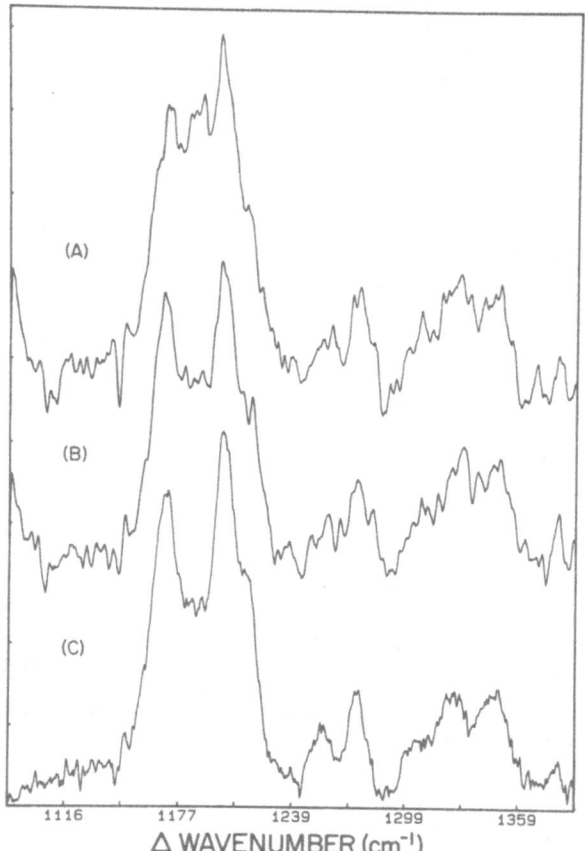

Figure 18: Two laser (590 nm probe, 570 nm pump) PTR[3] spectra of the fingerprint region of BR suspension in H_2O; A, 40 ps delay; B, zero delay; C, probe only.

band can be seen in the bands growing into the fingerprint region while the twisting of the hydrocarbon chain is reflected in the appearance of new bands in the HOOP region [33]. Although the time scales for both types of molecular transformations are generally the same (i.e., < 60 ps), further work is required to determine whether these changes occur simultaneously or sequentially. Data on the protonation of the Schiff base linkage has not yet been conclusively analyzed.

PTR[3] studies of molecular dynamics in the BR system are only beginning, but it is already clear that the molecular parameters now open to study provide a unique opportunity to characterize such complex biochemical reactions in exceptional detail. The versatility of the pump-probe configuration for recording TR[3] spectra is the source of that opportunity.

Conclusions:

The experimental capabilities of TR^3 spectroscopy for measuring real-time dynamics in liquid phase reactions at room temperature are beginning to provide the type of structural and conformational information previously reserved for gas phase reactions. The data described here illustrate some of the major characteristics of these techniques and their applications in chemistry and biochemistry.

Acknowledgement:

I would like to gratefully acknowledge members of my research group who have worked on TR^3 experiments, especially Dr. L. Dosser, Ms. J. Pallix, Mr. D. Gilmore, Dr. P. Killough, Dr. I. Grieger, Dr. G. Rumbles, Mr. T.L. Brack, and Mr. D. Blanchard. I am also very appreciative of the collaborative efforts of Mr. J. Moore and Professor D. Phillips in our work on anthraquinones and of Dr. L. Siemenkowski in the preparation of bacteriorhodopsin samples. I am also indebted to Mr. T.L. Brack and Mrs. M. Hosta in the preparation of this work. Work described here was supported by grants from the National Science Foundation and the National Institutes of Health and by Syracuse University and the University of Arizona.

References

1. J.F. Durama and A.W. Mantz in Fourier Transform Infrared Spectroscopy, 2 (J.R. Ferraro and L.J. Basile, eds.) Academic Press, New York, N.Y. (1979), p. 1-72.

2. E.B. Wilson, J.C. Decius, and P.C. Cross Molecular Vibrations McGraw Hill, New York, N.Y. (1955).

3. A.C. Albrecht, J. Phys. Chem. 33 (1960) p. 169.

4. 'Time-Resolved Raman Spectroscopy', G.H. Atkinson in Advances in Infrared and Spectroscopy, IX, Chapter 1 (R.J.H. Clark and R.E. Hester, eds.) Heyden and Sons, Inc. (1982) p. 1-62.

5. 'Time-Resolved Raman Spectroscopy', G.H. Atkinson in Advances in Laser Spectroscopy, I, Chapter 8 (B.A. Garetz and J.R. Lombardi, eds.) Heyden and Sons, Inc. (1982) p. 155-175.

6. Time-Resolved Vibrational Spectroscopy, National Science Foundation Workshop Report (G.H. Atkinson, ed.) (1983).

7. Time-Resolved Vibrational Spectroscopy (G.H. Atkinson, ed.) Academic Press, New York, N.Y., (1983).

8. Time-Resolved Vibrational Spectroscopy: Proceedings of the US/Japan Seminar (G.H. Atkinson and S. Maeda, eds.) Gordon and Breach, New York, N.Y. (1986).

9. P.M. Campion and A.C. Albrecht, Ann. Rev. Phys. Chem. 33 (1982) p. 353.

10. D. Blazej and W.L. Peticolas, J. Chem. Phys. 72 (1980) p. 3134.

11. T.A. Ward, G. Levin, and M. Szwarc, J. Am. Chem. Soc. 97 (1975) p. 258.

12. L.R. Dosser, J.B. Pallix, G.H. Atkinson, H.C. Wang, G. Levin, and M. Szwarc, Chem. Phys. Letters 62 (1979) p. 555.

13. H. Shindo and G.H. Atkinson in Time-Resolved Vibrational Spectroscopy (G.H. Atkinson, ed.) Academic Press, New York, N.Y. (1983), p. 191.

14. N.K. Bridge and G. Porter, Proc. Roy. Soc. London A 244 (1958) p. 259.

15. J.N. Moore, G.H. Atkinson, D. Phillips, P.M. Killough, and R.E. Hester, Chem. Phys. Lett. 107 (1984) p. 381.

16. J.N. Moore, G.H. Atkinson, D. Phillips, P. Killough, and R.E. Hester in Time-Resolved Raman Spectroscopy (D. Phillips and G.H. Atkinson, eds.) Gordon and Breach, New York, N.Y. (1985)

17. D.A. Gilmore and G.H. Atkinson in Time-Resolved Vibrational Spectroscopy (G.H. Atkinson, ed.) Academic Press, New York, N.Y. (1983) p. 151.

18. G.H. Atkinson and L.R. Dosser, J. Chem. Phys. 72 (1980) p. 2195.

19. G.H. Atkinson, D.A. Gilmore, L.R. Dosser, and J.B. Pallix, J. Phys. Chem. 86 (1982) p. 2305.

20. W. Siebrand, J. Chem. Phys. 47 (1967) p. 2411.

21. W. Siebrand in The Triplet State (A.B. Zohlan, ed.) Cambridge University Press, Cambridge, England (1967) p. 31.

22. G.H. Atkinson, J.B. Pallix, T.B. Freedman, D.A. Gilmore, and R. Wilbrandt, J. Am. Chem. Soc. 103 (1981) p. 5069.

23. G.H. Atkinson and J.B. Pallix in Laser Chemistry 3 (A. Zewail, ed.) Harwood Academic Publishers, Great Britain (1983) p. 321-332.

24. D. Oesterhelt and W. Stoeckenius, Proc. Natl. Acad. Sci. USA 70, 289 (1973).

25. W. Stoeckenius, R.H. Lozier, and R.A. Bogomolni, Biochim. Biophys. Acta 505, 215 (1979).

26. W. Stoeckenius and R.A. Bogomolni, Ann. Rev. Biochem. 51, 587 (1982).

27. R.H. Lozier, R.A. Bogomolni, and W. Stoeckenius, Biophy. J. 15, 955 (1975).

28. G.H. Atkinson, I. Grieger, and G. Rumbles in Time-Resolved Vibrational Spectroscopy (A. Laubereau and M. Stockburger eds.) Springer Verlag, Berlin (1985).

29. I. Grieger, and G.H. Atkinson, Biochemistry 24 5660 (1985).

30. M. Stockburger, W. Klusmann, H. Gattermann, G. Massig, and R. Peters, Biochemistry 18 4886 (1979).

31. T.L. Brack, G. Rumbles, I. Grieger, D. Blanchard, L. Siemenkowski, and G.H. Atkinson, Biophys. J. (submitted).

32. G. Rumbles, T.L. Brack, D. Blanchard, and G.H. Atkinson, Opt. Comm. (submitted).

33. T.L. Brack, D. Blanchard, L. Siemenkowski, and G.H. Atkinson (unpublished results).

HIGH SENSITIVITY, TIME-RESOLVED ABSORPTION SPECTROSCOPY BY INTRACAVITY LASER TECHNIQUES

George H. Atkinson
Department of Chemistry and
Optical Sciences Center
University of Arizona
Tucson, Arizona 85721

Absorption measurements obtained by intracavity laser spectroscopy (ILS) are described which encompass exceptionally high detection sensitivites while maintaining the characteristics associated with quantitative, time-resolved spectrometry. The optical physics underlying ILS and the instrumentation utilized in obtaining these spectra are described. Absorption spectra recorded by ILS for stable absorbers, transient intermediates formed during reactions, and van der Waals complexes formed at low-temperature in supersonic jet expansions are presented.

Introduction

Absorption spectroscopy has made fundamental contributions to the determination of molecular structure and kinetic properties in both the gas and condensed phase. The rapidly developing interest in reaction dynamics has placed increased emphasis on the detection of transient, reaction intermediates. With the advent of versatile laser systems, however, absorption spectroscopy recorded by classical methods is often not the detection technique of choice since it is significantly less sensitive than techniques such as laser-induced fluorescence (LIF). This disadvantage is particularly important in dynamical studies involving intermediates at low, instantaneous concentrations. The major advantage of absorption spectroscopy derives from its wide applicability to all types of species and experimental conditions. For example, significant changes in sample absorbtivities occur under experimental conditions where emission is efficiently quenched. Absorption spectroscopy could be utilized far more effectively in the study of reaction dynamics if its detection sensitivity was increased while preserving its capability for providing information on molecular structure, kinetic behavior, and quantitative concentrations.

   This paper describes the measurement of absorption spectra by intracavity laser techniques which utilize the properties of a laser resonator cavity to increase the sensitivity of absorption measurements by as much as six orders of magnitude over conventional absorption

207

P. M. Rentzepis and C. Capellos (eds.), Advances in Chemical Reaction Dynamics, 207–228.
© *1986 by D. Reidel Publishing Company.*

spectrometry. This enhanced sensitivity derives from the competition
between the gain in the laser medium and the wavelength-dependent
losses of the laser resonator. By placing the sample of interest
inside the laser resonator, the absorbtivity of the sample becomes one
of the intracavity losses. When the laser reaches threshold, the
losses attributable to the intracavity absorber can be superimposed on
the wavelength-dispersed spectrum of the laser output. The competition
that occurs within the laser system (medium plus resonator) in order to
reach threshold operation leads to enhanced detection of the sample
absorptions. Absorption spectroscopy with such high sensitivity can be
performed under conditions which maintain the information on structure, kinetics, and concentration normally anticipated in absorption
measurements.

The principles of intracavity laser spectroscopy (ILS) and the
application of ILS to the study of stable and transient molecular
species in the gas phase are presented here as a tutorial review
intended to provide an overview of activity in the field. This is not
an exhaustive review in which either a critical evaluation or even a
compilation of existing work is presented. Most of the examples used
here are taken from work in my own laboratory including some very
recent results only now submitted for publication in detail elsewhere.
These data are representative of the type of experimental capabilities
currently available.

ILS instrumentation

The primary instrumental requirements associated with ILS pertain to
the optical resonator of the laser in which the absorber is placed and
especially its mode structure. Both cw and pulsed lasers have been
used for ILS [1]. Most instrumental approaches have been designed to
monitor relatively broad-bandwidth absorption and therefore, the ILS
laser has been operated without tuning elements, relying instead on the
gain properties of the laser medium itself to determine the spectral
bandwidth. These choices of instrumentation have been dictated by
interest in monitoring molecular absorption features such as rotational
bandheads. Since it is important to keep the bandwidth of the ILS
laser significantly larger than that of the intracavity absorber,
untuned laser have been the system of choice. When atomic lines or
other narrow bandwidth absorption features (e.g., single rotational
molecular lines) are to be detected, tuned laser systems can also be
effectively utilized in ILS.

The ILS instrumentation involved in the work described here is
based on the modified optical resonator cavity of a cw jet stream, dye
laser. This apparatus is shown in Figure 1 together with the instrumentation required to photolytically-generate transient intermediates.
The optical cavity of the ILS laser is defined by mirrors M_1, M_2, and
M_3. The last mirror has been positioned to create a 1 m extended
optical cavity into which the sample cell is placed. The dye laser is
pumped by an argon ion laser operating at powers typically 1-2 times
that required for the threshold conditions of the dye laser (i.e., 2-3

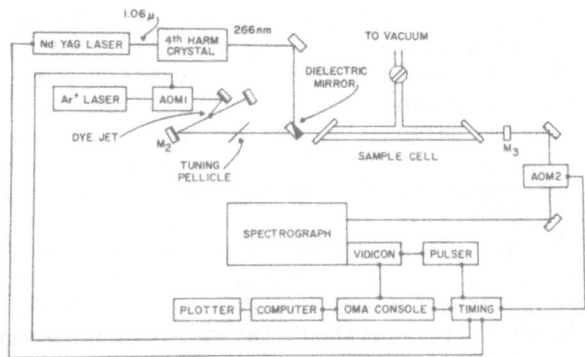

Figure 1: Instrumentation used for intracavity laser spectroscopy. Nd:YAG laser used separately to photolytically generate transient absorbers. See text for detailed discussion.

W all lines or 514.5 nm). The pumping radiation passes through an acousto-optic modulator (AOM1) prior to reaching the dye laser in order to modulate the pumping intensity. When a pulsed radio frequency voltage is applied to AOM1, the intensity of pumping radiation is reduced below that required to keep the dye laser above threshold and as a result, lasing action is terminated. As the voltage applied to AOM1 is reduced, the pumping power regains its maximum value in < 0.5 µs and the dye laser recovers full power within 1-2 µs. This general timing sequence is shown schematically in Figure 2. The dye laser radiation itself passes through a second AOM2 before reaching a detection system. The transmission of AOM2 is synchronized with that of AOM1 in order to control the time interval during which dye laser radiation is detected and thereby to determine the generation time, t_g (Figure 2).

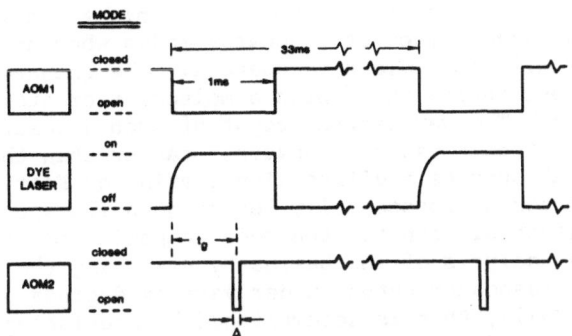

Figure 2: Time sequences illustrating the use of AOM1 to control dye laser activity and AOM2 to sample the output of the ILS dye laser. Also see Figure 1.

Spectral dispersion of the dye laser output is obtained with a monochromator. Multichannel detectors have been successfully used to monitor the dispersed ILS signal by placing the detector face in the

focal plane of the monochromator. The spectral region observed at any
one time, together with the resolution of the ILS spectrum itself, is
determined by the dispersion of the monochromator and the density of
pixels on the multichannel detector. Typically, a 1 m monochromator
operating with a 316 line/mm grating in an echelle mounting (7th
order), together with an intensified vidicon camera, can provide a
resolving power of about 90,000:1 [2]. Recently designed double-pass
monochromators operating with reticon detectors have demonstrated resolution of >800,000:1 [3].

Attention must be given to the optical elements that appear within
the dye laser cavity in order to avoid interferometric effects. Wedged
optical elements or efficient antireflective coatings are required to
obtain dye laser outputs without intensity modulations arising from
intracavity interferometers. A pellicle (uncoated) is commonly used
inside the resonator in order to maintain the dye laser at a central
maximum, but it does not significantly narrow the bandwidth of the
laser. The role of the extracavity Nd:YAG laser in Figure 1 will be
described in the section dealing with the time-resolved ILS detection
of transients.

Principles of ILS

The optical properties of a laser resonator, including those of the
lasing medium, have been analyzed extensively for many purposes. For
this discussion, it is important only to note that the wavelength-
dispersed profile of the output of the laser will be used to characterize the properties of a specific laser resonator under a well-
defined set of experimental parameters (e.g., pump power, laser dye
bandwidth, optical reflectivities, etc.). The mode competition within
the laser resonator is the prime factor determining the output profile
once the general experimental parameters have been chosen. Tunable dye
lasers have been of major interest in this field since they typically
contain a large number of resonator modes when untuned and they are
experimentally convenient to operate. The untuned dye laser provides
an excellent device into which a molecular or atomic absorber can be
introduced. The wavelength output of such a laser is thus modified or
"tuned" by the intracavity absorber rather than the conventional optical devices such as a diffraction grating or Fabry-Perot etalon.

In order to successfully record such ILS spectra, it is necessary
to quantitatively control the mode competition within the laser resonator. The analysis of ILS begins by examining the spectral profile of
the laser resonator output under varying degrees of mode competition.
Experimentally, this is accomplished by modulating the input pumping
power delivered to the dye laser with an acousto-optic modulator (AOM1
in Figure 1). When the output of the dye laser is also modulated prior
to detection by a second modulator (AOM2 in Figure 1), a selectable
portion of the mode competition can be monitored by controlling the
time delay between the two AOM devices (Figure 2). The resulting
spectral profiles are presented in Figure 3 for a variety of delay or
generation times (t_g). The distinct changes of the output profile

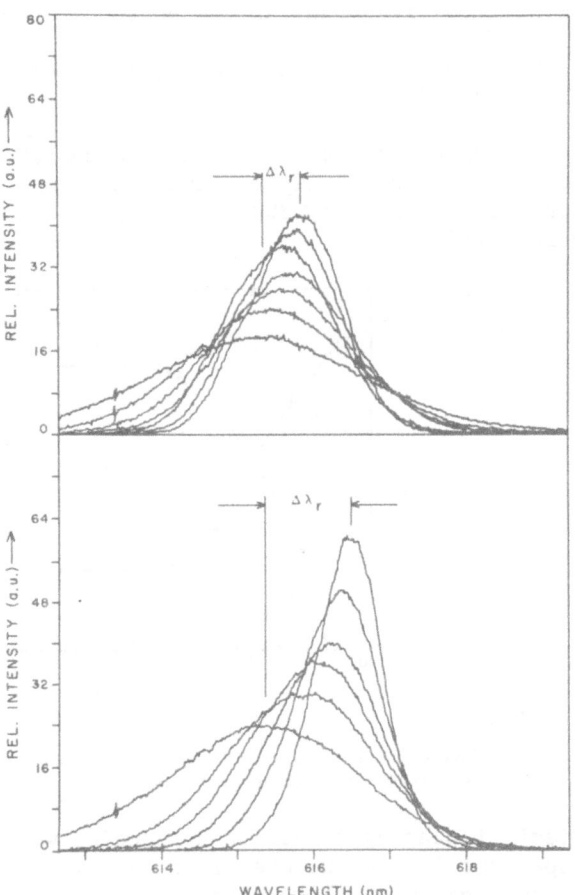

Figure 3: Generation spectra of a quasi-cw dye laser recorded for different t_g values and 514.5 nm argon-ion laser pumping powers. Top panel: 2 W pumping power: t_g = 10, 20, 30, 40, 60, 80, and 100 μsec for weakest to strongest curves, respectively; $\Delta\lambda_r$ = 0.26 nm. Bottom panel: 4 W pumping power; t_g = 10, 20, 30, 40, 60, and 100 μsec for weakest to strongest curves, respectively; $\Delta\lambda_r$ = 0.58 nm.

height and width as well as the pronounced shift of wavelength maxima to the red have been quantitatively described in terms of expression (1) [4]. The limited period over which the spectral profile follows

$$I(\sigma,t) = \frac{I'_0}{\Delta\sigma_0} \left(\frac{\gamma t_g}{\pi}\right)^{1/2} \exp\left[-\left(\frac{\sigma-\sigma_0}{\Delta\sigma_0}\right)^2 \gamma t_g\right] \qquad (1)$$

where: $I'_0 \equiv$ total laser intensity,
$t_g \equiv$ generation time (Fig. 2),
$\gamma \equiv$ losses in the optical cavity,
$\sigma \equiv$ spectral frequency (cm^{-1}),

$\sigma_0 \equiv$ spectral frequency at the center of the laser gain profile, and

$\Delta\sigma_0 \equiv$ bandwidth related to the total number of initial cavity modes.

this behavior is determined primarily by the mechanical stability of the resonator cavity, especially the dye jet. In a jet stream dye laser this period is typically 300 to 400 μs as demonstrated by the data in Figure 4 [4].

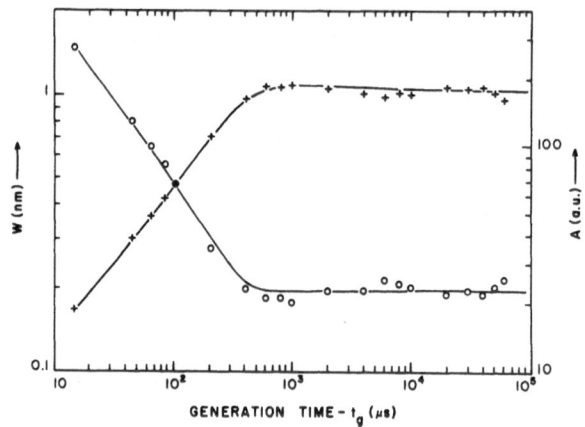

Figure 4: Spectral width (0) and amplitude (+) of generation spectra as a function of generation time (t_g).

The effect of placing an absorbing species with a well-defined absorption spectrum inside the dye laser cavity can now be determined on the basis of expression (1). When the absorbtivity of the intra-cavity absorber is small and its bandwidth is significantly narrower than the bandwidth of the ILS laser, then the spectral profile of the resulting dye laser output undergoes minimal perturbation. There exist thresholds, of course, for both absorbtivity and absorber bandwidth beyond which the dye laser output is significantly modified and the ILS information is not a good representation of the absorber properties. Examples of both cases are given in Figure 5 for intracavity absorption arising from the transient HCO [4]. For the first case, the absorption features due to the intracavity absorber obey a Beer-Lambert relation-ship. The overall, quantitative description of the ILS laser profile with an absorber present is given by expression (2) [4].

$$I(\sigma,t) = \frac{I'_0}{\Delta\sigma_0} \left(\frac{\gamma t_g}{\pi}\right)^{1/2} \exp\left[-\left(\frac{\sigma-\sigma_0}{\Delta\sigma_0}\right)^2 \gamma t_g - \alpha(\sigma)\frac{\ell}{L}c\,t_g\right] \quad (2)$$

where $\alpha(\sigma)$ is the absorption coefficient describing the intracavity loss, ℓ/L is the ratio of the absorption pathlength to the total length of the laser cavity, and c is the velocity of light. Quantitative absorption data can be extracted from this expression when:

$$\left[\alpha(\sigma)\,\frac{\ell}{L}\,c\,t_g\right] < 1 \tag{3}$$

and it is under these conditions that the ILS measurements described here are made.

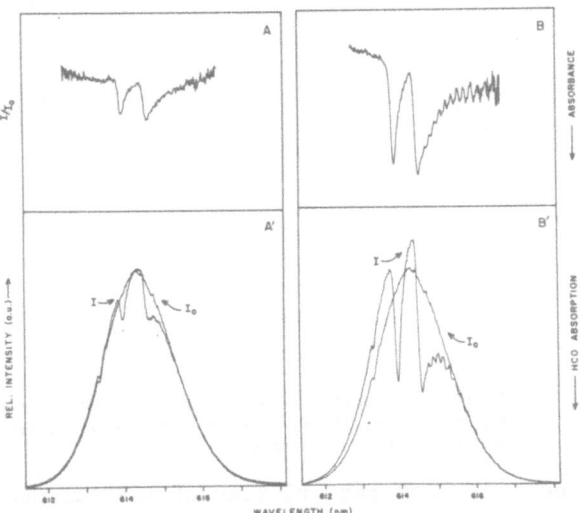

Figure 5: ILS absorption spectra for small (A) and (A') and large (B) and (B') concentrations of HCO.

Results

Absorption measurements have been made by ILS for stable molecules in bulk samples [5-7], transient intermediates generated by photolysis [8-14], atoms in thermolyzed beams [15,16], metastable species in discharges [17], and van der Waals (vdW) complexes formed in low temperature supersonic jet expansions [18,19]. The breadth of these applications is a reflection of the versatility of the fundamental optical physics underlying ILS. The high sensitivity of ILS makes it feasible to detect low absorber concentrations and exceptionally small absorbtivities. Indeed, some of the smallest absorbtivities ever reported have been obtained by ILS [6,7,18]. One quantitative measure of the detection sensitivity of ILS can be obtained through the effective pathlength for absorption. For a t_g of 250 µs and an intracavity sample cell length of 0.5 m, the effective absorption pathlength is >30 kilometers.

Stable molecules

Absorption transitions in the visible for H_2O [6] and O_2 [7] are

readily detectible by ILS even though these absorbtivities are extreme-
ly weak. The $b^1\Sigma_g^+ - A^3\Sigma_g^-$ transition or γ bands in O_2 near 628 nm are
observed in the ILS spectrum presented in Figure 6. The γ bands of O_2
were quantitatively fit with expression (2) and the results were used
to obtain line strength and collisional self-broadening data [6].

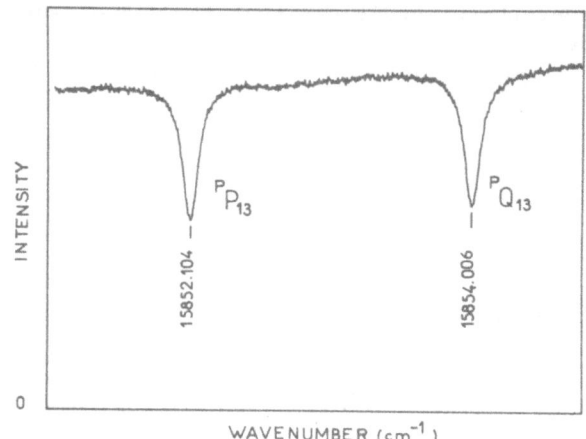

Figure 6: A portion of the oxygen γ band for pure O_2 at 0.847 atm
ℓ_{eq} = 3.64 km, 296 K. The spectrum is not corrected from the
photodiode array response.

The quantitative characteristics of ILS for measuring absorber
concentrations directly also were examined for NO_2 using a flashlamp-
pumped dye laser and vidicon camera detection [5]. Several NO_2 absorption features in the 590-593 nm region were selected for study. The
integrated band intensities of these NO_2 features were then measured as
a function of the NO_2 sample pressures over more than an order of
magnitude (Figure 7). The linearity of this dependence was an experimental demonstration that relative concentrations of intracavity
absorbers can be measured directly from ILS as long as the sample
absorbtivity remains small (i.e., expression (3)). Above a certain
value, nonlinear absorbtivity occurs as illustrated by the change in
slope in Figure 7.

The data for both O_2 and NO_2 demonstrate that ILS provides the
same general characteristics exhibited in Beer-Lambert relationships
describing classical absorption spectrophotometry. ILS can be used in
applications completely analogous to those of classical absorption if
the linear response range is utilized. The use of a multichannel
detector (i.e., using the multiplex wavelength advantage) with linear
response is, of course, a critical experimental improvement over photographic detection which itself requires some calibration.

Transient molecules and radicals

The initial detection of transient species by ILS was reported in 1973
[8] when a pulsed, flashlamp-pumped dye laser was used to record the

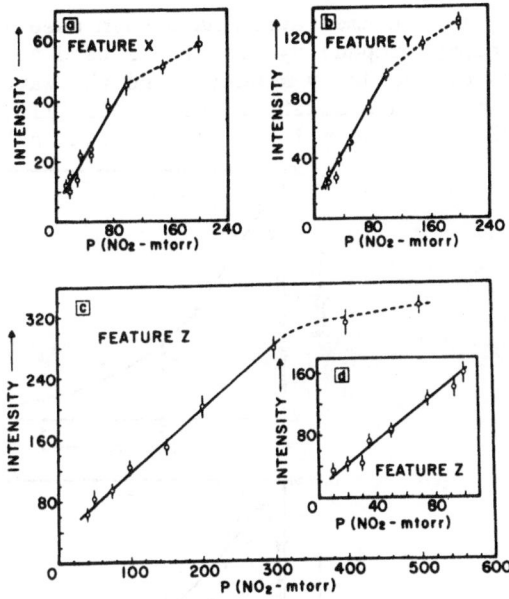

Figure 7: Quantitative ILD (590 nm region): The integrated intensities of spectral features labeled X, Y, and Z for the 590 nm region are plotted versus intracavity NO_2 pressure.

absorption spectrum of the HCO and NH_2 radicals. Both radicals were photolytically generated when a capacitively-driven nitrogen discharge produced a visible-ultraviolet continuum. The parent molecules used as precursor were acetaldehyde and formaldehyde for HCO and ammonia for NH_2. The pulse width of the flashlamp-pumped dye laser (0.3 μs) used for ILS determined the effective generation time. Detection was obtained on photographic plates after the output of the ILS laser transversed a 2 m grating spectrograph. These early results demonstrated the high sensitivity for absorption measurements that are inherent in ILS by detecting both HCO and NH_2 at concentrations several orders of magnitude lower than previously reported and with an optical pathlength of 0.75 m. Photographic detection did not facilitate rapid, quantitative ILS measurements, but it did provide the basis for high resolution spectroscopy. No discrepencies where found in the positions of absorption features measured by ILS [8]. The incorporation of multichannel detectors (as demonstrated for ILS of static NO_2 [5]) provided the linearity of response and time-resolution needed to obtain quantitative data for transients.

The spectroscopy and decay kinetics of the formyl radical, HCO, generated in the laser photolysis of acetaldehyde were studied by ILS [8-14]. The HCO radical is difficult to detect by nonabsorptive methods because its excited $^1A''$ state is predissociated. The quantum yield of fluorescence from $^1A''$ acetaldehyde is very low and as a result, LIF has not been used effectively to monitor HCO even though much of its absorption spectroscopy lies in the visible where powerful laser sources are available.

Two types of instrumentation have been used in ILS studies of HCO. A pulsed (flashlamp-pumped) dye laser was used to record HCO absorption in measurements of the decay kinetics and the wavelength dependence of HCO formation in the acetaldehyde photolysis [11,12]. The decay kinetics of HCO can be seen in the data shown in Figure 8 for two

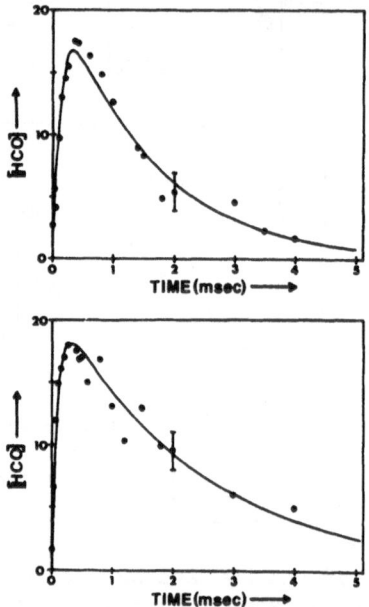

Figure 8: The time dependent concentration of HCO (0,0,0) following the excitation of acetaldehyde/argon (top panel) and acetaldehyde/cyclohexane (bottom panel) mixtures. The mixing ratios were 0.2 Torr acetaldehyde to 2.0 Torr of added gas.

sample conditions (with either argon or cyclohexane as an added gas) and with flashlamp excitation of acetaldehyde. The photolyzing flash-lamp was placed adjacent to the sample cell shown in Figure 1 (no Nd:YAG was used in this work). The decay rates can be attributed to the number of HCO collisions with either argon or cyclohexane. The collisional cross-section for each added gas was measured from these data [12]. The solid curves shown in Figure 8 are best fits to a kinetic mechanism describing the formation and decay of HCO as first order processes. The capability to detect HCO also makes it feasible to determine the energetic onset for HCO formation in the dissociation of acetaldehyde [11]. The photolysis source was a flashlamp-pumped dye laser which was used to conjunction with nonlinear optics to generate tunable ultraviolet radiation in the 260 to 340 nm range. The ultra-violet photolysis radiation was optically-coupled into the resonator cavity with a dielectrically-coated intracavity mirror. The relative quantum yield for the production of HCO (0,0,0) from acetaldehyde is presented in Figure 9 together with a low-resolution absorption spec-trum of acetaldehyde. The wavelength of the ILS laser was tuned to overlap absorption transitions originating in only the HCO (0,0,0) level, a capability which is expanded upon in results described below.

The sharp energetic onset for HCO (0,0,0) formation signals the population of excited states which efficiently dissociate via a radical channel.

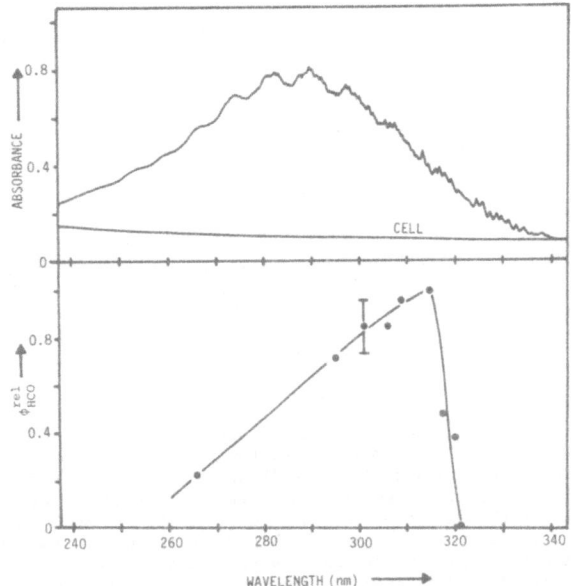

Figure 9: Upper panel: Low resolution absorption spectrum taken on a Cary 118 spectrometer for a 1000 Torr cm sample. Lower panel: Relative HCO (0,0,0) quantum yields (ϕ_{HCO}^{rel}) versus excitation wavelength. (Refer to text for details.)

A more complete view of the absorption spectroscopy and decay kinetics of HCO can be obtained by using a cw dye laser for ILS which has a well-controlled degree of intracavity mode competition [13,14]. The instrumentation used is shown in Figure 1. The Nd:YAG laser was used with nonlinear optics to generate the 266 nm radiation required to dissociate acetaldehyde into HCO. In order to obtain real-time kinetic data, the photolysis event must be synchronized with the timing sequence of the generation time. The schematic representation of the overall timing sequence is presented in Figure 10 where the real delay time, t_d, is defined. These definitions are consistent with those described in the instrumentation section (Figure 2).

The absorption spectroscopy of HCO (0,0,0) is presented in Figures 11 and 12 [13]. The excellent signal-to-noise ratios in these spectra permit even the rotational broadening caused by the shortening of excited-state lifetimes to be readily observed in the P(8) to P(14) linewidths (Figure 12). Ro-vibronically hot HCO can also be observed as illustrated by the spectra shown in Figure 13 where absorption bands from HCO (0,0,0) are also observed. These overlapping bands can be separated by recording ILS spectra at different delay times where the relative concentrations of the excited and ground-state HCO are different. An example of these data is presented in Figure 14.

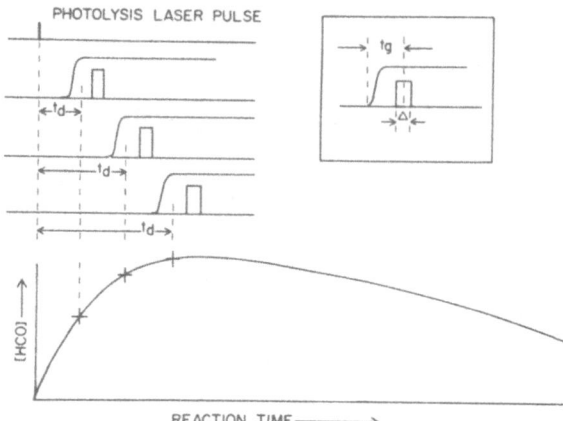

Figure 10: Timing sequence for ILS spectroscopy. The pulsed photolysis radiation at 266 nm initiates the timing. The initial rise in dye laser radiation exiting mirror M_3 (Fig. 1) is shown in the insert to define the beginning of the generation time, t_g. The termination of t_g is defined to be in the middle of the pulse used to operate AOM2 vidicon (shown in insert as a square wave pulse) which has a duration Δ. The time delay relative to the photolysis pulse that has kinetic significance is defined as t_d for three different values. The t_g for each of the three examples remains the same. The three t_d values are used to obtain data points on the curve schematically representing HCO kinetic shown at the bottom.

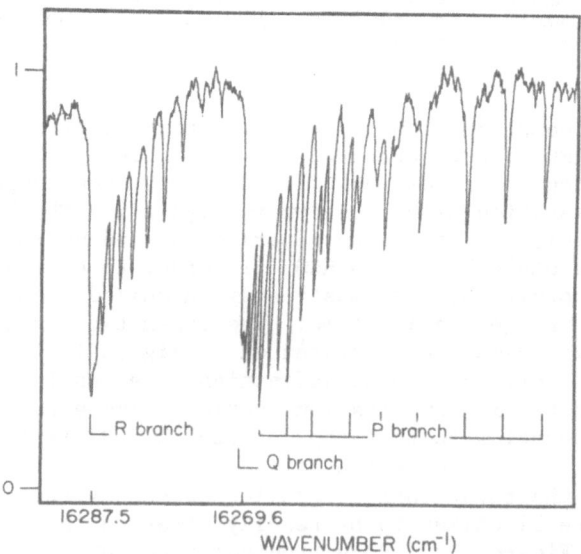

Figure 11: ILS absorption spectrum of the (0,9,0) ← (0,0,0) transition of HCO following the 266 nm photolysis of 0.1 Torr CH_3CHO and 10 Torr Ar. t_g = 12.5 µs and t_d = 20 µs.

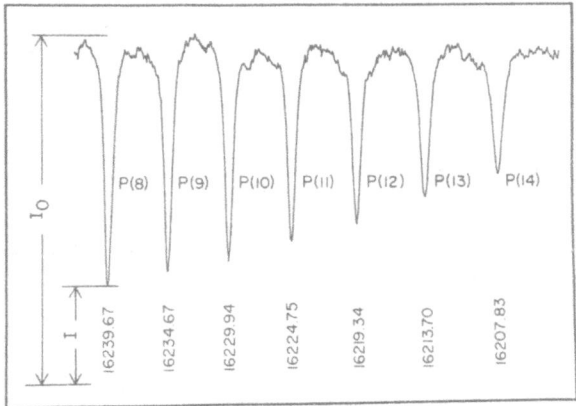

Figure 12: ILS absorption spectrum of the P(8) to P(14) lines in the (0,9,0) ← (0,0,0) transition of HCO formed in the 266 nm photolysis of 0.1 Torr CH_3CHO and 10 Torr Ar. The definitions of I_0 and I are also shown. The I_0/I values for the P(8), P(9), and P(10) lines were averaged to obtain the relative concentrations of HCO (0,0,0) discussed in the kinetic analysis.

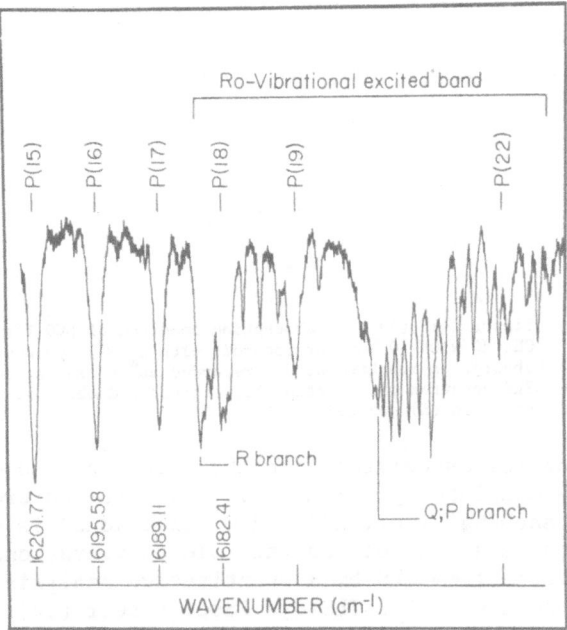

Figure 13: ILS absorption spectrum of several P-branch lines in the (0,9,0) ← (0,0,0) transition and an unidentified, vibrationally excited level of HCO formed in the 266 nm photolysis of 0.1 Torr CH_3CHO and 10 Torr Ar. The P-branch lines from HCO (0,0,0) and a part of the $\tilde{P}$, Q, and R branches for the vibrationally excited HCO are labelled. t_g = 12.5 μs and t_d = 20 μs.

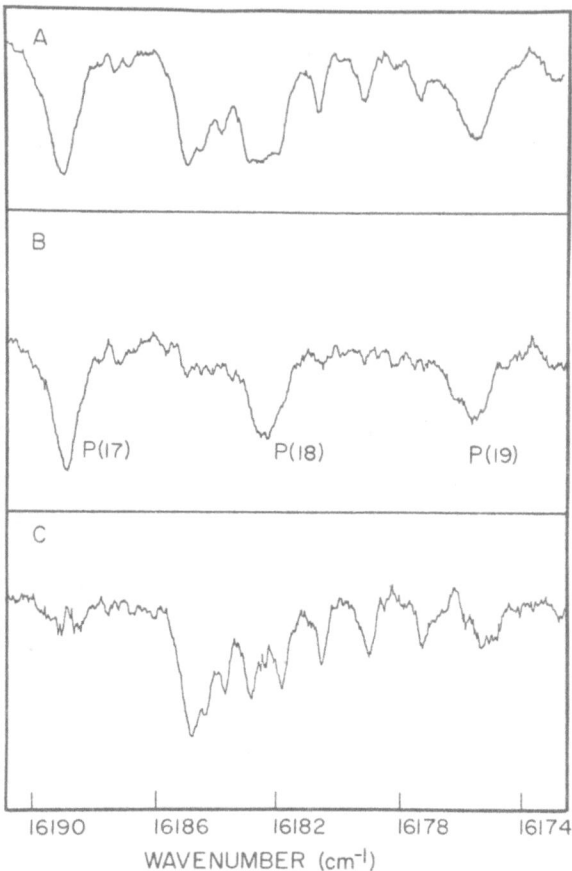

Figure 14: (A) ILS absorption spectrum of HCO for a sample of 0.1 Torr
CH_3CHO with 10 Torr Ar recorded with t_g = 20 μs and t_d = 10 μs. (B) ILS
absorption spectrum of HCO recorded as in (A) except t_d = 20 μs. (C) The
ILS spectrum of vibrationally excited HCO obtained by ratioing the spectra
shown in (A) and (B).

The kinetics associated with the formation and decay of HCO
(0,0,0) were examined by ILS as well. The second-order decay resulting
from self-quenching (i.e., HCO + HCO collisions) is shown in Figure 15.
The formation rates as well as the role of vibrationally-excited HCO
can be extracted from ILS by a quantitative analysis of the wavelength-
dispersed spectrum of the ILS dye laser itself [14]. Computational
models can be derived from expressions (1) and (2) for several different conditions [14]. The deviations from the modelled behavior can
be attributed to kinetics of HCO (0,0,0) formation rates. By monitoring absorption transitions originating in vibrationally-excited levels,
the contributions of HCO (0,1,0), HCO (0,0,1), and HCO (0,2,0) have
been elucidated. ILS absorption spectra of two vibrationally-excited

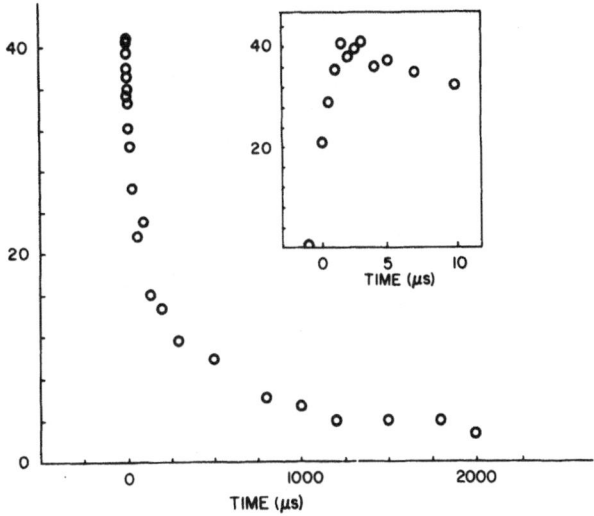

Figure 15: The time-dependent concentation of HCO (0,0,0) in absorbance units, α(σ), measured for the 266 nm photolysis of 10 Torr CH₃CHO. The data are plotted on two time scales in order to illustrate the short- and long-time behavior. Data are convoluted with the 1 μs time-resolution of the intracavity measurements made in this work.

HCO levels are shown in Figure 16. For the case of 266 nm excitation of acetaldehyde, about one-half of the HCO formed appears in vibration-ally-excited levels while the remaining is formed directly in the vibrationaless level [14]. The time dependence observed by ILS is different for each of the vibrational levels of HCO detected.

Metastable intermediates in discharges

The real-time measurement of absorption spectroscopy in highly reactive environments such as those created by discharges or plasmas has long been utilized to detect metastable intermediates that play important roles in chemistry. Perhaps no example of reactive discharges has had more commercial impact than that involving silane chemistry through its role in the deposition of silicon films. The gas phase chemistry assoc-iated with these deposition processes remains surprisingly unknown.

ILS was used recently to measure the absorption spectroscopy in microwave-driven discharges of SiH₄ under conditions which result in the deposition of silicon films. Of particular interest was the detec-tion of SiH₂ which is widely thought to act as a principal intermediate in the preparation of high quality amorphorous silicon. The ILS spec-trum obtained during deposition is presented at the bottom of Figure 17, together with assignments of rotational lines [17]. The absorp-tion bands of SiH₂ are labelled as are features due to excited state - excited state absorption in the carrier gas argon and in H₂. The identification of the metastable absorption are readily made by obtain-ing ILS spectra of discharges of a pure material as is illustrated at

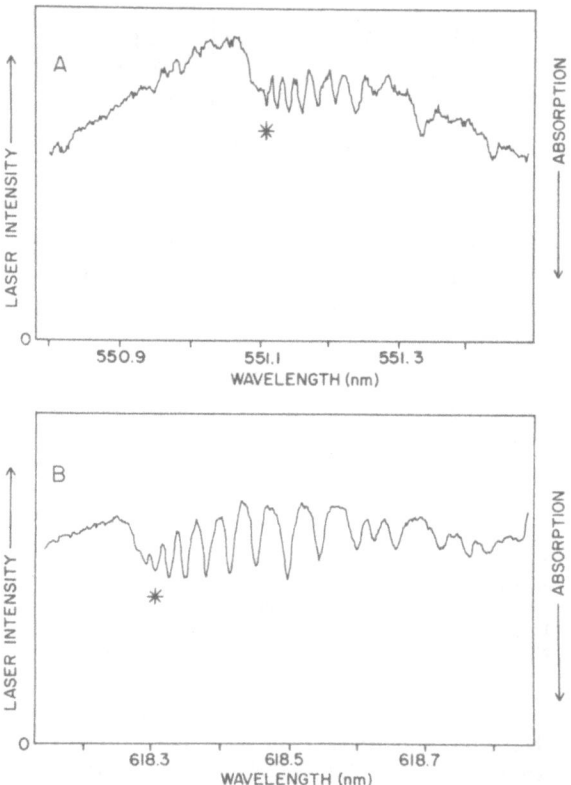

Figure 16: ILS spectra of the PQ bandheads of vibrationally-excited HCO.
(A) $(0,13,0) \leftarrow (0,1,0)$ absorption band (t_g = 10 μs, 2.3 Torr of
acetaldehyde and laser fluence at 266 nm ≅ 5 mj). (B) $(0,9,1) \leftarrow (0,0,1)$
band (t_g = 12 μs, 500 mTorr of acetaldehyde, and laser fluence at
266 nm ≅ 5 mj). Spectral features monitored in kinetic studies are
indicated by *.

the top of Figure 17 for argon and H_2 [17]. The ILS absorption spectrum of SiH_2 was recorded for much lower concentrations of the radical
than previously reported [20] and yet contains all of the same spectroscopic information. When recorded at higher spectral resolution, the
ILS spectrum should provide the basis for an improved structural analysis of SiH_2 relative to those currently available [20]. As importantly, ILS makes it feasible to monitor the relative concentrations of
SiH_2 in real-time while the discharge deposits the silicon film and
thereby, provides a spectroscopic method for optimizing characteristics
of the deposition process. The high sensitivity, time-resolved properties of ILS is the basis for these measurements.

Supersonic jet expansions

The formation of weakly-bound van der Waals (vdW) complexes has been

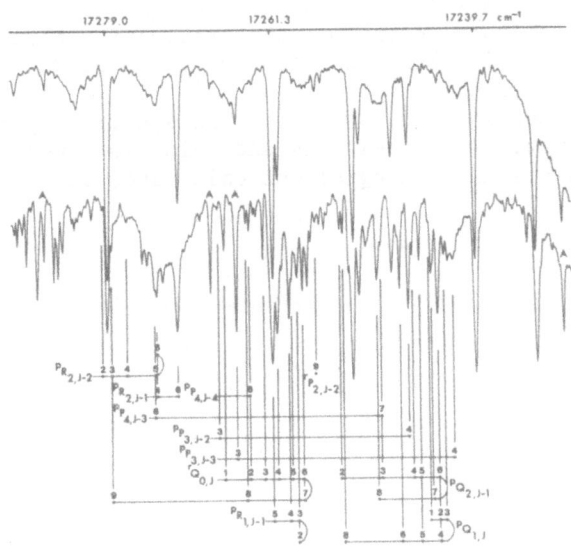

Figure 17: ILS absorption spectrum of silylene, SiH_2, in the region of the 579.6 nm $(0,2,0)'$ - $(0,0,0)''$ absorption spectrum of SiH_2. The SiH_2 was produced by microwave discharge dissociation of 4% silane in argon at a total pressure of 0.7 Torr. Assignment markers are for all bands in this region listed by Dubois in the Depository for Unpublished Data, National Science Library, NRC, Ottawa. The top spectrum indicates transitions due to absorption by electronically excited H_2 and was obtained from a microwave discharge of 13% H_2 in argon at a total pressure of 0.18 Torr.

accomplished in rare-gas matrices at low temperature and has led to extensive studies of their structures and reactivities. The development of supersonic jet technology supplanted much of that effort by making it feasible to prepare vdW complexes and ultracold (< 10 K rotational temperatures) molecules in the gas phase. Based on the rapid expansion of a high pressure gas (typical 10 to 100 atm) through a small orifice (typically 100 μ) into an evacuated chamber (e.g., 10^{-5} torr), supersonic jet expansions can be used to generate a wide range of vdW complexes from noble gas and molecules as well as metal clusters and ultracold polyatomic molecules. The principle experimental methods for detecting these species in jets have been based on phenomena occurring subsequent to absorption in the species itself. LIF or multiphoton ionization, for example, have been successfully used. Some direct absorption work has been done, but due to the exceptionally small pathlengths for absorption in normal jets (typically 100 μ), modifications were made in the nozzle design to obtain planar expansions with dimensions of several centimeters [21-23]. Recently, ILS has been used with jet expansions to record the absorption spectra of ultracold NO_2 [18] and vdW complexes of I_2 with He, Ar, Kr, and Xe [19].

The instrumentation is the same as that presented in Figure 1 except that the nozzle region of the jet expansion is incorporated inside the dye laser resonator (the extracavity photolysis laser is also not used). The ILS absorption spectrum of NO_2 is shown in Figure 18 for two rotational temperatures [18]. The pathlength of the circular jet nozzle used was 75 μ. The spectroscopic information in the form of wavenumber positions and relative intensities of the rotational bands are in good agreement calculated values.

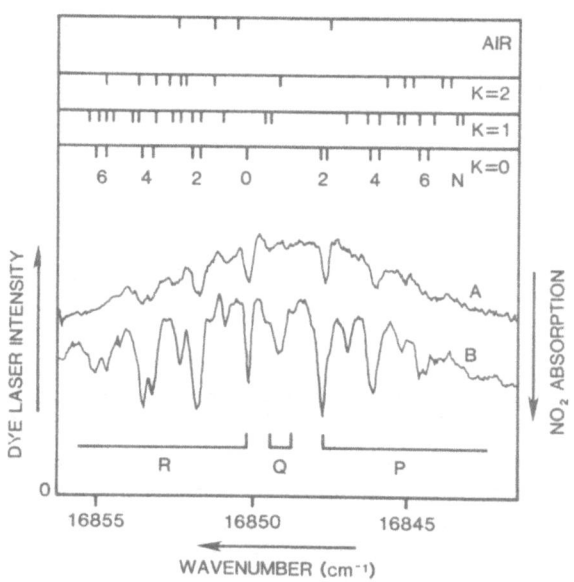

Figure 18: Central portion of dispersed dye laser radiation containing the 594 nm band of jet-cooled NO_2 under different expansion conditions: (A) ≈ 4 K and (B) ≈ 15 K.

The same type of instrumentation was used to record the absorption spectra of I_2, I_2He, and I_2Ar shown in Figure 19 for two sets of expansion conditions [19]. These are the first such absorption spectra of vdW complexes formed in a circular jet. The pathlength of absorption was 500 μ or less. The availability of absorption data is especially important since it provides a direct measurement of relative ground-state populations. With some additional information on transition probabilities for absorption, these ground-state populations measure the relative efficiencies for forming vdW complexes. These same data are essential for determining excited-state dynamics in vdW complexes. An example is provided by the I_2He and I_2Ar cases. Both ILS absorption and LIF spectra can be recorded for the same expansion containing I_2He and I_2Ar as illustrated by the data presented in Figure 20. The relative intensities of the I_2He and I_2Ar features are reversed on going from the ILS to the LIF spectra. This difference is a reflection of changes in the dissociation mechanism of the two vdW complexes. The

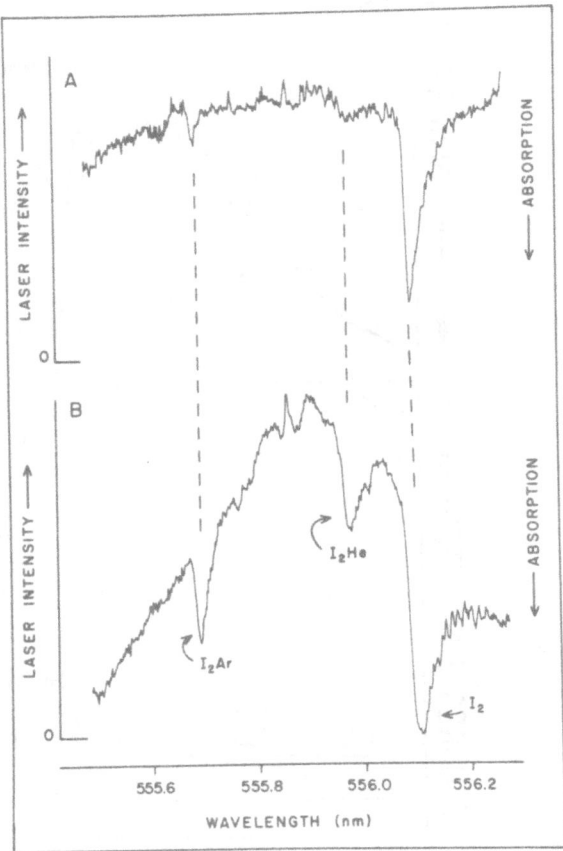

Figure 19: ILS absorption spectra of I_2 and I_2He, and I_2Ar van der Waals complexes. These B → X (21,0) transitions were monitored 1 cm from the orifice of a 0.5 mm pulsed nozzle. The 240 psi backing pressure of He was seeded with 1% Ar and 10 ppm I_2. (A) t_g = 10 μs and (B) t_g = 100 μs.

ILS spectrum clearly demonstrates that the I_2Ar complex is more effectively formed under these expansion conditions and yet the LIF signal from I_2Ar is significantly smaller than that from I_2He. The dynamical decay of I_2Ar apparently undergoes dissociation by more than one pathway, at least one of which does not generate a product seen in LIF. This analysis of the excited state decay mechanism of I_2Ar can be demonstrated with absorption data such as those provided by ILS [19].

Conclusion

The performance characteristics of ILS and the experimental techniques which control them are now well understood. It provides exceptionally high sensitivity, time-resolved detection via absorption spectroscopy.

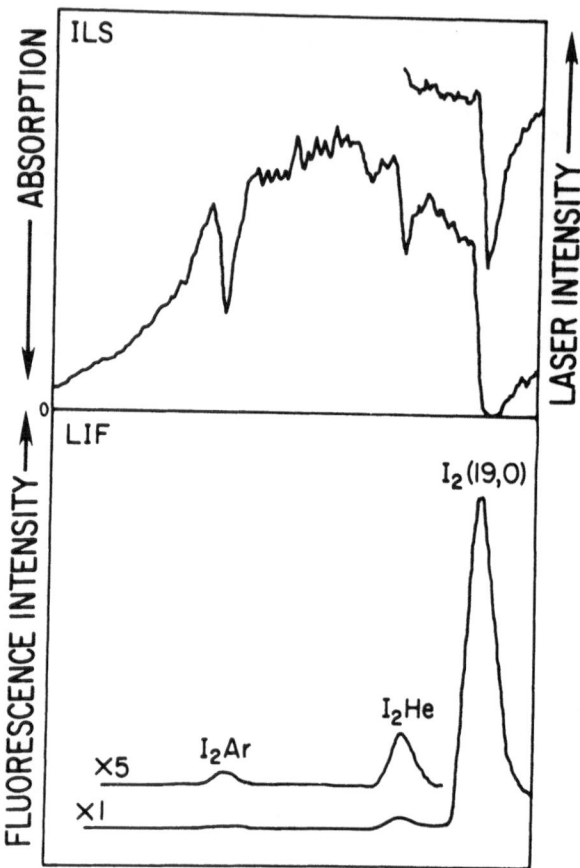

Figure 20: Absorption and LIF spectra of the same expansion. Expansion conditions are the same as in Figure 19. B → X (19,0) transitions of I_2, I_2He, and I_2Ar are shown. Note the lower relative detection sensitivity of LIF for I_2Ar than for I_2He or I_2.

It can be utilized to obtain quantitative information on absorber concentrations and can be adapted for the absorbtive detection of species with transient populations or weakly-bound complexes formed at low temperatures over 10^{-5} m pathlengths. Although ILS has already been the basis for a wide variety of studies in spectroscopy and chemical dynamics, its recent emergence as a mature experimental method should result in greatly expanded applications.

Acknowledgements

I would like to express my sincere appreciation to the students and researchers in my laboratory who have worked on ILS. These include Mr. T.N. Heimlich, Dr. R. Gill, Mr. W.D. Johnson, Prof. M. Schulyer, Prof.

M. Schuh, Mr. T.L. Brack, Dr. N. Goldstein, Mr. W. Maynard, and Dr. J. O'Brien. I am especially indebted to Prof. F. Stoeckel for his collaboration on several projects involving ILS and for many stimulating discussions. I am also indebted to Dr. N. Goldstein and Mrs. M. Hosta for their assistance in preparing this paper. This research has been supported by grants from the Research Corp., the Petroleum Research Fund of the American Chemical Society, the U.S. Army Research Office, and the National Science Foundation. Support was also provided by Syracuse University and the University of Arizona.

References

1.   T.D. Harris "Laser Intracavity-Enhanced Spectroscopy", in Ultrasensitive Laser Spectroscopy, (D. Klinger ed.) Academic Press, New York) 1983.

2.   N. Goldstein, T.L. Brack, and G.H. Atkinson, Chem. Phys. Lett. 116 223 (1985).

3.   F. Stoeckel (unpublished results)

4.   F. Stoeckel and G.H. Atkinson, Appl. Optics 24 3591 (1985).

5.   G.H. Atkinson, T.N. Heimlich, and M.W. Schuyler, J. Chem. Phys. 66 5005 (1977).

6.   F. Stoeckel, M.A. Melieres, and M. Chenevier, J. Chem. Phys. 76 2191 (1982).

7.   M. Chenevier, M.A. Melieres, and F. Stoeckel, Opt. Commun. 45 385 (1983).

8.   G.H. Atkinson, A.H. Laufer, and M.J. Kurylo, J. Chem. Phys. 59 350 (1973).

9.   J.H. Clark, C.B. Moore, and J.P. Reilly, Intern. J. Chem. Kinetics 10 427 (1978).

10. J.P. Reilly, J.H. Clark, C.B. Moore, and G.C. Pimentel, J. Chem. Phys. 69 4381 (1978).

11. R.J. Gill and G.H. Atkinson, Chem. Phys. Lett. 64 426 (1979).

12. R.J. Gill, W.D. Johnson, and G.H. Atkinson, Chem. Phys. 58 29 (1981).

13. F. Stoeckel, M.D. Schuh, N. Goldstein, and G.H. Atkinson, Chem. Phys. 95 135 (1985).

14. N. Goldstein and G.H. Atkinson, Chem. Phys. (in press).

15. P. Kumar, G.O. Brink, S. Spence, and H.S. Lakkaraju, Opt. Commun. 32 (1980) 129.

16. G.O. Brink and S.M. Heider, Opt. Lett. 6 (1981) 366.

17. J. O'Brien and G.H. Atkinson, Chem. Phys. Lett. (submitted for publication).

18. N. Goldstein, T.L. Brack, and G.H. Atkinson, Chem. Phys. Lett. 116 223 (1985).

19. N. Goldstein and G.H. Atkinson, J. Chem. Phys. (submitted for publication).

20. I. Dubois, Can. J. Phys. 46 (1968) 2485.

21. A. Amirav, M. Sonnensheim, and J. Jortner, Chem. Phys. 88 199 (1984).

22. A. Amirav, U. Even, and J. Jortner, Chem. Phys. Letters 94 545 (1983).

23. A. Amirav, U. Even, and J. Jortner, Chem. Phys. Letters 95 295 (1983).

DYNAMICS OF EXCITED STATES OF AN ATOM INTERACTING WITH STRONG LASER
FIELDS

Constantine Mavroyannis
Division of Chemistry
National Research Council of Canada
Ottawa, Ontario, Canada K1A 0R6

ABSTRACT. We have considered the dynamic effects arising from the
interaction of a three-level atom in the "V" configuration with two
strong laser fields, whose initially populated modes are resonance with
the two atomic transition frequencies. One-, two- and three-photon
processes have been considered and the corresponding spectral functions
have been calculated in the limit of high photon densities of both laser
fields. Numerical calculations have been made for selected values of
the Rabi frequencies of the two laser fields, and the computed results
have been graphically presented and discussed. Comparison has been made
between the results derived when the laser fields are quantized, and
photon correlations have been considered in the limit of high photon
densities with those obtained when the laser fields are treated
classically; the latter treatment is identical to those derived by the
dress-atom method. It is found that the appearance of new sidebands,
for which there is no classical analog, reveals the boson character of
the photon fields when the laser fields are quantized taking into
account photon correlations in the limit of high photon densities. The
merit of using the latter method is that it yields results describing
both the classical as well as the quantum nature of the photon fields
involved.

INTRODUCTION

With the advent of the tunable dye laser, theoretical and experimental
interest has been directed towards the interactions of atoms and
molecules with strong resonant electromagnetic fields and nonlinear
optical processes. A strong laser field is materialized when the rela-
tion $\Omega^2 >> \gamma_0^2$ is fulfilled, where Ω is the Rabi frequency defined as (1)

$$\Omega = \vec{\mu}_{12} \cdot E \qquad (1)$$

and where the laser field is treated classically, while when the field
is quantized

P. M. Rentzepis and C. Capellos (eds.), Advances in Chemical Reaction Dynamics, 229–249.
© *1986 by D. Reidel Publishing Company.*

$$\Omega^2 = \omega_p^2 f_{12}(\tfrac{1}{2}+\bar{n}) \ . \tag{2}$$

In Eqs. (1) and (2), $\vec{E}$ is the strength of the laser field, ω_p is the atomic plasma frequency, and $\bar{n}$ is the photon density of the laser field, while f_{12} and $\vec{\mu}_{12}$ denote the oscillator strength at resonance and the dipole moment for the transition $|1>\leftrightarrow|2>$ in question, respectively. γ_0 describes the spontaneous emission probability per unit time for the decay process $|2>\rightarrow|1>$ at the frequency ω_{21} and is defined as

$$\gamma_0 = \frac{4}{3} (\omega_{21}/c)^3 |\vec{\mu}_{12}|^2 \ , \tag{3}$$

which is exactly the Einstein A coefficient (1); units in which $\hbar=1$ are used throughout.

Physically, the relation $\Omega^2 >> \gamma_0^2$ implies that the radiative lifetime of the excited state in question is much greater than that due to the induced absorption (power broadening). Hence, the excited state of the atom can interact several times with the laser field before spontaneously emitting a photon. The result of this dynamic interaction is the production sidebands, whose lifetimes differ from those of the excited states of the atom. In such a case, the incident resonant field may be considered to be very strong and the atomic transition to be completely saturated.

The excitation spectrum of the light scattered by a single two-level atom driven at resonance by a strong laser field consists of three peaks: a central peak at the excitation frequency and two symmetrically located sidebands. This so-called dynamic Stark effect has been theoretically predicted by Mollow (2) and observed by Schuda et al. (3) and by others (4-9). The three peaks are described by Lorentzian lines whose radiative widths are equal to $\gamma_0/2$, $3\gamma_0/4$ and $3\gamma_0/4$, respectively, where $\gamma_0/2$ is the natural width of a photon spontaneously emitted from a single atom. The ratio of the central peak height to the heights of the sidebands is 3:1. Several theoretical and experimental investigations on the subject in question have since been conducted (10-30).

The first part of this lecture will be devoted to the dynamics of the excited state of a single atom interacting with strong laser fields. As an example, we will consider the resonance fluorescence spectra arising from the interaction of a three-level atom in the "V" configuration with two strong laser fields, where each of the fields couples resonantly the ground state to an excited state as indicated by thick lines in Figure 1. The frequency modes ω_a and ω_b of the two laser fields are assumed to be highly populated. They are defined as $\omega_a=\omega_{21}$ and $\omega_b=\omega_{31}$, where $\omega_{21}=\omega_2-\omega_1$ and $\omega_{31}=\omega_3-\omega_1$ are the two atomic transition frequencies, respectively. The parity of the ground state $|1>$ is assumed to be different from those of the excited states $|2>$ and $|3>$; hence, the electronic transition $|1>\leftrightarrow|2>$ and $|1>\leftrightarrow|3>$ are electric dipole allowed while the transition $|2>\leftrightarrow|3>$ is electric dipole forbidden. The electron states $|1>$, $|2>$ and $|3>$ are simultaneously coupled to the remaining modes of the electromagnetic field (vacuum or single field), these being initially empty. Hence, the signal field is considered to be the weak perturbing field describing the radiative decay process $|2>\rightarrow|1>$ and $|3>\rightarrow|1>$, which are denoted by wiggly lines in Figure 1.

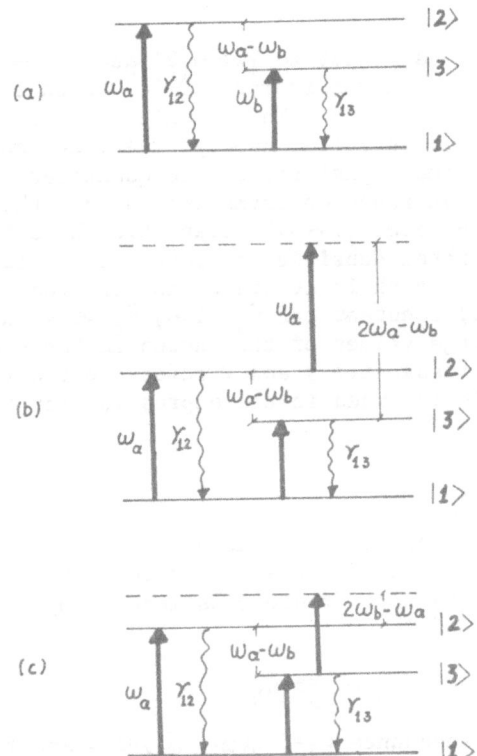

Figure 1. (a) Energy level diagram of a three-level atom in the "V" configuration. Thick lines indicate the laser fields operating between the levels $|1\rangle \leftrightarrow |2\rangle$ and $|1\rangle \leftrightarrow |3\rangle$. Wiggly lines denote radiative decay processes. (b) Third-order nonlinear frequency generation in the region $\omega = \pm(\omega_b - 2\omega_a)$ and (c) in the region $\omega = \pm(\omega_a - 2\omega_b)$.

Using the atomic system shown in Figure 1, we will consider the following processes: Firstly, one-photon processes (30,31), which occur at the frequencies $\omega = \omega_a = \omega_{21}$ and $\omega = \omega_b = \omega_{31}$, respectively; secondly, two-photon processes (32,33) which occur in the range of frequencies $\omega = \pm\omega_{ab} = \pm(\omega_a - \omega_b)$ as depicted in Figure 1. These excitations result from interference effects arising through indirect interactions between the two laser fields (32,33). Since the required frequency range is equal to the difference between the frequencies of the two laser fields, $\omega = \pm(\omega_a - \omega_b)$, the spectra for these excitations may then be called interference or Raman spectra (32) describing physical processes where one photon with frequency ω_a is absorbed while another photon with frequency ω_b is emitted and vice versa. Lastly, we will consider three- photon processes (33-35), which arise from the third-order nonlinear optical mixing of the frequencies ω_a and ω_b of the two laser fields to produce the frequencies (34,35) $\omega = \omega_b - 2\omega_a$ and $\omega = \omega_a - 2\omega_b$. This is shown in part b and c of Figure 1, respectively.

ONE-PHOTON PROCESSES

To study one-photon processes due to $|1> \leftrightarrow |2>$ and $|1> \leftrightarrow |3>$ transitions, a model Hamiltonian has been considered (30-34) which describes the atomic system shown in Figure 1a. This Hamiltonian is expressed in the second quantization representation, namely, the free and interacting electron and the laser and signal fields are quantized. Using this Hamiltonian and the Green function formalism (36,37) the excitation spectra due to the one-photon $|1> \leftrightarrow |2>$ transition have been considered in the limit of high photon densities of both laser fields. The high photon density limit for both laser fields is achieved when the ratios $(\bar{n}_a/V) \to$ constant, $(\bar{n}_b/V) \to$ constant for $\bar{n}_a >> 1 \to \infty$, $\bar{n}_b >> 1 \to \infty$ and $V \to \infty$. Here $\bar{n}_a$ and $\bar{n}_b$ denote the average values of the photon number operators of the laser fields a and b, respectively and V refers to the volume container of the system. This is included in the expression for the atomic plasma frequency, namely,

$$\omega_p^2 = 4\pi e^2/mV , \tag{4}$$

where m and -e are the electron mass and charge, respectively; ω_p appears in the expression (2) for the Rabi frequency. In the high photon density limit, the Rabi frequencies induced by the laser fields a and b take the form (30-34)

$$\Omega_a^2 \approx \omega_p^2 f_1 \bar{n}_a , \qquad \Omega_b^2 \approx \omega_p^2 f_2 \bar{n}_b , \tag{5}$$

respectively where at resonance $f_1 = f_{12}(\omega_{21}/\omega_a) = f_{12}$ and $f_2 = f_{13}(\omega_{31}/\omega_b) = f_{13}$ represent the oscillator strengths for the corresponding transitions $|1> \leftrightarrow |2>$ and $|1> \leftrightarrow |3>$. In order to derive the final results in the limit of high photon densities, the hierarchy of the Green functions is truncated by employing a decoupling aproximation where photon-photon correlations from each laser field are taken fully into consideration (31).

Excitation Spectra

We make use of the notation

$$X = (\omega - \omega_a)/\gamma_0 ,$$

$$\eta_a = \Omega_a/\gamma_0 , \qquad \eta_b = \Omega_b/\gamma_0 , \qquad \eta^2 = (\eta_a^2 + \eta_b^2)/2 , \tag{6}$$

where X is the reduced detuning frequency for the one photon peak and η_a and η_b are the relative Rabi frequencies of the laser fields a and b, respectively. For the sake of convenience, it is assumed that the decay rates for the transitions $|1> \leftrightarrow |2>$ and $|1> \leftrightarrow |3>$ are equal, i.e., $\gamma_{21}^0 = \gamma_{31}^0 = \gamma_0$; they are given by Eq. (3). Then the expression for the absorption coefficient, which describes the one-photon transition $|1> \leftrightarrow |2>$ in the limit of high photon densities of both laser fields, (apart from a constant factor) is defined by the spectral function (31)

$$P_1(\omega) = \frac{(\bar{n}_1 - \bar{n}_2)}{2\pi\gamma_0} I(X) , \tag{7}$$

where $\bar{n}_1$ and $\bar{n}_2$ denote the electron population of the states $|1>$ and $|2>$, respectively. $I(X)$ is the relative intensity representing the excitation spectra for the transition in question and is given by

$$I(X) = \left(1-\eta_b^2/8\eta^2\right)\frac{\tfrac{1}{2}}{X^2+\tfrac{1}{4}} + \frac{1}{8}\left[L(X,\eta_a\sqrt{2})+L(X,-\eta_a\sqrt{2})\right] +$$

$$+ \tfrac{1}{4}\left[L(X,\eta_b/\sqrt{2})+L(X,-\eta_b/\sqrt{2})\right] + \frac{\eta_b^2}{8\eta^2}\left[L,(X,\eta)+L(X,-\eta)\right]$$

$$+ \frac{\eta_a^2}{16\eta^2}\left[L(X,2\eta)+L(X,-2\eta)\right] . \tag{8}$$

The functions $L(X,\lambda)$ and $L(X,-\lambda)$ describe the shape of the sidebands at the relative frequencies $X=\lambda$ and $X=-\lambda$, respectively, and are defined as

$$L(X,\pm\lambda) = \frac{\tfrac{1}{4}\mp(X\mp\lambda)1/4\lambda}{(X\pm\lambda)^2+9/16} . \tag{9a}$$

The first term on the right-hand side (rhs) of Eq. (8) describes the central peak, which is a Lorentzian line peaked at $X=0$, and which has a spectral width of the order of $\gamma_0/2$. The intensity of the central peak depends on the factor $1-(\eta_b^2/8\eta^2)$, where $\eta_b^2/8\eta^2$ is an interference effect due to the presence of the laser field b. The remaining terms on the rhs of Eq. (8) describe four pairs of Lorentzian peaks centered at $X=\pm\eta_a\sqrt{2}$, $\pm\eta_b/\sqrt{2}$, $\pm\eta$ and $\pm2\eta$ and having spectral widths of the order of $3\gamma_0/4$. The expression (8) for $I(X)$ implies that in the limit of high photon densities and strong fields where $\eta_a^2\gg1$ and $\eta_b^2\gg1$, the relative intensities of the pairs of sidebands at $X=\pm\eta_a\sqrt{2}$ and $\pm\eta_b/\sqrt{2}$ are constant and independent of the values of η_a and η_b. However, the pairs of sidebands at $X=\pm\eta$ and $\pm2\eta$ depend on the values of the ratios η_b^2/η^2 and η_a^2/η^2, respectively. In the limiting case, when both relative Rabi frequencies are equal, namely, when $\eta_a=\eta_b=\eta$ then for $\eta^2>1$, the maximum relative intensities of the peaks described by Eq. (8) are independent of η provided that η is large enough so as to avoid overlap between the peaks of the system (31).

In an earlier attempt (30), we have considered the same problem by making use of the Green function method, where the hierarchy of the Green functions has been truncated by using a decoupling approximation. In this procedure some Green's functions have been prematurely decoupled so that photon correlations for the laser field a have been considered in the high photon density limit but not those corresponding to the laser field b. The final expression for the relative intensity obtained by this approach is given by Eq. (29) in ref. (30), which cannot be studied analytically and whose excitation spectra can be only obtained by numerical computation using a computer. In order to derive Eq. (8),

we have followed the correct approach suggested recently (38), where
instead of decoupling the Green functions in question prematurely, the
equations of motion for the Green functions are derived by means of the
Hamiltonian of the system and then the resulting higher-order Green
functions are decoupled. By this approach, both laser fields are
treated on the same footing as far as photon correlations are concerned.
The final result is derived in a closed form (31), and it is correct in
the high photon density limit for both laser fields. Numerical results
obtained from each method are illustrated graphicaly in figures 2-4,

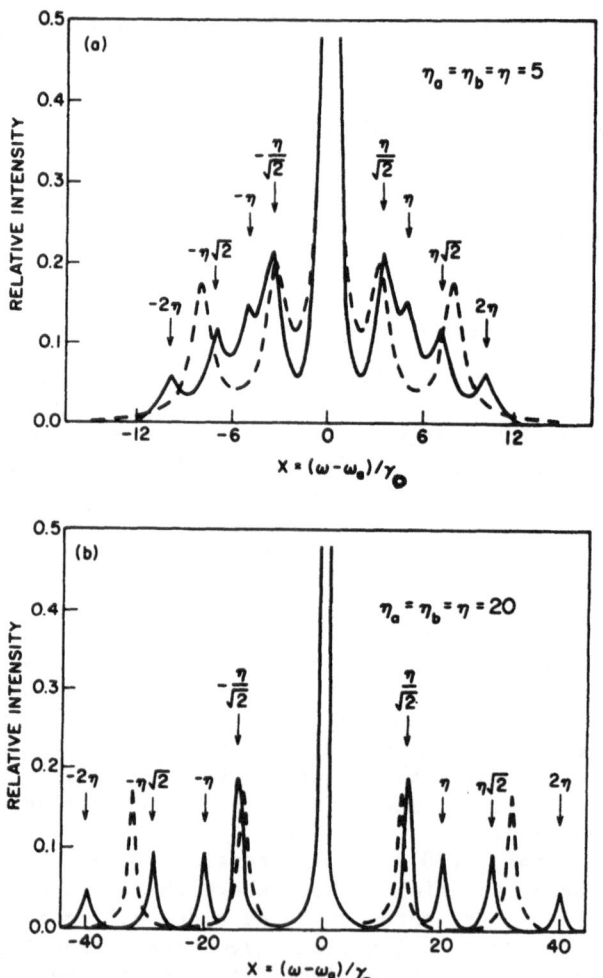

Figure 2. The relative intensity of the fluorescent light I(X) is
plotted versus $X=(\omega-\omega_a)/\gamma_0$ for different values of η. (a) $\eta=\eta_a=\eta_b=5$ and
(b) $\eta=\eta_a=\eta_b=20$. Solid lines are derived from Eq. (8) while dashed lines
are obtained from Figures 2b and 2c of ref. (30).

where the relative intensities derived from Eq. (8) and Eq. (29) of ref. (30), respectively, are plotted versus the relative frequency $X=(\omega-\omega_a)/\gamma_0$ for selective values of the relative Rabi frequencies η_a and η_b. Solid lines in Figures 2-4 refer to the results derived from Eq. (8) while dashed lines refer to those obtained in ref. (30).

Figures 2a and 2b illustrate the spectra for values $\eta=\eta_a=\eta_b=5$ and 20, respectively. In Figure 2a, the value of $\eta=5$ is not large enough to separate the peaks so that there is an overlap in the relative intensities of the peaks while for $\eta=20$ in Figure 2b, the peaks are well

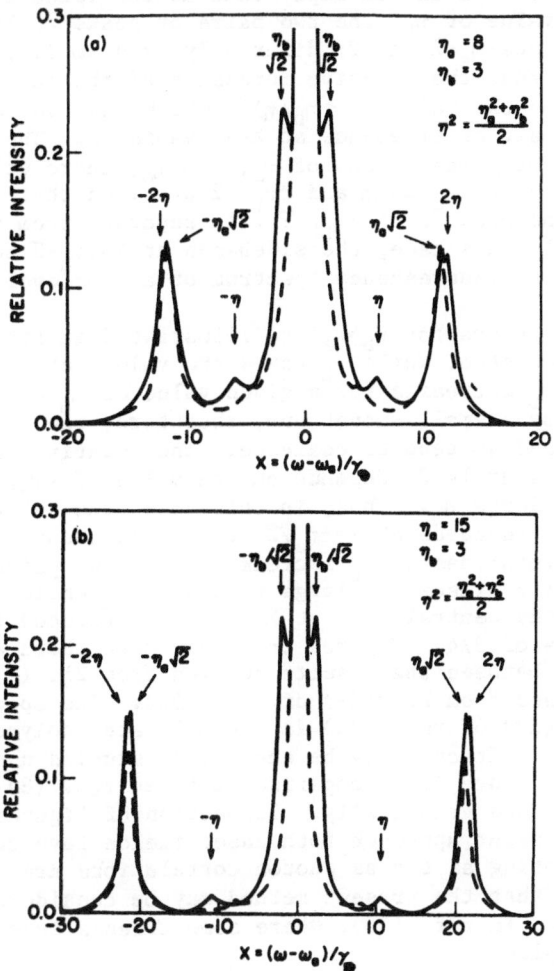

Figure 3. As in Figure 2 but $\eta_b=3$ is kept constant while η_a is varied. (a) $\eta_a=8$ and (b) $\eta_a=15$. Solid lines are derived from Eq. (8) while dashed lines are obtained from Figures 3b and 3c of ref. (30).

separated and their relative intensities remain constant and independent of η for values of $\eta > 20$. There are four pairs of sidebands, which are peaked at the frequencies $X = \pm \eta / \sqrt{2}$, $\pm \eta$, $\pm \eta \sqrt{2}$ and $\pm 2\eta$ as is illustrated by the solid lines; the dashed lines give only two pairs of sidebands. In Figure 2b, the heights of the sidebands at the frequencies $X = \pm \eta / \sqrt{2}$, $\pm \eta$, $\pm \eta \sqrt{2}$ and $\pm 2\eta$ are found to be 1/5, 1/10, 1/10 and 1/20 of that of the central peak while the corresponding height for the sidebands given by the dashed lines is 1/6. The linewidths of all sidebands are equal to $3\gamma_0/4$.

In Figures 3a and 3b, we consider the case where the relative Rabi frequency $\eta_b = 3$ is kept constant while η_a takes the values of $\eta_a = 8$ and 15, respectively. Figures 3a and 3b imply that as the value of η_a increases for a given value of η_b, the two pairs of peaks at $X = \pm \eta_a \sqrt{2}$ and $\pm 2\eta$ coalesce rapidly since $\eta_a \sqrt{2}$ and 2η difer only by a small amount. Equation (8) indicates that the relative intensity of the pair of sidebands at $X = \pm \eta$ depends on the value of $(\eta_b / \eta)^2$, which goes to zero as η_a increases; hence, the pair of sidebands at $X = \pm \eta$ vanishes. Thus, for a given value of η_b and for large values of η_a, $\eta_a \gg \eta_b$, there will be only two pairs of sidebands at $X = \pm \eta_a \sqrt{2} \approx \pm \eta$ and $\pm \eta_b / \sqrt{2}$ provided that η_b is large enough so that the peaks at $X = \pm \eta_b / \sqrt{2}$ will separate from the central peak at $X = 0$. In this case, the sidebands at $X = \pm \eta_a \sqrt{2}$ become identical to those of the fluorescence spectrum of a two-level system (2–20).

The fluorescence spectra for $\eta_b > \eta_a$ are illustrated in Figues 4a and 4b where $\eta_a = 3$ is kept constant while η_b takes the values of $\eta_b = 10$ and 15, respectively. As η_b increases for a given value of η_a, the values of $\eta_b / \sqrt{2}$ and η differ by a small amount and, therefore, the two pairs of sidebands at $X = \pm \eta_b / \sqrt{2}$ and $\pm \eta$ tend to coalesce. The relative intensity of the pair of sidebands at $X = \pm 2\eta$ depends on the value of $(\eta_a / \eta)^2$, which becomes very small for large η_b. Thus, in this case, the relative intensity of the pair of sidebands at $X = \pm \eta_a \sqrt{2}$ remains constant while that of the pair at $X = \pm 2\eta$ diminishes as η_b increases. The two pairs of peaks at $X = \pm \eta_b \sqrt{2}$ and $\pm \eta$ will coalesce for large values of η_b while the relative intensity of the central peak at $X = 0$ will be reduced from $1 - (\eta_b^2 / 8\eta^2)$ to the value of 3/4. Figures 2–4 imply that there are characteristic differences between the results derived from Eq. (8) (solid lines) and those obtained from Eq. (29) of ref. (30). The spectral function given by Eq. (29) of ref. (30) is a complicated polynomial of the parameters involved. It can only be accurately studied numerically through a computer and, hence, the comparison between Eqs. (8) and (29) of ref. (30) has to be done graphically. Inspection of Figures 2–4 and the fact that in the present approach both laser fields have been treated on the same footing as far as photon correlations are concerned, lead to the conclusion that the present method may be considered as a improvement on that used in ref. (30), where some Green's functions have been prematurely decoupled.

The fluorescence spectra due to the transition $|1> \leftrightarrow |3>$ can be obtained from Eq. (8) if ω_a and η_a are replaced everywhere by ω_b and $\eta_b (\omega_a \leftrightarrow \omega_b$, $\eta_a \leftrightarrow \eta_b)$, respectively, and if $\bar{n}_2$ is replaced $\bar{n}_3 (\bar{n}_2 \leftrightarrow \bar{n}_3)$ as

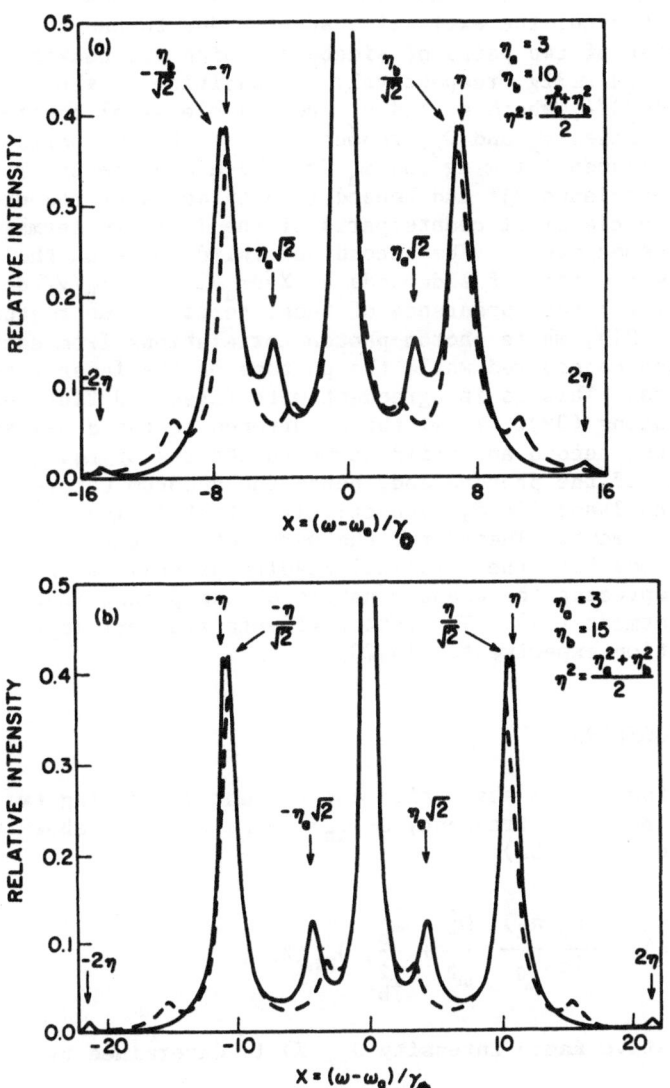

Figure 4. As in Figure 2 but η_a=3 is kept constant while η_b is varied.
(a) η_b=10 and (b) η_b=15. Solid lines are derived from Eq. (8) while
dashed lines are obtained from Figures 4b and 4c of ref. (30).

well. The fluorescence spectra for the transition $|1\rangle\leftrightarrow|3\rangle$ consist of
nine Lorentzian lines peaked at the frequencies ω_b, $\omega_b\pm\Omega_b\sqrt{2}$, $\omega_b\pm\Omega_a/\sqrt{2}$,
$\omega_b\pm\eta\gamma_0$ and $\omega_b\pm2\eta\gamma_0$, respectively. The spectral width of the central
peak is $\gamma_0/2$ while those for the sidebands are equal to $3\gamma_0/4$.

It has been shown (31) that when the classical description for both laser fields is used, the excitation spectra due to the transition $|1> \leftrightarrow |2>$ consist of two pairs of sidebands which are peaked at the frequencies $\omega_a \pm g$ and $\omega_a \pm 2g$, respectively, in addition to the central peak. Here $g^2 = (g_a^2 + g_b^2)/4$, where g_a and g_b are the classical counterparts of the Rabi frequencies Ω_a and Ω_b, respectively. The two pairs of sidebands at the frequencies $\omega_a \pm g$ and $\omega_a \pm 2g$, which are identical to those derived by Cohen-Tannoudji and Renaud (22) by means of the dress-atom method, are the classical counterparts of the last two terms on the rhs of Eq. (8), respectively. The second and third terms on the rhs of Eq. (8) describing the pair of sidebands at $X = \pm \eta_a \sqrt{2}$ and $\pm \eta_b / \sqrt{2}$ do not have classical analogs; the appearance of these terms is due to the decoupling procedure (31), where photon-photon correlations from each laser field have been considered while the photons of the laser fields obey Bose statistics. This is in agreement with Dirac's definition of interference of photons (39). Thus, the occurrence of these new sidebands described by the second and third terms on the rhs of Eq. (8) reveal the Bose character of the photons and, hence the quantum nature of the photons of each laser field, respectively, which is lost in the classical treatement. Therefore, the merit of the present method is that it describes both the classical results as well as the quantized ones, which represent the quantum nature of the photon lost in the classical treatment (31). The latter effects are very important in photon corelation experiments (15,20).

TWO-PHOTON PROCESSES

The contribution to the absorption coefficient describing two-photon Raman spectra near the frequency $\omega \approx \omega_{ab} = \omega_a - \omega_b$, which is shown in Figure 1, is found to be (32,33)

$$P_2(\omega) = \frac{(\bar{n}_3 - \bar{n}_2)}{16\pi\gamma_0} \left(\frac{\Omega_a^2}{\omega_a^2} + \frac{\Omega_b^2}{\omega_b^2} \right) J_{ab}(X) , \tag{9b}$$

where the relative Raman intensity $J_{ab}(X)$ is determined by

$$J_{ab}(X) = \frac{1 + 3\mu/4}{X^2 + 1} + \frac{1}{2} \left[L(X, \eta_a/\sqrt{2}) + L(X, -\eta_a/\sqrt{2}) \right]$$

$$+ \frac{1}{2} \left[L(X, \eta_b/\sqrt{2}) + L(X, -\eta_b/\sqrt{2}) \right] + \frac{(1-\mu)}{2} \left[L(X, \eta) + L(X, -\eta) \right]$$

$$+ \frac{\mu}{16} \left[L(X, 2\eta) + L(X, -2\eta) \right] \tag{10}$$

In Eqs. (9b) and (10), $\bar{n}_2$ and $\bar{n}_3$ are the electron populations of the states $|2>$ and $|3>$, respectively, while

$$X = (\omega - \omega_{ab})/\gamma_0 = (\omega - \omega_a + \omega_b)/\gamma_0 , \qquad \mu = \eta_a^2 \eta_b^2 / 2\eta^4 , \tag{11}$$

being the reduced Raman frequency. η_a, η_b, η and the function $L(X, \pm\lambda)$ are defined by Eqs. (6) and (9a), respectively. The expression (10) describes the relative intensity in the region of positive frequencies $\omega\approx\omega_{ab}$ and is valid in the limit of high photon densities of both laser fields. The first term on the rhs of Eq. (10) describes the main peak, which is a Lorentzian line peaked at the frequency $X=0$ or at $\omega-\omega_{ab}=\omega_a-\omega_b$, and which has a radiative width of the order of γ_0. The relative intensity of this peak is equal to $1+3\mu/4$, where the term $3\mu/4$ is due to the presence of the strong fields. In the limit when $\mu\to0$, this peak is identical to the corresponding peak in the presence of weak fields (40,41).

The last four terms in the square brackets on the rhs of Eq. (10) describe four pairs of Lorentzian lines peaked at the frequencies $X=\pm\eta_a/\sqrt{2}$, $\pm\eta_b/\sqrt{2}$, $\pm\eta$ and $\pm2\eta$, respectively, and having radiative widths of the order of $3\gamma_0/4$. The two pairs of peaks at the frequencies $X=\pm\eta_a/\sqrt{2}$ and $\pm\eta_b/\sqrt{2}$ are induced by the laser fields, respectively, and their relative intensities are roughly constant and equal to 1.34. The pairs of peaks at $X=\pm\eta$ and $\pm2\eta$ are induced simultaneously by both laser fields and their reltaive intensities depend strongly on the value of η_a and η_b.

When both laser fields have equal intensities and $f_1\approx f_2$, namely, when $\eta_a=\eta_b=\eta$, then $\mu=\frac{1}{2}$, and the two pairs at $X=\pm\eta_a/\sqrt{2}$ and $\pm\eta_b/\sqrt{2}$ coincide while the relative intensities of the pairs of peaks at $X=\pm\eta$ and $\pm2\eta$ become roughly equal to 0.33 and 0.08, respectively. In this case, the ratio of the relative intensities of the main peak at $X=0$ to those pairs described by the remaining terms in Eq. (10) is given by 1.38: 1.34:0.33:0.08. This result is illustrated in Figure 5, where $\eta_a=\eta_b=\eta=10$. We next consider the case when the laser fields have different relative intensities $\eta_a\neq\eta_b$.

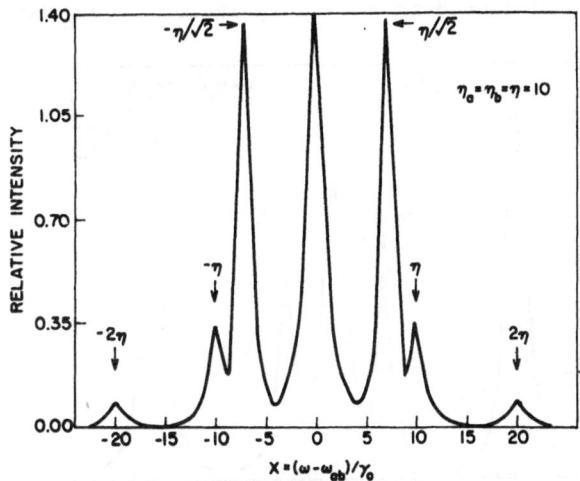

Figure 5. The relative intensity of the fluorescent light $J_{ab}(X)$ determined by Eq. (10) is plotted versus the relative frequency $X=(\omega-\omega_{ab})/\gamma_0$ for equal relative Rabi frequencies $\eta_a=\eta_b=\eta=10$.

Figure 6 illustrates the spectra when $\eta_a=2\eta_b=10$ with $\eta=7.9$ and $\mu=0.32$, where there is the main peak at X=0 and four pairs of sidebands and

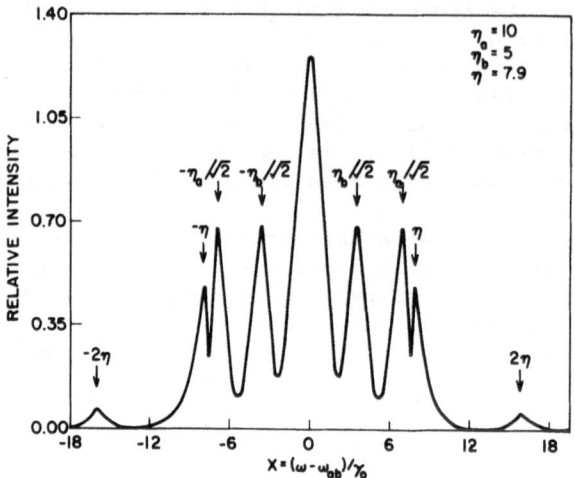

Figure 6. As in Figure 5 but for $\eta_a=10$, $\eta_b=5$ and $\eta=(\eta_a^2+\eta_b^2)^{\frac{1}{2}}/\sqrt{2}\approx7.9$.

where the ratios of the intensity of the main peak to those of the side-bands are equal to 1.24:0.67:0.46:0.05. When the intensity of one of the lasers becomes much larger than the other then μ takes its minimum value. For instance, when $\eta_a>>\eta_b$ then $\mu\to0$ and $\eta\to\eta_a/\sqrt{2}$, the pair of peaks at $X=\pm\eta_a/\sqrt{2}$ and $\pm\eta$ coincide while the intensity of the main peak at X=0 is reduced to 1 and the pair of sidebands at $X=\pm2\eta$ vanishes. Such an example is depicted in Figure 7, where $\eta_a=10\eta_b=50$ then

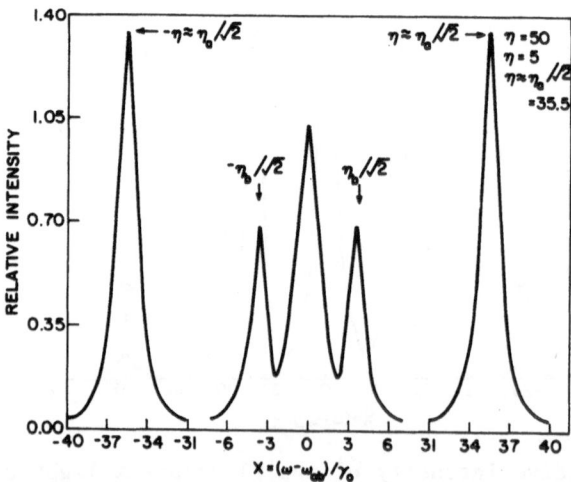

Figure 7. As in Figure 5 but for $\eta_a=50$, $\eta_b=5$ and $\eta\approx\eta_a/\sqrt{2}\approx35.5$.

$\eta \approx \eta_a / \sqrt{2} \approx 35.5$ and $\mu=0.02$; the ratios of the intensities of the main peak at $\bar{X}=0$ to those of the sidebands are given by $1:1.32:0.67$, which indicates that the pair of peaks at $X=\pm\eta_a / \sqrt{2} \approx \pm\eta$ has a larger relative intensity than that of the main peak.

As in the case of the one-photon spectra (31), there is no classical analog for the two pairs of peaks occurring at the frequencies $X=\pm\eta_a / \sqrt{2}$ and $\pm\eta_b / \sqrt{2}$, which are given by the second and third terms on the rhs of Eq. (10), respectively. Each pair of these sidebands describes the Bose character of the photons of the corresponding laser field, which results when photon correlations are taken into account through the decoupling approximation. Thus, the occurrence of these pairs of sidebands reveals the boson character and, therefore, the quantum nature of the photons of both laser fields which is lost in the classical treatment. The remaining last two terms on the rhs of Eq. (10) representing the pair of sidebands at $X=\pm\eta$ and $\pm2\eta$ are the quantum mechanical counterparts of those peaked at $X=\pm g$ and $\pm2g$, respectively, which are obtained when the laser fields are treated classically or, equivalently, through the dress-atom approach (42).

The spectral function $P_2(\omega)$ and, consequently, the relative intensity $J_{ab}(X)$ defined by Eqs. (9b) and (10), respectively, is positive when $\bar{n}_3 > \bar{n}_2$ and negative when $\bar{n}_2 > \bar{n}_3$, where $\bar{n}_2$ and $\bar{n}_3$ are the electron popula- tion of the states $|2\rangle$ and $|3\rangle$, respectively. Negative spectral function or relative intensity implies that the physical process of stimulated emission (amplification) takes place at the frequencies in question. Since the stronger the laser field the larger the induced electron population and considering that the electron states $|2\rangle$ and $|3\rangle$ are populated by the laser fields a and b, respectively, we conclude that when $\eta_a > \eta_b$ then $\bar{n}_2$ is expected to be larger than $\bar{n}_3$, provided that both states have equal radiative lifetimes. Thus, when the conditions $\eta_a > \eta_b$ and $\gamma_{12}^0 = \gamma_{31}^0 = \gamma_0$ are fulfilled, then $\bar{n}_2 > \bar{n}_3$ and, consequently, the spectral function becomes negative indicating that amplification (Raman gain) is anticipated to take place. In the opposite case, which occurs when $\eta_b > \eta_a$ and $\eta_{12}^0 = \gamma_{13}^0 = \gamma_0$ then $\bar{n}_3 > \bar{n}_2$, the spectral function is positive and the Raman loss process should be materialized.

THIRD-ORDER NONLINEAR MIXING OF FREQUENCIES

In ref. (34), use has been made of a model Hamiltonian, which is correct beyond the rotating wave approximation and which describes the atomic system shown in Figure 1a. Then the Green function formalism has been used to calcualte the excitation spectra for the process shown in Figure 1c, where the nonlinear mixing of the frequencies ω_a and ω_b of the two laser fields produces the frequency $\omega=\omega_a -2\omega_b$. In the limit of high photon densities of both laser fields, the spectral function describing the excitation spectra near the frequency $\omega \approx \omega_a -2\omega_b$ is determined by Eq. (30) of ref. (34) in the form

$$S_{12,21}(\omega) = \frac{(\bar{n}_3-\bar{n}_2)(\frac{\Omega_b}{4\omega_b})^2 (\chi-\frac{1}{2}\gamma)}{2\pi(z^2-\frac{1}{2}\Omega_b^2)} \left[1 + \frac{\Omega_a^2 N(z^2)}{4(z^2-2\Omega_b^2)D(z^2)}\right], \quad (12)$$

where

$$z^2 = (x-\gamma)(x-\tfrac{1}{2}\gamma) , \qquad x = \omega-\omega_a+2\omega_b , \quad (13)$$

$$N(z^2) = (z^2-\Omega_b^2)\{(z^2-2\Omega_b^2)[z^2(z^2 - \tfrac{5}{2}\Omega_b^2+2\Omega_a^2)+\Omega_b^2(2\Omega_b^2-\Omega_a^2)]$$

$$+ \tfrac{1}{2}\Omega_a^2\Omega_b^2(\Omega_a^2 - \tfrac{5}{2}\Omega_b^2)-\tfrac{1}{2}\Omega_b^6\}+\Omega_b^2\Omega_a^2(\Omega_b^2-\tfrac{1}{4}\Omega_a^2) , \quad (14)$$

$$D(z^2) = z^2[(z^2-\tfrac{1}{2}\Omega_b^2)(z^2-\Omega_b^2+2\Omega_a^2)(z^2-3\Omega_b^2-\tfrac{1}{2}\Omega_a^2)+\Omega_a^2\Omega_b^2(z^2+2\Omega_a^2)]$$

$$- \tfrac{1}{4}\Omega_b^4\Omega_a^2(\Omega_a^2+\Omega_b^2)$$

$$= (z^2-\tfrac{1}{2}\Omega_b^2)\{z^2[(z^2-\Omega_b^2+2\Omega_a^2)(z^2-3\Omega_b^2-\tfrac{1}{2}\Omega_a^2)+\Omega_a^2\Omega_b^2]$$

$$+\Omega_a^2\Omega_b^2(2\Omega_a^2+\tfrac{1}{2}\Omega_b^2)\}+\tfrac{1}{4}\Omega_b^4\Omega_a^4 \quad (15)$$

and γ is the scattering function defined by Eq. (5) of ref. (34), whose imaginary part is equal to $\gamma_0=\mathrm{Im}\gamma$ given by Eq. (3). The excitation spectra near the frequency $\omega \approx \omega_b-2\omega_a$ depicted in Figure 1b can be derived from Eq. (12) if the index a is replaced everywhere by b, i.e., $\Omega_a \leftrightarrow \Omega_b$, $\omega_a \leftrightarrow \omega_b$ and $\bar{n}_2 \leftrightarrow \bar{n}_3$. If we introduce the relative detuning frequency

$$X = (\omega-\omega_a+2\omega_b)/\gamma_0 , \quad (16)$$

then the expression for the relative intensity $I(X)$ is defined

$$\text{Relative Intensity} = \frac{-2\pi\gamma_0}{(\bar{n}_3-\bar{n}_2)} (\frac{4\omega_b}{\Omega_b})^2 \mathrm{Im}S_{12,21}(\omega) \quad (17)$$

where $\mathrm{Im}S_{12,21}(\omega)$ is the imaginary part of the function $S_{12,21}(\omega)$ defined by Eq. (12). The procedure used to compute the relative intensity (17) from Eq. (12) is similar to that described in the literature (35,43,44). For the sake of convenience, the excitation spectra when $\eta_a<\eta_b$, $\eta_a=\eta_b$ and $\eta_a>\eta_b$ will be considered separately (33-35).

Excitation Spectra for $\eta_a<\eta_b$.

Figures 5a and 5b illustrate the computed spectra for values of the ratio η_a/η_b equal to 1/10 and 1/50, respectively. The spectra

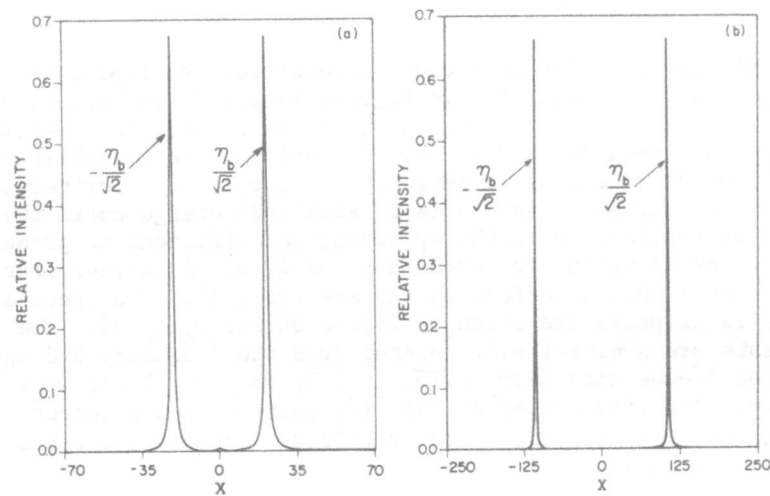

Figure 8. The relative intensity of the fluorescent light computed from Eq. (17) is plotted versus $X=(\omega-\omega_a+2\omega_b)/\gamma_0$ for various values of the relative Rabi frequencies $\eta_a<\eta_b$. (a) $\eta_a=3$, $\eta_b=30$. (b) $\eta_a=3$, $\eta_b=150$.

consist of a doublet peaked at the relative frequencies $X=\pm\eta_b/\sqrt{2}$ or, equivalently, at $\omega=\omega_a-2\omega_b\pm\Omega_b/\sqrt{2}$. The two peaks have equal intensities, which remain constant and independent of the value of the ratio η_a/η_b provided that $\eta_a/\eta_b<1$. The spectral widths of the peaks are of the order of $3\gamma_0/4$.

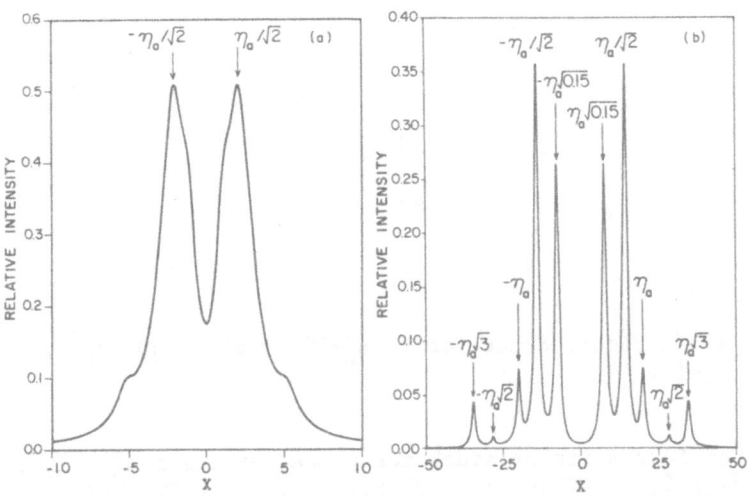

Figure 9. As in Figure 8 but for $\eta_a=\eta_b$. (a) $\eta_a=\eta_b=3$. (b) $\eta_a=\eta_b=20$.

Excitation Spectra for $\eta_a = \eta_b$.

When both laser fields have equal intensities and $f_1 \approx f_2$, then the computed spectra are illustrated in figures 9a and 9b for values of $\eta_a = \eta_b = 3$ and 20, respectively. Figure 9a indicates that for $\eta_a = \eta_b < 3$, the main peak at the frequency X=0 at $\omega = \omega_a - 2\omega_a$ splits into a doublet peaked at $X = \pm\eta_a/\sqrt{2}$, whose intensity decreases as η_a increases. Furthermore, as the value of η_a increases, new pairs of sidebands emerge until the value of $\eta_a = \eta_b = 10$ is reached, where the splitting is sufficient to produce five pairs of peaks which are completely resolved. A further increase of the value of η_a has no effect on the spectra (35). The spectra of the five pairs of peaks are shown in figure 9b for $\eta_a = \eta_b = 20$. The five pairs of peaks are symmetrically located from the frequency X=0 and peaked at the frequencies $X = \pm\eta_a\sqrt{0.15}$, $\pm\eta_a/\sqrt{2}$, $\pm\eta_a$, $\pm\eta_a\sqrt{2}$ and $\pm\eta_a\sqrt{3}$, respectively. The line shapes of the five pairs of peaks become steeper as the value of η_a increases but their relative intensities remain practically constant.

Excitation Spectra for $\eta_a > \eta_b$.

When $\eta_a > \eta_b$, we consider two cases where the ratio η_a/η_b takes the value of 10 and 50; the computed spectra are illustrated in Figures 10a and 10b and 11a and 11b, respectively. The spectra in Figures 10a and

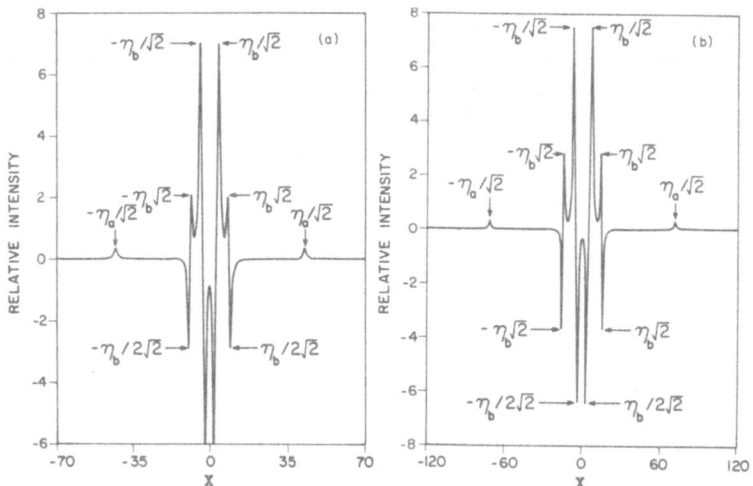

Figure 10. As in Figure 8 but for $\eta_a/\eta_b = 10$. (a) $\eta_a = 60$, $\eta_b = 6$. (b) $\eta_a = 100$, $\eta_b = 10$.

10b consist of four pairs of sidebands peaked at the frequencies $X = \pm\eta_b/2\sqrt{2}$, $\pm\eta_b/\sqrt{2}$, $\pm\eta_b\sqrt{2}$ and $\pm\eta_a/\sqrt{2}$, respectively. The pair of sidebands at $X = \pm\eta_b\sqrt{2}$ having positive intensities vanishes as the values of

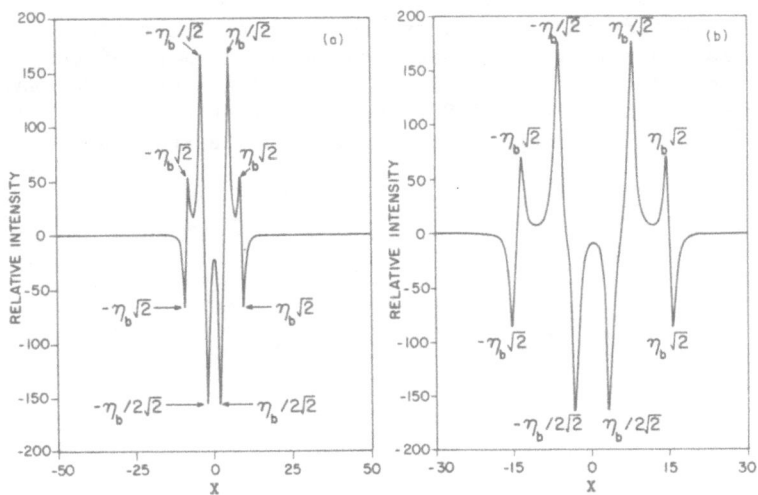

Figure 11. As in Figure 8 but for η_a/η_b=50. (a) η_a=300, η_b=6. (b) η_a=500, η_b=10.

η_a and η_b decrease and become less than those shown in Figures 10a and 11a, but the ratios η_a/η_b remain consant (35) and are equal to 10 and 50, respectively. In Figure 10a and 10b, the relative intensities of the sidbands at X=$\pm\eta_b/\sqrt{2}$ and $\pm\eta_a/\sqrt{2}$ are positive indicating absorption; those at X=$\pm\eta_b/2\sqrt{2}$ are negative implying amplification, while the two pairs of sidebands at X=$\pm\eta_b\sqrt{2}$ have positive and negative components and, thus, describe the mixed process of absorption-amplification. The net relative intensity of the two pairs of sidebands at X=$\pm\eta_b\sqrt{2}$ is negative indicating that the physical process of stimulated emission (amplification) prevails. The pairs of sidebands at X=$\pm\eta_b\sqrt{2}$ are analogous to the absorption-amplification line shapes observed for a two-level system by Wu et al. (8) and to those recently predicted to occur in the resonance fluorescence spectra of three-level atoms (43,44). The pair of sidebands at X=$\pm\eta_b/\sqrt{2}$ has a vanishing small relative intensity, and the pair disappears for value of the ratio η_a/η_b greater than 10 as shown in Figures 11a and 11b.

Comparison between Figures 10 and 11 implies that the ratio I_{50}/I_{10} of the relative intensities of the sidebands occurring at the value of the ratio η_a/η_b=50 to the corresponding ones at η_a/η_b=10 is

$$I_{50}/I_{10} \approx (50/10)^2 = 25 .$$

This implies that the relative intensities of the sidebands vary as $(\eta_a/\eta_b)^2$, namely,

$$I_{(\eta_a/\eta_b)} = \frac{1}{\kappa} (\eta_a/\eta_b)^2 , \qquad (18)$$

where κ is a constant and, therefore, an enhancement of the relative intensity of the sidebands occurs as the value of the ratio η_a/η_b increases. In order to verify that such an enhancement of the relative intensity takes place, we consider the excitation spectra for values of the ratio η_a/η_b equal to 40 and 80, the results of which are illustrated in Figures 12a and 12b, respectively. Figures 12a and 12b imply that

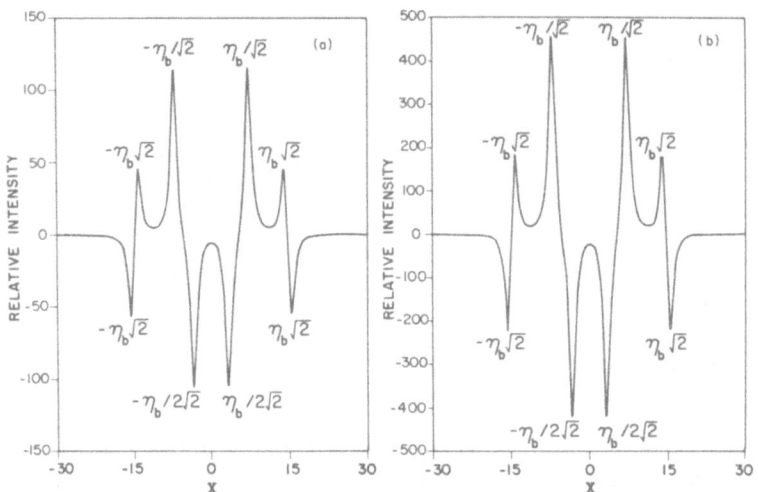

Figure 12. As in Figure 8 but for $\eta_a > \eta_b$. (a) $\eta_a=400$, $\eta_b=10$, $\eta_a/\eta_b=40$. (b) $\eta_a=800$, $\eta_b=10$, $\eta_a/\eta_b=80$.

the ratio I_{η_1}/I_{η_2} of the intensities of the sidebands appearing at the same frequencies but at two different values of the ratios $\eta_1=(\eta_a/\eta_b)$ and $\eta_2(\eta_1 \neq \eta_2)$ varies as the squares of the ratio $(\eta_1/\eta_2)^2$, i.e.,

$$I_{\eta_1}/I_{\eta_2} \approx (\eta_1/\eta_2)^2 \tag{19}$$

It is pointed out that the net relative intensity of the sidebands at $X = \pm \eta_b \sqrt{2}$ is not only negative but also increases as the ratio η_a/η_b increases. This is an important property as far as the amplification process is concerned, which is achieved without having a population inversion.

CONCLUDING REMARKS

Dynamic effects of the excited state of a single atom occurring in the presence of strong laser fields have been outlined. One-, two- and three-photon processes have been considered for a three-level atom interacting with two strong resonant laser fields. The main dynamic effects of the excited state of a three-level atom due to the presence of two strong resonant laser fields for the processes under consideration are the following: (i) The creation of pairs of sidebands whose lifetimes differ from that of the central peak which occurs at the excitation frequency. (ii) the number of pairs of sidebands as well as their relative intensities depend upon the strength of the laser fields involved. (iii) The relative intensities of the sidebands are either positive or negative describing the physical processes of absorption (attenuation) and stimulated emission (amplification), respectively, which occur at the corresponding frequencies. (iv) In the process of third-order nonlinear mixing of frequencies, the relative intensities of the sidebands are found to vary as $(\eta_a/\eta_b)^2$ for $(\eta_a/\eta_b) > 1$; hence, the relative intensities are immensely enhanced as the ratio η_a/η_b increases and (v) the appearance of sidebands which have positive and negative components describing the mixed process of absorption-stimulated emission. The net relative intensity of the pairs is negative and increases as the ratio η_a/η_b increases for $\eta_a/\eta_b > 1$ indicating that amplification has prevailed and has been enhanced without having a population inversion.

Comparison has been made between the results obtained when the laser fields are quantized taking into account photon correlations in the limit of high photon densities with those derived when the laser fields are treated classically, which are identical to those obtained when the dress-atom approach is used. The sidebands, whose appearance has no classical analog, describe the boson character of the photons of the laser fields. Their existence suggests that the results-obtained when the fields are quantized in the limit of high photon densities- describe the classical as well as the quantum nature of the photons, the latter being lost in the classical treatment.

REFERENCES

1. Allen, L. and Eberly, J.H.: 1975, Optical Resonance and Two-level Atoms, Wiley, New York.

2. Mollow B.R.: 1979, Phys. Rev. 188, p. 1969.

3. Schuda, F., Stroud, C.R., Jr., and Hercher, M.: 1974, J. Phys. B, 7, L198.

4. Walther, H.: 1975, in Proc. Second Inter. Conf. on Laser Spectroscopy, eds. Haroch, S., Pebory-Peyrouls, J.C., Hänsch, T.W., and Haris, S.E., Springer Verlag, Berlin.

5. Wu, F.Y., Grove, R.E., and Ezekiel, S.: 1975, Phys. Rev. Lett. 35, p. 1426.

6. Hartig, W., Rusmussen, W., Scieder, R., and Walther, H.: 1976, Z. Phys. A278, p. 205.

7. Grove, R.E., Wu, F.Y., and Ezekiel, S.: 1977, Phys. Rev. A, 15, p. 227.

8. Wu, F.Y., Ezekiel, S., Ducloy, M., and Mollow, B.R.: 1977, Phys. Rev. Lett. 38, p. 1077.

9. Ezekiel, S. and Wu, F.Y.: 1979, Quantum Electron. 8, p. 978.

10. Mollow, B.R.: 1975, Phys. Rev. A, 12, p. 1919.

11. Carmichael, H.J., and Walls, D.F.: 1975, J. Phys. B, 8, p. L77.

12. Hassam, S.S., and Bullough, R.K.: 1975, J. Phys. B, 8, p. L147.

13. Agarwal, G.S.: 1976, Phys. Rev. Lett., 37, p. 1383.

14. Eberly, J.H.: 1976, Phys. Rev. Lett., 37, p. 1387.

15. Kimble, H.J., and Mandel, L.: 1976, Phys. Rev. A, 13, p. 2123.

16. Mollow, B.R.: 1976, Phys. Rev. A, 13, p. 758.

17. Carmichael, H.J., and Walls, D.R.: 1976, J. Phys. B, 9, p. 1199.

18. Mollow, B.R.: 1977, Phys. Rev. A, 15, p. 1023.

19. Whitley, R.M., and Stroud, C.R.: 1976, Phys. Rev. A, 14, p. 1498.

20. Kimble, H.J., and Mandel, L.: 1977, Phys. Rev. A, 15, p. 689.

21. Cohen-Tannoudji, C.: 1977, Frontiers in Laser Spectroscopy, Les Houches Summer School (1975) Section 25, eds. Balian, R., Haroche, S., and Liberman, S., North-Holland, Amsterdam, pp. 1-104.

22. Cohen-Tannoudji, C., and Reynaud, S.: 1977, J. Phys. B, 10, p. 345, 2311.

23. Salomaa, R.: 1977, J. Phys. B, 10, p. 3005.

24. Avan, P., and Cohen-Tannoudji, C.: 1977, J. Phys. B, 10, p. 171.

25. Agarwal, G.S.: 1978, Phys. Rev. A, 18, p. 1490.

26. Mavroyannis, C.: 1979, Mol. Phys. 37, p. 1175.

27. Mavroyannis, C.: 1979, Optics Commun. 29, p. 80.

28. Mavroyannis, C. and Sharma, M.P.: 1980, Physica A, 102, p. 431.

29. Mavroyannis, C.: 1980, Can. J. Phys. 58, p. 957.

30. Sharma, M.P., Villaverde, A.B., and Mavroyannis, C.: 1980, Can. J. Phys. 58, p. 1570.

31. Mavroyannis, C.: 1984, J. Math. Phys. 25, p. 2780.

32. Mavroyannis, C.: 1983, Mol. Phys. 48, p. 847.

33. Mavroyannis, c.: 1984, J. Phys. Chem. 88, p. 4868.

34. Mavroyannis, C.: 1981, Can. J. Phys. 59, p. 1917.

35. Mavroyannis, C., Woloschuk, K.J., Hutchinson, D.A., and Downie, C.: 1982, Can. J. Phys. 60, p. 245.

36. Zubarev, D.N.: 1974, Noneqluilibrium Statistical Thermodynamics, Plenum, New York, Chapter 16.

37. Mavroyannis, C.: 1975, The Green Function Method, in Physical Chemistry, eds. Eyring, H., Jost, W., and Henderson, D., Academic Press, New York, Vol. XI A, Chapter 8.

38. Mavroyannis, C.: 1983, Phys. Rev. A, 27, p. 1414.

39. Dirac, P.A.M.: 1958, The Principles of Quantum Mechanics, Oxford U.P., Oxford, 4th ed., pp. 7-10.

40. Bloembergen, N.: 1967, Am. J. Phys. 35, p. 989.

41. Levenson, M.D., and Song, J.J.: 1980, Coherent Raman Spectroscopy, in Coherent Nonlinear Optics, Recent Advances, eds. Field, M.S., and Letokhov, V.S., Springer-Verlag, Berlin, p. 293.

42. Agrawal, G.S., and Jha, S.S.: 1979, J. Phys. B, 12, p. 2655.

43. Woloschuk, K.J., Hontzeas, S., and Mavroyannis, C.: 1982, Can. J. Phys. 60, p. 968.

44. Woloschuk, K.J., Hontzeas, S., and Mavroyannis, C.: 1982, Optics Commun. 42, p. 77.

DYNAMICS OF EXCITED STATES OF TWO IDENTICAL ATOMS INTERACTING WITH STRONG LASER FIELDS

Constantine Mavroyannis
Division of Chemistry
National Research Council of Canada
Ottawa, Ontario, Canada K1A OR6

ABSTRACT. Dynamic effects occurring in the excited states of two identical atoms in the presence of a strong laser field have been considered. Emphasis has been given to dynamic effects arising from stimulated one- and three-photon processes for a system consisting of two identical three-level atoms interacting simultaneously with a strong laser field and a weak signal field. The atoms consist of an upper excited state and two lower states; they interact through their dipole-dipole interaction and radiate to each other as well. The spectral function for the signal field, which describes the stimulated one- and three-photon proceses of the symmetric and antisymmetric modes arising from the dipole-dipole and cooperative radiative interactions between the two atoms, has been calculated and compared with that of the single atom. The computed spectra are presented graphically for different values of the Rabi frequencies, dipole-dipole interactions, and for different detunings.

INTRODUCTION

Consider that there are two identical two-level atoms at a distance R apart and that at a given time, the first atom is excited and the second is in its ground state. The first atom can emit a photon, which is then absorbed by the second atom. In the initial state of the system, the first atom is excited and the second atom is in its ground state. The atoms are supposed to be close together at a distance $R < \lambda$, where $\lambda = c/\omega_b$ is the wavelength of the emitted radiation at the resonant transition frequency $\omega_{21} = \omega_b$ from the excited to the ground state of the atoms. Then this state of the atoms can be equally well represented by a superposition of states which are symmetric and antisymmetric with respect to the interchange of the atoms; these states are analogous to the singlet and triplet states of two spins. Only the symmetric state is capable of emitting a photon and decaying into the ground state of the two unexcited atoms, which is also symmetric. The radiative decay rate of the symmetric state is γ_0, which is twice that of the isolated atom $\gamma_0/2$ or, equivalently, the lifetime of the symmetric state is one-

251

P. M. Rentzepis and C. Capellos (eds.), Advances in Chemical Reaction Dynamics, 251–266.

half that of the isolated atom. The antisymmetric state is stable and
the photon is trapped between the two atoms. Dicke (1) was first to
emphasize the cooperative nature of the spontaneous emission from a
system of identical atoms and named it superradiance. The corresponding
symmetric states of the system were called superradiant states.

In the process of superradiant emission, the atoms are coupled
together by their common radiation field and, hence, they decay coopera-
tively. In this case, the intensity emitted by N atoms is proportional
to N^2 instead of N, as in the ordinary emission by isolated atoms.
Therefore, superradiance is a fundamental effect and its existence has
been verified experimentally (2). The effects of the dipole-dipole
interaction (3-5) and retardation (6) in the resonant interaction of two
identical atoms have been considered. The structure of the resonance
fluorescence spectrum arising from the cooperative effects of two and
many atoms has also been investigated (7-15).

In the presence of a strong resonant laser field operating between
the two levels of each of the two atoms at a distance $R < \lambda$, the excitation spectrum of the symmetric modes consists of Lorentzian lines describing the central peak and two pairs of sidebands, whose radiative
widths are of the order of γ_0, $3\gamma_0/2$ and $5\gamma_0/2$, respectively (9,13).
The decay rates γ_0 and $3\gamma_0/2$ are twice the corresponding ones arising
from an isolated atom interacting with the strong laser field (13). The
energy shift and the spectral width of the last pair of sidebands are
found to be two and five times larger than those for the isolated atom,
respectively. The probability of occurrence of this pair of sidebands
depends on parameters such as the ratio of the dipole-dipole interaction
energy and the Rabi frequency, as well as on the ratio of the square of
the spontaneous emission probability and the square of the Rabi frequency. Hence, the existence of this pair of sidebands is due entirely to
the cooperative behavior of the two atoms (13,15). The resonance
fluorescence spectrum of the symmetric modes of many atoms should
consist of pairs of sidebands occurring at the harmonics of the Rabi
frequency in addition to the usual three peaks of the isolated atom
(7-15).

The two-atom resonance fluorescence spectrum of the antisymmetric
modes consists of the central peak and a pair of sidebands (9). The
central peak has a delta function distribution indicating the stability
of the mode in question while the pair of sidebands is described by
Lorentzian lines which have spectral widths equal to $\gamma_0/2$ (9). When the
two atoms are taken far apart, $R \gg \lambda$, the excitation spectrum for the
symmetric and antisymmetric modes becomes identical to that of the
isolated atom interacting with the laser field. Explicit expressions
for the dipole-dipole interaction energy V_{AB}, the Rabi frequency Ω and
γ_0 can be found elsewhere (9,13,15). The laser field is considered as a
strong field when the condition $\Omega^2 \gg \gamma_0^2$ is satisfied; the reader is
referred to the first part of the lecture for further details and
literature. Excitation spectra of two-identical three-level atoms at
high photon densities have been also considered (16).

The second part of this lecture will be devoted to the description
of stimulated one-photon and three-photon processes arising from the
cooperative and dipole-dipole interactions between the two three-level

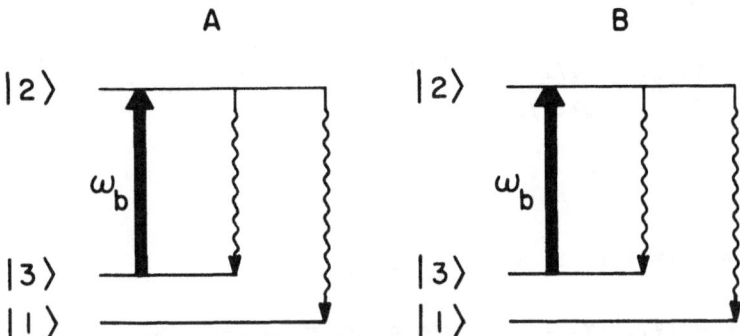

Figure 1. Energy level diagram of two identical three-level atoms, A and B, respectively. Thick lines indicate the strong laser field operating between the states $|2\rangle\leftrightarrow|3\rangle$ while wiggly lines describe the radiative decay $|2\rangle\rightarrow|3\rangle$ and $|2\rangle\rightarrow|1\rangle$, respectively.

atoms shown in Figure 1. Each atom in Figure 1 consists of an upper excited state $|2\rangle$ and two lower states $|1\rangle$ and $|3\rangle$, where the degeneracy of the ground state has been removed by an internal process or by applying an electric or a magnetic field. The strong laser field depletes the metastable state $|3\rangle$ by bringing electrons into the excited state $|2\rangle$ from where the electrons decay through the signal field into the lower ground states of the system. The two identical atoms in Figure 1 interact through their dipole-dipole interaction and radiate to each other as well. Hence, due to the cooperative interaction between the atoms, new states are generated which are symmetric and antisymmetric with respect to the interchange between the atoms. We shall describe the excitation spectrum due to the decay process for the symmetric and antisymmetric modes of the atoms depicted in Figure 1.

STIMULATED ONE-PHOTON SPECTRA

Using the Green function formalism in the limit of high photon densities, the excitation spectrum for the one-photon process of the symmetric modes has been calculated (17) when the atoms are close together at a distance R<λ, for which the spontaneous transition probabilities of the isolated atoms for the decay processes $|2\rangle\rightarrow|1\rangle$ and $|2\rangle\rightarrow|3\rangle$ as well as the cooperative ones between the atoms are equal to γ_0. For the sake of convenience, we will discuss the excitation spectrum of the symmetric modes in the absence and in the presence of the dipole-dipole interaction between the atoms separately.

Excitation Spectrum in the Absence of Dipole-Dipole Interactions

When the dipole-dipole interaction V_{AB} between the atoms is very weak, namely, in the limit when $|V_{AB}|<<\gamma_0$, then the relative intensity for the symmetric modes is given by (17)

$$F_{12}^{(\pm)}(X) = \frac{1/2}{X^2 + 1/4} - \sum_{j=1}^{4} \frac{N_j^0 \Gamma_j^0 - (X - C_j^0) M_j^0}{(X - C_j^0)^2 + (\Gamma_j^0)^2} \text{ , for } \xi = \nu = 0 \text{ ,} \tag{1}$$

where

$$X = (\omega - \omega_{21})/\gamma_0 \text{ , } \quad \eta_b = \Omega_b/\gamma_0 \text{ ,} \tag{2}$$

$$\xi = |V_{AB}|/\gamma_0 \text{ , } \quad \nu = \delta_b/\gamma_0 = (\omega_{23} - \omega_b)/\gamma_0 \text{ .} \tag{3}$$

Here X is the relative frequency and ω_{21} and ω_b are the atomic transition frequency $|2> \rightarrow |1>$ and the frequency of the laser field, respectively. Ω_b and δ_b denote the Rabi frequency and the detuning of the laser field b while η_b and ν are the corresponding relative Rabi frequency and relative detuning; ξ designates the relative dipole–dipole interaction between the atoms A and B. The numerical values of the parameters C_j^0, N_j^0, Γ_j^0, M_j^0 and N_j^0/Γ_j^0 for j=1–4 are given in Table 2 of ref. (17) for given values of η_b and $\xi = \nu = 0$. The results obtained from Eq. (1) will be presented here only graphically.

The first term on the right–hand side (rhs) of Eq. (1) describes the main peak, which is a Lorentzian line peaked at the signal frequency $\omega = \omega_{21}$ (X=0) and which has a spectral width equal to $\gamma_0/2$. The lifetime of this peak is equal to $2/\gamma_0$ which is twice the value of that arising from the corresponding term in Eq. (37) of ref. (18) for the isolated atom; units with ħ=1 are used throughout. It is shown in ref. (17)that for the stimulated $|1> \leftrightarrow |2>$ transition for the two–atom one–photon process, the decay rate of the main peak at X=0 is given by $\gamma_0/2$ for both the symmetric and antisymmetric modes, while for the corresponding spontaneous two–atom one–photon process (9,13,16), the decay rates are equal to $2\gamma_0$ and zero for the symmetric and antisymmetric modes, respectively. The last term on the rhs of Eq. (1) represents four sidebands, which have asymmetric Lorentzian profiles and which are peaked at $X = C_j^0$ and have spectral widths of the order of Γ_j^0 for j=1–4.

Using the numerical data of Table 2 of ref. (17), the relative intensity $F_{12}^{(\pm)}(X)$ given by Eq. (1) is plotted versus the relative frequency X in Figures 2a–2f for values of the relative Rabi frequency η_b equal to η_b=3,5,7,10,15 and 20, respectively. Dashed and solid lines represent the spectra of the single atom and two–atom symmetric modes, respectively. The dashed lines in Figure 2a–2f calculated from Eq. (37) of ref. (18) describe the main peak at X=0 and a pair of sidebands peaked at $X = \pm \eta_b/\sqrt{2}$; the relative intensity of the main peak and that of the pair of sidebands takes positive and negative values indicating absorption (attenuation) and stimulated emission (amplification) of the signal field at the corresponding frequencies, respectively.

In the absence of the dipole–dipole interaction between the atoms ($\xi=0$) at resonance ($\nu=0$), the difference between the single atom and the

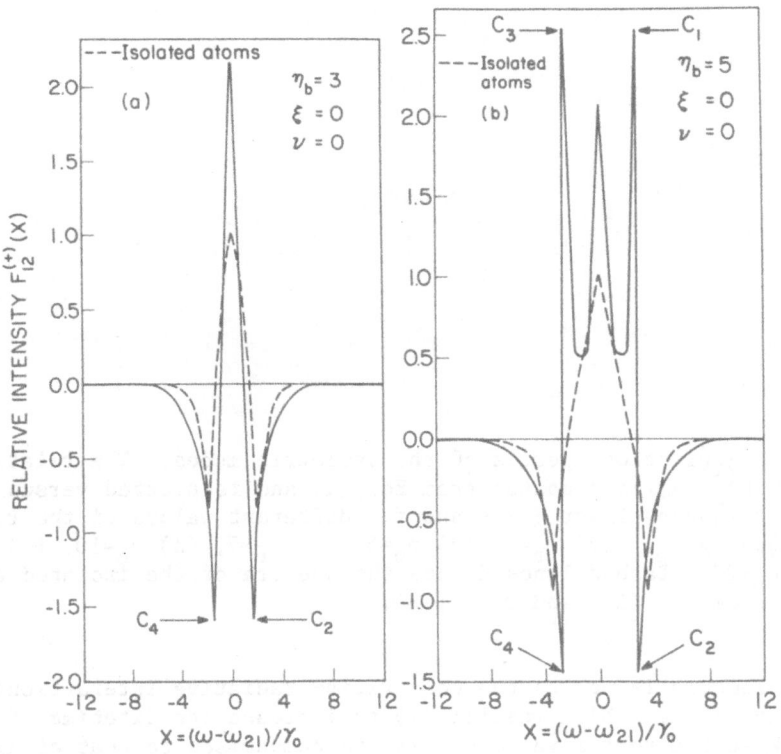

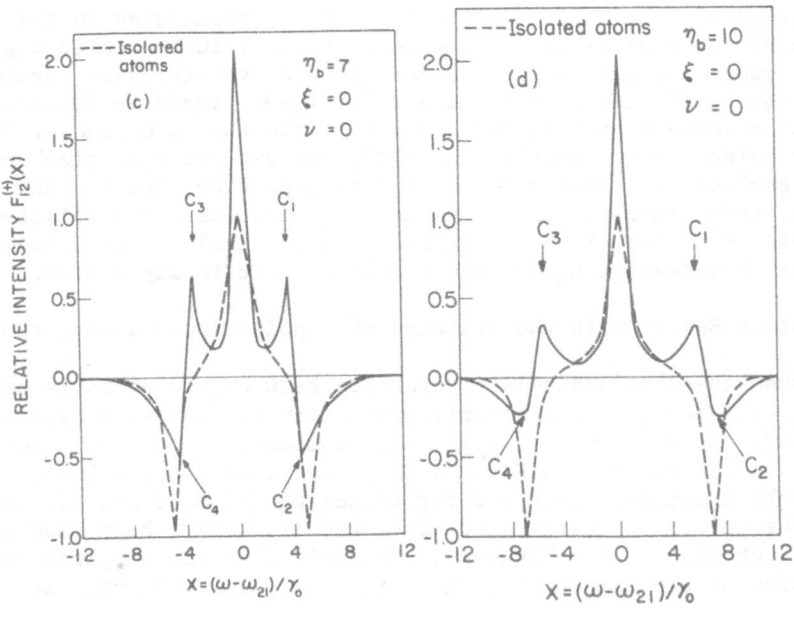

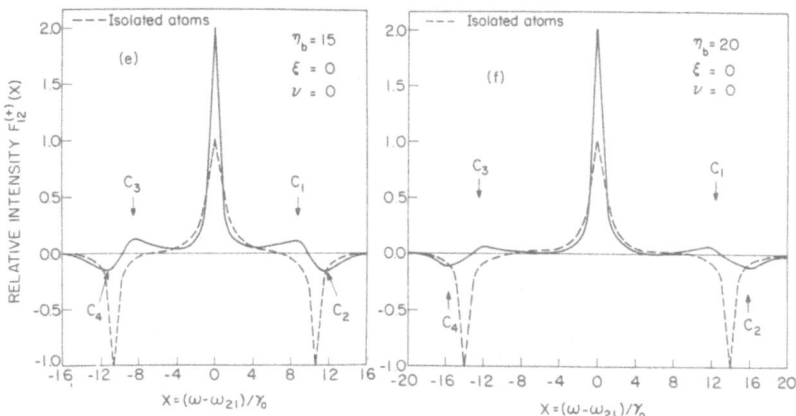

Figure 2. Excitation spectra of the symmetric modes. The relative intensity $F_{12}^{(+)}(X)$ is computed from Eq. (1) and is plotted versus the relative frequency X for $\xi=\nu=0$ and for different values of the relative Rabi frequency η_b. (a) $\eta_b=3$, (b) $\eta_b=5$, (c) $\eta_b=7$, (d) $\eta_b=10$, (e) $\eta_b=15$ and (f) $\eta_b=20$. Dashed lines denote the spectra of the isolated atom computed from Eq. (37) and ref. (18).

two-atom spectra is due to the cooperative radiative interactions between the atoms, which result: i) to increase the lifetime of the main peak at X=0 to the value of $2/\gamma_0$ in comparison to that of $1/\gamma_0$ for the isolated atom; ii) to split the one pair of sidebands of the single atom spectrum into two pairs of sidebands; iii) to vary the relative intensities of the two pairs of sidebands as a function of the relative Rabi frequency η_b, and iv) to induce strong asymmetries in the line-shapes of the side-bands. These results are illustrated in figures 2a-2f, where the two pairs of sidebands are symmetrically located from X=0. For small values of η_b, i.e., for $\eta_b=3$, Figure 2a implies that the pair of sidebands at X=±1.378 coalesces with the main peak at X=0. At higher values of η_b, namely, for $\eta_b>5$, the two pairs of sidebands are well resolved and their relative intensities take positive and negative values, respectively. As the value of η_b increases the relative intensities of the sidebands decrease, and, finally, for values of η_b greater than twenty, $\eta_b>20$, the sidebands practically vanish.

Excitation Spectrum in the Presence of Dipole-Dipole Interactions ($\xi>1$)

When the dipole-dipole interaction between the atoms is appreciable, i.e., when $\xi>1$, the relative intensity $F_{12}^{(+)}(X)$ is determined by Eqs. (27) and (28) of ref. (17). The computed spectra at resonance ($\nu=0$) are illustrated in Figures 3a-3d for a constant value of the relative Rabi frequency $\eta_b=10$ and for values of $\xi=2,5,10$ and 15, respectively. The values of the parameters C_j for $j=1-4$ have been tabulated in Table 1 of ref. (17). Figures 3a-3d imply that for $\nu=0$, the relative intensity of the main peak at X=ξ, in comparison with that at X=ξ=0 in

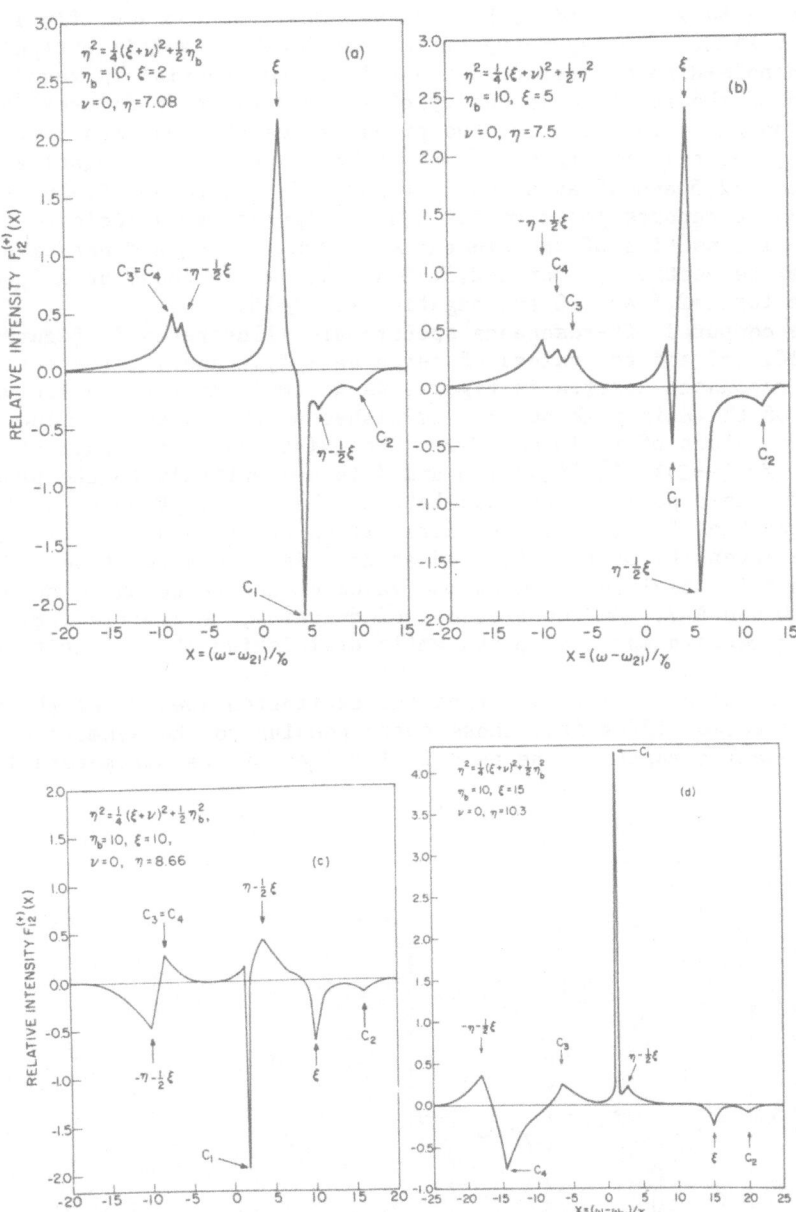

Figure 3.   As in Figure 2 but $F_{12}^{(+)}(X)$ is computed from Eq. (27) of ref. (17) for $\eta_b = 10$, zero detuning $\nu = 0$ and for different values of ξ.   (a) $\xi = 2$, (b) $\xi = 5$, (c) $\xi = 10$ and (d) $\xi = 15$.

Figure 2d, takes its maximum positive value of $\xi\approx4$, diminishes for
values of ξ between 4 and 5, becomes negative for $\xi>5$ and, finally, vanishes for $\xi>20$. The asymmetry of the main peak at $X=\xi$ is entirely due
to the dipole-dipole interaction, and it is substantial in magnitude.
The sign of the relative intensity of the sidebands in Figures 3a-3d
depends on the value of ξ for the given values of $\eta_b=10$ and $\nu=0$. For
instance, the relative intensity of the peak at $X=C_1$ is negative for
values of $\xi=2,5$ and 10 as shown in Figures 3a-3c, respectively, while in
Figure 3d it becomes positive for $\xi=15$. Figures 3a-3d indicate that the
relative intensities of the sidebands at $X=C_2$ and C_3 are negative and
positive, respectively, for $\xi=2,5,10$ and 15, while those at $X=C_4$ are
positive for $\xi=2,5$ and 10 and negative for $\xi=15$.

The computed off-resonance spectra are illustrated in Figures 4a-4c
for $\eta_b=10$, $\xi=2$ and for values of detuning $\nu=5,10$ and 20, respectively.
The off-resonance spectra in Figures 4a-4c imply that the relative intensity of the main peak at $X=\xi$ diminishes as the value of ν increases
for given values of $\eta_b=10$ and $\xi=2$. The existence of the pair of sidebands at $X=\frac{1}{2}(\nu-\xi)\mp\eta$ in Figures 3 and 4 is due entirely to the presence
of the dipole-dipole interaction between the atoms. Figures 4a-4c
illustrate that for $\nu\neq0$, the spectral lines are very asymmetric. The
relative intensity of the sideband at $X=C_1$ is always negative and
diminishes in absolute value as the value of the detuning ν increases
for $\eta_b=10$ and $\xi=2$. In the presence of detuning, asymmetries arise
enhancing certain spectral peaks while diminishing the intensity of
others.

It is shown in ref. (17) that the excitation spectra of the antisymmetric modes differ from those corresponding to the symmetric ones
only by small changes in the numerical values of the parameters involved

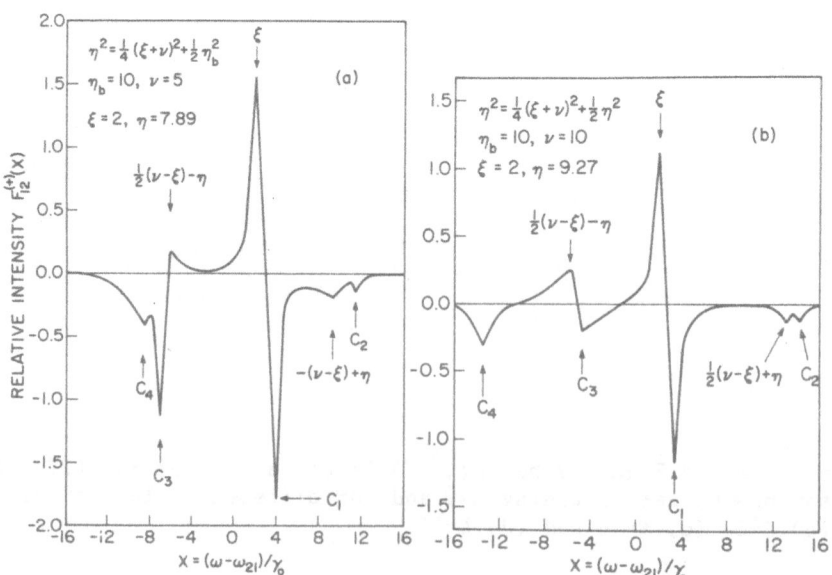

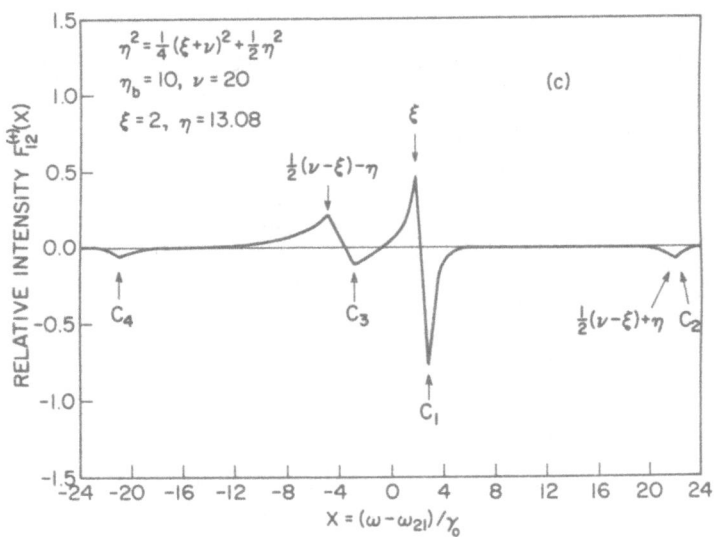

Figure 4. As in Figure 3 but for $\eta_b=10$, $\xi=2$ and different values of detuning ν. (a) $\nu=5$, (b) $\nu=10$ and (c) $\nu=20$.

which, from the physical point of view, do not play any significant role; we refer to ref. (17) for further details.

STIMULATED THREE-PHOTON PROCESSES

When the atoms are close together at distances $R<\lambda$, the relative intensity describing three-photon excitations of the symmetric modes for the system of atoms shown in Figure 1 has been found to be (19)

$$J_{12}^{(+)}(Y) = - \sum_{j=1}^{4} \frac{N_j \Gamma_j - (Y-\Omega_j)M_j}{(Y-\Omega_j)^2 + \Gamma_j^2} \,, \tag{4}$$

where the relative frequency Y is defined as

$$Y = (\omega-\omega_{21}+2\omega_b)/\gamma_0 \,, \tag{5}$$

while the numerical values for the parameters $\Omega_j, N_j, \Gamma_j, M_j$ and Γ_j/M_j for $j=1=4$ are tabulated in Tables I–IV of ref. (19). The expression (4) describes the physical process, where a photon of the signal field with frequency ω_{21} is absorbed while two photons of the laser field are emitted and vice versa.

At resonance and in the absence of the dipole-dipole interactions between the atoms, namely, when $\xi=\nu=0$, the parameters appearing in Eq.

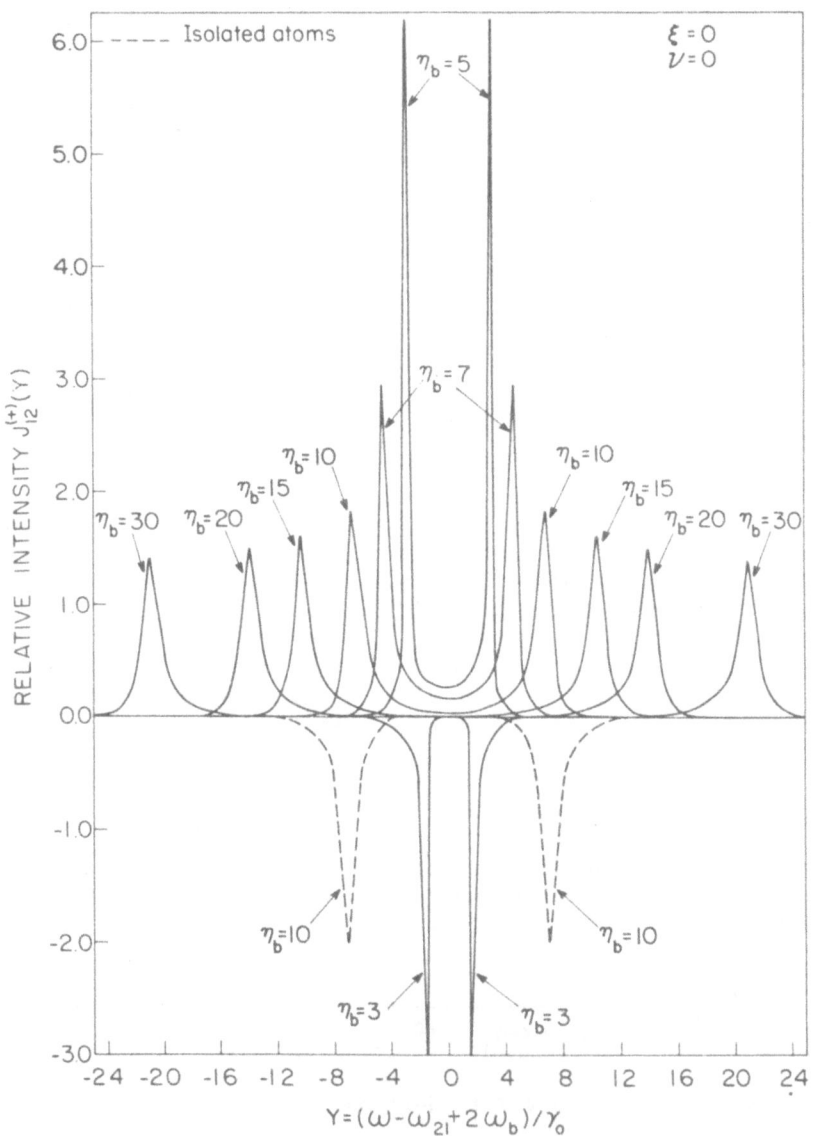

Figure 5. Three-photon spectra of the symmetric modes. The relative intensity $J_{12}^{(+)}(Y)$ determined by Eq. (4), where the parameters are given in Table I of ref. (19), is plotted versus the relative frequency Y for $\xi = \nu = 0$ and for different values of the relative Rabi frequency $\eta_b = 3-30$. Dashed lines denote the spectra of the isolated atoms computed from Eq. (38) of ref. (18). The relative intensity of the doublet for $\eta_b = 3$ is cut-off at the value -3.0 (units) while its real value is -12.881 (units as indicated in Table I of ref. (19).

(4) are given in Table I of ref. (19) for various values of the relative Rabi frequency η_b=3-100 while the corresponding spectra are depicted in Figure 5. The data from Table I of ref. (19) imply that the function $J_{12}^{(+)}(Y)$ is described by two doublets which are peaked at the frequencies $Y=\pm\Omega_1(\Omega_2=-\Omega_1)$ and $Y=\pm\Omega_3(\Omega_4=-\Omega_3)$ and which have spectral widths of the order of $\Gamma_1=\Gamma_2$ and $\Gamma_3=\Gamma_4$, respectively. In this case, each pair of sidebands is symetrically located from $Y=0$. However, for $\xi=\nu=0$ and $\eta_b=$ 3-100, only one of the doublets has an appreciable intensity while the intensity of the other is vanishingly small. Thus, in this case, the function $J_{12}^{(+)}(Y)$ describes only one pair of sidebands whose relative intensity is negative or positive depending on the value of the relative Rabi frequency η_b, for instance, for η_b=3 the relative intensity of the doublet at $Y=\Omega_1=-\Omega_2=\pm1.661$ takes the negative value of -12.881 (units) indicating that amplification of the signal field takes place. For values of η_b equal and greater to 5, i.e., for η_b>5, the relative intensity of the doublet becomes positive while its magnitude decreases as the value of η_b increases; it changes slowly for high values of η_b until it becomes nearly constant for values of η_b between 50 and 100. This behaviour of the intensity of the doublet is illustrated in Figure 5 by solid lines while dashed lines depict the doublet for η_b=10 describing the spectra of the single atom, which are determined from Eq. (38) and Figure 4 of ref. (18). Figure 5 indicates that the relative intensity of the doublet for the single atom case at resonance (ν=0) takes always constant and negative values independently of η_b while the corresponding intensity for the two-atom doublet varies with the value of η_b and becomes a constant for values of η_b between 50 and 100; it also changes from negative to positive values for values of η_b>5. Since ξ=0 indicates the absence of dipole-dipole interactions between the atoms, the variation of the intensity of the doublet with respect to η_b as well as the induced asymmetries in Figure 5 are due entirely to the cooperative interaction between the atoms. This effect is similar to that found for the stimulated one-photon spectra.

The values of the parameters for the off-resonance ($\nu\neq0$) case are given in Table II and IV of ref. (19) for η_b=10 and for the values of ξ=0 and ξ=2, respectively, while the corresponding spectra are illustrated in Figures 6a and 6b. In this case, the positions of the sidebands become completely asymmetric. Figure 6a implies that for ν=5, the relative intensity of the peak at $Y=\Omega_3=-7.721$ is positive and equal to 7.455 (units) becoming abruptly negative for a very small change in the value of Y because of the large value of the asymmetry M_3=0.332 (units), while the relative intensity of the peak at $Y=\Omega_4$=4.888 is negative. For values of ν=10 and 20, the intensities of the corresponding peaks take negative values as indicated in Figure 6a. Comparison between Figure 3a with the corresponding one for the isolated atom given by Figure 4 of ref. (18) implies that the spectra in the two cases are completely different, and the different behaviour is attributed to the cooperative radiative interaction between the two atoms. Figure 6b indicates that there are three peaks for ν=5, where the relative intensity of one of them is positive while the remaining two peaks have negative relative intensities. For ν=10 and 20, there are two and three peaks, respectively, whose relative intensities take negative values. Hence, the

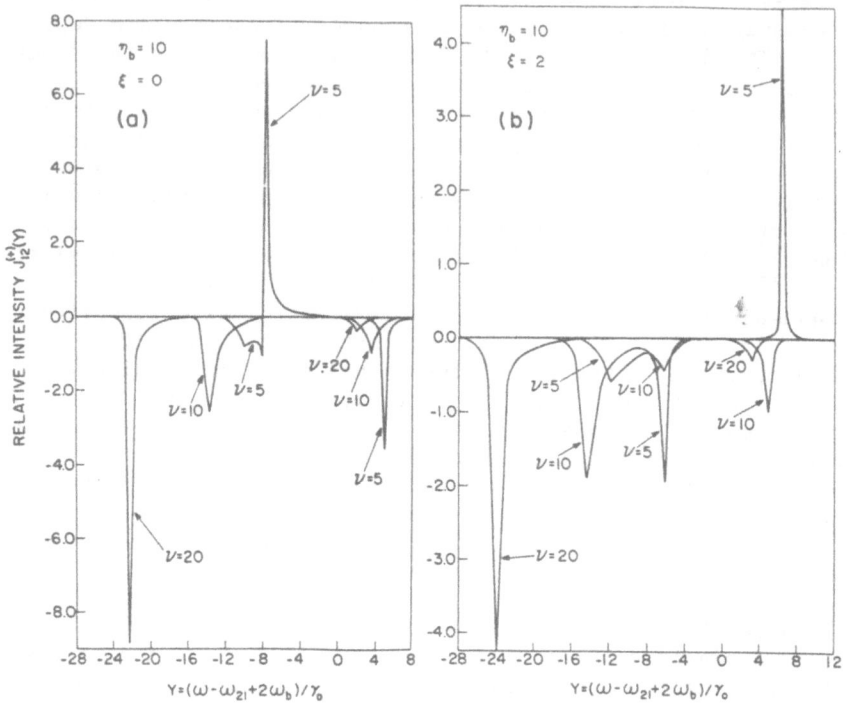

Figure 6. As in Figure 5 but for η_b=10, detunings ν=5, 10 and 20 and for different values of ξ. (a) ξ=0 and (b) ξ=2. The data are taken from Tables II and IV of ref. (19) for ξ=0 and ξ=2, respectively.

effect of finite detuning is to induce asymmetries enhancing some peaks while diminishing the intensity of others; the presence of the detuning changes the sign of the intensities of some peaks as well.

In the presence of the dipole–dipole interaction between the atoms, i.e., when ξ>1, the parameters appearing in Eq. (4) for η_b=10 and at resonance (ν=0) are given in Table III of ref. (19) while the corresponding spectra are depicted in Figure 7. The data in Table III of ref. (19) and Figure 7 imply that at least three of the four peaks designated by Eq. (4) have appreciable relative intensities. The sign and magnitude of these relative intensities as well as the positions and asymmetries of the peaks depend entirely on the values of ξ, namely, on the strength of the dipole–dipole interaction between the atoms. To make a comparison, we consider the case, where η_b=10 and ν=0, then for ξ=0, the relative intensity of the doublet for the one–atom (dashed lines) and two–atom case (solid lines) in Figure 5 takes a negative and positive value, respectively. On the other hand, in Figure 7 for $\xi\neq0$, the relative intensity of one of the peaks takes a positive value while for the other two peaks the relative intensities take negative values

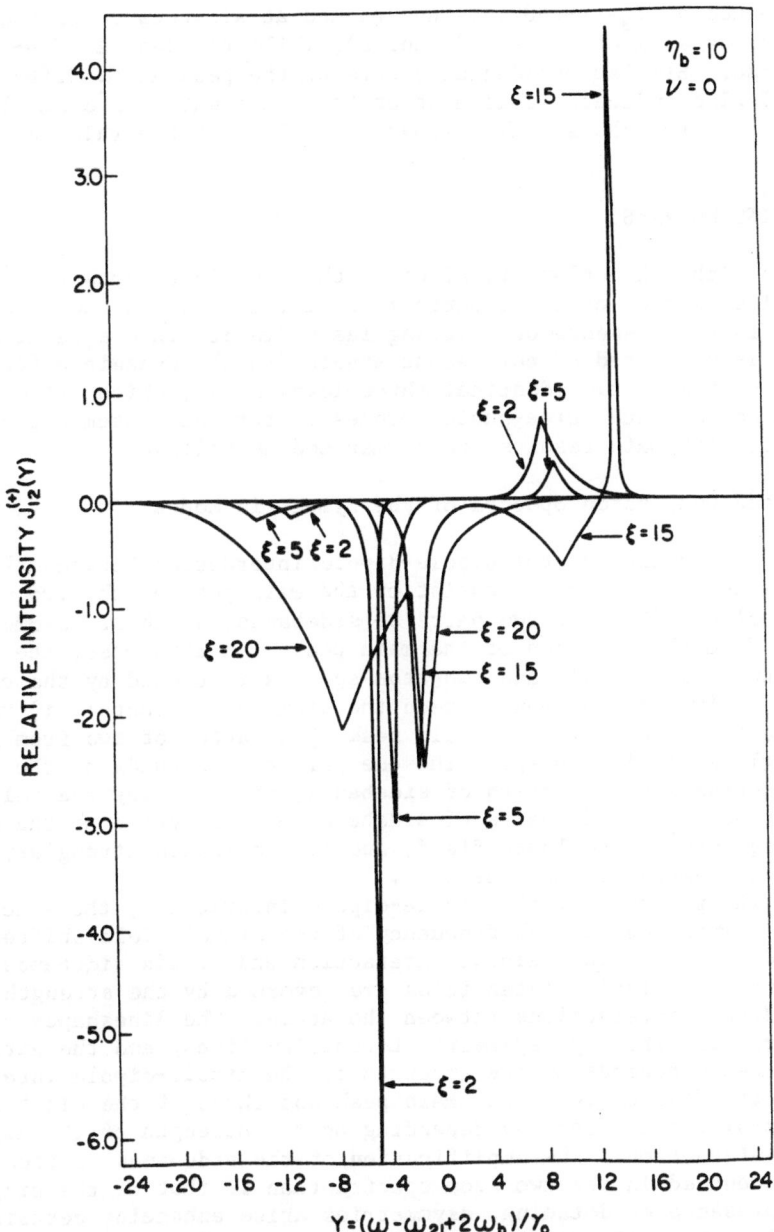

Figure 7. As in Figure 5 but the data are taken from Table III of ref. (19) for η_b=10, ν=0 and for different values of ξ=2,5,15 and 20.

for a given value of ξ. As the value of ξ increases, the value of M_1 increases to the extent that for ξ=20, the sideband, which is supposed

to be peaked at $\Omega_3=-19.682$, is not peaked at all because of the high
value of the asymmetry $M_3=1.59$ (units), while the peak at $\Omega_2=-7.438$ is
very broad. Similar broadening occurs of the peak at $\Omega_1=9.883$ for $\xi=15$.
Figure 7 also indicates that most of the peaks which have small relative
intensities are substantially broad regardless of the value of ξ.

CONCLUDING REMARKS

To begin with, an outline is given of the dynamic effects that occur in
the excited states of two identical two-level atoms in the absence as
well as in the presence of a strong laser field. Then a detailed discussion is presented of our recent studies on the dynamic effects of the
excited states of two identical three-level atoms, which arise in the
stimulated one- and three-photon processes for the system depicted in
Figure 1. The main results are summarized as follows.

Stimulated One-Photon Spectra of the Symmetric Modes

In the absence of the dipole-dipole interaction between the atoms
at resonance, the spectra consist of the main peak at the frequency of
the signal field and of two pairs of sidebands, which are symmetrically
located from the position of the main peak. In this case, the difference between the single and two-atom spectra is caused by the cooperative radiative interaction between the atoms and amounts: i) to
increase the lifetime of the main peak by a factor of two from that of
the single atom; ii) to split the one pair of sidebands of the single
atom spectrum into two pairs of sidebands; iii) to vary the relative
intensities of the two pairs of sidebands as a function of the relative
Rabi frequency of the laser field, and iv) to induce strong asymmetries
in the lineshapes of the sidebands.
In the presence of the dipole-dipole interaction, the spectra consist of a main peak at the frequency of the signal field shifted by the
magnitude of the dipole-dipole interaction and of six sidebands, whose
positions and relative intensities are governed by the strength of the
dipole-dipole interactions between the atoms. The lineshapes of all
peaks are described by asymmetric Lorentzian lines, and the extent of
the asymmetry depends on the strength of the dipole-dipole interaction.
The relative intensity of the main peak and those of the sidebands are
either positive or negative depending on the strength of the dipole-
dipole interaction. The amplification of the sidebands is found to be
more pronounced in the two-atom spectra than in that of the single atom.
In the presence of detuning, asymmetries arise enhancing certain
spectral peaks while diminishing the intensity of others.

Stimulated Three-Photon Spectra of the Symmetric Modes

At resonance and in the absence of the dipole-dipole interaction,
the spectra consist of a doublet whose peaks are symmetrically located
from the frequency $\omega=\omega_{21}-2\omega_b$, while the magnitude and sign of the relative intensity of the peaks depend on the value of the relative Rabi

frequency η_b. For $\eta_b = 3$, the relative intensity of the doublet takes negative values, and it becomes positive for $\eta_b > 5$. As the value of η_b increases, the magnitude of the relative intensity of the doublet decreases, and it becomes nearly constant for values of η_b between 50 and 100. The single atom spectra consist of a doublet having a constant negative relative intensity independent of η_b. The difference between the single and two-atom spectra is attributed to the cooperative radiative interaction between the two atoms. In the presence of detunings, the symmetry in the positions and relative intensities of the two sidebands vanishes. In the presence of the dipole–dipole interaction and at resonance, the spectra consist of three sidebands whose position, lineshape, and sign, as well as magnitude of the relative intensities depend entirely on the strength of the dipole–dipole interactions between the atoms.

For the spectra of the antisymmetric modes, it is concluded in refs. (17) and (19) that no significant changes occur in the spectra of the stimulated one- and three-photon excitations, which arise from the symmetric and antisymmetric combination of the two atoms, respectively. Of course, this conclusion is correct within the limits of the model and approximations that have been made in our study.

REFERENCES

1. Dicke, R.H.: 1954, Phys. Rev. 93, p. 99.

2. Feld, M.S. and MacGillivray, J.C.: 1980, Coherent Nonlinear Optics, Recent Advances, eds. Feld, M.S., and Letokhov, V.S., Springer-Verlag, New York, pp. 7–54.

3. Stephen, M.J.: 1964, J. Chem. Phys. 40, p. 669.

4. Hutchinson, D.A. and Hameka, H.F.: 1964, J. Chem. Phys. 41, p. 2006.

5. MacLachlan, A.D.: 1964, Mol. Phys. 8, p. 409.

6. Milonni, P.W., and Knight, P.L.: 1974, Phys. Rev. A, 10, p. 1096.

7. Senitzky, I.R.: 1972, Phys. Rev. A, 6, p. 1171.

8. Agarwal, G.S., Brown, A.C., Narducci, L.M., and Vetri, B.: 1977, Phys. Rev. A, 15, p. 1613.

9. Mavroyannis, C.: 1978, Phys. Rev. A, 18, p. 185.

10. Amin, A.S. and Cordes, J.G.: 1978, Phys. Rev. A, 18, p. 1298.

11. Senitzky, I.R.: 1978, Phys. Rev. Lett. 40, p. 1334.

12. Agarwal, G.S., Narducci, L.M., Feng, D.H. and Gilmore, R.: 1979, Phys. Rev. Lett. 42, p. 1260.

13. Mavroyannis, C.: 1980, Optics Commun. 33, p. 42.

14. Agarwal, G.S., Saxena, R., Narducci, L.M., Feng, D.H. and Gilmore, R.: 1980, Phys. Rev. A, 21, p. 257.

15. Mavroyannis, C.: 1980, Phys. Rev. A, 22, p. 1129.

16. Mavroyannis, C.: 1979, Physica 99A, p. 435; 1982, Ibid., 110A, p. 431; 1984, Ibid., 125A, p. 398; 1984, Ibid., 127A, p. 407.

17. Mavroyannis, C.: 1985, J. Math. Physics, submitted.

18. Mavroyannis, C.: 1985, J. Math. Phys. 26, p. 1093.

19. Mavroyannis, C.: 1985, J. Math. Phys., submitted.

STIMULATED ONE-PHOTON AND MULTIPHOTON PROCESSES IN STRONGLY DRIVEN MULTILEVEL ATOMIC SYSTEMS

Constantine Mavroyannis

Division of Chemistry, National Research Council of Canada
Ottawa, Ontario, Canada K1A OR6

ABSTRACT

We derive and discuss the behaviour of several multilevel atomic systems relevant to the problem of stimulated one-photon and multiphoton processes under the simultaneous action of strong laser fields and weak signal fields. The systems considered here consist of a single two-level atom, whose level degeneracies of the ground and excited states have been removed by internal process or by applying an external electric or magnetic field, respectively. In this system, one of the two ground state levels of the atom is a metastable one and its electron population is depleted through the action of either one laser field (three-level atom) or two laser fields (four-level atom), respectively. The strong laser fields deplete the metastable state by bringing the electrons into the excited states from where the electrons decay through the signal fields into the lower ground states of the system. For such systems, the fluorescent spectra of the signal field are described by the stimulated processes of one-photon, three-photon, and two-photon Raman, respectively. The dynamic aspects of such processes are discussed in detail. It is shown that the intensities of the sidebands for the processes in question take negative values indicating that a strong amplification of the signal field occurs at the corresponding frequencies. The computed spectra for the processes in question are presented graphically for different values of the Rabi frequencies and detunings. The derived results are discussed in detail and compared with the observed spectra.

P. M. Rentzepis and C. Capellos (eds.), Advances in Chemical Reaction Dynamics, 267–287.
© *1986 by D. Reidel Publishing Company.*

INTRODUCTION

Considerable interest has been recently directed towards the study of the amplification of sidebands occurring in strongly driven two-level atomic systems without population inversion. Mollow (1) was first to consider the optical amplification arising in a two-level atomic system interacting with a strong pump field and a weak signal field simultaneously. His treatment (1) is a semiclassical one where the pump field and the signal field are both treated classically while the electron states are quantized. Mollow (1) has shown that at high pump intensities the spectral function takes negative values indicating amplification (stimulated emission) of the signal field. It has been shown (1) that the amplification of the signal field occurs at the expense of the pump field, whose rate of attenuation increases as the rate of amplification of the signal field increases. Mollow's theoretical predictions have been confirmed experimentally by Wu et al. (2) in a Na atom beam experiment. Galbraith et al. (3) have extended Mollow's treatment to calculate double-resonance line shapes for arbitrary angular momentum states of molecules.

The amplification of weak sidebands by a strongly driven two-level vapor system without population inversion has been observed in microwave (4,5), rf. (6), and optical (7,8) transitions. Tam (9) was first to observe large sideband amplifications in self-focused light beams in alkali-metal-atom vapors at high pump intensities. The observed strong sideband amplification has been interpreted (9) as an off-resonance stimulated Raman scattering.

The purpose of this lecture is to discuss the excitation spectra arising from a three-level atom interacting with a strong laser field and a weak signal field simultaneously. The limit of the strong laser field operating between two atomic states is realized when the Rabi frequency induced by the laser field in question is much higher than the spontaneous emission probability describing the decaying process between the two atomic states.

STIMULATED PROCESSES IN STRONGLY DRIVEN THREE-LEVEL ATOMIC SYSTEM

We consider a three-level atom whose nondegenerate energy levels are depicted in Figure 1. The atom consists of an upper excited state and two lower states, where the degeneracy of the ground state has been removed by applying an internal or an external electric or a magnetic field. The energies of the lower ground state $|1>$, the upper ground state $|3>$ and the excited state $|2>$ are denoted by ω_1, ω_3 and ω_2, respectively, and the transition frequencies $\omega_{ij}=\omega_i-\omega_j$ with $i,j=1,2$ and 3, where units in which $\hbar=1$ are used throughout. The parity of the state $|2>$ is assumed to be different from those of $|1>$ and $|3>$, hence, the electron transitions $|1> \leftrightarrow |2>$ and $|2> \leftrightarrow |3>$ are electric dipole allowed while the transition $|1> \leftrightarrow |3>$ is electric

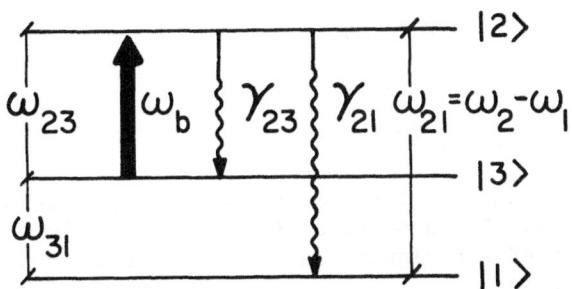

Figure 1. Energy-level diagram of a three-level atom. The solid
line indicates the laser field operating between the states $|2>$ and
$|3>$, $|2> \leftrightarrow |3>$. Wiggly lines describe the radiative decays $|2> \leftrightarrow |3>$
and $|2> \leftrightarrow |1>$, respectively.

dipole forbidden. The atom is pumped between the levels $|3>$ and $|2>$,
$|2> \leftrightarrow |3>$, by a strong laser field whose frequency mode ω_b is initial-
ly populated and is near resonance with the transition frequency $\omega_{23} = \omega_2 - \omega_3$. The atomic states are simultaneously coupled to the remaining
modes of the electromagnetic field (signal or vacuum field), those
being initially empty.

In Figure 1, the strong laser field operating between the states
$|2> \leftrightarrow |3>$ is indicated by the thick line, while the signal field,
which is assumed to be a weak perturbing field, describes the radia-
tive decays $|2> \rightarrow |3>$ and $|2> \rightarrow |1>$, respectively, indicated by wiggly
lines. Since the transition $|1> \leftrightarrow |3>$ is electric dipole forbidden
for parity considerations the electronic state $|3>$ may be considered
as a metastable state and, therefore, it is initially populated while
the states $|1>$ and $|2>$ are initially empty. The electron population
of the state $|3>$ is depleted by the action of the laser field which
excites the electrons into the state $|2>$, where the electrons emit
photons through the signal field and decay into the states $|3>$ and
$|1>$ simultaneously. The energy shifts (Lamb shifts) induced by the
signal field are discarded as being negligibly small in comparison to
the corresponding Rabi frequency induced by the laser field.

Excitation Spectra (Quantized Laser Field)

A model Hamiltonian has been considered (10) where all free and
interacting fields, namely, the electron, the laser and signal fields
are quantized and describe the atomic system depicted in Figure 1.
Then using this Hamiltonian and initial conditions with state $|3>$
occupied, and the Green function formalism (11,12) in the limit of
high photon densities of the laser field, the spectral function for
the signal field has been calculated (10) in the form

$$P_{12}(\omega) = P^{(1)}(\omega) + P^{(3)}(\omega) + R^{(2)}(\omega) , \tag{1}$$

where the functions

$$P^{(1)}(\omega) , \quad P^{(2)}(\omega) , \quad R^{(2)}(\omega) ,$$

describe stimulated one-photon, three-photon and two-photon Raman processes, respectively. The expression (1) for $P_{12}(\omega)$ defines the absorption coefficient of the signal field describing the transition $|1> \rightarrow |2>$ while the function

$$P_{21}(\omega) = P_{12}(-\omega)$$

describes the corresponding emission process $|2> \rightarrow |1>$ and can be derived from Eq. (1) if ω is replaced everywhere by $-\omega$. Hence, positive and negative values of the function $P_{12}(\omega)$ indicate that the physical processes of absorption (attenuation) and stimulated emission (amplification) of the signal field take place at the frequencies in question, respectively. In deriving the expression (1), the hierarchy of the Green functions has been truncated through a decoupling scheme, which takes into account photon-photon correlations in the limit of high photon densities of the laser field (10,13-17). For the sake of convenience, the functions that appear on the right-hand side (rhs) of Eq. (1) will be discussed separately.

Stimulated One-Photon Process

The expression for the absorption coefficient describing stimulated one-photon process is found to be (10)

$$P^{(1)} = \frac{(\bar{n}_3 - \bar{n}_1)}{2\pi\gamma_+^0} F_{12}(X) , \tag{2}$$

where $\bar{n}_1$ and $\bar{n}_3$ denote the average values of the electron density operators of the states $|1>$ and $|3>$, respectively, and

$$\gamma_+^0 = (\gamma_{21}^0 + \gamma_{23}^0)/2 \tag{3}$$

with γ_{21}^0 and γ_{23}^0 representing the spontaneous radiative transition probabilities for the transition $|2> \rightarrow |1>$ and $|2> \rightarrow |3>$ and defined by

$$\gamma_{21}^0 = \frac{4}{3}(\omega_{21}/c)^3 |\vec{\mu}_{21}|^2 , \quad \gamma_{23}^0 = \frac{4}{3}(\omega_{23}/c)^3 |\vec{\mu}_{32}|^2 . \tag{4}$$

In Eq. (4), $\vec{\mu}_{21}$ and $\vec{\mu}_{32}$ are the matrix elements of the electric dipole moment operators for the transitions $|1> \leftrightarrow |2>$ and $|2> \leftrightarrow |3>$, respectively, and c is the speed of light. The spectral function $F_{12}(X)$ in Eq. (2) is determined by

$$F_{12}(X) = 1/(X^2+1) - \tfrac{1}{4}L(X,\nu_b,\eta) - \tfrac{1}{4}L(X,\nu_b,-\eta) , \tag{5}$$

where the functions $L(X, \nu_b, \pm\eta)$ describe the shape of the sidebands at the relative frequencies $X = -\frac{1}{2}\nu_b \pm \eta$, respectively, and are defined as

$$L(X, \nu_b, \pm\eta) = \frac{\frac{1}{2}(1 \pm \nu_b/2\eta) \pm (X + \frac{1}{2}\nu_b \mp \eta)/2\eta}{(X + \frac{1}{2}\nu_b \mp \eta)^2 + \frac{1}{4}} . \tag{6}$$

In Eqs. (5) and (6) use has been made of the following notation

$$X = (\omega - \omega_{21})/\gamma_+^0 , \qquad \nu_b = (\omega_{23} - \omega_b)/\gamma_+^0 , \tag{7}$$

$$\eta_b = \Omega_b/\gamma_+^0 , \qquad \eta^2 = \frac{1}{4}\nu_b^2 + \frac{1}{2}\eta_b^2 , \tag{8}$$

$$\Omega_b^2 = \omega_p^2 f_b(\frac{1}{2} + \bar{n}_b) \approx \omega_p^2 f_b \bar{n}_b , \qquad f_b = f_{23}(\omega_{23}/\omega_b) , \tag{9}$$

$$\omega_p^2 = 4\pi e^2/mV , \qquad \bar{n}_b = \langle \beta_b^+ \beta_b \rangle . \tag{10}$$

In Eqs. (5)-(10), X is the reduced frequency of the signal field and Ω_b is the Rabi frequency of the laser field. ν_b and η_b represent the relative detuning and the relative Rabi frequency of the laser field, respectively, while η denotes the total relative frequency shift arising from the Rabi frequency and the detuning of the laser field. The function f_{23} is the oscillator strength for the transition $|2\rangle \leftrightarrow |3\rangle$, ω_p is the plasma frequency, m is the electron mass, $-e$ is the electron charge and V is the volume of the sample container. $\bar{n}_b$ is the average number of photons of the laser field, which in the high photon density limit considered here, takes the value $\bar{n}_b \gg 1$. Within this convention the radiative width appears in units of γ_+^0.

The spectral function $F_{12}(X)$ given by Eq. (5) describes the excitation spectra near the reduced frequency X. The first term on the rhs describes the central peak of the signal field, which is a Lorentzian line with maximum relative intensity equal to unity ($h_c = 1$) at X=0 and halfwidth equal to γ_+^0. The second and third terms respectively describe a pair of sidebands with maximum relative intensities at zero detunings ($\nu_b = 0$) equal to $-1(h_\pm = -1)$ which are peaked at $X = \pm\eta$; the halfwidth of each sideband is $\frac{1}{2}\gamma_+^0$. The shape of the sidebands is determined by Eq. (6), which are asymmetric Lorentzian lines peaked at $X = \pm\eta - \frac{1}{2}\nu_b$ with negative relative intensities, and the shape of the lines depends strongly on the value of the detuning ν_b. The ratio of the maximum height h_c of the central peak at X=0 to those of the sidebands $h_\pm$ at $X = \pm\eta - \frac{1}{2}\nu_b$, respectively, is given by

$$h_c/h_\pm = -1/(1 \pm \nu_b/2\eta) . \tag{11}$$

Since the intensity of the central peak is positive while those
of the sidebands are negative, the spectral function $F_{12}(X)$ given by
Eq. (5) describes the physical process of absorption (attenuation) at
$X=0$ and stimulated emission (amplification) at $X=\pm\eta-\frac{1}{2}\nu_b$ with $\nu_b=0$ as
well as with $\nu_b\neq0$, respectively. As an illustration, the function
$F_{12}(X)$ denoted as the relative intensity = $F_{12}(X)$ is plotted versus
the reduced relative frequency X in Figs. 2a–d for a constant value
of $\eta_b=10$ and various values of detunings ν_b. Figure 2a illustrates

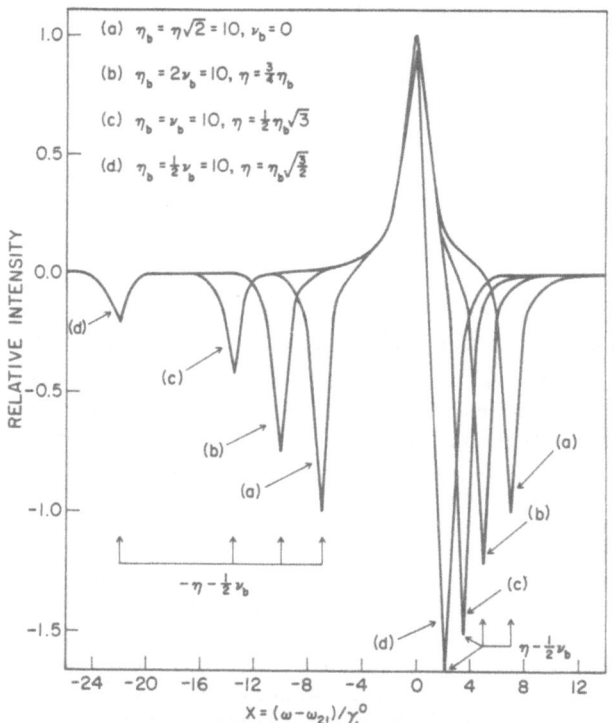

Figure 2. One–photon spectra. The relative intensity = $F_{12}(X)$ is
computed from the rhs of Eq. (5) and is plotted versus the relative
frequency $X=(\omega-\omega_{21})/\gamma_+^0$ for the relative Rabi frequency $\eta_b=10$ and
various detunings. (a) $\nu_b=0$, (b) $\nu_b=5$, (c) $\nu_b=10$ and (d) $\nu_b=20$.

the spectra at resonance ($\nu_b=0$) and for $\eta_b=10$, where the relative
intensity of the central peak at $X=0$ and those of the sidebands at
$X=\pm\eta_b/\sqrt{2}$ are equal but opposite in sign indicating that signal field
absorption and amplification will take place at the corresponding
frequencies. Figures 2b–d depict the spectra for values of ν_b equal

to 5, 10 and 20, respectively. It is shown from Figs. 2b-d and Eq. (6) that as the value of the detuning increases, the intensities of the sidebands at $X=\eta-\frac{1}{2}\nu_b$ and $X=-\eta-\frac{1}{2}\nu_b$ increase and decrease, respectively, provided that $\nu_b/2\eta<1$, which is always true for $\eta_b>1$. In the limit for values of $\nu_b>\eta_b$ for $\nu_b/2\eta<1$, $h_c\to1$, $h_-\to0$ while h_+ takes its maximum negative value $h_+\to-2$ at $X=\eta-\frac{1}{2}\nu_b$. A schematic representation of the splitting of the excited state $|2>$ is shown in Fig. 3.

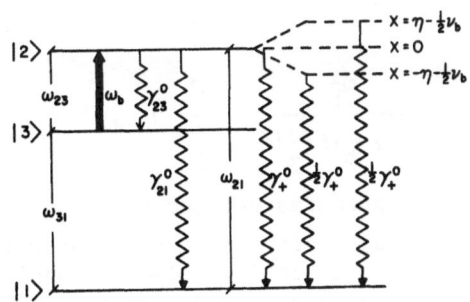

Figure 3. Schematic representation of the energy splitting at the relative frequency $X=(\omega-\omega_{21})/\gamma_+^0$.

Stimulated Three-Photon Processes

The second term on the rhs of Eq. (1) represents stimulated three-photon absorption and it is found to be (10)

$$P^{(3)}(\omega) = \frac{(\bar{n}_3-\bar{n}_1)\Omega^2_b}{8\pi\gamma_+^0 d^2_{12}(\omega)} J_{12}(Y) , \tag{12}$$

where the propagator $d_{12}(\omega)$ is defined by

$$d_{12}(\omega) = \omega-\omega_{21}+i\gamma_+^0 , \tag{13}$$

and the spectral function $J_{12}(Y)$ is determined by

$$J_{12}(Y) = -L(Y,\nu_b,\eta)-L(Y,\nu_b,-\eta) , \tag{14}$$

where the function $L(Y,\nu_b,\pm\eta)$ is derived from Eq. (6) when X is replaced everywhere by $Y, X\leftrightarrow Y$. In Eqs. (12) and (14), Y is defined as

$$Y = (\omega-\omega_{21}+2\omega_b)/\gamma_+^0 , \tag{15}$$

and denotes the reduced frequency for the three-photon process. The spectral function $J_{12}(Y)$ represents stimulated three-photon absorption spectra, which are generated near the frequency $\omega=\omega_{21}-2\omega_b>0$, where two photons of the laser field with frequency $2\omega_b$ are absorbed while a photon of the signal field with frequency ω_{21} is emitted. In Eq. (12), the factor

$$\Omega_b^2/4d_{12}^2(\omega) \tag{16a}$$

may be taken as

$$\frac{\Omega_b^2}{4d_{12}^2(\omega)} = \frac{\Omega_b^2}{4(\omega-\omega_{21}+i\gamma_+^0)^2} \approx \frac{\Omega_b^2}{4(-2\omega_b+i\gamma_+^0)^2} \approx \left(\frac{\Omega_b}{4\omega_b}\right)^2 , \tag{16b}$$

where ω has been replaced by its approximate value at $\omega\approx\omega_{21}-2\omega_b$. The factor given by the expression (16a) implies that the spectral function $J_{12}(Y)$ in Eq. (12) describes stimulated third-order nonlinear spectra near the frequency Y. At resonance $\nu_b=0$, and for $\omega_{21}=\omega_a$, the expressions (12) and (14) are identical to the corresponding ones given by Eqs. (11) and (12) in ref. (18), which describe the spectra near the frequency $\omega=\omega_a-2\omega_b$.

The spectral function $J_{12}(Y)$ given by Eq. (14) describes a doublet with maximum intensities $i_\pm$ at the frequencies $Y=\pm\eta-\frac{1}{2}\nu_b$ equal to

$$i_\pm = -2(1\pm\nu_b/2\eta) , \tag{17}$$

respectively, and halfwidths equal to $\frac{1}{2}\gamma_+^0$. The intensities of the doublet are always negative and, hence, the physical process of amplification is expected to take place at the frequencies $Y=\pm\eta-\frac{1}{2}\nu_b$ and the shape of the doublet depends on the value of the ratio $\nu_b/2\eta$. The spectra are illustrated in Figures 4a-d, where the function $J_{12}(Y)$ is plotted as $J_{12}(Y)$=relative intensity versus the relative frequency Y for the Rabi frequency $\eta_b=10$ and various detunings ν_b. The resonance case, $\nu_b=0$, is depicted in Figure 4a, which is identical to Figure 2a in ref. (18) when $\omega_{21}=\omega_a$. In this case the maximum relative intensity of the doublet remains constant and equal to -2. Figures 4b-d illustrate the spectra for values of the detunings $\nu_b=5$, 10 and 20, respectively. It is shown that the shape of the doublet in Figures 4b-d depends strongly on the value of the ratio $\nu_b/2\eta$. A schematic representation of the energy splitting which occurs at the relative frequency Y is depicted in Figure 5.

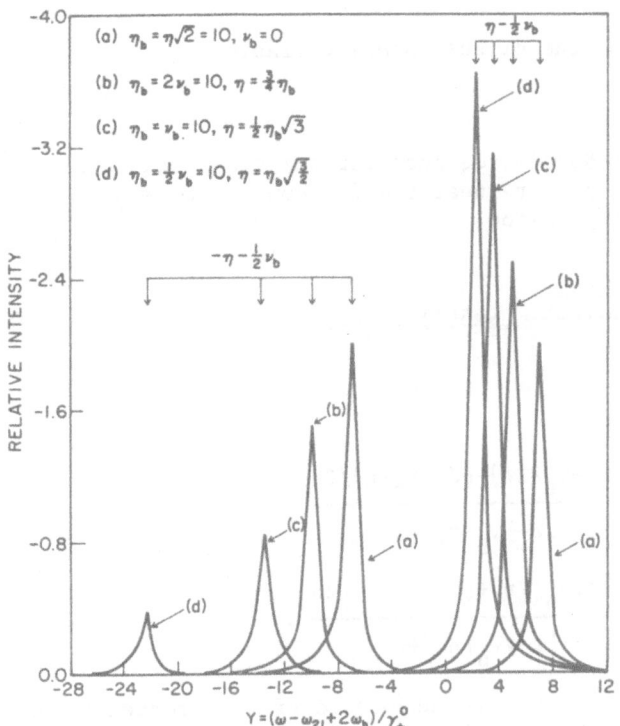

Figure 4. Three-photon spectra. The relative intensity=$J_{12}(Y)$ is computed from the rhs of Eq. (14) and is plotted versus the relative frequency $Y=(\omega-\omega_{21}+2\omega_b)/\gamma_+^0$ for the relative Rabi frequency $\eta_b=10$ and various detunings. (a) $\nu_b=0$, (b) $\nu_b=5$, (c) $\nu_b=10$ and (d) $\nu_b=20$.

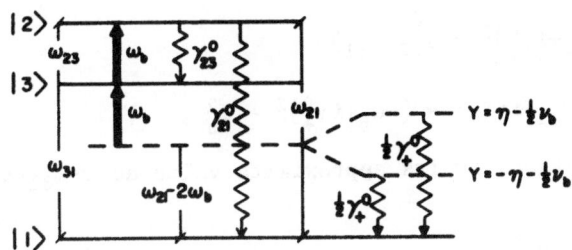

Figure 5. Schematic representation of the energy splitting at the relative frequency $Y=(\omega-\omega_{21}+2\omega_b)/\gamma_+^0$.

Stimulated Two-Photon Raman Process

If we introduce the dimensionless variable

$$Z = (\omega - \omega_{31})/\gamma_+^0 \tag{18}$$

which denotes the reduced frequency for the two-photon excitations, then the excitation spectra near the frequency $\omega = \omega_{31} \approx \omega_{21} - \omega_b$ are determined by the expression

$$R^{(2)}(\omega) = \frac{(\bar{n}_3 - \bar{n}_1)\Omega_b^2}{4\pi\gamma_+^0 d_{12}^2(\omega)} \left[-\pi\delta(Z) + I_{12}(Z)\right] , \tag{19}$$

where

$$I_{12}(Z) = -\tfrac{1}{2} \frac{\tfrac{1}{2}(1-\nu_b/2\eta) - (Z - \tfrac{1}{2}\nu_b - \eta)/2\eta}{(Z - \tfrac{1}{2}\nu_b - \eta)^2 + \tfrac{1}{4}}$$

$$-\tfrac{1}{2} \frac{\tfrac{1}{2}(1+\nu_b/2\eta) + (Z - \tfrac{1}{2}\nu_b + \eta)/2\eta}{(Z - \tfrac{1}{2}\nu_b + \eta)^2 + \tfrac{1}{4}} . \tag{20}$$

The excitations in the neighbourhood of the frequency $\omega = \omega_{31} \approx \omega_{21} - \omega_b$ are generated by interference effects arising between the signal field and the laser field and represent the physical process where, simultaneously, a photon of the signal field with frequency ω_{21} is absorbed while a photon of the laser field with frequency ω_b is emitted. Thus, the spectra described by the function $R^{(2)}(\omega)$ given by Eq. (19) may be attributed to the absorption of the Raman frequency $\omega = \omega_{31} \approx \omega_{21} - \omega_b$. Then $d_{12}(\omega)$ in Eq. (19) may be approximated by

$$d_{12}^2(\omega) = (\omega - \omega_{21} + i\gamma_+^0)^2 \approx d_{12}^2(\omega_{31})$$

$$= (\omega_{31} - \omega_{21} + i\gamma_+^0)^2 \approx (-\omega_b + i\gamma_+^0)^2 \approx \omega_b^2 , \tag{21}$$

where ω has been replaced by its approximate value at $\omega \approx \omega_{31} \approx \omega_{21} - \omega_b$. The factor

$$\Omega_b^2/2d_{12}^2(\omega) \approx \Omega_b^2/2\omega_b^2 , \tag{22}$$

in Eq. (19) implies that $R^{(2)}(\omega)$ describes stimulated third-order nonlinear spectra near the frequency $\omega = \omega_{31} \approx \omega_{21} - \omega_b$.

The expression (19) describes the stimulated Raman spectra near the relative frequency Z and it is valid in the limit when $\bar{n}_b \gg 1$.

The first term on the rhs of Eq. (19) describes the main peak, which has a delta-function distribution at the frequency Z=0 indicating the stability of the mode in question. This is in agreement with Breit's (19) suggestion denoting the absence of spontaneous emission for atoms having a common upper level and two different lower levels (20-23). The function $I_{12}(Z)$ given by Eq. (20) describes a pair of sidebands, which are peaked at the frequencies $Z=\frac{1}{2}\nu_b \pm \eta$ and have radiative halfwidths equal to $\frac{1}{2}\gamma_+^0$, respectively. Thus, although the main peak at Z=0 is stable (non-radiative), the two sidebands decay radiatively with a lifetime of the order $2/\gamma_+^0$. The shape of the sidebands depends strongly on the value of the ratio $\nu_b/2\eta$ while their intensities are always negative, which implies that the physical process of amplification occurs at the frequencies $Z=\frac{1}{2}\nu_b \pm \eta$ and their maximum values $I_\pm$ are given by

$$I_\pm = -\left(1\mp\nu_b/2\eta\right) . \tag{23}$$

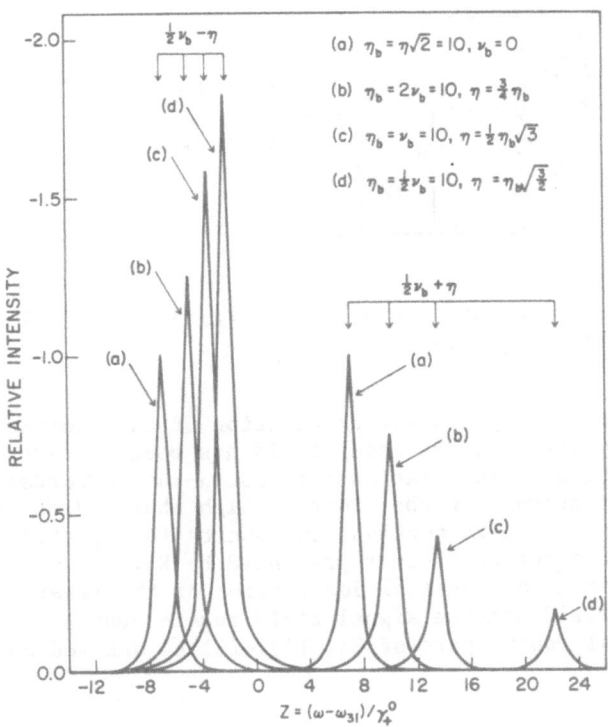

Figure 6. Two-photon or Raman spectra. The relative intensity= $I_{12}(Z)$ is computed from the rhs of Eq. (20) and is plotted versus the relative frequency $Z=(\omega-\omega_{31})/\gamma_+^0$ for the relative Rabi frequency η_b=10 and various detunings. (a) ν_b=0, (b) ν_b=5, (c) ν_b=10 and (d) ν_b=20.

The Raman spectra are illustrated in Figures 6a–d, where the function $I_{12}(Z)$ is plotted as the relative intensity=$I_{12}(Z)$ versus the relative frequency Z for the Rabi frequency η_b=10 and for various values of the detuning ν_b=0,5,10 and 20, respectively. Figure 6a implies that at resonance, ν_b=0, the doublet has equal negative intensities; for finite values of the detuning ν_b, as the value of ν_b increases the negative intensity of the peak at $Z=\frac{1}{2}\nu_b+\eta$ decreases while the corresponding one at $Z=\frac{1}{2}\nu_b-\eta$ increases as is depicted in Figures 6b–d. Finally as ν_b takes values greater than η_b, $\nu_b>\eta_b$, but the inequality $\nu_b/2\eta<1$ must be always satisfied; the intensity of the sideband at $Z=\frac{1}{2}\nu_b+\eta$ becomes negligibly small, $I_+\to0$ while the corresponding one at $Z=\frac{1}{2}\nu_b-\eta$ takes its maximum negative value, $I_-\to2$. The splitting of the state $|3>$ at the relative frequency Z is shown schematically in Figure 7.

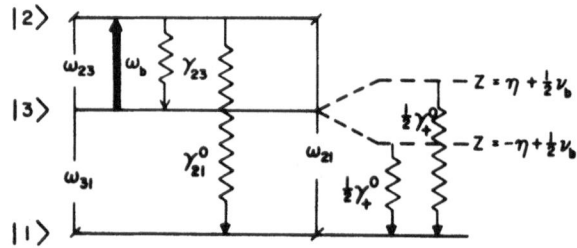

Figure 7. Schematic representation of the energy splitting at the relative frequency $Z=(\omega-\omega_{31})/\gamma_+^0$.

As mentioned previously, in the calculation of the excitation spectra described by Eq. (1), the laser field has been considered in the second quantization representation and photon–photon correlations have been taken into account in the limit of high photon densities of the laser field. In the next section, the excitation spectra describing the same processes as those represented by Eq. (1) will be discussed by considering a classical description of the laser field while the electron field and the signal field remain quantized; namely, the classical counterpart of Eq. (1) will be derived and discussed.

CLASSICAL DESCRIPTION OF THE LASER FIELD

In order to treat the laser field classically, we consider the same Hamiltonian as before but with the laser field taken in the interaction representation while the term describing the laser-atom interaction is replaced by its classical counterpart with a coupling constant g_b, which is the corresponding classical counterpart of the quantum field expression for the Rabi frequency Ω_b defined by Eq. (9). Using this Hamiltonian and a procedure, which is similar to that used before, the spectral function describing the excitation spectra has been calculated (24), and is given by

$$P^c_{12}(\omega) = P_c^{(1)}(\omega) + P_c^{(3)}(\omega) + R_c^{(2)}(\omega) . \tag{24}$$

The expression (24) is the classical counterpart of Eq. (1) and defines the absorption coefficient of the signal field describing the physical processes under investigation.

One-Photon Process

The spectral function describing one-photon process is found to be (24)

$$P_c^{(1)}(\omega) = \frac{(\bar{n}_3 - \bar{n}_1)}{2\pi\gamma_+^0} F^c_{12}(X) , \tag{25}$$

where

$$F^c_{12}(X) = 2/(X^2+1) - L(X, \nu_b, \xi) - L(X, \nu_b, -\xi) , \tag{26}$$

$$\xi_b = g_b/\gamma_+^0 , \qquad \xi^2 = \tfrac{1}{4}(\xi_b^2 + \nu_b^2) , \tag{27}$$

and the shape function $L(X, \nu_b, \pm\xi)$ can be obtained from Eq. (6) if η is replaced everywhere by ξ, $\eta \leftrightarrow \xi$. In Eqs. (25)-(27), ξ_b is the relative classical Rabi frequency while ξ defines the total relative shift arising from the classical Rabi frequency ξ_b and the detuning ν_b of the laser field.

Comparison between Eq. (26) for $F^c_{12}(X)$ and Eq. (5) for $F_{12}(X)$ implies that, apart from the difference in the expressions between η and ξ defined by Eqs. (8) and (27), respectively, the function $F^c_{12}(X)$ is larger by a factor of two than that of $F_{12}(X)$, provided that the correspondence principle is applicable, namely, $g_b \leftrightarrow \Omega_b$ or, equivalently, $\xi_b \leftrightarrow \eta_b$. As an illustration, the functions $F_{12}(X)$ and $F^c_{12}(X)$ denoted as relative intensities are plotted in Figure 8 versus the relative frequency X for the constant value of $\eta_b = \xi_b = 10$ and different values of detuning ν_b. The computed spectra described by the functions $F_{12}(X)$ and $F^c_{12}(X)$ are depicted in Figure 8 by solid and dashed lines, respectively; the spectra for $F_{12}(X)$ (solid lines) are

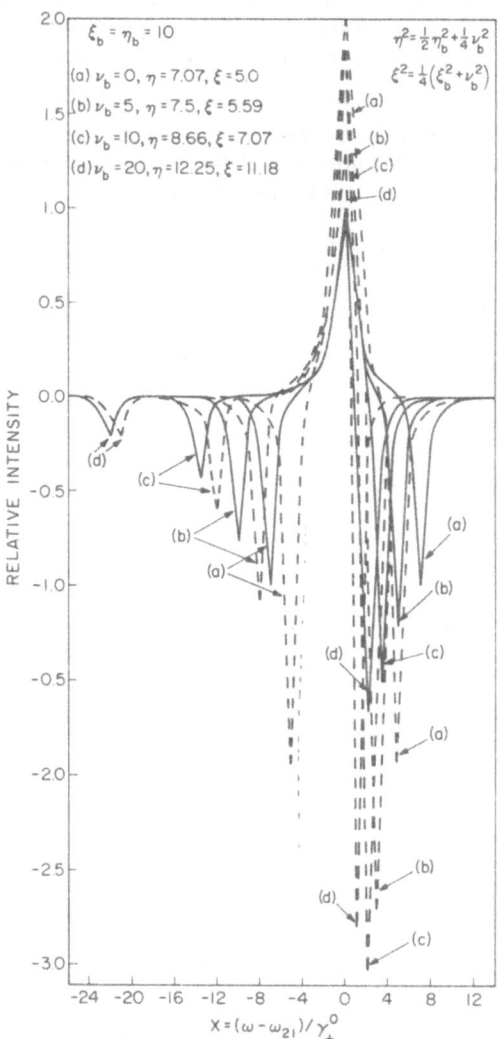

Figure 8.   One-photon spectra.   The relative intensities $F_{12}(X)$ and
$F^C_{12}(X)$ are computed from the rhs of Eqs. (5) and (26), respectively,
and are plotted versus the relative frequency $X=(\omega-\omega_{21})/\gamma^0_+$ for the
relative Rabi frequency $\eta_b=10$ and its classical counterpart $\xi_b=10$ and
various detunings.   (a) $\nu_b=0$, (b) $\nu_b=5$, (c) $\nu_b=10$ and (d) $\nu_b=20$.
Solid and dashed lines denote the spectra described by the shape
functions $F_{12}(X)$ and its classical counterpart $F^C_{12}(X)$, respectively.

identical to those given in Figure 2. The spectra in Figure 8 consist of the main peak at the frequency X=0 of the signal field and a pair of sidebands which are peaked at $X=-\frac{1}{2}\nu_b \pm \eta$ (solid lines) and at $X=-\frac{1}{2}\nu_b \pm \xi$ (dashed lines), respectively. The frequency shifts between the positions of the sidebands described by $F_{12}(X)$ (solid lines) and those by $F_{12}^C(X)$ (dashed lines) are equal to $\pm(\eta-\xi)$, while the corresponding maximum relative intensities are determined by the expressions

$$i_\pm = -(1\pm\nu_b/2\eta) \text{ and } i_\pm^C = 2(1\pm\nu_b/2\xi) \ ,$$

respectively.

Three-Photon Process

The spectral function $P_c^{(3)}(\omega)$ is determined by (24)

$$P_c^{(c)}(\omega) = \frac{(\bar{n}_3 - \bar{n}_1)}{8\pi\gamma_+^0 d_{12}^2(\omega)} J_{12}^C(Y) \ , \tag{28}$$

where

$$J_{12}^C(Y) = -L(Y,\nu_b,\xi)-L(Y,\nu_b,-\xi) \tag{29}$$

and the reduced frequency Y is given by Eq. (15) while the function $L(Y,\nu_b,\pm\xi)$ can be derived from Eq. (6) after the replacements $X\leftrightarrow Y$ and $\eta\leftrightarrow\xi$ have been made.

The spectral function $J_{12}^C(Y)$ defined by Eq. (29) for the three-photon process describes a pair of sidebands which are peaked at the frequencies $Y=-\frac{1}{2}\nu_b \pm \xi$, respectively. The frequency shifts between the peaks of the doublets described by the functions $J_{12}(Y)$ and $J_{12}^C(Y)$ are equal to $\pm(\eta-\xi)$, while the corresponding maximum relative intensities of the doublets are determined by the expressions

$$i_\pm = -2(1\pm\nu_b/2\eta) \text{ and } i_\pm^C = -2(1\pm\nu_b/2\xi) \ ,$$

respectively. Hence, the spectra described by the function $J_{12}^C(Y)$ may be obtained from those computed from Eq. (14) and depicted in Figure 4 if the positions of the peaks are shifted by $\pm(\eta-\xi)$ while their corresponding intensities by $\pm\nu_b(1/\eta-1/\xi)$, respectively.

Two-Photon Raman Process

The excitation spectra near the Raman frequency $\omega=\omega_{31}\approx\omega_{21}-\omega_b$ are described by the spectral function

$$R_c^{(2)}(\omega) = \frac{(\bar{n}_3 - \bar{n}_1)g_b^2}{4\pi\gamma_+^0 d_{12}^2(\omega)} \, I_{12}^c(Z) \, , \tag{30}$$

where

$$I_{12}^c(Z) = - \frac{\tfrac{1}{2}(1-\nu_b/2\xi)-(Z-\tfrac{1}{2}\nu_b-\xi)/2\xi}{(Z-\tfrac{1}{2}\nu_b-\xi)^2+\tfrac{1}{4}} - \frac{\tfrac{1}{2}(1+\nu_b/2\xi)+(Z-\tfrac{1}{2}\nu_b+\xi)/2\xi}{(Z-\tfrac{1}{2}\nu_b+\xi)^2+\tfrac{1}{4}} \, . \tag{31}$$

The expressions (30) and (31) are the classical counterparts of Eqs. (19) and (20), respectively. The shape function $I_{12}^c(Z)$ describes the two-photon Raman spectra near the frequency Z when the laser field is treated classically. The spectra consist of two sidebands which are peaked at the frequencies $Z=\tfrac{1}{2}\nu_b\pm\xi$ and have radiative widths of the order $\tfrac{1}{2}\gamma_+^0$. The intensities of the doublet take negative values, which indicate that amplification of the signal field is anticipated to take place at the frequency $Z=\tfrac{1}{2}\nu_b\pm\xi$, and their maximum values $I_\pm^c$ are determined by

$$I_\pm^c = -2(1\mp\nu_b/2\xi) \, , \tag{32}$$

which, apart from the differences between η and ξ, are larger by a factor of two (in absolute values) than the corresponding ones given by Eq. (20). Hence, Eq. (30) for $I_{12}^c(Z)$ describes spectra similar to those described by Eq. (20) for $I_{12}(Z)$, which are depicted in Figure 6. The frequency shifts between the peaks of the doublets described by the functions $I_{12}(Z)$ (Figure 6) and $I_{12}^c(Z)$ are equal to $\pm(\eta-\xi)$ while the maximum relative intensities of the corresponding peaks are given by Eq. (23) and Eq. (32), respectively. Thus, the spectra described by $I_{12}^c(Z)$ can be easily obtained from those depicted in Figure 6 for $I_{12}(Z)$.

SUMMARY

We have considered the excitation spectra for a three-level atom as shown in Figure 1, where a strong laser field operates near resonance between the states $|2\rangle$ and $|3\rangle$, $|2\rangle\leftrightarrow|3\rangle$, while the signal field, which is assumed to be a weak perturbing field, describes the radiative decays $|2\rangle\rightarrow|3\rangle$ and $|2\rangle\rightarrow|1\rangle$, respectively. The state $|3\rangle$ has the same parity as the state $|1\rangle$ and, hence the state $|3\rangle$ as a metastable one is initially populated while the remaining states $|1\rangle$ and $|2\rangle$ are empty. The electron population of the state $|3\rangle$ is depleted by the action of the laser which excites the electrons into the state $|2\rangle$, where the electrons emit photons through the signal field and decay into the states $|3\rangle$ and $|1\rangle$ simultaneously.

Using the Green function method, we have calculated the excitation spectra of the signal field in the limit of high photon densities of the laser field for the following processes:

i) One photon process. The spectral function is given by Eq. (2), where the function $F_{12}(X)$ is defined by Eqs. (5) and (6) and is graphically presented in Figures 2a–d; the splitting of the state $|2\rangle$ at the frequency X is schematically depicted in Figure 3. The intensity of the central peak at X=0 or at $\omega=\omega_{21}$ is positive indicating signal-field absorption while those of the pair of sidebands at $X=\pm\eta-\frac{1}{2}\eta_b$ are negative implying the stimulated emission of the signal field. The one-photon process discussed here is the three-level analog to that for the two level system predicted by Mollow (1) and confirmed experimentally by Wu et al. (2). It is pointed out that for the three-level system under investigation the sum of the intensities is equal to

$$h_+ + h_- = -2 \ , \quad \text{for } \nu_b = 0 \ ,$$

and

$$h_+ + h_- = -2/\left(1-\nu_b^2/4\eta^2\right) \ , \quad \text{for } \nu_b \neq 0 \ ,$$

while the intensity of the central peak is $h_c=1$. For the two-level system (1,2), stimulated emission takes place for certain values of photon densities but the gain is rather small; for instance, the observed maximum gain (2) was approximately 0.4% with a resonant laser-field intensity of 130 mW/cm². Thus the results of the present study are encouraging suggesting that the amplification of the signal field in a three-level system might occur more easily and greater gains might be achieved than for the corresponding two-level system. Of course this could be only verified experimentally.

ii) Stimulated three-photon process. The expression (12) describes the stimulated three-photon process near the frequency $\omega=\omega_{21}-2\omega_b$, where one photon of the signal field is absorbed while two photons of the laser field are emitted simultaneously. The shape function $J_{12}(Y)$ is defined by Eq. (14) and describes a doublet whose intensity is always negative implying amplification while the shape of the doublet depends on the value of the ratio $\nu_b/2\eta$. The calculated spectra are shown in Figures 4a–d while the splitting that occurs at the frequency Y is depicted schematically in Figure 5.

iii) Stimulated two-photon Raman process. The spectral function for the stimulated Raman spectra at $\omega=\omega_{31}\approx\omega_{21}-\omega_b$ is determined by the expressions (19) and (20). The spectra consist of a main peak at $\omega=\omega_{31}$ or Z=0, which has a delta-function distribution, and a pair of sidebands peaked at $Z=\frac{1}{2}\nu_b\pm\eta$ which have negative intensities implying amplification at the frequencies in question; the halfwidths of the sidebands is $\frac{1}{2}\gamma_+^0$. The spectra of the doublet are shown in Figures 6a–d while the splitting of the state $|3\rangle$ is depicted schematically

in Figure 7. The spectra for the stimulated Raman process discussed
here have, qualitatively, the three main characteristic properties
that have been observed experimentally by Tam (9) namely: 1) The
sidebands appear at frequencies Z≠0 and, in particular, at $Z=\frac{1}{2}\nu_b\pm\eta$;
2) the sidebands have always negative intensities indicating amplification and 3) the factor $(\bar{n}_3-\bar{n}_1)\Omega_b^2$ in Eq. (19) implies that the
relative intensity of the sidebands depends on the photon density $\bar{n}_b$
of the laser field as well as on the atomic density N. Thus, at
suitable atomic density, photon density of the laser field and detuning will generate strong sideband amplification at the frequencies
$Z=\frac{1}{2}\nu_b\pm\eta$. Hence, the stimulated Raman spectra discussed here agree at
least qualitatively with the observed spectra by Tam (9). For a
quantitative comparison one has to have accurate experimental data
regarding the parameters involved.

Using a classical description of the laser field, we have
considered the same processes as before, namely, one-, three- and
two-photon Raman spectra arising from the $|1\rangle\leftrightarrow|2\rangle$ transition for the
atomic system shown in Figure 1. The spectral function for the
one-photon process is defined by Eq. (25), where the function $F_{12}^C(X)$
as the relative intensity is determined by Eq. (26); $F_{12}^C(X)$ is the
classical counterpart of the function $F_{12}(X)$ defined by Eq. (5).
Numerical results computed from the functions $F_{12}(X)$, Eq. (5), and
$F_{12}^C(X)$, Eq. (26), are presented graphically in Figure 8 by solid and
dashed lines, respectively. Resonance ($\nu_b=0$) and off-resonance
($\nu_b\neq0$) spectra are depicted in Figure 8 for constant values of η_b and
ξ_b, namely, for $\eta_b=\xi_b=10$. The spectra in Figure 8 describe: (i) the
main peak at the frequency X=0 of the signal field, whose relative
intensity is always positive and (ii) one pair of sidebands peaked at
the frequencies $X=-\frac{1}{2}\nu_b\pm\eta$ (solid lines) and $X=-\frac{1}{2}\nu_b\pm\xi$ (dashed lines),
respectively, whose relative intensities are always negative. The
frequency shifts between the positions of the sidebands described by
the function $F_{12}(X)$ (solid lines) and $F_{12}^C(X)$ (dashed lines) are equal
to $\pm(\eta-\xi)$ while the corresponding maximum intensities are given by
$i_\pm=-(1\pm\nu_b/2\eta)$ and $i_\pm^C=-2(1\pm\nu_b/2\xi)$, respectively. Figure 8 indicates
that at resonance ($\nu_b=0$) the maximum intensities of the peaks described by the classical expression $F_{12}^C(X)$ (dashed lines) are larger
by a factor of two than those described by the function $F_{12}(X)$ (solid
lines). However, in both treatments, the signal field is attenuated
at the frequency X=0 while it is strongly amplified at the frequencies $X=-\frac{1}{2}\nu_b\pm\eta$ and $X=-\frac{1}{2}\nu_b\pm\xi$, respectively.

The expression (28) describes three-photon processes near the
frequency Y. The shape function $J_{12}^C(Y)$ is determined by Eq. (29) and
describes a doublet peaked at the frequencies $Y=-\frac{1}{2}\nu_b\pm\xi$ with maximum
intensities equal to $i_\pm^C=-2(1\pm\nu_b/2\xi)$, respectively, and spectral
widths of the order of $\frac{1}{2}\gamma_\pm^0$. The shape function $J_{12}^C(Y)$ is the classical counterpart of the function $J_{12}(Y)$ given by Eq. (14) and they

differ only as far as the definition of ξ and η are concerned, which are determined by Eqs. (27) and (8), respectively. Hence, the spectra described by the function $J_{12}^{c}(Y)$ can be obtained from those computed from $J_{12}(Y)$, which are shown in Figure 4, by appropriately changing the positions and the intensities of the doublet.

The spectral function $R_{c}^{(2)}(\omega)$ describing the two-photon Raman process is determined by Eq. (30), which is the classical counterpart of the function $R^{(2)}(\omega)$ defined by Eq. (19). The expression (19) consists of two terms one of which is the delta-function $\delta(Z)$, while the other is the shape function $I_{12}(Z)$ given by Eq. (20). The expression (30) consists only of the shape function $I_{12}^{c}(Z)$ defined by Eq. (31), which is the classical counterpart of Eq. (20). The function $I_{12}^{c}(Z)$ describes a doublet peaked at the frequencies $Z = \frac{1}{2}v_{b} \pm \xi$ with maximum intensities equal to $I_{\pm}^{c} = -2(1 \mp v_{b}/2\xi)$, which are, apart from the different definitions of ξ and η, larger by a factor of two than the corresponding ones defined by Eq. (23) namely, $I_{\pm} = -(1 \mp v_{b}/2\eta)$. Hence, the spectra described by the function $I_{12}^{c}(Z)$ can be easily obtained from those described by the function $I_{12}(Z)$, which are depicted in Figure 6. The presence and the absence of the delta-function $\delta(Z)$ in Eqs. (19) and (30), respectively, is due to the quantum nature of the quantized field, and it is attributed to the different way by which the quantized field and the classical field split the excitation spectrum (24).

The results of the present study are consistent with those derived in recent studies (10,13,14,16,24), where comparison has been made between the results obtained when the laser fields are quantized and photon-photon correlation are taken into account in the limit of high photon densities of the laser fields and when the laser fields are treated classically. As has been discussed in the literature (10,13,14,16,24), both treatments have their own merits and should be used whenever they are appropriate. However, only experimental observations will reveal with certainty which of the two treatments provides a more appropriate description of the problem under investigation.

The present study has been recently extended (16) by considering the fluorescent spectra of the signal field arising from stimulated one-photon, three-photon and two-photon Raman processes for an atomic system consisting of a four level atom interacting with two strong laser fields and a weak signal field simultaneously; the interested reader is referred to the literature (16) for details.

REFERENCES

1. Mollow, B.R.: 1972, Phys. Rev. A, 5, p. 2217.

2. Wu, F.Y., Ezekiel, M., Ducloy, M. and Mollow, B.R.: 1977, Phys. Rev. Lett., 3, p. 1077.

3. Galbraith, H.W., Dubs, M. and Steinfeld, J.I.: 1982, Phys. Rev. A, 26, p. 1528.

4. Senitzky, B., Gould, G. and Cutler, S.: 1963, Phys. Rev., 130, p. 1460.

5. Senitzky, B. and Cutler, S.: 1964, Microwave J., 1, p. 62.

6. Bonch-Bruevich, A.M., Khodovoi, V.A. and Chigir, N.A.: 1974, Zh. Eksp. Teor. Fiz., 67, p. 2069 (:1975, Sov. Phys. JETP, 40, p. 1027).

7. McCall, S.L.: 1974, Phys. Rev. A, 9, p. 1515.

8. Gibbs, H.M., McCall, S.L. and Venkatesan, T.N.C.: 1976, Phys. Rev. Lett., 36, p. 1135.

9. Tam, A.C.: 1979, Phys. Rev. A, 19, p. 1971.

10. Mavroyannis, C.: 1985, J. Math. Phys., 26, p. 1093.

11. Zubarev, D.N.: 1974, Nonequilibrium Statistical Thermodynamics, Plenum, New York, Chapter 16.

12. Mavroyannis, C.: 1975, The Green Function Method, in Physical Chemistry, eds. Erying, H., Jost, W., and Henderson, D., Academic Press, New York, Vol. XI A, Chapter 8.

13. Mavroyannis, C.: 1983, Phys. Rev. A, 27, p. 1414.

14. Mavroyannis, C.: 1984, J. Math. Phys., 25, p. 2780.

15. Mavroyannis, C.: 1984, J. Phys. Chem., 88, p. 4868.

16. Mavroyannis, C.: 1985, Phys. Rev. A, 31, p. 1563.

17. Mavroyannis, C.: 1985, J. Chem. Phys., 82, p. 3563.

18. Mavroyannis, C.: 1983, Optics. Commun., 46, p. 323.

19. Breit, G.: 1933, Rev. Mod. Phys., 5, p. 91.

20. Schenzle, A. and Brewer, R.G.: 1975, Laser Spectroscopy, eds. Haroche, S., Pebay-Peyroula, J.C., Hänsch, T.W. and Harris, S.E., Springer Verlag, New York, p. 420.

21. Chow, W.W., Scully, M.O. and Stoner, J.O.: 1975, Phys. Rev. A., 11, p. 1380.

22. Herman, R.M., Grotch, H.H., Kornblith, R. and Elerly, J.H.: 1975, Phys. Rev. A, 11, p. 1389.

23. Milonni, P.W.: 1976, Phys. Rep. 25, p. 53.

24. Mavroyannis, C.: 1985, J. Math. Phys., submitted.

Ab-Initio Multireference Determinant Configuration Interaction (MRD-CI)
and CASSCF Calculations on Energetic Compounds

Joyce J. Kaufman, P. C. Hariharan, S. Roszak, C. Chabalowski, M.
van Hemert[*], M. Hotokka[Δ] and R. J. Buenker[*]
Department of Chemistry
The Johns Hopkins University
Baltimore, Maryland 21218 USA

ABSTRACT. The molecular decomposition pathways of $>C - NO_2$ and $>N - NO_2$ bonds are one of the key primary steps in initiation of explosives.
To study molecular decomposition pathways it is necessary to use ab-initio MRD-CI (multireference determinant-configuration interaction) or CASSCF (complete active space multiconfiguration SCF) calculations.
Our preliminary beyond Hartee-Fock calculations had indicated that the wave function of nitromethane even in its ground electronic state at equilibrium geometry had two-determinant character with additional determinants becoming important along the $CH_3 - NO_2$ dissociation pathway. There is even more pronounced multideterminant character in the excited states. Our MRD-CI and CASSCF calculations on nitromethane and MRD-CI calculations on RDX and nitrobenzene confirm this multi-determinant character. The MRD-CI calculations on RDX were carried out both with localized orbitals and with canonical orbitals in the region of the $>N-NO_2$ bond being dissociated. The localized orbital approach was very promising and saved an order of magnitude of computer time.

[*]Visiting scientist, The Johns Hopkins University; permanent address:
Gorleaus Laboratories, Post Office Box 9502, 2300 RA, Leiden, The
Netherlands.

[Δ]Visiting scientist, The Johns Hopkins University; permanent address:
Institutionen For Fysikalisk Kemi, Department of Physical Chemistry',
ABO AKEDEMI Porthansgatan 3-5, SF-20500 Abo 50, Finland.

[*]Visiting scientist, The Johns Hopkins University; permanent address:
Theoretische Chemie, Bergiscle Universitat, Gesamthochschule Wuppertal,
Gauss-Strasse 20, 5600 Wuppertal 1, West Germany.

P. M. Rentzepis and C. Capellos (eds.), Advances in Chemical Reaction Dynamics, 289–309.
© *1986 by D. Reidel Publishing Company.*

1. INTRODUCTION

The electronic potential energy surfaces along the dissociation pathways of energetic compounds play a decisive role in determining
subsequent events leading to decomposition and finally to detonation.
The dissociation of $>C - NO_2$ and $>N - NO_2$ bonds of nitroexplosives
are one of the key primary steps in initiation of explosives. The
molecular decomposition pathways of the nitroexplosives and the initiation and subsequent steps in detonation would then be completely
intertwined. To study molecular decomposition pathways it is necessary
to use ab-initio MRD-CI (multireference determinant-configuration
interaction) or CASSCF (complete active space multiconfiguration SCF)
calculations.

Quantum chemical calculations beyond Hartree-Fock (which are
single determinant SCF calculations) such as MC-SCF (multiconfiguration
SCF) and/or MRD-CI (multireference determinant-configuration
interaction) calculations are necessary to describe correctly the
potential energy surfaces of dissociation pathways. Compounds containing an $-NO_2$ group are even more complicated since $R - NO_2$ compounds are
not describable properly as a single determinant wave function even at
their equilibrium geometries in their ground electronic states. Our
earliest beyond Hartree-Fock calculations on nitromethane, CH_3NO_2, had
indicated this multideterminant behavior.[1] Nitromethane has a two-
determinant wave function even in its ground electronic state at
equilibrium geometry. As nitromethane dissociates along the $H_3C - NO_2$
bond, even more determinants became significant.[2] The electronically
excited states have a mixture of a number of different determinants in
different regions along the $H_3C - NO_2$ dissociation pathway.[2]

When dissociating a $>C - NO_2$ or $>N - NO_2$ bond in a large
nitroexplosive, it is logistically unfeasible computerwise to allow
excitations from all occupied molecular orbitals. A number of years
ago we derived and implemented an effective CI Hamiltonian technique.[3]
In this method, the effect of all of the occupied molecular orbitals
from which excitations are not allowed are folded into the effective CI
Hamiltonian. To within the 16 digit accuracy of the computer, this
technique gives the same results as carrying out frozen core CI calculations in which excitations are not allowed from "frozen" occupied
molecular orbitals. The great advantage of the effective CI
Hamiltonian technique is that it is not necessary to transform the
integrals over atomic orbitals to integrals over molecular orbitals for
the occupied molecular orbitals which have been folded into the effective CI Hamiltonian. It is this transformation which is the computer
time and computer space limiting part of a CI calculation. Using this
effective CI Hamiltonian it is very feasible to calculate ionization
potentials, electron affinities and excitation spectra for

large molecules based on conventional delocalized molecular orbitals
using the canonical highest occupied and lowest virtual molecular
orbitals. However, for dissociation of a group from a large molecule
it is necessary to include explicitly in the CI calculations all of the
occupied molecular orbitals in that region of interest (as well as the
virtual orbitals in that region of space). We put forth the suggestion
at the 1980 Canadian Theoretical Conference[4] to use the effective CI
Hamiltonian method for molecular decomposition by transforming to
localized orbitals and using explicitly the localized molecular orbitals (both occupied and virtual) in the region of interest.

We implemented that localized CI approach and used it[5] for MRD-CI
calculations studying the $>N - NO_2$ decomposition pathway of RDX

RDX—molecular sketch.

Just as with CH_3NO_2, the wave function of RDX even at its equilibrium geometry was not a single determinant wave function. RDX has
three nitro groups. Thus there are multideterminant contributions
corresponding to the three nitrogroups. The potential energy surfaces
of excited states along the $>N - NO_2$ decomposition pathway of RDX
contain a multitude of determinants.

For our initial MRD-CI study of the $>C - NO_2$ decomposition pathway
of nitrobenzene, we used a slightly different approach. Since it is
not unambiguous for an aromatic system how best to transform to localized orbitals, we used all of the canonical occupied molecular orbitals
and virtual orbitals with major contributions in that region of space.
Again, even in the ground state at equilibrium geometry nitrobenzene
has a multideterminant wave function and the excited states contain
major contributions from a multitude of determinants.

2. METHODS

2.1. MRD-CI

The MRD-CI (in this case, explicitly multireference double excitation configuration interaction) method used was that of Peyerimhoff and Buenker[6] which was converted and adapted at the Johns Hopkins University to the CYBER 175,[∇] CDC 7600,[∇] CRAY 1M (CTSS)[*] and CRAY XMP-12[*] (COS). The general strategy of these MRD-CI calculations was first to run a small CI calculation (several hundred to ≈ a thousand singly and doubly excited configurations relative to the SCF ground state). From the results of this smaller CI calculation, up to 20 most important configurations were picked as reference configurations. Then the singly and doubly excited configurations were generated relative to the reference configurations. An interaction selection energy threshold was selected. The energy contribution of each configuration (from the selected symmetry adapted functions, SAF's) was calculated by a perturbative procedure and all configurations contributing greater than a certain energy threshold were included explicitly in the larger MRD-CI wave function to be solved. After the CI calculations were carried out, the CI energy from the selected SAF's was augmented by E (extrapolated) perturbation calculations from the summation of energy lowerings for rejected SAF's at two or more thresholds. This was then followed by a Davidson-type correction to account for the linked double excitations:

$$E(\text{full CI}) = E(\text{Ext}) + (1 - \sum_{p}^{\text{ref}} c_p^2)[E(\text{Ext}) - E(\text{Ref})]$$

E(Ref) = Energy for only reference configurations.

This Davidson type correction has the effect of making the MRD-CI calculation essentially size-consistent. (The definitions of the various energies used in the present article are consistent with those used by Peyerimhoff and Buenker.[6])

2.1.1. <u>Nitromethane.</u> The nitromethane calculations were carried out on a CYBER 175 and a CDC 7600.

[∇]C. Chabalowski and P. C. Hariharan, The Johns Hopkins University, 1982, with the collaboration of R. J. Buenker on visits to the Johns Hopkins University, 1981-1982.

[*]M. van Hemert and P. C. Hariharan, The Johns Hopkins University, 1985.

[*]V. Saunders, M. van Hemert and P. C. Hariharan, The Johns Hopkins University, 1985.

From the results of the small CI calculation, the nineteen most important configurations were picked as reference configurations. These were mostly the out-of-plane (π-type) orbitals with respect to the plane of the NO_2 group and in-plane (σ-type) orbitals in the C-N bond (if the NO_2 group was perpendicular to two of the H atoms of the CH_3 group). The virtual space was chosen to be complementary. Then all single and double excitations were generated relative to these nineteen reference configurations. For the $CH_3 - NO_2$ decomposition pathway, this typically generated ~850,000 configurations. An interaction selection threshold of 0.000050 hartree was used. From the total number of symmetry adapted functions (SAF's) which were generated for each C-N distance (~850,000), approximately 4000 to 4200 SAF's were retained.

The MRD-CI procedure described above was followed.

Within the size of the core memory and disk space available to a user on the CYBER 175 and CDC 7600, on which these MRD-CI calculations on CH_3NO_2 were carried out, it was feasible for us only to carry out the MRD-CI calculations with 66 atomic basis functions. A 9^s5^p atomic basis set contracted to 4^s2^p was used on the heavy atoms and a 5^s1^p contracted to 2^s1^p was used on the H.

2.1.2. <u>RDX</u>. The RDX calculations were carried out on a CDC 7600. RDX is quite a large molecule and on a CDC 7600 it was computationally intractable, both in terms of computer memory and time requirements, to carry out CI calculations for the entire molecule (although the new vector supercomputers with large core alleviate this problem to an extent dependent on core and peripheral resources). By localizing and focusing on the $>N - NO_2$ decomposition pathway, much important insight can be gained about how many and what kinds of potential energy surfaces are involved, and to what types of product species they lead.

First, ab-initio MODPOT/VRDDO/MERGE[7-9] SCF calculations were carried out on RDX at its equilibrium geometry,[10] and then at various $>N_{(1)} - N_{(4)}O_2$ distances. These calculations were carried out with our own rapid and effective ab-initio Gaussian program, MOLASYS, into which we had long since incorporated as options desirable computational strategies for ab-initio calculations on large molecules: MODPOT - ab-initio effective core model potentials which permit the calculation of valence electrons only, explicitly yet accurately; VRDDO - a charge-conserving, integral size prescreening evaluation which decides whether an integral (or block of integrals) should be calculated explicitly (especially effective for spatially extended systems); and MERGE - to save and reuse common skeletal integrals.

We had shown when we first described this ab-initio MODPOT/VRDDO method by extensive testing on 75 molecules carrying out ab-initio calculations using all-electron basis sets, the VRDDO procedure, the

ab-initio MODPOT procedure and the combined ab-initio MODPOT/VRDDO
procedure, that there was excellent agreement of the ab-initio
MODPOT/VRDDO results with the results of ab-initio all-electron calculations using the same valence atomic basis sets (and the same core
atomic basis sets for the ab-initio effective core model potential).
The gross atomic populations agree to within 0.02 e, the orbital
energies agree to within 0.02 a.u., and agreement of the potential
energy surfaces for isomer total energy differences or along a potential energy surface was even better, 0.001 to 0.0001 a.u. What is just
as significant is the savings in computer time. For nitrobenzene,
which was the test case, the ab-initio MODPOT/VRDDO method was almost
an order of magnitude faster than the corresponding ab-initio all-
electron calculations. For larger molecules the ab-initio MODPOT/VRDDO
method saves more than an order of magnitude depending on the size and
shape of the molecule. Additional savings of greater magnitude accrue
from use of the MERGE technique.

The Boys procedure[11,12] was applied to the unitary transformation
from the canonical Hartree-Fock orbitals of the closed shell RDX
molecule to localized orbitals. The occupied and virtual orbitals are
localized independently. Mulliken population of localized orbitals has
been used as a criterion of the localization. Then smaller size
(several hundred to ~ a thousand) CI calculations were carried out
along the $\frac{C}{C}$>N - NO_2 decomposition pathway of RDX based first on canonical delocalized orbitals located primarily on the >N - NO_2 fragment and
then on strictly orthogonal localized orbitals on that fragment for
both the ground state and for three excited states of RDX.

From the results of this smaller CI calculation, the nineteen most
important configurations were picked as reference configurations. For
the >N - NO_2 decomposition for delocalized occupied molecular orbitals
of the order of 3 to 1000 symmetry adapted configurations (SAF's) were
generated, and for localized orbitals of the order of 37,500 SAF's were
generated. An interaction selection threshold of 0.000050 hartree was
used. From the total number of symmetry adapted functions (SAF's)
which were generated for each $N_{(1)}$ - $N_{(4)}$ distance approximately 2050
(localized) to 3350 (delocalized) SAF's were retained explicitly for
the MRD-CI calculations.

2.1.3. Nitrobenzene. Since it is not unambiguous how to choose the
best way to localize molecular orbitals for an aromatic system, the
MRD-CI calculations were carried out based on delocalized canonical
molecular orbitals centered in the region of the $\frac{C}{C}$>C - NO_2 bond being
dissociated.

Ab-initio SCF calculations were carried out for nitrobenzene
along the >C - NO_2 dissociation pathway. There were a total of 50
molecular orbitals (32 occupied and 18 virtual). For this study, we
used conventional delocalized molecular orbitals and for the MRD-CI
calculations took into account explicitly all of the occupied molecular

orbitals in the region of the $>C - NO_2$ bond (15 occupied molecular orbitals) and folded the other 17 occupied orbitals into an effective CI Hamiltonian and used 18 virtual orbitals. So far we have investigated the $^1A_1$, $^1A_2$, $^3A_2$, $^1B_1$, $^3B_1$, $^1B_2$, $^3B_2$ states.

2.2. CASSCF

The CASSCF method used was that of Roos and Siegbahn[13] - which at the Johns Hopkins University was converted and adapted[#] to the CRAY-1M. The CASSCF method is an MCSCF procedure which uses the concept of a complete active space. Within the core memory and disk space available to a user at the time these calculations were run, it was feasible for us to carry out the CASSCF calculations along the CH_3NO_2 decomposition pathway with a 9^s5^p basis contracted to 4^s2^p on the heavy atoms and a 4^s contracted to 3^s basis on H; 49 basis functions. Since then, we have been able to carry out larger basis set calculations, $9^2 5^p 1^d$ on the heavy atoms contracted to $4^s2^p1^d$ and a 4^s1^p basis contracted to 2^s1^p on the H; 82 basis functions for the equilibrium geometry of CH_3NO_2.

A plane of symmetry perpendicular to the plane of the NO_2 group was used in the calculation. Of the seven active orbitals, three symmetric and two antisymmetric orbitals were in the active space and two symmetric orbitals were in the virtual space.

3. RESULTS AND DISCUSSION

3.1. Nitromethane

3.1.1. MRD-CI Calculations on the Ground and Three Electronically Excited States Along the $CH_3 - NO_2$ Decomposition Pathway of Nitromethane. A plot of the five highest occupied and five lowest unoccupied molecular orbital energies (labeled with their symmetry character) vs. $CH_3 - NO_2$ distance is presented in Figure 1. At equilibrium geometry the first HOMO is out-of-plane (π-like) on O_1, O_2 and the first LUMO is out-of-plane (π^*-like) on N, O_1, O_2. The second HOMO is C-N(σ) and the second LUMO is C-N(σ^*). At a C-N distance of ~3.8 a.u., the first HOMO becomes C-N(σ) and the first LUMO becomes C-N(σ^*), while the second HOMO is now out-of-plane (π-like)

[#]M. Hotokka and P. C. Hariharan, The Johns Hopkins University, 1984.

on O_1, O_2 and the second LUMO is out-of-plane (π^*-like) on N, O_1, O_2. This plot indicates that there is a major crossing of molecular orbitals, both in the occupied space and in the virtual space.

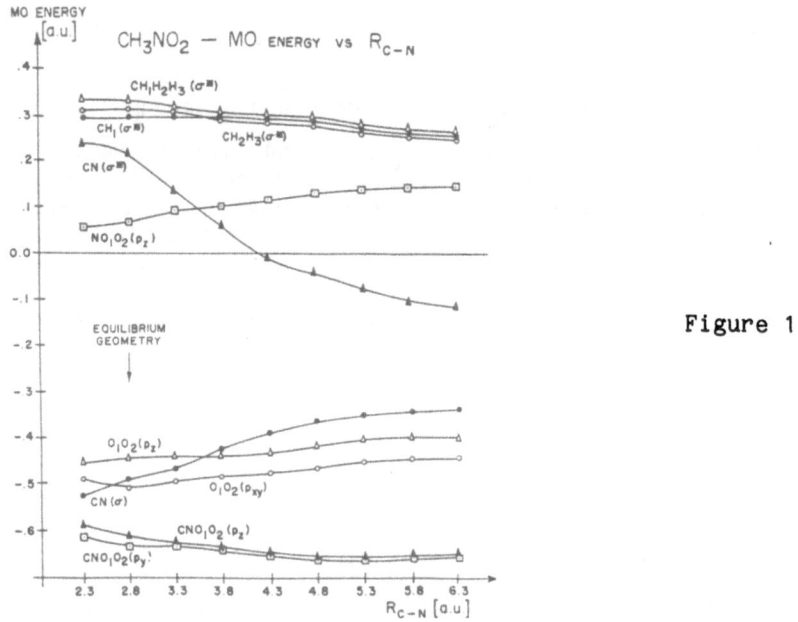

Figure 1

The total number of configurations generated, the selection and the number of configurations included explicitly in the MRD-CI calculations are presented in Table I. (All tables are collected at the end of this article).

A plot of the SCF and various levels of CI energies vs. CH_3NO_2 distance for the ground state are plotted in Figure 2.

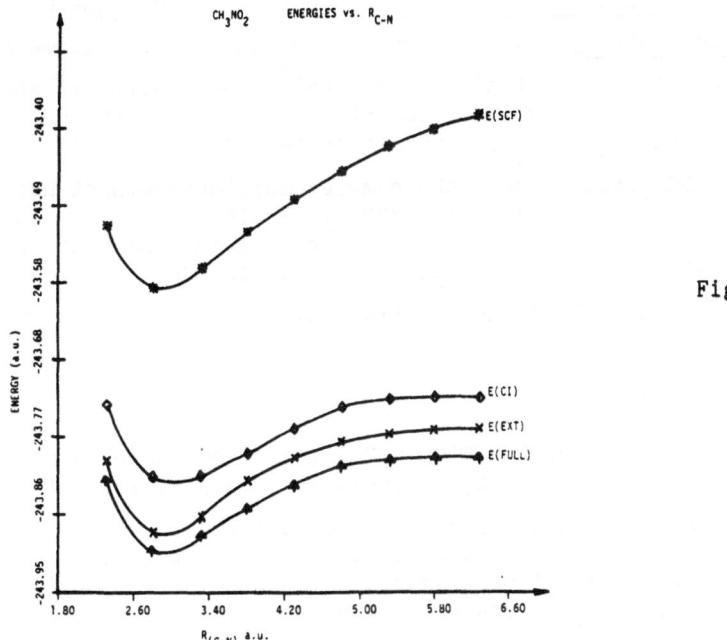

Figure 2

In Table II are presented the coefficients and the explicit chemical character of the most important configurations in the ground state. It can be seen that even in the ground electronic state that the $(\pi$-like$)^2 \to (\pi^*$-like$)^2$ $[H_1(0_1 0_2: p_z)^2 \to L_1(NO_1 0_2: p_z)^2]$ remains a contributor all the way out to the decomposition asymptote. A number of additional contributions also become important along the CH_3 - NO_2 dissociation pathway. Major among these are sigma-type:

$$(C-N: \sigma)^2 \to (C-N: \sigma^*)^2 \qquad [H_1(C-N: \sigma)^2 \to L_1(C-N: \sigma^*)^2]$$

and $$(C-N: \sigma) \to (C-N: \sigma^*) \qquad [H_1(C-N: \sigma) \to L_2(C-N: \sigma^*)]$$

There are a wide variety of configurations contributing to the electronically excited states all along the CH_3 - NO_2 decomposition pathway. (For details of the configurations and their coefficients see reference 2).

These results indicate the complexity of multideterminant electronic configurations to be expected along the $>C-NO_2$ decomposition pathways of all nitrocompounds, including the larger energetic nitrocompounds.

3.1.2. CASCCF Calculations on the Ground State Along the $CH_3 - NO_2$ Decomposition Pathway.

Again, even at the equilibrium geometry, the contribution of the $(\pi\text{-like})^2 \rightarrow (\pi^*\text{-like})^2$ contribution is apparent. Also, there is a complexity of contributions from additional configurations along the $CH_3 - NO_2$ decomposition pathway.

The CASSCF results show the complex multideterminant nature of the ground state as indicated by the MRD-CI results.

Again there are a wide variety of configurations contributing to the electronically excited states all along the $CH_3 - NO_2$ decomposition pathway. (For details of the configurations and their coefficients see reference 2).

3.2. RDX - MRD-CI

The ab-initio MODPOT/VRDDO/MERGE SCF calculations were carried out for RDX as a function of $>N_{(1)}-N_{(4)}$ distance. The ab-initio effective core model potential (MODPOT) basis set gave 66 molecular orbitals, of which 42 molecular orbitals are doubly occupied. A plot of even just the 9 highest occupied and 5 lowest unoccupied molecular orbitals of RDX as a function of $N_{(1)} - N_{(4)}$ distance, Figure 3, indicates that there is considerable level crossing which further complicates the problem.

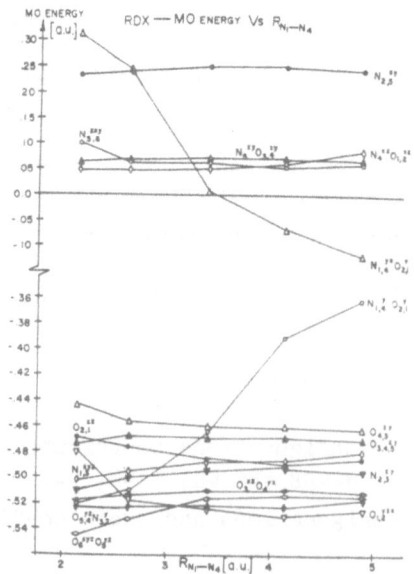

Figure 3

The RDX orbitals were well localized upon Boys localization. Occupied orbitals are practically localized on the two center bonds or on lone pairs on the oxygen atoms. Virtual orbitals are localized on the two center bonds or, in some, cases, on the three center nitrogroups.

Since we have a set of localized orbitals, the extraction of orbitals belonging to the chosen part of the molecule is a very easy task. Twelve occupied orbitals are involved in the electronic description of this area. In the case of virtual orbitals, we have chosen eight orbitals. Because of the inadequate description of the virtual orbitals from the MODPOT basis (which is minimal), we have extended this space by high lying C-H orbitals. This choice gives us the 20 active orbitals space with 12 doubly occupied.

MRD-CI calculations have been performed for different choices of reference functions and have been carried out for the ground and three low lying excited states. Table III presents results for thé ground state at equilibrium geometry. The first three columns contain results for the localized-orbitals space (20 active orbitals), while the next two contain calculations for a bigger (34 orbitals), but intuitively chosen canonical delocalized space. This bigger space has been selected from the canonical orbitals set for a good description of the

$$N_{(1)} - N_{(4)} \Big\langle \begin{array}{c} O_{(1)} \\ O_{(2)} \end{array}$$

part. The results indicate that the orbital space for the localized orbitals is a good choice of space in the "localized" case, where the calculated correlation energy is even bigger than in the "delocalized" space. It indicates a very good choice of localized orbitals, that seems to lead to the best saturated space in the region of

$$N_{(1)} - N_{(4)} \Big\langle \begin{array}{c} O_{(1)} \\ O_{(2)} \end{array}$$

within our 66 orbital space. The reduction of the number of "active" orbitals has the consequence of decreasing the computational time by an order of magnitude; because of the excellent localized orbitals, the reduction in the size of the space needed for the CI calculations has resulted in a decrease of CPU time of almost an order of magnitude compared to conventional CI calculations using canonical orbitals. There is a very systematic (and corresponding to chemically intuitive) choice which is possible within a localized space.

Table IV, which presents the coefficients and molecular orbital character of the ground electronic state of RDX along the $N_{(1)} - N_{(4)}$ dissociation pathway, indicates the complexity of the multiconfiguration character. Determinants in addition to those important at the

equilibrium distance become important along the $N_{(1)} - N_{(4)}$ dissociation
pathway.

Remarks on the correlation energy and multideterminant character
for excited states remain similar to those for the ground state, with
even more multideterminant character in the excited states.

3.3. Nitrobenzene - MRD-CI

As with the dissociation of other $>C - NO_2$ or $>N - NO_2$ compounds, there
is a crossing of molecular orbitals along the dissociation pathway
(Figure 4).

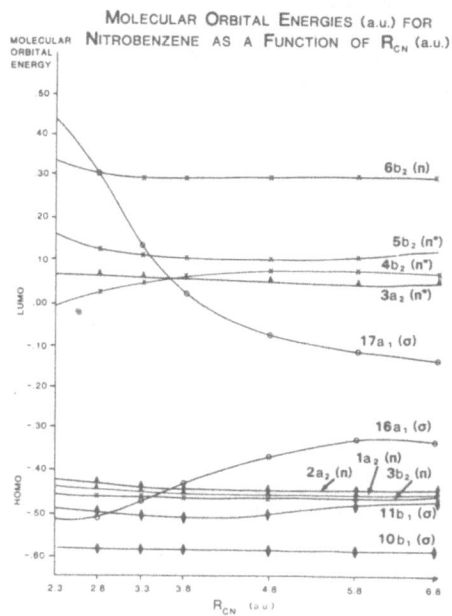

Figure 4

Our results on the $^1A_1$, $^1A_2$, $^3A_2$, $^1B_1$, $^3B_1$, $^1B_2$ and $^3B_2$ (Figure 5)
show a wealth of structure in the potential energy surfaces for the
various electronic states of nitrobenzene as a function of $>C - NO_2$
distance. Both the $^1A_2$ and $^3A_2$ states are predissociative and both
change dominant configuration at least twice along their potential
energy surfaces. The $^1B_1$ state is predissociative and changes dominant
configuration at least once along its potential energy surface. The

$^3B_1$ state is very slightly predissociative changing dominant configuration at least once along its potential energy surface. The $^1B_2$ and $^3B_2$ surfaces are predissociative each changing dominant configuration at least twice along their potential energy surfaces. We are continuing these investigations to identify the structures of the higher roots.

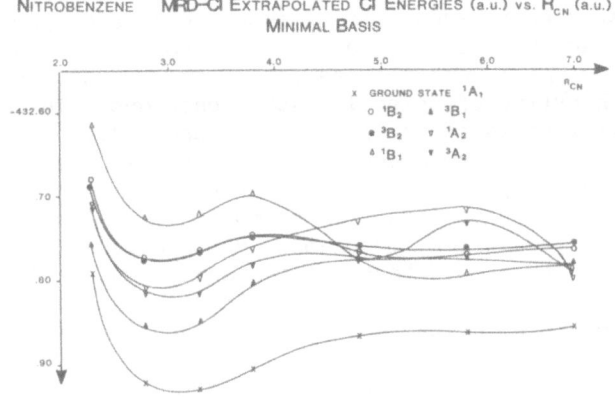

Figure 5

4. CONCLUSIONS

Both the MRD-CI and CASSCF calculations on CH_3NO_2 verify our earlier preliminary GVB and MCSCF/CI results of the multideterminant character of even the ground electronic state of CH_3NO_2 at its equilibrium geometry and indicate the character of the additional configurations that become important along the $CH_3 - NO_2$ decomposition pathway of nitromethane. This will be a general phenomenon for the $>C - NO_2$ decomposition pathways of all nitrocompounds, including the higher energetic nitrocompounds.

The MRD-CI calculations of the decomposition pathway of the excited states of CH_3NO_2 also indicate the complexity of electronic configurations to be expected along the decomposition pathways of excited states of $>C - NO_2$ compounds.

The MRD-CI calculations along the decomposition pathway of nitromethane were carried out with medium size atomic basis sets (including some polarization functions) limited by the available memory and disk restrictions. However, the basis set used for the MRD-CI calculations was sufficiently large that the identification and characterization of the multideterminant character and the conclusions will remain unchanged.

Likewise, the CASSCF calculations along the $CH_3 - NO_2$ decomposition pathway were carried out with medium size atomic basis sets. A recent calculation at the CH_3NO_2 equilibrium geometry was made with a larger basis set including some polarization functions on both the heavy atoms and on hydrogen. Additional disk space now available will permit even larger basis set CASSCF calculations. However, again the basis sets used for the CASSCF calculations along the CH_3 - NO_2 decomposition pathway were sufficiently large that the identification of the multideterminant character and the conclusions will remain unchanged.

The MRD-CI calculations presented for the RDX molecule can be looked upon as a first step towards solving some correlation problems. The localization approach has several advantages: not only does it allow one to choose the proper orbital space, but it is also very helpful in the later handling of molecular wave functions. The major advantage is the ability to adopt intuition based upon chemical bonds, lone pairs, etc. In spite of our encouraging results, we are aware of problems which can arise in particular cases. Highly symmetrical systems or special chemical compounds (for example those with strong conjugation effects) can cause many problems. Nevertheless, even poor localization (localization on more than two centers, coupling between distant chemical groups, etc.) is still useful in distinguishing localized regions in a molecule which may not be as obvious as in the case of RDX.

The treatment of correlation connected with a portion of a molecule has the same requirements as a treatment of whole molecules with the same number of electrons as the portion being considered. The main advantage seems to be translation from the big molecule problem (RDX 42 electrons) to the moderate problem ($>N$ - NO_2 part, 12 electrons). Especially when CI methods are used, it can help to solve the size consistency problem.

For these test calculations on RDX, only a minimal (but well-balanced) MODPOT basis set was used. We are aware that this is not a sufficient basis set representation. We will be carrying out similar calculations with larger basis sets on the newer supercomputers to which we now have access.

The use of localized orbitals in an MRD-CI calculation holds promise especially for investigating the potential energy surfaces for decomposition pathways of small groups dissociating from large molecules (or adding to large molecules). The size of calculations we carried out in this present research was the limit of what was feasible with the core memory and disc space available to a user on the CDC 7600.

The MRD-CI calculations on nitrobenzene indicate a wealth of structure in the potential energy surfaces especially of the excited states. All of the excited states exhibit predissociative character some changing dominant configuration several time along the $>C$ - NO_2 dissociation pathway. Thus all of these excited states can dissociate to products with kinetic energy which may be significant in initiation of processes leading to detonation.

NITROMETHANE

Table I

MULTIREFERENCE DETERMINANT CI   (MRD-CI)
DETAILS OF CONFIGURATION INTERACTION CALCULATIONS

REFERENCE CONFIGURATION = 19   (ALL SINGLES AND DOUBLES)
ELECTRONIC STATES EXPLORED = 4
TOTAL SAF'S GENERATED = 845,020
INTERACTION (SELECTION) THRESHOLD = 0.000050 HARTREE

$R_{(C-N)}$ A.U.	2.3	2.8	3.3	3.8	4.3	4.8	5.3	5.8	6.3
SAF'S (SELECTED)	4647	4449	4075	4780	4616	4715	4137	4082	4071

E(CI): SAF'S SELECTED

E (EXTRAPOLATED): CALCULATED FROM SUMMATION OF ENERGY LOWERINGS FOR REJECTED SAF'S AT 2 OR MORE THRESHOLDS.

$$E(FULL\ CI) = E(EXT) + (1 - \sum_{P}^{REF} c_P^2)[E(EXT) - E(REF)]$$

E(REF) = ENERGY FROM ONLY REFERENCE CONFIGURATIONS

Table II
CH_3NO_2 STATE 1 (GROUND STATE) MRD-CI

r(C-N)	DESCRIPTION	c^2
2.3	GS	.90
	$H1(0_1 0_2: p_z)^2 \rightarrow L1(NO_1 0_2: p_z)^2$	.02
2.8	GS	.89
	$H1(0_1 0_2: p_z)^2 \rightarrow L1(NO_1 0_2: p_z)^2$	.02
3.3	GS	.89
	$H1(0_1 0_2: p_z)^2 \rightarrow L1(NO_1 0_2: p_z)^2$	.02
MO'S SWITCH		
3.8	GS	.87
	$H2(0_1 0_2: p_z)^2 \rightarrow L2(NO_1 0_2: p_z)^2$	.01
	$H1(C-N: \sigma)^2 \rightarrow L1(C-N: \sigma*)^2$	.01
	$H1(C-N: \sigma) \rightarrow L1(C-N: \sigma*)$	.01
4.3	GS	.82
	$H1(C-N: \sigma)^2 \rightarrow L1(C-N: \sigma*)^2$	.04
	$H1(C-N: \sigma) \rightarrow L1(C-N: \sigma*)$	.03
	$H2(0_1 0_2: p_z)^2 \rightarrow L2(NO_1 0_2: p_z)^2$	.01
4.8	GS	.75
	$H1(C-N: \sigma)^2 \rightarrow L1(C-N: \sigma*)^2$	.07
	$H1(C-N: \sigma) \rightarrow L1(C-N: \sigma*)$	.06
	$H2(0_1 0_2: p_z)^2 \rightarrow L2(NO_1 0_2: p_z)^2$	.01
5.3	GS	.65
	$H1(C-N: \sigma)^2 \rightarrow L1(C-N: \sigma*)^2$	.12
	$H1(C-N: \sigma) \rightarrow L1(C-N: \sigma*)$	.11
	$H2(0_1 0_2: p_z)^2 \rightarrow L2(NO_1 0_2: p_z)^2$	.01
5.8	GS	.56
	$H1(C-N: \sigma)^2 \rightarrow L1(C-N: \sigma*)^2$	.16
	$H1(C-N: \sigma) \rightarrow L1(C-N: \sigma*)$	.15
	$H2(0_1 0_2: p_z)^2 \rightarrow L_2(NO_1 0_2: p_z)^2$	.006
6.3	GS	.49
	$H1(C-N: \sigma)^2 \rightarrow L1(C-N: \sigma*)^2$	.20
	$H1(C-N: \sigma)^2 \rightarrow L1(C-N: \sigma*)$	.15
	$H2(0_1 0_2: p_z)^2 \rightarrow L2(NO_1 0_2: p_z)^2$	.005

Table III

RDX MRD-CI   CORRELATION ENERGY (A.U.) FOR THE GROUND STATE OF RDX FOR A DIFFERENT NUMBER OF REFERENCE FUNCTIONS IN THE CASES OF LOCALIZED AND CANONICAL ORBITAL SPACES

| | LOCALIZED ORBITALS | | | CANONICAL ORBITALS | |
| | I | II | III | IV | V |
	2 REF. FUNCT.	19 REF. FUNCT. (A)	19 REF. FUNCT. (B)	2 REF. FUNCT.	19 REF. FUNCT
E_{CI}	-171.436907	-171.443561	-171.450047	-171.302694	-171.336571
E_{EXT}	-171.445164	-171.458082	-171.461869	-171.385781	-171.381572
E_{FULL}	-171.470216	-171.470412	-171.468546	-171.392018	-171.388501
CPU TIME (SECS.) CDC 7600	52.6	498.6	646.0	197.0	6745.7

(A)   THE CHOICE OF REFERENCE FUNCTIONS SUITABLE FOR THE 4 LOWEST STATES

(B)   THE CHOICE OF REFERENCE FUNCTIONS ESPECIALLY SUITABLE FOR THE GROUND STATE

E_{SCF} = -171.216514

Table IV

RDX   CI   LOCALIZED ORBITALS

RUN 1

$R_{N(1) - N(4)}$ (a.u.)

| | 2.13 | | 2.63 | | 3.38 | | 4.13 | | 4.88 |
Coefficients (c²) Greater Than 0.005	Excitations	Coefficients (c²) Greater Than 0.005	Excitations	Coefficients (c²) Greater Than 0.005	Excitations	Coefficients (c²) Greater Than 0.005	Excitations	Coefficients (c²) Greater Than 0.005	Excitations
0.91488	Ground State	0.87880	Ground State	0.84084	Ground State	0.76969	Ground State	0.66449	Ground State
0.00597	$3,5 \cdot (13)^2$	0.00576	$(3)^2 \cdot (14)^2$	0.02009	$(12)^2 \cdot (13)^2$	0.06911	$(12)^2 \cdot (13)^2$	0.14787	$(12)^2 \cdot (15)^2$
0.00584	$(4)^2 \cdot (14)^2$	0.00496	$(4)^2 \cdot (16)^2$	0.00757	$4,11 \cdot (14)^2$	0.00698	$4,12 \cdot (13)^2$	0.01272	$7,12 \cdot (13)^2$
0.00531	$(6)^2 \cdot (15)^2$	0.00548	$(5)^2 \cdot (15)^2$	0.00741	$1,6 \cdot (14)^2$	0.00692	$8,12 \cdot (13)^2$	0.01270	$6,12 \cdot (13)^2$
				0.00527	$(8)^2 \cdot (15)^2$	0.00584	$2,10 \cdot (14)^2$	0.00887	$11,12 \cdot (13)^2$
				0.00501	$(4)^2 \cdot (14)^2$	0.00570	$1,7 \cdot (14)^2$	0.00521	$12 \cdot 13$

5. ACKNOWLEDGEMENTS

This research was supported by the Office of Naval Research, Power Programs under Contract N00014-80-C-0003 and also especially by the Army Research Office through ONR.

The authors would like to thank Dr. Richard S. Miller, ONR, and Dr. David Squire, ARO, for their belief that we could successfully carry out what Dr. Squire called "high-risk research" namely, ab-initio MRD-CI calculations for the $>C - NO_2$ and $>N - NO_2$ decomposition pathways of large nitroexplosive molecules. While the calculations reported here are all somewhat preliminary in that one would like to have larger atomic basis sets, more active electrons, etc., they represented the maximum size calculations that could be carried out within the computer core and disk resources that were available to us at the time. With the advent of larger CRAY-XMP and CRAY-2 computers, larger calculations will now be feasible.

The authors are appreciative of the cooperation and earlier collaboration at the Johns Hopkins University of Professor Robert J. Buenker on conversion of the MRD-CI program to the CYBER 175 and CDC 7600.

The CH_3NO_2 and RDX MRD-CI calculations were run on the CDC 7600 at BRL. The authors would like to thank Dr. Larry Puckett, Assistant Director, US Army Ballistics Research Laboratory, Aberdeen Proving Ground, Aberdeen, MD, for arranging the use of the BRL CDC 7600.

The authors are also appreciative to Dr. Martin Guest and Dr. Victor M. Saunders of the Atomic and Molecular Theory Group at Daresbury, England for the CRAY (COS) version of the MRD-CI Program. They would especially like to thank Dr. Victor Saunders for his collaboration at the Johns Hopkins University in putting this version of the MRD-CI program on the NRL CRAY XMP-12.

The authors are appreciative to Professors Bjorn Roos and Per Siegbahn for making available to us the CASSCF program.

The CASSCF calculations were carried out on a CRAY-1M at Cray Research, Incorporated, at Chippewa Falls, Wisconson. The authors are most appreciative to Cray Research, Inc., and to Dr. S. Chen for arranging the use of the CRAY-1M computer time, and to Drs. C. Hsiung, P. M. Johnson and G. Spix for their outstanding help on the CRAY-1M system itself.

The MRD-CI calculations on nitrobenzene were run on the CRAY XMP-12 at NRL.

6. REFERENCES

1. C. Chabalowski, P. C. Hariharan, Joyce J. Kaufman and R. J. Buenker, 'Ab-Initio Multireference CI Calculations on CH_3NO_2 Confirm Earlier Preliminary GVB and MCSCF/CI Results that HNO_2 and CH_3NO_2 Have Multiconfiguration Ground As Well As Electronically

Excited States Even at Equilibrium Geometry, A Symposium Note,'
Int. J. Quantum Chem. S17, 643-644 (1983).

2. Joyce J. Kaufman, P. C. Hariharan, C. Chabalowski and M. Hotokka,
   'Multireference Determinant CI Calculations and CASSCF
   Calculations on the CH_3 - NO_2 Decomposition Pathway of
   Nitromethane,' An invited paper presented at the Sanibel
   International Symposium on Quantum Chemistry, Solid State Theory,
   Many-Body Phenomena and Computational Quantum Chemistry,
   Marineland, Florida, March 1985. In press, Int. J. Quantum Chem.
   Symposium Issue.

3a. R. Rafenetti, The Johns Hopkins University, 1973.

 b. R. Rafenetti and H. J. T. Preston, The Johns Hopkins University,
    1974.

4. Joyce J. Kaufman, 'Ab-Initio MODPOT/VRDDO/MERGE Calculations and
   Electrostatic Molecular Potential Maps: A. Large Carcinogens,
   Drugs and Biomolecules, B. Mechanism of Polymerization
   Initiation,' An invited plenary lecture presented at the
   Symposium on Theory of Complex Systems of Chemical and Biological
   Interest, Banff, Alberta, Canada, June 1980.

5a. Joyce J. Kaufman, P. C. Hariharan, C. Chabalowski, S. Roszak and
    A. Laforgue, 'Ab-Initio CI and Coupled Cluster Calculations on
    Energetic Compounds,' An invited paper presented at the Sanibel
    International Symposium on Quantum Chemistry, Solid-State Theory,
    Many-Body Phenomena and Computational Quantum Chemistry, Palm
    Coast, Florida, March 1984.

 b. Joyce J. Kaufman, P. C. Hariharan and S. Roszak, 'Ab-Initio
    MODPOT/VRDDO/MERGE Multireference Determinant Configuration
    Interaction (MRD-CI) Calculations for the >N-NO_2 Decomposition
    Pathway of RDX Based on Localized Orbitals,' An invited paper
    presented at the Fifth International Congress of Quantum
    Chemistry, Montreal, Canada, August 1985.

6. C. Petrongolo, R. J. Buenker and S. D. Peyerimhoff, 'Nonadiabatic
   Treatment of the Intensity Distribution in V-N Bonds of Ethylene,'
   J. Chem. Phys. 76, 3655-3667 (1982).

7. Joyce J. Kaufman, H. E. Popkie and P. C. Hariharan, 'New Optimal
   Strategies for Ab-Initio Quantum Chemical Calculations on Large
   Drugs, Carcinogens, Teratogens and Biomolecules.' An invited
   lecture presented at the Symposium on Computer Assisted Drug
   Design, Division of Computers in Chemistry at the American
   Chemical Society National Meeting, Honolulu, Hawaii, April 1979.
   In Computer Assisted Drug Design, Eds. E. C. Olson and R. E.
   Christoffersen, ACS Symposium Series 112, Am. Chem. Soc.,
   Washington, D. C., 1979, pp. 415-435.

8. Joyce J. Kaufman, P. C. Hariharan and H. E. Popkie, 'Symposium Note: Additional New Computational Strategies for Ab-Initio Calculations on Large Molecules,' Int. J. Quantum Chem. S15, 199-201 (1981).

9. Joyce J. Kaufman, 'Reliable Ab-Initio Calculations for Energetic Species,' An invited lecture presented at the NATO Advanced Study Institute on Fast Reactions and Energetic Systems, Preveza Beach, Greece, July 1980. In Fast Reactions In Energetic Systems, Eds., C. Capellos and R. F. Walker, NATO Advanced Institute Series, D. Reidel Publishing Company, Boston, Massachusetts, 1980, pp. 569-609.

10. P. C. Hariharan, W. S. Koski, Joyce J. Kaufman, R. S. Miller and A. H. Lowrey, 'Ab-Initio MODPOT/VRDDO/MERGE Calculations on Large Energetic Molecules. II. Nitroexplosives: RDX and α-, β- and δ-HMX,'* Int. J. Quantum Chem. S16, 363-375 (1982).

*Paper XVI in the Series 'Molecular Calculations With the Non-Empirical Ab-Initio MODPOT/VRDDO/MERGE Procedures.'

11. J. M. Foster and S. F. Boys, 'Canonical Configurational Interaction Procedure,' Revs. Mod. Phys. 32, 300-311 (1960).

12. T. L. Gilbert, 'Self-Consistent Equations For Localized Orbitals in Polyatomic Systems.' In Molecular Orbitals in Chemistry, Physics and Biology, Eds. P. O. Lowdin and B. Pullman, Academic Press, New York, 1964, pp. 405-420.

13a. B. O. Roos, P. R. Taylor and P. E. M. Siegbahn, 'A Complete Active Space Method (CASSCF) Using A Density Formulated Super - CI Approach,' Chem. Phys. 48, 157-173 (1980).

   b. P. E. M. Seigbahn, A. Heiberg, B. O. Roos and B. Levy, 'The Present Limits in Accurancy in Atomic Calculations of Small Systems,' Phys. Scr. 21, 328-334 (1980).

   c. P. E. M. Siegbahn, J. Almlöf, A. Heiberg and B. O. Roos, 'The Complete Active Space SCF (CASSCF) Method in a Newton-Raphson Formulation With Application to the HNO Molecule,' J. Chem. Phys. 74, 2384-2396.

Crystal Structures of Energetic Compounds: Ab-Initio
Potential Functions and Ab-Initio Crystal Orbitals

Joyce J. Kaufman, P.C. Hariharan, S. Roszak, J.M. Blaisdell,
A.H. Lowrey*

Department of Chemistry   and Richard S. Miller
The Johns Hopkins University   ONR
Baltimore, Maryland 21218   800 North Quincy Street
USA   Arlington, Virginia 22217

ABSTRACT. Detonation pressures and detonation velocities are governed
by the crystal densities (gms./cc.) of explosives. Our approach to
predicting optimal crystal-packing and crystal-structure parameters is
based on ab-initio potential functions from nonempirical ab-initio
calculations of smaller molecular aggregates (monomers, dimers,
trimers, etc.). The total SCF interaction energies are partitioned into
the different components, and then these components are fit
individually to functional forms or when necessary recalculated or
estimated explicitly for certain interaction components for each
different unit cell dimension change. The CRYSTAL-JHU program, given
the crystal symmetry, allows us to vary and optimize the crystal-
structure parameters. The agreement of our calculated unit cell
dimensions of nitromethane (CH_3NO_2) and of RDX with experiment was
excellent, within 1 to 2.8%.
    Most recently, we have derived and implemented and tested a
program POLY-CRYST for ab-initio SCF calculations on crystals and
polymers.

1. INTRODUCTION

To be able to predict crystal densities even for completely
hypothetical compounds is of interest. While there are empirical
methods to predict crystal densities, these methods depend only on
the number and type of fragments or groups. These empirical methods
are incapable of predicting differences in crystal densities between
various position isomers of the same molecule. These empirical methods
for crystal densities are also utterly incapable of predicting crystal-
structure arrangements.

*Fellow by Courtesy, The Johns Hopkins University, Permanent Address:
NRL, Washington DC 20375.

P. M. Rentzepis and C. Capellos (eds.), Advances in Chemical Reaction Dynamics, 311–326.
© *1986 by D. Reidel Publishing Company.*

Thus we embarked several years ago on a project to calculate optimal crystal-packing and crystal-structure parameters based on ab-initio SCF calculations plus calculations of dispersion energy contributions. There had been previous calculations on crystals using empirically derived potential functions. Most often these calculations were carried out for the lattice energies at the experimentally determined crystal structure. However, these empirical methods depend on a very large number of empirical parameters which must be determined from the experimental crystal structures of similar molecules. For new hypothetical molecules, such experimental data are lacking.

Hence, we initiated a major program to predict crystal densities and crystal-structure parameters from potential functions derived from ab-initio calculations.

Our approach, aimed at evaluating the intermolecular interactions, such as in polymer propagation and in molecular crystals, and at analyzing the optimal crystal packing is based on nonempirical ab-initio calculations for smaller molecular aggregates (monomers, dimers, trimers, etc.), partitioning the total SCF interaction energies into the different components and then fitting these components individually to functional forms or when necessary recalculating or estimating explicitly for certain interaction components for each different unit cell dimension change.

For this current research we derived and implemented a complete program, including higher-order terms where necessary, and including special features necessary only for calculation on crystals[1,2]. For this current research we also derived and implemented a complete CRYSTAL-JHU program which, given the crystal symmetry, allows us to vary and optimize the crystal-structure parameters[3,4].

Since this is a pioneering study using such potential functional forms fitted to energy partitioned a -initio all-electron SCF or ab-initio MODPOT/VRDDO/MERGE SCF wave functions, we investigated carefully such aspects as the comparison of the various intermolecular interaction terms calculated from the potential functions compared to those calculated by ab-initio all-electron wave functions. We also investigated for the CRYSTAL-JHU program how many surrounding unit cells had to be included for energy convergence.

For our first test on predicting unit cell dimensions for an energetic nitrocompound, we investigated the crystal of nitromethane, CH_3NO_2. An experimental crystal structure had previously been determined. We calculated the intermolecular interactions for CH_3NO_2 at more than 50 geometries, partitioned the interaction energies, and fit these to functional forms[4]. We then minimized the crystal energy with respect to unit cell dimensions. The agreement of our calculated unit cell dimensions with experiment was excellent.

Thus, we fulfilled successfully our goal of being able to optimize theoretically crystal unit cell dimensions to within ~2-3% using potentials derived from our partitioned ab-initio MODPOT/VRDDO calculations.

Furthermore, our method fits the partitioned ΔE_{SCF} results to atom-class - atom-class potential functions, which enables a general library of such potential functions to be built up for the molecular classes of interest.

2. METHODOLOGY

2.1 Intermolecular Interactions

The stabilization energy (per mole of molecules) ΔE_A in the crystal or in a molecular complex system with a number of other molecules or molecular parts can be estimated reliably from:

$$\Delta E_A = {}^1\!/_2 \sum_{B}^{\text{all other molecules}} \Delta E_{AB} + {}^1\!/_3 \sum_{C \neq B}^{\text{all other molecules}} \Delta E_{ABC} + \dots$$

where ΔE_{AB}, ΔE_{ABC} are the two- and three-body interaction energies.

The value of this approach is that these ab-initio atom-class - atom class potential functions can be derived from smaller molecules containing the necessary structural features.

2.1.1 Nonempirical Two Body SCF Interaction Energy. The nonempirical two-body interaction energy

$$\Delta E_{AB}^{SCF} = E_{EL}^{(1)} + E_{EX}^{(1)} + E_{IND,LE}^{(2)} + E_{IND,CT}^{(2)}$$

has been decomposed[1,2,4] into the electrostatic $E_{EL}^{(1)}$, exchange $E_{EX}^{(1)}$, classical long range $E_{IND,LE}^{(2)}$, and short range $E_{IND,CT}^{(2)}$ induction contributions.

2.1.2 The Electrostatic Term. In addition, the electrostatic term has been partitioned into short-range penetration $E_{EL,PEN}^{(1)}$ and a long-range multipole term $E_{EL,MTP}^{(1)}$ estimated within atomic multipole approximation. For this term we use, q, μ, θ - the cumulative atomic monopoles, dipoles, and quadrupoles derived directly from the ab-initio wave function.

We calculate this electrostatic term up through quadrupoles.

The electrostatic long-range multipole terms are one of the dominant contributions to intermolecular interaction energies between molecules which have strong intramolecular charge redistribution. Our treatment of this contribution includes explicitly the influence of anisotropy. We have recently shown[5] that the electrostatic term is the

dominant contribution to anisotropy. Since we calculate this term
explicitly, we take into account the anisotropy of intermolecular
interactions.

2.1.3 <u>Short Range Contributions</u> $E_{EL,PEN}^{(1)}$, $E_{EX}^{(1)}$, and $E_{IND,CT}^{(2)}$

All short range contributions from ab-initio intermolecular SCF
calculations, i.e., $E_{EL,PEN}^{(1)}$, $E_{EX}^{(1)}$, and $E_{IND,CT}^{(2)}$ are then fitted to a
functional of the type:

$$E = \sum_{a \epsilon A} \sum_{b \epsilon B} (\alpha_{ab} + \beta_{ab} R_{ab}^{-1}) \exp(-\gamma_{ab} R_{ij})$$

where R_{ab} denotes the interatomic distances between atoms belonging to
both interacting molecules A and B, and α, β, and γ constants are
fitted to reproduce the nonempirical ab-initio values of the
interaction energy components.

2.1.4 <u>Long Range Two-Body Induction Term</u>. Since long range induction
energy terms are not additive (but only the electric field due to
surrounding molecules is additive) the long range two-body induction
term $E_{IND,LE}^{(2)AB}$ is approximated as:

$$E_{IND,LE}^{(2)AB} = -1/2[\sum_{a \epsilon A} \alpha_a (\vec{E}_a^{B \to A})^2 + \sum_{b \epsilon B} \alpha_b (\vec{E}_b^{A \to B})^2]$$

where α_a and α_b denote the atomic polarizabilities (fitted to reproduce
nonempirical ab-initio values of $E_{IND,LE}^{(2)AB}$ where $\vec{E}_a^{B \to A}$ or $\vec{E}_b^{A \to B}$ correspond
to the electric field calculated within the atomic multipole expansion:

$$\vec{E}_b^{A \to B} = \sum_{b B} [q_b \vec{R} |R|^{-3} - \vec{\mu}_b |R|^{-3} + 3(\vec{\mu} \cdot \vec{R}) \vec{R} |R|^{-5} - 2(\overline{\theta} \cdot \vec{R}) |R|^{-5} +$$

$$+ 5\vec{R}(\vec{R} \cdot \overline{\theta} \cdot \vec{R}) |R|^{-7}$$

2.1.4.1 <u>Atomic Multipole Approximation of Electric Field</u>. To evaluate
the two- and three-body induction energy from approximate formula based,
on atomic polarizabilities α, one has to calculate an estimate for the
values of electric field $(\vec{E}_a^{B \to A})^2$ or $(\vec{E}_b^{A \to B})^2$ created by one molecule on

the atom of other interacting molecules. This can be achieved within
the atomic multipole approximation:

$$\vec{E}_a^{B \to A} = \sum_{b \in B} [q_b \vec{R} |R|^{-3} - \vec{\mu}_b |R|^{-3} + 3(\vec{\mu}_b \cdot \vec{R}) \vec{R} |R|^{-5} - 2(\overline{\Theta} \cdot \vec{R}) |R|^{-5} +$$

$$+ 5 \vec{R} (\vec{R} \cdot \overline{\Theta} \cdot \vec{R}) |R|^{-7}$$

This approximation was tested by calculating the exact value of
the electric field and comparing with the corresponding estimates in
monopole $(\vec{E}_q)^2$, dipole $(\vec{E}_{q\mu})^2$, and quadrupole $(\vec{E}_{q\mu\Theta})^2$ approximation.
The results showed that atomic quadrupoles are essential to obtain
satisfactory estimates of the electric field value.

2.1.4.2 Atomic Polarizabilities. We obtain a consistent set of
atomic polarizabilities α by fitting these to the explicitly calculated
values of $E_{IND,LE}^{(2)}$ from the partitioned ab-initio SCF calculations
against the electric field contributions.

2.1.5 Three Body-Interaction Energy. According to recent studies,[6,7]
the three-body interaction energy is dominated by the corresponding
long range induction term:

$$\Delta E_{ABC} \approx E_{IND,LE}^{(2)ABC} = -\sum_{a \in A} \alpha_a (\vec{E}_a^{B \to A} \ \vec{E}_a^{C \to A})$$

which is estimated in our approach using the same α values as in the
$E_{IND,LE}^{(2)AB}$ two-body term.

2.1.6 The Second Order Two-Body Dispersion Contribution

2.1.6.1 Semi-Theoretical Evaluation of Dispersion Energy. Up until
now, we have estimated this semi-theoretically from the expression:

$$E_{IND,LE}^{(2)AB} = -\sum_{a \in A} \sum_{b \in B} c_{aa} c_{bb} \ R_{ab}^{-6}$$

where c_{aa} constants were fitted to reproduce the nonempirical
dispersion energy values from studies of C. Huiszoon and F. Mulder[8], B.
Jeziorski and M. van Hemert,[9] and H. Lischka.[10]

2.1.6.2 Nonempirical Evaluation of Dispersion Energy. We have now
written a more rigorous program for calculating the dispersion energies
based on perturbation-variation approach.

2.1.6.2.1 <u>Second Order Perturbation Expression</u>. The dispersion term not included in the ΔE_{SCF} interaction energy could be evaluated from the classical second order perturbation expression:

$$E_{DISP}^{(2)} = 4 \sum_{i \in A}^{occ} \sum_{j \in B}^{occ} \sum_{k \in A}^{vac \; \infty} \sum_{\ell \in B}^{vac \; \infty} \langle ik|j\ell \rangle^2 \; / \; (E_i + E_j - E_k - E_\ell)$$

where sum-over-states includes all singly excited configurations $i \to k$ and $j \to \ell$.

Expanding molecular integrals $\langle ik|j\ell \rangle$ in LCAO MO expression:

$$\langle ik|j\ell \rangle = \sum_{r \in A}^{AO} \sum_{s \in B}^{AO} \sum_{t \in A}^{AO} \sum_{v \in B}^{AO} A_{ir} B_{js} A_{kt} B_{\ell v} \langle rt|sv \rangle$$

In all previously published calculations of this kind the monomer wave functions A_{ir}, A_{kt}, B_{js}, and $B_{\ell v}$ have always been evaluated in monomer basis sets. However, this yields significantly underestimated dispersion energy values as the number of vacant MO's (and therefore, singly excited configurations) is usually too small (especially in minimal and even extended basis sets).

2.1.6.2.2 <u>Variation-Perturbation Approach</u>. Much better results could be obtained in the variation-perturbation approach proposed by Jeziorski and van Hemert[9] where the monomer wave-functions have been evaluated in dimer basis set. It could be demonstrated (Sz. Roszak - who is a visiting scientist with our group at JHU) that the above mentioned approach is <u>eqivalent</u> to the use of formula with molecular integrals $\langle ik|j\ell \rangle$ expanded in dimer basis set AB:

$$\langle ik|j\ell \rangle = \sum_{r \in AB}^{AO} \sum_{s \in AB}^{AO} \sum_{t \in AB}^{AO} \sum_{v \in AB}^{AO} A_{ir} B_{js} A_{kt} B_{\ell v} \langle rs|tv \rangle$$

Such an approach allows consideration of many more vacant orbitals and leads to obtaining better dispersion energy estimates than within the classical formula.

Both formulas have been coded and the first (classical) formula has been tested against values given by Kochanski.[11]

2.2 Crystal-JHU

The CRYSTAL-JHU program calculates:

$$E_A = 1/2 \sum_B^{\text{lattice}} E_{AB}$$

optimizing 12 parameters determining the crystal structure within the assumed space group using the ΔE_{AB} described above.

2.3 POLY-CRYST - A Program For Ab-Initio Crystal Orbitals and Polymer Orbitals[12]

2.3.1 <u>Major Features</u>. This technique permits one to calculate ab-initio quantum chemical crystal orbitals and polymer orbitals making use of the translational symmetry in a crystal and translational and/or translational-rotational symmetry in a polymer.

Main features of the ab-initio SCF portion of the crystal orbital part of the program method:

> Fully ab-initio
> Full use of translational symmetry
> Analagous to molecular Hartree-Fock-Roothaan method
> No theoretical problems in approaching Hartree-Fock limit
> > (All numerical difficulties can have arbitrarily strict convergence criteria)
> General in 1, 2 or 3 dimensions
> Differences from the ab-initio SCF molecular method:
> Periodic Gaussians replace Gaussians in basis set
> Additional quantum number present in orbital description
> Gross atomic orbital populations and total overlap populations in molecule reflect influence of surrounding molecules

In the basic POLY-CRYST program itself we have meshed in as an options the desirable computational strategies we had derived over the years for ab-initio calculations on large molecules[13-15]. These strategies include: ab-initio effective core model potentials (MODPOT) which enable calculations on valence electrons only explicitly, yet accurately; a charge conserving integral size prescreening evaluation (which we named VRDDO - variable retention of diatomic differential overlap) which decides if an integral (or an block of integrals) is larger than a predetermined threshold and thus should be calculated, especially effective for spatially extended molecules and molecular systems; and an efficient MERGE technique to save and reuse common skeletal (or other) integrals. Thus for the calculation of the integrals for construction of the Fock matrix in the basic POLY-CRYST program we already calculate all integrals explicitly that will be larger than a certain threshold. Our particular VRDDO method of prescreening cuts down enormously on the number of integrals which must be calculated.

2.3.2 Problem Definition. As an electron moves in the periodic potential of the crystal, the difference between unit cells is only due to the phase difference. This can be introduced into our formalism along with the periodicity by introducing phase factors of $e^{-i\vec{k}\cdot\vec{R}}$ into our functions. It is not convenient to deal directly with a continuous variable in our calculations. In practice we evaluate the crystal properties on a mesh of discrete $\vec{k}$ points which can be as fine as necessary to display all the features of the band structure.

The form of our periodic basis function is now:

$$\sum_{\vec{R}} e^{-i\vec{k}\cdot\vec{R}} \chi_\mu(\vec{r} + \vec{R})$$

where $\vec{R}$ runs in principle over all lattice vectors of the crystal. Only cells with interactions above our threshold must be included.

The form of the answer is:

$$\phi_{n\vec{k}}(r) = \frac{1}{\sqrt{N}} \sum_{\vec{R}}^{N} \sum_{\mu}^{B} e^{-i\vec{k}\cdot\vec{R}} c_{n\mu\vec{k}} \chi_\mu(\vec{r} + \vec{R})$$

where N is the number of cells and B is the size of the basis set in the primary unit cell. The $\vec{k}$ dependence is indicated by subscripts on ϕ and c. The variation within $\vec{k}$ is smooth. This band holds 2N electrons, so the normalization of $\frac{1}{\sqrt{N}}$ gives the correct number per unit cell.

2.3.3 Hartree-Fock Crystal Orbital Equations and Matrix Elements. The Hartree-Fock crystal orbital equations are:

$$\vec{H}_{n\vec{k}} \vec{C}_{n\vec{k}} = E_{n\vec{k}} \vec{S}_{\vec{k}} \vec{C}_{n\vec{k}}$$

in matrix form.

2.3.4 Logic of Crystal Orbital Program.
Step 1.   Generation of lattice vectors from cell parameters and cell selection criteria.
Step 2.   Calculation of real-space one-electron integrals for all cases where one function is in the primary unit cell.
Step 3.   Calculation of real-space two-electron integrals for all cases where one function is in the primary unit cell.

Step 4.   For each $\vec{k}$-point, obtain the integrals for the periodic basis
           functions. The pieces from different cells must be collated
           and phase factors introduced.

Step 5.   Input initial coefficients and $\vec{k}$-occupation.
Steps 6-9 are SCF cycle
Step 6.   Form density matrix by 3-dimensional numerical integration
           over $\vec{k}$.

Step 7.   Form Fock matrix for each $\vec{k}$.
Step 8.   Diagonalize each Fock matrix.
Step 9.   Obtain new coefficients, test for convergence and do
           bookkeeping.

2.3.5 Additional Program Features

2.3.5.1 Population Analysis. The results of the population analyses
give the gross atomic charges on the atoms and the total overlap
populations TOP's between atoms (both intra- and intermolecular).
These TOP's indicate whether the intramolecular bonding is strengthened
or weakened when a molecule is placed in a crystal or polymer. These
TOP's also indicate the strength of intermolecular bonding.

2.3.5.2 Properties Package. These calculate atomic and molecular
multipole moments which are important in intermolecular interaction.

2.3.5.3 Possible Long-Range Corrections. Ab-initio SCF calculations
on molecular solids differ somewhat in character from ab-initio
calculations on covalently bound solids such as inorganic ionic solids
and metallic solids.
      In neutral molecular solids, however, there are at most only small
intramolecular charge redistributions, and thus the long-range behavior
of ionic or metallic solids is not expected to be a problem for
molecular solids. We plan to investigate this problem, and if long-
range terms have to be included, we will explore using strategies
similar to the multipole-multipole and multipole-induced multipole
techniques we developed for ab-initio atom-class - atom class potential
functions.[4]

3. RESULTS AND DISCUSSIONS

3.1 CRYSTAL-JHU

Our test calculations gave encouraging unit cell parameters agreement
with experiment (N_2 - within 7%; CO_2 - within 1%).
      Nitromethane, CH_3NO_2, was the first energetic molecule for which
we have tried to optimize the crystal unit cell dimensions.
      We carried out ab-initio MODPOT/VRDDO SCF calculations for the
CN_3NO_2 dimer for 46 different orientations and intermolecular

We carried out ab-initio MODPOT/VRDDO SCF calculations for the
CN_3NO_2 dimer for 46 different orientations and intermolecular
geometries of the CH_3NO_2 dimer. These ab-initio intermolecular SCF
calculations were corrected for the bases set superposition error
BSSE[16,17].

[BSSE, the basis set superposition error, is due to not having
very large completely saturated atomic basis sets. The majority of the
BSSE arises from inadequate description of the inner shells which are
heavily energy weighted - but these inner shells play virtually no role
in chemistry or in intermolecular interactions. Our ab-initio MODPOT
method turns out to have very little BSSE compared to conventional ab-
initio all-electron calculations including inner shells. Thus, our
MODPOT method is well suited for calculations of intermolecular
interactions.]

The results from CRYSTAL-JHU for the optimized unit cell
parameters for CH_3NO_2[4] are in excellent agreement with
experiment[18], a(2.7%), b(1.3%), and c(3.7%) [a = 5.1832 A expt. (5.039
A calc.); b = 6.2357 A expt. (6.153 A calc.); c = 8.5181 A expt. (8,198
A calc.)].

Next, we investigated optimizing unit cell parameters for the
nitroexplosive RDX

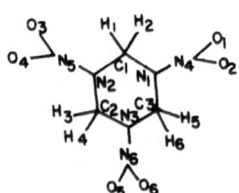

RDX—molecular sketch.

For RDX we used atom-class - atom-class potential functions derived
from CH_3NO_2 for the short range potentials, calculated the
electrostatic and induction terms explicitly using our calculated wave
function of the isolated RDX molecule, plus the semi-theoretical
estimate of the dispersion contribution. The great advantage of our
method for calculating intermolecular interaction energies is that for
the larger nitroexplosives it is only necessary to calculate the wave
function of the single isolated molecule (from which to calculate
explicitly the electrostatic and induction energies) and is not
necessary to calculate the wave function of pairs or triads of these
large molecules at many different geometries.

We used the experimental crystal symmetry of the RDX
molecule,[19] and our ab-initio MODPOT/VRDDO calculations for the RDX

molecule[20]. Our CRYSTAL-JHU results for optimizing the unit cell parameters of RDX were very encouraging. The agreement with experiment was most gratifying, to better than 1%.

3.2 POLY-CRYST

The POLY-CRYST program has been tested for H_2 in a linear chain and, as a severe test of the numerical convergence of the program at large distances, for H_2 in a fictitious crystal with one H_2 per unit cell and a lattice translation of 50 Å.

For the latter problem we used 27 unit cells and 27 $\vec{k}$ points and our customary ab-initio basis set for H used in our ab-initio MODPOT/VRDDO calculations on molecules.

The total energy per H_2 molecule in this crystal should be just a little lower than the total energy of the isolated H_2 molecule. The E_{HOMO} of H_2 in the crystal should be a little higher and the energy of the E_{LUMO} should be a little lower than the corresponding eigenvalues for the isolated H_2 molecule.

The results for the isolated H_2 molecule and the POLY-CRYST results on the H_2 crystal described above (even in this severe case with a lattice translation of 50 Å) in total energies (Table I) and in eigenvalues (more detailed results in reference 12) show that the above behavior is followed.

We also tested the POLY-CRYST program for a linear chain of H_2 molecules, 9 cells, 3 Å apart. In figure 1, E_{HOMO} and E_{LUMO} are plotted as a function of $\vec{k}$.

TABLE 1.

POLY-CRYST TEST

H_2 27 UNIT CELLS

27 $\vec{k}$ POINTS

LATTICE TRANSLATION LENGTH 50 Å

Energies (a.u.)	H_2 Isolated Molecule	H_2 POLY-CRYST 27 Unit Cells 27 $\vec{k}$ Points		
			k_x	
		0	+1	−1
E_{HOMO}	−0.586167	−0.592234	−0.589279	−0.589279
E_{LUMO}	+0.612240	+0.612240	+0.612241	+0.612241
E_{TOTAL}	−1.121794	−1.12256		

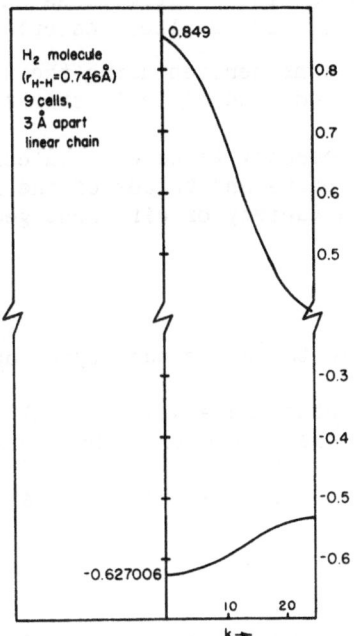

H₂ molecule
(r_{H-H}=0.746Å)
9 cells,
3 Å apart
linear chain

Figure 1

4.   CONCLUSIONS

4.1   CRYSTAL-JHU

What is very valuable is that, since we derive atom-class - atom-class
potentials, these results now enable us to start to build up a library
of potentials, such as involving compounds with NO_2 groups.

The largest amount of computational research which goes into an
approach of this sort is the initial building up of the library of
atom-class - atom-class potential functions. We are investigating
further the optimal basis set balance, integral thresholds, etc., for
intermolecular interactions involving nitrocompounds.

The CRYSTAL-JHU method was proven capable of optimizing the unit
cell parameters of CH_3NO_2 in excellent agreement with experiment using
the atom-class - atom-class potential functions for the short range
potentials derived from a large number of intermolecular calculations
on CH_3NO_2 plus explicit calculation of the electrostatic and induction
terms derived from the wave function of an isolated CH_3NO_2 molecule
plus a semi-theoretical estimate of the dispersion term energy.

What is even more encouraging is that for RDX the CRYSTAL-JHU
method for optimizing unit cell parameters gave excellent agreement

with experiment, using the atom-class – atom-class short range potentials derived from CH_3NO_2, plus explicit calculation of the electrostatic and induction terms derived using the wave function of an isolated RDX molecule plus a semi-theoretical estimate of the dispersion energy.

Hence, it appears not necessary to have to calculate the electronic wave functions of pairs and triads of the large nitroexplosive molecules in a variety of different geometrical arrangements.

4.2 POLY-CRYST

The POLY-CRYST will enable one to answer such types of pertinent questions as:
a. How do the occupied and virtual molecular orbital energy levels and the bond strengths of a molecule change from the isolated molecule when the molecule is in a crystal?
b. What is the bonding between molecules in a crystal?

5. ACKNOWLEDGEMENTS

This research was supported by the Office of Naval Research, Power Programs Branch under Contract N00014-80-C-0003

6. REFERENCES

1a. W. A. Sokalski, P. C. Hariharan, V. Lewchenko, and Joyce J. Kaufman, 'MOLASYS-INTER: A Program to Calculate Ab-Initio Partitioned Energy Contributions from Ab-Initio (or Ab-Initio MODPOT/VRDDO) SCF Wave Functions; s and p Orbitals. Also Includes Calculation of Electric Field at a Molecule from Surrounding Molecules and Procedures for Calculation of Dispersion Energy,' The Johns Hopkins University, 1979-present.
  b. A less extensive version of this program was written and used as early as 1979. W. A. Sokalski, P. C. Hariharan, Joyce J. Kaufman and C. Petrongolo, 'Basis Set Superposition Effect on Difference Electrostatic Molecular Potential Contour Maps,' Int. J. Quantum Chem. 18, 165 (1980).

2. W. A. Sokalski, P. C. Hariharan, V. Lewchenko, and Joyce J. Kaufman, 'GIPSY-INTER: A Program to Calculate Ab-Initio Partitioned Energy Contributions from Ab-Initio (or Ab-Initio MODPOT/VRDDO) SCF Wave Functions: s, p, d, and f Orbitals. Also Includes Calculation of Electric Field at a Molecule from Surrounding Molecules and Procedures for Calculation of Dispersion Energy,' The Johns Hopkins University, 1982-present.

3. A. H. Lowrey, W. A. Sokalski, P. C. Hariharan, and Joyce J. Kaufman. CRYSTAL-JHU, THe Johns Hopkins University, 1980- present.

4. W. A. Sokalski, S. Roszak, P. C. Hariharan, W. S. Koski, Joyce J.
    Kaufman, A. H. Lowrey and R. S. Miller. 'Crystal Structure Studies
    Using Ab-Initio Potential Functions From Partitioned Ab-Initio
    MODPOT/VRDDO SCF Energy Calculations I. N_2 and CO_2 Test Cases.
    II. Nitromethane,' Int. J. Quantum Chem. S17, 375-391 (1983).

5. W. A. Sokalski, P. C. Hariharan, and Joyce J. Kaufman, 'On the
    Nature of Anisotropy of Nonempirical Atom-Atom Potentials,'
    manuscript to be submitted for publication.

6. E. Clementi, M. Kistenmacher, W. Kolos, and S. Romano, 'Non-
    Additivity in Water-Ion-Water Interactions,' Theo. Chim. Acta 55,
    257-266 (1980).

7. W. Kolos and B. Jeziorski, 'Perturbation Approach to The Study of
    Weak Intermolecular Interactions,' Mol. Interact. 3, 1-46 (1982).

8a. C. Huiszoon and F. Mulder, 'Long-Range C, N and H Atom-Atom
    Potential Parameters from Ab-Initio Dispersion Energies for
    Different Azabenzene Dimers,' Mol. Phys. 38, 1497-1506 (1979).

  b. C. Huiszoon and F. Mulder, 'Long-Range C, N and H Atom-Atom
    Potential Parameters from Ab-Initio Dispersion Energies for
    Different Azabenzene Dimers,' Mol. Phys. 40, 249-251 (1980).
    (Errata).

9. B. Jeziorski and M. van Hemert, 'Variation Perturbation Treatment
    of the Hydrogen Bond Between Water Molecules', Mol. Phys. 31, 713-
    729 (1970).

10. H. Lischka, 'Ab-Initio Calculations on Intermolecular Forces.
    Effect of Electron Correlation on the Hydrogen Bond in the HF
    Dimer,' J. Am. Chem. Soc. 96, 4761-4766 (1974).

11. E. Kochanski, 'Evaluation of Intermolecular Energy Between Two
    Hydrogen Molecules near van der Waals Minimum, from a Perturbative
    Procedure,' J. Chem. Phys. 58, 5823-5831 (1973).

12a. J. M. Blaisdell, W. A. Sokalski, P. C. Hariharan, Joyce J. Kaufman
    and R. S. Miller. 'POLY-CRYST: A Program for Ab-Initio
    MODPOT/VRDDO Calculations on Polymers, Crystals, and Solids,' A
    paper presented at the Sanibel International Symposium on Quantum
    Chemistry, Solid-State Theory, Many-Body Phenomena and
    Computational Quantum Chemistry, March 1984. In press, Int. J.
    Quantum Chem., Symposium Issue.

  b. J. M. Blaisdell, W. A. Sokalski, P. C. Hariharan and Joyce J.
    Kaufman, 'POLY-CRYST - A Program for Ab-Initio Crystal Orbitals
    and Polymer Orbitals,' Presented at the International Conference
    on the Theory of the Structures of Non-Crystalline Solids,

Bloomfield Hills, Michigan, June 1985. In press, J Non-Crystalline Solids.

13. Joyce J. Kaufman, H. E. Popkie and P. C. Hariharan, 'New Optimal Strategies For Ab-Initio Quantum Chemical Calculations On Large Drugs, Carcinogens, Teratogens And Biomolecules.' An invited lecture presented at the Symposium on Computer Assisted Drug Design, Division of Computers in Chemistry at the American Chemical Society National Meeting, Honolulu, Hawaii, April 1979. In: Computer Assisted Drug Design, eds, E. C. Olson and R. E. Christoffersen, ACS Symposium Series 112, Am. Chem. Soc., Washington DC, 1979, pp. 415-435.

14. Joyce J. Kaufman, P. C. Hariharan and H. E. Popkie, 'Symposium Note: Additional New Computational Strategies for Ab-Initio Calculations on Large Molecules,' Int. J. Quantum Chem. S15, 199-201 (1981).

15. Joyce J. Kaufman, 'Ab-Initio Calculations Incorporating Desirable Options For Large Molecules And Solids: A Competitive, Completely General Alternative To Local Density Methods For Many Systems.' An invited paper presented at the Symposium Local Density Approximations in Quantum Chemistry and Solid State Theory, Copenhagen, Denmark, June 1982. In: Local Density Approximations in Quantum Chemistry and Solid State Physics, eds. J. P. Dahl and J. Avery, Plenum Press, NY, 1984 pp. 815-827.

16. J. M. Foster and S. F. Boys, 'Canonical Configuration Interaction Procedures,' Revs. Mod. Phys. 32, 300-311 (1960)

17. T. L. Gilbert 'Self-Consistent Equations For Localized Orbitals in Polyatomic Systems' in Molecular Orbitals In Chemistry, Physics And Biology, Eds. P. O. Lowdin and B. Pullman, Academic Press, NY, 1964, pp 405-420.

18. S. F. Trevino, E. Prince and C. R. Hubbard, 'Refinement of the Structures of Solid Nitromethane,' J. Chem. Phys. 73, 2096 - 3000 (1980).

19. C. S. Choi and E. Prince, 'The Crystal Structure of Cyclotrimelhyleno-trinitramine,' Acta Cryst. Sect. B28, 2857-2862 (1972).

20. P. C. Hariharan, W. S. Koski, Joyce J. Kaufman, R. S. Miller and A. H. Lowrey, 'Ab-Initio MODPOT/VRDDO/MERGE Calculations on Large Energetic Molecules. II. Nitroexplosives: RDX and α-, β- and δ-HMX,' Int. J. Quantum Chem. S16, 363-375 (1982).

THEORETICAL STUDIES OF THE C_4 MOLECULE

James P. Ritchie,[&] Harry F. King,[*] and William S. Young[*]
Mail Stop B214
Los Alamos National Laboratory
Los Alamos, New Mexico 87545

ABSTRACT. Optimized geometries and relative energies for three states of the C_4 molecule have been obtained from single-reference configuration interaction (SRCI) calculations. The $^1\Sigma_g^+$ acetylenic form correlates with two ground state $^1\Sigma_g^+$ C_2 molecules, from which it can be formed without activation. The $^1\Sigma_g^+$ state, however, is calculated to lie approximately 25 kcal above the $^3\Sigma_g^-$ state. At the SRCI level, a rhombic form is calculated to lie 1.1 kcal below the triplet form; consideration of the Davidson correction reduces this difference to 0.4 kcal, while more complete basis sets are expected to increase the difference only by about 0.2 kcal. Consideration of these effects and difference in zero-point energy leads to a final estimated splitting of 1.2 kcal, favoring the rhombus. To aid the determination of the ground state, preliminary estimates of the lowest optical transitions were obtained from SRCI calculations and vibrational frequencies were obtained from SCF calculations. Comparison of the calculated results with experimentally obtained spectra suggest the possibility that both the linear triplet and the rhombus may have already been observed.

1. INTRODUCTION

We have undertaken a theoretical study of small carbon clusters to determine what role, if any, they may play in soot formation. Theoretically, small carbon clusters have been the subject of several investigations.[1-4] Experimentally, C_2,[5] and C_3[6] are well-characterized and often-observed species in combustion and explosive processes. It

---

[&]Author to whom correspondence should be addressed.
[*]Permanent address: Department of Chemistry, State University of New York at Buffalo, Buffalo, NY 14214

P. M. Rentzepis and C. Capellos (eds.), Advances in Chemical Reaction Dynamics, 327–338.

is also known that carbon in the form of soot is also frequently produced in combustion and explosive processes, usually with deleterious
effects. C_4 was studied first because it is a simple dimer of C_2, the
most abundant molecular form of carbon. Furthermore, C_4 may have an
acetylenic ($\cdot$C≡C-C≡C$\cdot$, 1) or cummulenic (^{3}:C=C=C=C:, 2) structure. The
relative energies of these forms are important for determining the
structure of sp hybridized allotropes of carbon, which may likewise
occur in acetylenic ($\cdot$(C≡C)$_n$$\cdot$) or cummulenic (^{3}:(C=C)$_n$:) forms.

Our aim is to calculate the thermodynamic and kinetic stability of
C_4, as well as to calculate spectroscopic constants to aid in the identification and characterization of C_4. Observations attributed to C_4
have been previously reported.[7,8]

Additional interest arises from reports that a rhombic structure
(3) is calculated to be the ground state for C_4.[4]

$$\underline{1} \qquad\qquad\qquad \underline{2} \qquad\qquad\qquad \underline{3}$$

2. METHODS

Calculations were performed with Dunning's double zeta plus polarization basis set,[9] unless otherwise noted. The MOLECULE-SWEDEN codes
were used on a Cray computer.[10]

CI calculations for these species included all single and double
replacements from the reference function. For the rhombus, the reference function is the closed shell SCF configuration. For the triplet,
it is the RHF-SCF configuration; i.e.: $(core)5\sigma_g^2 4\sigma_u^2 \pi_u^4 \pi_g^2$. Because
of the biradical character of 1, two-configuration SCF optimized orbi-
tals were used in the CI treatment. The reference function for the CI
calculations is composed of those configurations from the two-configu-
ration SCF wavefunction; i.e.: $(core)5\sigma_g^2 \pi_u^4 \pi_g^4 4\sigma_u^0$ and $4\sigma_u^2 \pi_u^4 \pi_g^4 5\sigma_g^0$.
Optimized geometries were obtained from grid searches.

3. RESULTS AND DISCUSSION

3.1 Dimerization of C_2

A series of calculations were performed in which two ${}^1\Sigma_g{}^+$ C_2 molecules approached each other along a linear path. The two C_2 internuclear distances were fixed at 1.203 Å. To account properly for the biradical character of the two reactant molecules and the C_4 product, 4 electron in 4 orbital Complete Active Space SCF (CASSCF) calculations were performed at each point along the reaction path. For these calculations, the 3-21G basis set was employed.[11]

The results of these calculations showed that there is no barrier to the formation of ${}^1\Sigma_g{}^+$ C_4 along a linear approach of two C_2 molecules and that the reaction is 127 kcal exothermic when calculated in this fashion. These results are in keeping with the biradical character of the C_2 molecules. Thus, 1 represents a kinetically viable intermediate.

3.2 Linear and Rhombic Structures

Results from geometry optimization for 1, 2, and 3 are presented in Table I. For the ${}^1\Sigma_g{}^+$ and ${}^3\Sigma_g{}^-$ cases, slight deviations from linearity along either of the two bending modes resulted in higher energies, regardless of the level of theory used. For the rhombic case, deviations from planarity resulted also in higher energies at all levels of theory.

TABLE I. Optimized parameters for some C_4 structures. Results for ${}^3\Sigma_g{}^-$ and rhombic C_4 were obtained at the single-reference CI level. Results for the ${}^1\Sigma_g{}^+$ structure were obtained using a two-configuration reference function for the CI calculations. See text for details. Bond lengths in Å.

	${}^1\Sigma_g{}^+$	${}^3\Sigma_g{}^-$	Rhombus
$R(C_1\text{-}C_2)$	1.213	1.316	1.448
$R(C_2\text{-}C_3)$	1.398	1.297	1.500
Rel Energy (kcal)	0.0	-24.7	-25.8

The acetylenic nature of $\underline{1}$ is evidenced by the short C_1-C_2 bond length and the longer C_2-C_3 bond length. In comparison, butadiyne has C-C bond lengths of 1.218 and 1.384 Å.[12] A large amount of biradical character is indicated by coefficients of about 0.68 and -0.63 for the two configurations corresponding essentially to doubly occupied in-phase and out-of-phase combinations of sigma orbitals primarily on C_1 and C_4. The equilibrium acetylenic form of C_4 lies energetically above the triplet and rhombic equilibrium forms, as well as several other states; thus, it is only metastable.

The $^3\Sigma_g^-$ form was verified to be stable to bending at the CI level. At the optimized geometry, the coefficient of the reference function in the normalized CI wavefunction is -0.924. This state of C_4 has nearly equal bond lengths characteristic of a cummulene. For example, buta-triene has a C_1-C_2 bond length of 1.318 Å and a C_2-C_3 bond length of 1.283 Å.[13] The similar bond lengths in $\underline{2}$ and butatriene occur because the two unpaired electrons of $\underline{2}$ are in orthogonal orbitals having the same nodal properties as that of the highest doubly occupied pi orbital in butatriene. This trend is general and it is expected that the bond lengths of the $^3\Sigma_g^-$ form of C_n, where n is even, will be very similar to those of the analogous cummulene.

The rhombic form has been suggested as the ground state for C_4^4. This form is particularly interesting because it possesses a bond be-tween inverted sp^2 centers. Bonds between inverted sp^3 carbons have been previously investigated.[14] At the optimized geometry, the coef-ficient of the reference function in the normalized CI wavefunction is 0.927. The calculated bond lengths listed in Table I are slightly longer, as would be expected, than those obtained with the 6-31G* basis set of 1.425 and 1.457 for C_1-C_2 and C_2-C_3, respectively. The C_2-C_3 bond undergoes the greater change and its length is not at all charac-teristic of a typical double bond as it is represented in $\underline{3}$.

Because of the unique bond between inverted carbons, the effect of the inclusion of correlation was further examined. Figure 1 shows a map of the electron density obtained from a CI calculation less that obtained from an SCF calculation. The CI optimized geometry was used for both calculations. Electron density is removed from near the bond mid-points and placed closer to the atoms in the CI calculation. The bond between the inverted carbons as shown on the left in Fig. 1 has two symmetry-related regions along the internuclear axis of increased density between the carbons. Similar features are not observed for bonds along the periphery.

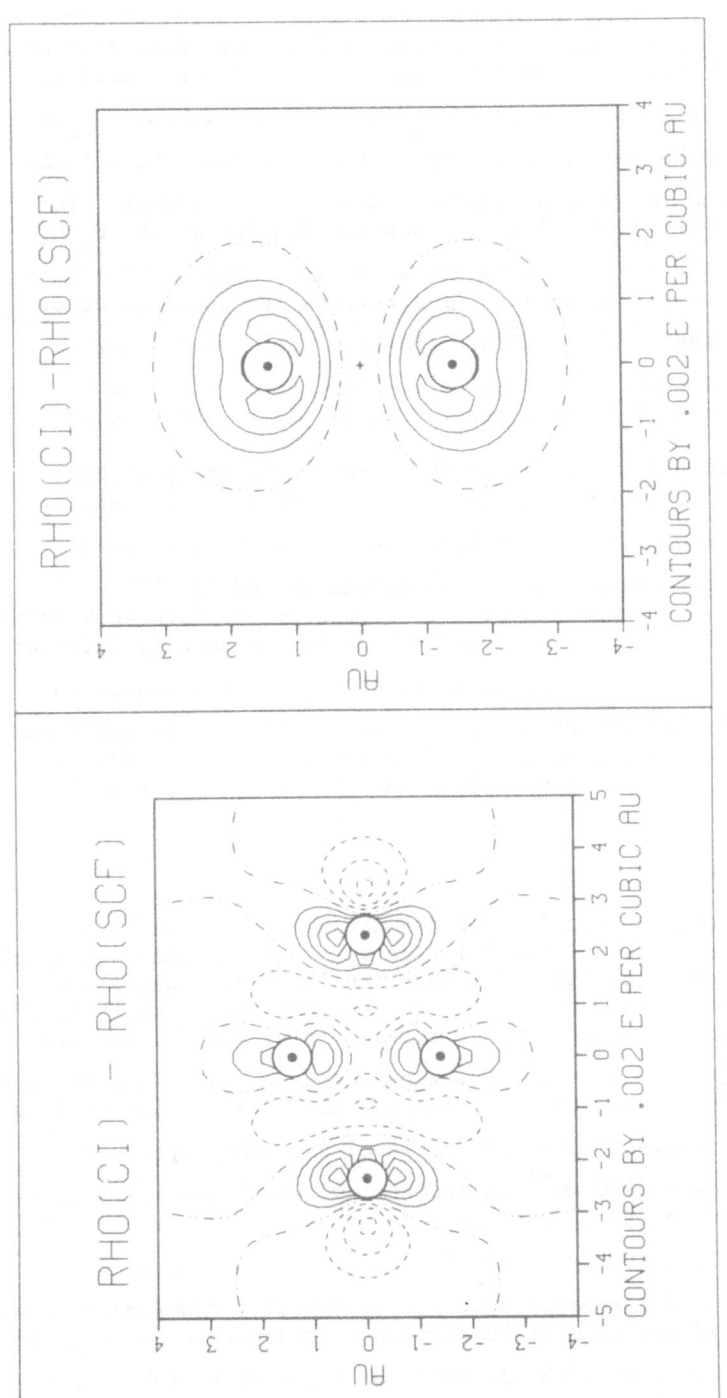

Figure 1. Electron density distribution obtained from CI calculations less that obtained by SCF calculations. Illustrated planes are in the plane of the molecule (left) and in the perpendicular plane containing the pi bond. Solid lines represent positive values, dashed lines represent negative values, and the combination dash-dot-dot lines represent zero as a value. Atom positions in the illustrated planes are indicated by a dark circle (●), while projections of atoms not in the plane are indicated with a plus (+).

Figure 2 shows density difference maps created by subtracting an electron density distribution obtained from a superposition of sphericalized SCF (double-zeta plus polarization basis) atoms from that of the molecular CI calculation. Deformation density maps were used in a discussion of bonding between inverted sp^3 hybridized carbons.[15] In contrast to the bond between inverted sp^3 carbons, a positive deformation density is found in the bond between inverted sp^2 carbons. Inclusion of CI, it will be recalled, reduces the density at the C_2-C_3 bond midpoint, so this difference is not a result of the CI wavefunctions used in this work. Lone pairs of electrons are apparent at C_1 and C_4. Peaks of about 0.07 e/au^3 are observed near the mid-points of bonds along the periphery. Interestingly, these peaks lie near if not directly on the internuclear axis indicating little if any bent bond character.

The deformation density found in the bonds along the periphery contrasts with that found between the inverted centers. The density deformation at the mid-point of the C_2-C_3 bond is about 0.03 e/au^3 - less than that for the other bonds, even though the $C2$-C_3 bond is represented as being of higher bond order. Displacements from this point along the C_1-C_4 vector result in larger values of the density deformation. Displacements from this point along the C_2-C_3 axis result in lesser values of the density deformation. Inspection of the deformation density in the plane containing the pi bond reveals electron density has been built-up along the internuclear axis and depleted from the atoms' p orbitals.

3.3 Spectroscopic Data

Calculated vibrational frequencies for the linear triplet and the rhombus are shown in Tables II and III. Results for the linear triplet species were obtained using the UHF method and the 6-31G* basis set. Results for the rhombus were obtained using the RHF method with the DZP basis set. Vibrational frequencies calculated in either fashion are usually about 9-11% too high for stretching modes.[16] Bending modes are usually thought to be more accurate, although the observed low bending frequency for C_3 is overestimated at the level of theory used here.[17]

Weltner observes a 2170 cm^{-1} IR absorption, which was previously assigned by Thompson to the stretching mode of an acetylenic C_4 species. In $\underline{2}$, the calculated frequency for this IR allowed mode of 1740 cm^{-1} seems in good agreement with the essentially cummulenic structure. Indeed, the IR spectrum of butatriene is reported as having absorptions in this region at 1708 and 1610 cm^{-1}; the only higher fre-

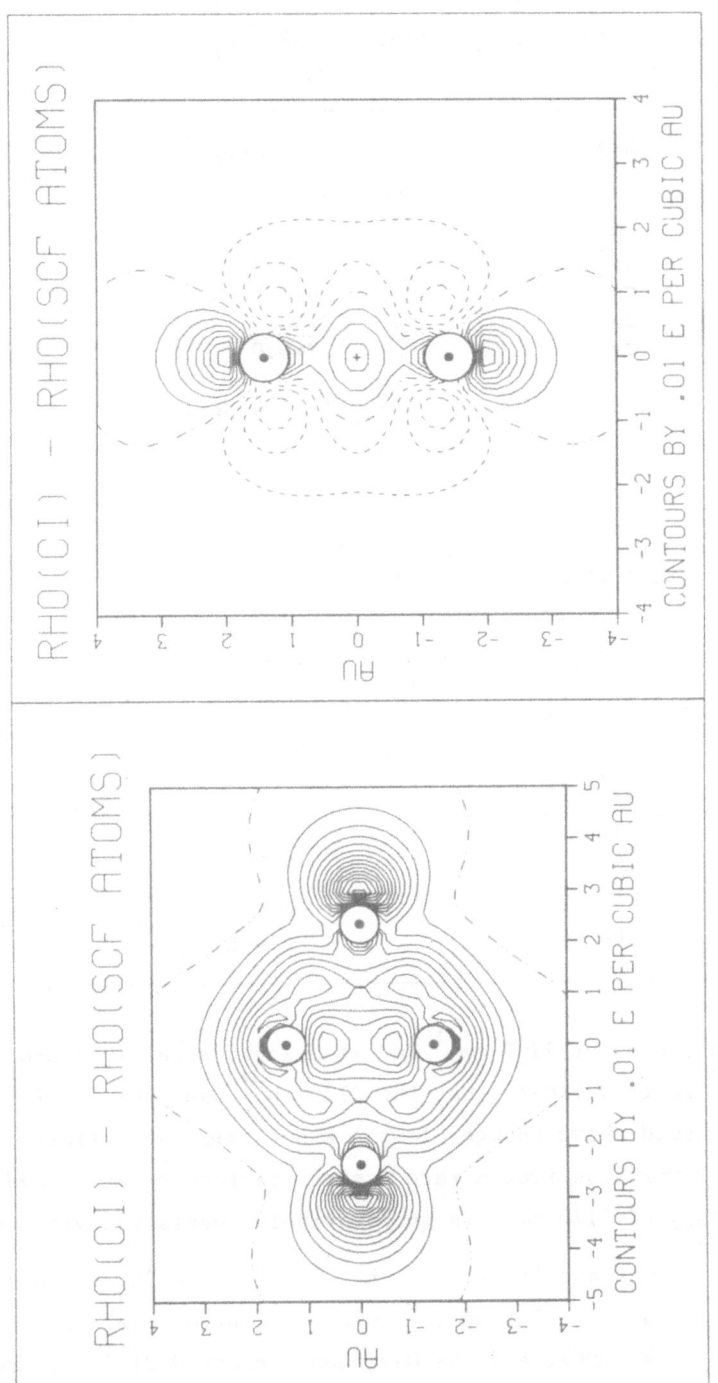

Figure 2. Electron density distribution obtained from a CI calculation less that obtained from a superposition of spherical atoms from SCF calculations. Plots are presented as in Figure. 1.

TABLE II. Calculated vibrational frequencies in cm^{-1} and resulting zero-point energy for $^3\Sigma_g^-$ C_4. The 6-31G* geometry was used.

Mode	Frequency
πu	209.3
πg	407.7
σg	1021.5
σu	1740.5
σg	2345.0
ZPE	9.1 kcal

TABLE III. Calculated vibrational frequencies in cm^{-1} and resulting zero-point energy for rhombc C_4. The DZP SCF optimized geometry was used.

Mode	Frequency
b_{3u}	345.2
b_{1u}	443.5
a_g	1061.3
b_{1g}	1086.7
a_g	1419.2
b_{2u}	1541.3
ZPE	8.4 kcal

quency absorption occurs at 2990 cm^{-1} and is almost certainly a C-H stretch.[18] Thus, it is unlikely that the 2170 cm^{-1} band results from $^3\Sigma_g^-$ C_4. As discussed above though, $\underline{1}$ is a kinetically accessible state, even at the very low frozen rare gas matrix temperatures, and a stretching frequency of 2170 cm^{-1} is qualitatively consistent with the calculated structure.

Low-lying optical transitions from the $^3\Sigma_g^-$ and the ^1A_g rhombus were also calculated and are listed in Table IV. Three levels of theory were used: SCF, SRCI, and the Davidson corrected CI.[19] There are two low-lying transitions for the $^3\Sigma_g^-$. Using Davidson corrected CI energies, the $^3\pi_u$ transition is predicted to lie below the $^3\Sigma_u^-$ transition. The $^3\pi_u$ state may even drop lower at higher levels of theory because of its multi-reference character. On the other hand, the $^3\Sigma_u^-$ state appears to have less multi-reference character.

TABLE IV. Low-lying optically allowed transitions in cm^{-1} at various levels of theory.

	SCF	SRCI	Davidson
From $^3\Sigma_g^-$ to:			
$^3\Pi_u$	30,556	20,049	15,488
$^3\Sigma_u^-$	14,729	19,797	21,291
From ^1A_g to:			
$^1B_{2u}$	27,317	21,592	19,765

Weltner observed two UV absorptions in rare gas matricies and assigned them to C_4. These bands were regarded as arising from a vibrational progression of an electronic transition from the $^3\Sigma_g^-$ state. The results in Table IV suggests the possibility that the observed 19569 cm^{-1} (19765 calculated) is the ^1A_g to $^1B_{2u}$ absorption of the rhombus, while the observed 21659 cm^{-1} (21291 calculated) results from the $^3\Sigma_g^-$ to $^3\Sigma_u^-$ transition. While it would be a fortunate coincidence for the calculations to be of the accuracy implied by the assigned transitions, it does seem quite plausible that the experiment results in more than one C_4 species. This interpretation requires that either or both of two possibilities be true: the rhombus and linear triplet lie extremely close in energy, as the calculations suggest, or that a non-thermodynamic mixture is trapped in the rare gas matrix. The latter could occur because the C_4 is formed from reactions of smaller carbon clusters at 4 to 25 K.

3.4 The Ground State of C_4

As indicated in Table I, at the SRCI level of theory the rhombus is favored by 1.1 kcal. Consideration of the Davidson correction lowers this preference to only 0.4 kcal. The calculated zero-point energies differ by 0.7 kcal, favoring the rhombus again. Inclusion of an extra set of diffuse d orbitals in the basis set followed by a SRCI calculation appears to favor the rhombus very slightly, by about 0.2 kcal. Starting with the 0.4 kcal figure, then adding the 0.2 kcal difference to account for an incomplete basis set and the 0.6 kcal (reduced by ~10% to account for the overestimation of vibrational frequencies at the SCF level) difference in zero-point energies, the rhombus would be favored by 1.2 kcal. This result is close to that of 2.5 kcal obtained from RMP4 calculations.[4] The absolute value of this splitting is not so important as realizing that the two structures are very close in energy.

4.   ACKNOWLEDGMENT

This work was sponsored jointly by the United States Department of the Energy and Department of the Army.

REFERENCES

1. K. S. Pitzer and E. Clementi, J. Am. Chem. Soc. 81, 4477 (1959).

2. E. Clementi, J. Am. Chem. Soc. 83, 4501 (1961).

3. D. W. Ewing and G. V. Pfeiffer, Chem. Phys. Letters 86, 365 (1982) and references therein.

4. R. A. Whiteside, R. Krishnan, D. J. Defrees, John A. Pople, and P. v.R. Schleyer, Chem. Phys. Letters 78, 358 (1981).

5. R. L. Altman, J. Chem. Phys. 32, 615 (1968) and references contained therein.

6. a) L. Gausset, G. Herzberg, A. Lagerqvist, and B. Rosen, Astrophys. J. 142, 45 (1965). b) W. Weltner and D. McLeod, J. Chem. Phys. 45, 3096 (1966).

7. W. R. M. Graham, K. I. Dismuke, and W. Weltner, Jr., Astrophys. J. 204, 301 (1976).

8. K. R. Thompson, R. L. DeKock, and W. Weltner, Jr., J. Am. Chem. Soc. 93, 4688 (1971).

9. T. H. Dunning, J. Chem. Phys. 53, 2823 (1970).

10. P. Seighbahn, B. Roos, P. Taylor, J. Almlof, M. Heiberg, and C. Bauschlicher.

11. J. S. Binkley, J. A. Pople, and W. J. Hehre, J. Am. Chem. Soc. 102, 939 (1980).

12. M. Tanimoto, K. Kuchitsu, and Y. Morino, Bull. Chem. Soc. Jap. 42, 2519 (1969).

13. Quoted by O. Bastiansen and M. Traetteberg, Tetrahedron 17, 147 (1962).

14. K. B. Wiberg, Acc. Chem. Res. 17, 379 (1984) and references therein.

15. J. E. Jackson and L. C. Allen, J. Am. Chem. Soc. 106, 591 (1984).

16. J. A. Pople, H. B. Schlegel, R. Krishnan, D. J. DeFrees, J. S. Binkley, M. J. Frisch, R. A. Whiteside, Int. Journal Quant. Chem.: Quantum Chemistry Symposium 15, 269 (1981).

17. D. H. Liskow, C. F. Bender, and H. F. Schaefer, III, J. Chem. Phys. 56, 5075 (1972).

18. W. M. Schubert, T. H. Liddicoct, and W. A. Lanka, J. Am. Chem. Soc. 76, 1919 (1954).

19. S. R. Langhoff and E. R. Davidson, Int. Journal Quant. Chem. 8, 61 (1974); J. Chem. Phys. 64, 4699 (1976).

FRACTAL ASPECTS OF HETEROGENEOUS CHEMICAL REACTIONS

Panos Argyrakis
Department of Physics
University of Crete
Iraklion, Crete
GREECE

Raoul Kopelman
Department of Chemistry
University of Michigan
Ann Arbor, Michigan 48109
U.S.A.

ABSTRACT

Heterogeneous reactions in which the rate is limited by the motion on the heterogeneous substrate differ significantly in their kinetics from diffusion-limited homogeneous reactions. We test a conjecture that combines the exploration space concept of the classical collision-rate-theory with the newly discovered paradigm of Brownian motion on low-dimensional (fractal-like) structures. We define a reduced time, which replaces real time in the classical rate equations. The reduced-time relations are tested by computer simulations, effectively proving the simple connection between a single-particle Brownian motion on a fractal structure and the dynamics of an ensemble of reacting Brownian particles. The real-time kinetic relations are governed by the random walk exponents, e.g. the effective spectral dimension which characterizes the time dependence of the number-of-distinct-sites-visited on a low-dimensional structure.

1. INTRODUCTION

There are a number of approaches to transport-limited-reactions, from the 19th century collision rate theory to the modern formulations of diffusion-controlled reactions.[1] While these theories apply, at best, only to homogeneous reactions, they all give an important common result: a rate constant K that is constant in time (at least for longer times). Can one generalize this result to heterogeneous reactions? No! We study here a simple conjecture that generalizes the homogeneous (3-dimensional) approach to heterogeneous (lower-dimensional) situations. The result is a time-depenent rate coefficient K(t). However, K is constant in reduced-time. This reduced-time has a simple and unique definition, thus allowing for unambiguous tests via computer simulations.

For the simple, homogeneous, elementary reaction,

$$A + A \rightarrow Products\uparrow, \tag{1}$$

P. M. Rentzepis and C. Capellos (eds.), Advances in Chemical Reaction Dynamics, 339–344.
© *1986 by D. Reidel Publishing Company.*

one has the well known rate equation,

$$-d\rho/dt = K\rho^2,$$ (2)

and its well known solution,

$$\rho^{-1} - \rho_o^{-1} = Kt,$$ (3)

where $\rho(t)$ is the instantaneous density (concentration) of the reactant A and $\rho_o = \rho(t = 0)$. It has been argued[2,3] that eqs. (2,3) should be generalized to

$$-d\rho/dS = K\rho^2,$$ (2')

$$\rho^{-1} - \rho_o^{-1} = KS,$$ (3')

where the "reduced time" S is proportional to the net space (S_t) swept out by the molecule A in time t. In homogeneous space S is <u>proportional</u> to t, that is S ~ t, and the only distinction between (3) and (3') is in the units of the <u>constant</u> K. However, this linear relation is not necessarily true for heterogeneous reaction spaces, and it has been shown that for fractal[2,3] and low-dimensional[4] spaces,

$$S \sim t^f , \quad f = d_s/2 < 1,$$ (4)

where d_s is an effective spectral dimension[5] and f is a fraction of unity. We note that, for Brownian motion on a lattice, S_t is the number of distinct sites visited by a randomly walking reactant molecule A after t steps. We thus relate the rate equation of an <u>ensemble</u> of <u>reacting</u> random walkers with the net space explored by a <u>single</u> random walker. We note that the misleadingly simple-looking reduced-time equations (2',3') give new, more complex, <u>real-time equations</u>,

$$-d\rho/dt = K(t)\rho^2 = K_o t^{-h}\rho^2 , \quad h = 1 - f$$ (2")

$$\rho^{-1} - \rho_o^{-1} = (K_o/f)t^f , \quad f = d_s/2$$ (3")

where the real-time rate coefficient is, in general,

$$\boxed{K(t) \sim dS/dt} .$$ (5)

We show, via computer simulations, the validity of eq. (5) for reactions on percolation networks.

2. ORIGIN OF CONJECTURE

Fig. (1a) shows the 19th century collision-rate concept. The molecule A, with cross section πr^2, sweeps out, in time t, a cylinder of length ℓ and volume $V = \pi r^2 \ell$. Obviously, $V \sim \ell \sim t$ and the rate constant K describes the volume swept out in unit time:

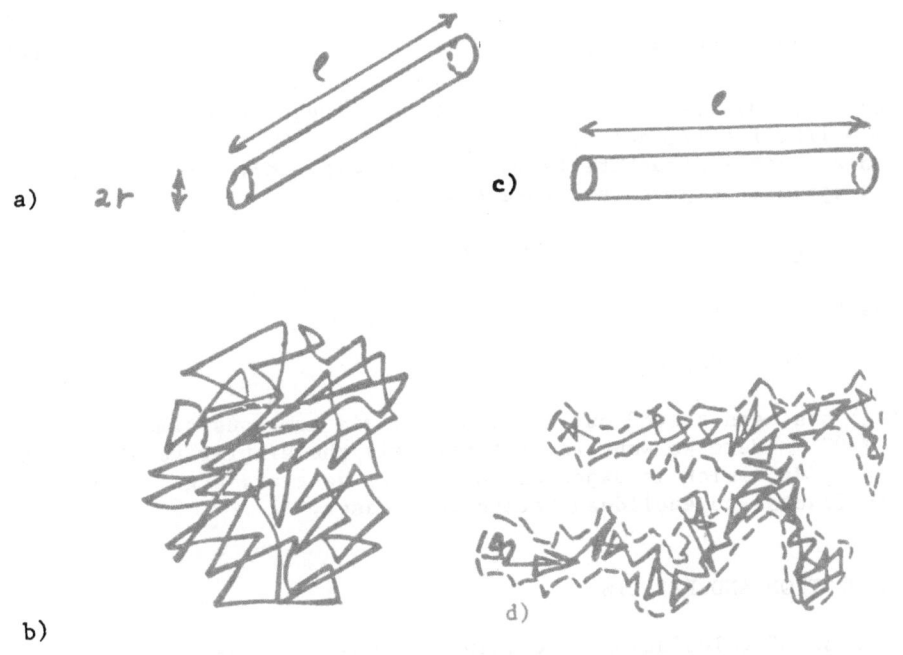

Fig.1:   a) Collision Theory Cylinder (3-dimensions).
         b) Brownian Motion Path (3-dimensions).
         c) Brownian Motion Path (1-dimension).
         d) Brownian Motion Path ("Fractal" Structure).
Note: For b) and d) the cross-section (2r) is the line-
         width.

$$K \sim dV/dt = \text{const.} \tag{6}$$

For Brownian motion in homogeneous 3-dimensional space (Fig. 1b),
Smoluchowski has shown[1] that, for longer times,

$$K \sim D = \text{const.}, \tag{7}$$

where D is the diffusion constant. Eq. (6) <u>cannot</u> be used if V is
taken as the diffusion sphere. However, we replace V with S_t, the
<u>net space</u> swept out by the randomly walking molecule A in time t (S_t)
is the volume of a self-penetrating snake with cross section πr^2).
Now, using eq. (5), we get[6] $S_t \sim t$ (t → ∞) and thus,

$$K(t) \sim dS_t/dt = \text{const.}, \tag{8}$$

in agreement with eq. (7).
    Fig. (1c) shows a 1-dimensional reaction, i.e., the molecule is
restricted to move on a (horizontal) line. The 19th century collision

model gives the same answer as in 3-dimensions, i.e., eq. (6). However, if we now assume Brownian motion on a line, we get[8] $S_t \sim \sqrt{t}$, and thus

$$K(T) \sim dS_t/dt \sim t^{-1/2} . \tag{9}$$

We note that this is a rigorous result[7] for longer time, and agrees with h = 1/2, f = 1/2 and d_s = 1.

On a fractal structure (symbolized by Fig. 1d), the effective space swept out is given by (see eq. 4):

$$S_t \sim t^f , \quad t \to \infty, \ f = d_s/2 < 1, \tag{10}$$

leading to

$$K(t) \sim dS_t/dt \sim t^{-h}, \ h = 1 - f . \tag{11}$$

Eq. (11) has been tested out successfully via numerous simulations.[3,8-10] Here we want to test directly the simpler relation (eq. 5), $K(t) \sim dS_t/dt$, which is expected to be more general, e.g., apply also to fractal-to euclidean "crossover" cases.[11]

3. SIMULATION AND RESULTS

The methods of calculation have been previously discussed in detail.[5,11] Specifically, improved Monte-Carlo algorithms are used to monitor the random walk properties, employing large lattices (4 x 10^6 sites) and extending over long times (2 x 10^5 steps). A large number of realizations is used (usually 1000 runs) for improved statistics.

The single particle dynamics has been extensively studied by us[5,11] recently. Detailed computer simulations of eq. (10) have provided very accurate values for f (and d_s), for 2-dimensional and 3-dimensional lattices. Our single walker results show that, at the exact critical percolation threshold:

$$d_s = 1.30 \pm 0.02 \quad (2\text{-dim.}),$$

$$d_s = 1.33 \pm 0.02 \quad (3\text{-dim.}). \tag{12}$$

These values provide the spectral dimensions of a percolating cluster exactly at criticality, i.e., at C_c = 0.59 (2-dim.) and C_c = 0.31 (3-dim.). Above the critical point there is a crossover to the euclidean behavior of a perfect lattice. Thus at C = 0.60 we find d_s = 1.36 (2-dim.) and at C = 0.32 we find d_s = 1.45 (3-dim.). For reacting particles the corresponding d_s values are:[8]

$$d_s = 1.35 \text{ at } C = 0.60 \quad (2\text{-dim.}),$$

$$d_s = 1.37 \text{ at } C = 0.32 \quad (3\text{-dim.}), \tag{13}$$

in satisfactory agreement with the d_s values of the single particle exploration. This leads us to plot the single particle exploration in terms of the reacting particle density for the corresponding time intervals, but in a time-independent plot. This is done in Fig.2, where we plot S_t for one particle <u>vs.</u> $(\rho^{-1} - \rho_o^{-1})$ for reacting particles, for 2-dim. lattices, for a variety of times and concentrations. We notice that all points fall close to the straight line x = y, thus verifying the conjecture (eq.3').

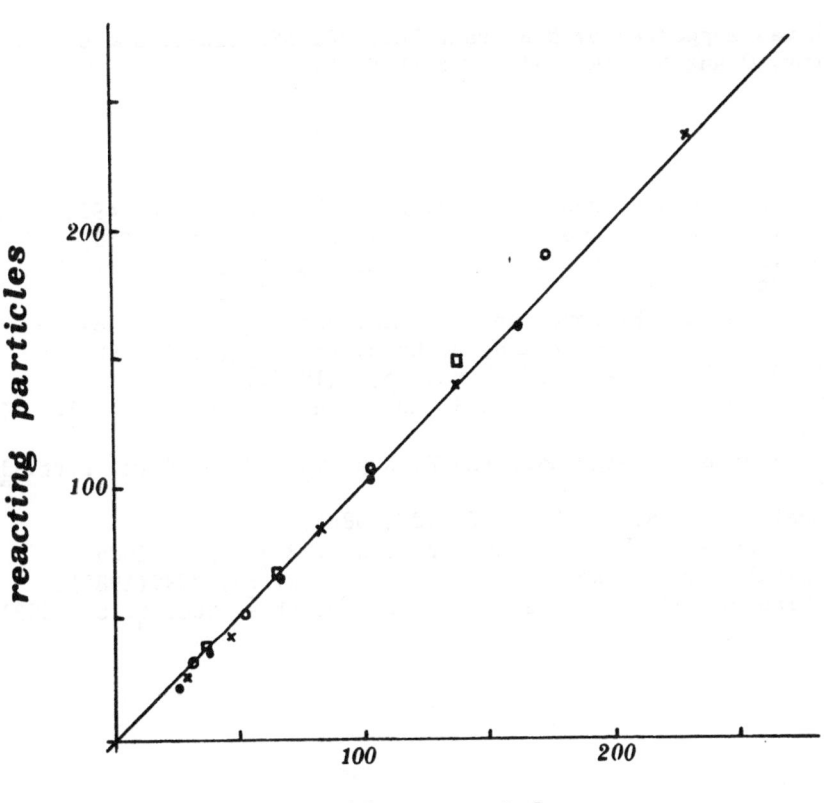

single particle

Fig.2: Single particle S_t vs.reacting particles $(\rho^{-1} - \rho_o^{-1})$ for 2 dim. lattices. Points are as follows: ●: C = 0.60, ✗: C = 0.65, ○: C = 0.70, ◘ : C = 0.80. The interval t = 1 - 2000 steps is covered. The results are averages of 1000 realizations on 300x300 lattices with cyclic boundary conditions, and ρ_o = 0.02. Random walks are on the largest cluster only and the "blind ant" model is employed.[12]

We thus show that one-particle motion indeed parallels reacting particle behavior; the same exponents are recovered from both cases. We note that we have effectively eliminated real time from our simulations (by using "reduced time"). The conclusion is that one-particle diffusion, and particularly the range of the walker (S_t) is of much importance to the understanding of the more complicated kinetics of ensembles of reacting random walkers.

ACKNOWLEDGEMENTS

This work was supported by NIH Grant No.2 R01 NS80016-17 and U.S. Dept. of the Army, Grant No. ARO-DAAG-20-83-K-0131.

REFERENCES

1) D.F. Calef and J.M. Deutsch, Ann. Rev. Phys. Chem. 34, 493(1983).
2) P.W. Klymko and R. Kopelman, J. Phys. Chem. 87, 4565(1983).
3) R. Kopelman, P.W. Klymko, J.S. Newhouse and L.W. Anacker, Phys. Rev. B 29, 3747(1984).
4) L.W. Anacker and R. Kopelman, J. Chem. Phys. 81, 6402(1984).
5) P. Argyrakis and R. Kopelman, J. Chem. Phys. 81, 1015(1984).
6) S. Chandrasekhar, Rev. Mod. Phys. 15, 1(1943).
7) L.W. Anacker, R.P. Parson and R. Kopelman, J. Phys. Chem. 84, 4758 (1985).
8) J.S. Newhouse, P. Argyrakis and R. Kopelman, Chem. Phys. Lett. 107, 48(1984).
9) R. Kopelman, J. Stat. Phys. 42, 185(1986).
10) J.S. Newhouse and R. Kopelman, Phys. Rev. B 31, 1677(1985).
11) P. Argyrakis and R. Kopelman, J. Chem. Phys. 83, 3099(1985).
12) C.D. Mitescu and J. Roussenq, Ann. Israel Phys. Soc. 5, 81(1983).

THE PROPAGATION OF TURBULENT FLAMES AND DETONATIONS IN TUBES

JOHN H.S. LEE
McGill University
Mechanical Engineering Department
817 Sherbrooke St. W.
Montreal, Quebec
Canada H3A 2K6

ABSTRACT. By artificially increasing the roughness of the wall of the tube using a wire spiral or a sequence of equally spaced orifice plates, it is possible to generate very intense large scale turbulence. Under these conditions it is possible to observe five different propagation regimes of combustion waves. The self quenching regime corresponds to a flame accelerating initially to a high velocity before quenching itself when the turbulent mixing rate exceeds the chemical reaction rate of the combustible mixture. The weak turbulent deflagration regime corresponds to flame speeds of the order of few tens of meters per second. The steady state velocity of this regime is achieved by the balance of the positive and negative effects of turbulence on the burning rate (ie., enhancement of mass and energy transport versus quenching due to mixing and flame stretch). In the sonic or choking regime, the flame speed corresponds closely to the sound speed in the burnt gases. The gasdynamic choking is brought about by the combined effect of friction and heat addition in a compressible pipe flow. The quasi-detonation regime corresponds to the low velocity detonation phenomenon in which the severe momentum losses give detonation velocity significantly below the normal Chapman-Jouguet value. The existence of this regime is based on the criterion $\lambda/d \lesssim 1$ where "λ" and "d" denote the detonation cell size and the orifice diameter respectively. The fifth regime of normal Chapman-Jouguet detonation occurs when $d/\lambda \gtrsim 13$ in accord with the result of the critical tube diameter problem. Qualitative discussions of the turbulent flame structure according to the ideas of Chomiak are given. A unified concept is advanced in that it is postulated that shear and turbulence play the essential roles not only in the propagation of deflagration, but in detonation as well. Auto-ignition by shock heating is assigned a lesser role, while the transverse turbulent shear layers generated by the triple shock configuration in the front of a cellular detonation are assumed to play the key role in the enhancement of rapid chemical reactions necessary for the propagation of a detonation wave. The principle argument being that since free radicals are in abundance in the reaction zone, it is more efficient to induce chemical reactions in the unburned gases by rapid turbulent mixing rather

345

P. M. Rentzepis and C. Capellos (eds.), Advances in Chemical Reaction Dynamics, 345–378.
© *1986 by D. Reidel Publishing Company.*

than to generate the free radicals by thermal dissociation in the shock. Thus the role of the shock front in detonation is to preheat the mixture leading to higher local diffusion rates and more important, to generate turbulent shear layers via triple shock collisions.

1. INTRODUCTION

Perhaps the most fundamental problem in combustion is to find out how fast things burn, or more precisely, how fast the combustion front propagates in the explosive medium. Generally speaking, the lower and the upper limit of propagation speed of a combustion wave corresponds to that of a laminar deflagration and a Chapman-Jouguet detonation respectively. For fuel-air mixtures at standard conditions, the laminar deflagration speed (relative to stationary coordinates) is of the order of a few meters per second, while the Chapman-Jouguet detonation velocity is about 1800 m/s. The mechanism of propagation of a laminar flame is due to the molecular diffusion of mass (free radicals) and heat from the reaction zone to the unburned mixture to initiate the chemical reaction. For a Chapman-Jouguet detonation, the precise nature of the propagation mechanism is still uncertain. However, according to the classical theory of Zeldovich, Döring and Von Neumann for the detonation structure, ignition is assumed to be due to the adiabatic compression of the unburned mixture by the shock wave that precedes the reaction zone. Laminar flame theory is now well developed and if the kinetics of the reactions and the transport coefficients of the mixture are known, the laminar flame speed can be computed theoretically. For Chapman-Jouguet detonations, the equilibrium detonation states can readily be determined (for gaseous explosives). However, no theory exists as yet whereby the dynamic detonation parameters (ie., initiation energy, limits, critical diameter, etc.) can be evaluated. This is due to the fact that the dynamic detonation parameters are linked directly to the detailed structure of the detonation front. Thus far it has not been possible to describe quantitatively the three dimensional cellular detonation structure as observed experimentally. However, important progress has been made in recent years in achieving direct correlations between the dynamic parameters with the detonation cell size which can be measured experimentally using the smoked foil technique. The advances in detonation theory have been summarized in a recent review [1].

In between the lower and the upper limit of laminar flames and Chapman-Jouguet detonations, there can exist a continuous spectrum of turbulent deflagration speeds which depend on the boundary conditions. The principal mechanism of propagation is due to convective turbulent transport even though molecular diffusion still plays the dominant role in the local fine scale rate processes. Unlike laminar flames and Chapman-Jouguet detonations, turbulent deflagration speeds are no longer fundamental quantities which depend only on the physical and chemical properties of the mixture. Turbulence is controlled by the specific initial and boundary conditions of the problem. Thus each problem should be treated individually and the ideas and results derived from the study of one particular situation should not be generalized in

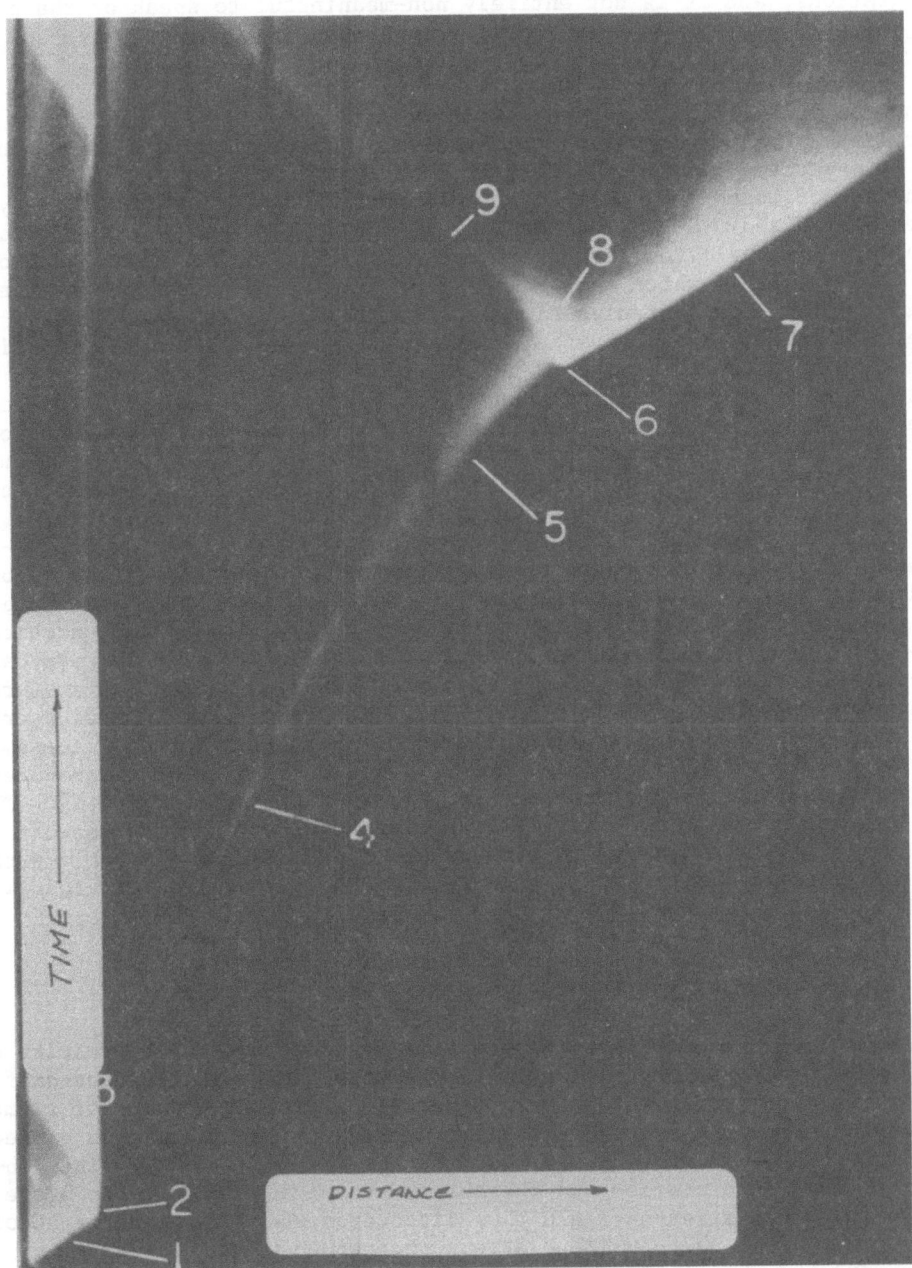

Fig. 1   Streak photograph of the transition from deflagration to detonation in $C_2H_2-O_2$ mixture, $P_O = 60$·torr in a smooth 25 mm
diameter tube.

principle. However, there are many universal features of turbulent deflagrations, and it is not entirely non-meaningful to speak of the propagation mechanisms of turbulent deflagrations in a general way, but we always bear in mind the problem specific nature of turbulence.

The object of this lecture is to discuss the influence of turbulence on combustion in gaseous explosive mixtures. The exact nature of turbulence in the combustion of condensed explosives is not known, but it is expected that it will play an important role also. The strong dependence of turbulence on initial and boundary conditions has already been emphasized. Hence, we shall limit our discussion to a particular problem. We shall consider the propagation of turbulence deflagrations in a long tube in which the walls may be "smooth" or roughened by the placement of flow obstacles along the length of the tube to increase the turbulence. The combustible mixture in the tube is initially quiescent and ignition is at one of the closed ends of the long tube. It is well known that upon ignition the initially laminar flame will accelerate and eventually will either transit to a Chapman-Jouguet detonation or reach some steady state turbulent deflagration speed. The problem is thus divided into two phases. The first phase is the transient acceleration phase of the flame, while the second phase is the final steady state turbulent deflagration or Chapman-Jouguet detonation regimes. The mechanisms of turbulent flame acceleration have already been discussed by the Author at the previous NATO Advanced Study Institute in 1980 [2] and also in a later review paper [3]. These flame acceleration mechanisms are fairly well understood on an individual basis, although their cooperative effects when the different mechanisms are present simultaneously are still not clear. However, quantitative descriptions of the flame acceleration process have met with little success and the numerical studies that have been carried out managed to reproduce only the qualitative features of the complex flow structure [4-6]. The present lecture will concentrate on a discussion of the final steady state regimes which cover the wide range of phenomena from weakly turbulent deflagrations (or wrinkled laminar flames) to quasi-detonations and Chapman-Jouguet detonations.

2. PROPAGATION REGIMES IN ROUGH TUBES

It is of importance to first review briefly the experimental results of turbulent deflagrations propagating in smooth and rough tubes of different sizes that have been obtained to date. The first studies of flame propagation in long smooth tubes were carried out by the French pioneers Mallard, Le Chatelier [7], Berthelot and Vielle [8] in 1881. They observed flame acceleration and the abrupt transition to a steady state supersonic combustion wave and thus discovered the phenomenon of detonation. A typical self-luminous streak photograph of an accelerating flame and the abrupt transition to a detonation wave in a smooth glass tube is shown in Fig. 1. In the earlier studies there appears to be little interest shown in the steady state high speed deflagrations that were obtained in less sensitive mixtures in which transitions to Chapman-Jouguet detonations were not obtained. Flame propagation in rough tubes was first studied by Chapman and Wheeler [9] in 1926. They showed that

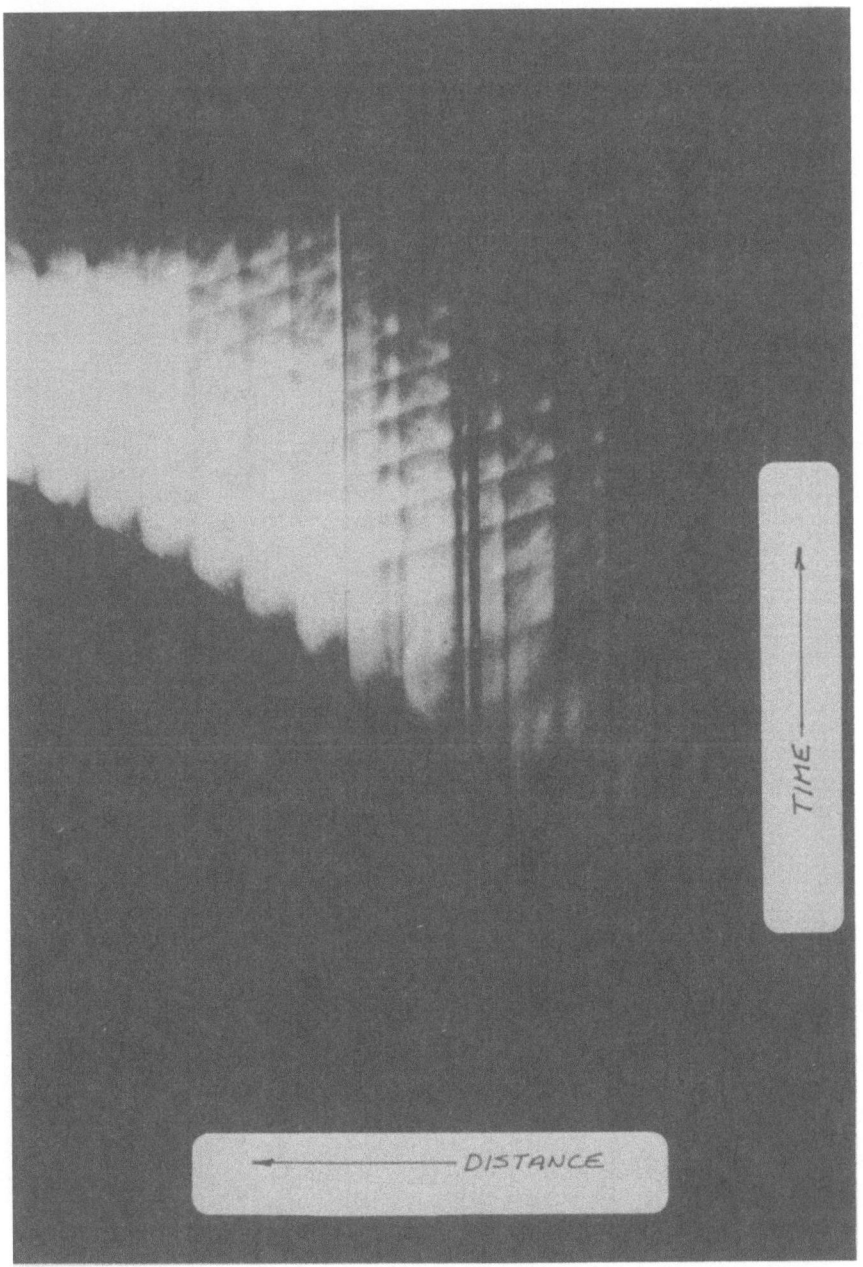

Fig. 2   Streak photograph of the acceleration to the steady state
          choking regime for a flame in CH_4-air mixtures in a 4 cm dia-
          meter tube with orifice plate obstacles spaced 10 cm apart.
          (Courtesy of H.Gg. Wagner)

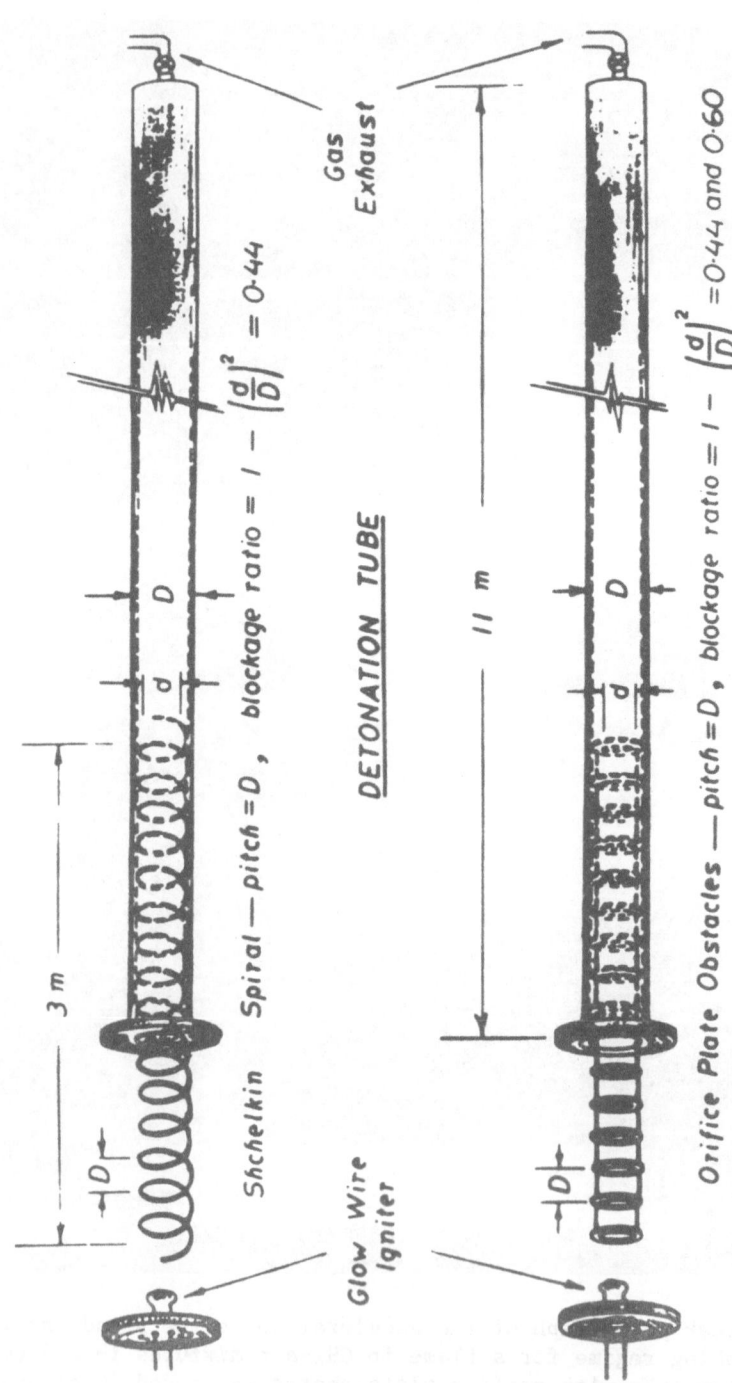

Fig. 3   Schematic of flame tube and flow obstacles.

turbulent deflagration speeds of about 420 m/sec were obtained for methane-air mixtures in a 5 cm diameter tube with the wall "roughened" by
the placement of a sequence of restricting rings (orifice plates) spaced
about one diameter apart along the tube. Schelkhin [10] studied the
propagation of flame and detonations in tubes where the wall is roughened by placing a wire spiral inside the tube. He observed a significant reduction in the transition distance from flame to detonation and
also reported steady state detonation velocities in the rough tube as
low as half their normal Chapman-Jouguet value. These low velocity
detonations (or quasi-detonations) were studied later by Guenoche and
Manson [11] and Brochet [12]. Recent interests in high speed turbulent deflagrations have prompted further studies of flame propagation
in rough tubes [13-22]. A typical streak photograph of the flame acceleration process and the final steady turbulent deflagration in a
CH_4-air mixture in a 4 cm diameter tube with a sequence of orifice
plates (3 cm diameter) spaced about 19 cm apart, is shown in Fig. 2.
In contrast to Fig. 1, it can be observed that the approach to the final
steady state flame speed is continuous with no distinct transition from
one regime (deflagration) to another (detonation). The final steady
state deflagration speed in Fig. 2 is about 770 m/s while in a smooth
tube, the turbulent deflagration speed for CH_4-air mixtures is typically
few tens of meters per second. It is interesting to note in Fig. 2 that
the slope of the trajectory of the deflagration front is more or less
parallel to the trajectories of the sound waves in the product gases indicating that the flow is sonic.

From the extensive epxerimental studies carried out at McGill over
the past decade, different regimes of flame (or detonation) propagation
have been identified. We shall describe in some detail the results that
have been obtained to date.

The McGill experiments were carried out in long steel tubes (11 m
to 19 m long) of diameters ranging from 5 cm to 30 cm. To "roughen" the
walls of the tubes, wire spirals as used by Schelkhin and orifice plates
as used by Chapman and Wheeler were employed. The orifice plates are
spaced about one diameter apart and the "wall roughness" is controlled
by the blockage ratio of the orifice plates (ie., BR = $1 - d^2/D^2$). Similarly for the Schelkhin spirals, the pitch is about one tube diameter and
the wall roughness is controlled by the wire diameter of the spiral. A
range of fuels (C_2H_4, CH_4, C_3H_8, and H_2) of different compositions in
air initially at atmospheric pressure and room temperature were used.
Flame speeds were measured by ionization probes spaced about 0.5 m apart
along the tube and pressure transducers were also used to record the
pressure time profiles at various locations along the tube. A schematic
of the experimental apparatus is given in Fig. 3.

The general behavior is as follows: Upon ignition the flame accelerates rapidly and after propagating past a number of obstacles it approaches a steady state velocity. It should be emphasized that the
steady state velocity referred to is based on an averaged value over a
distance of about half to one meter (depending on the tube diameter).
Thus the flame velocity is averaged over at least three or four obstacle
plates even for the largest diameter tube (ie., 30 cm). The local combustion phenomena in the obstacle field is extremely complex and it is

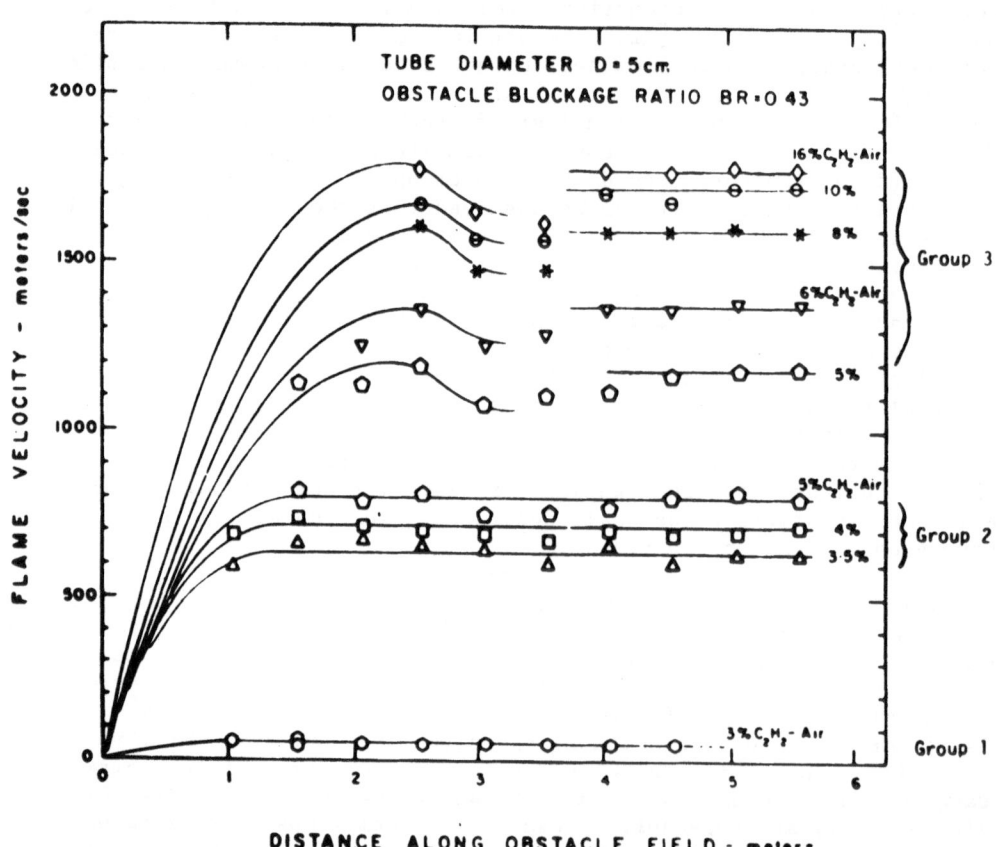

Fig. 4   Flame velocity history in the obstacle field.

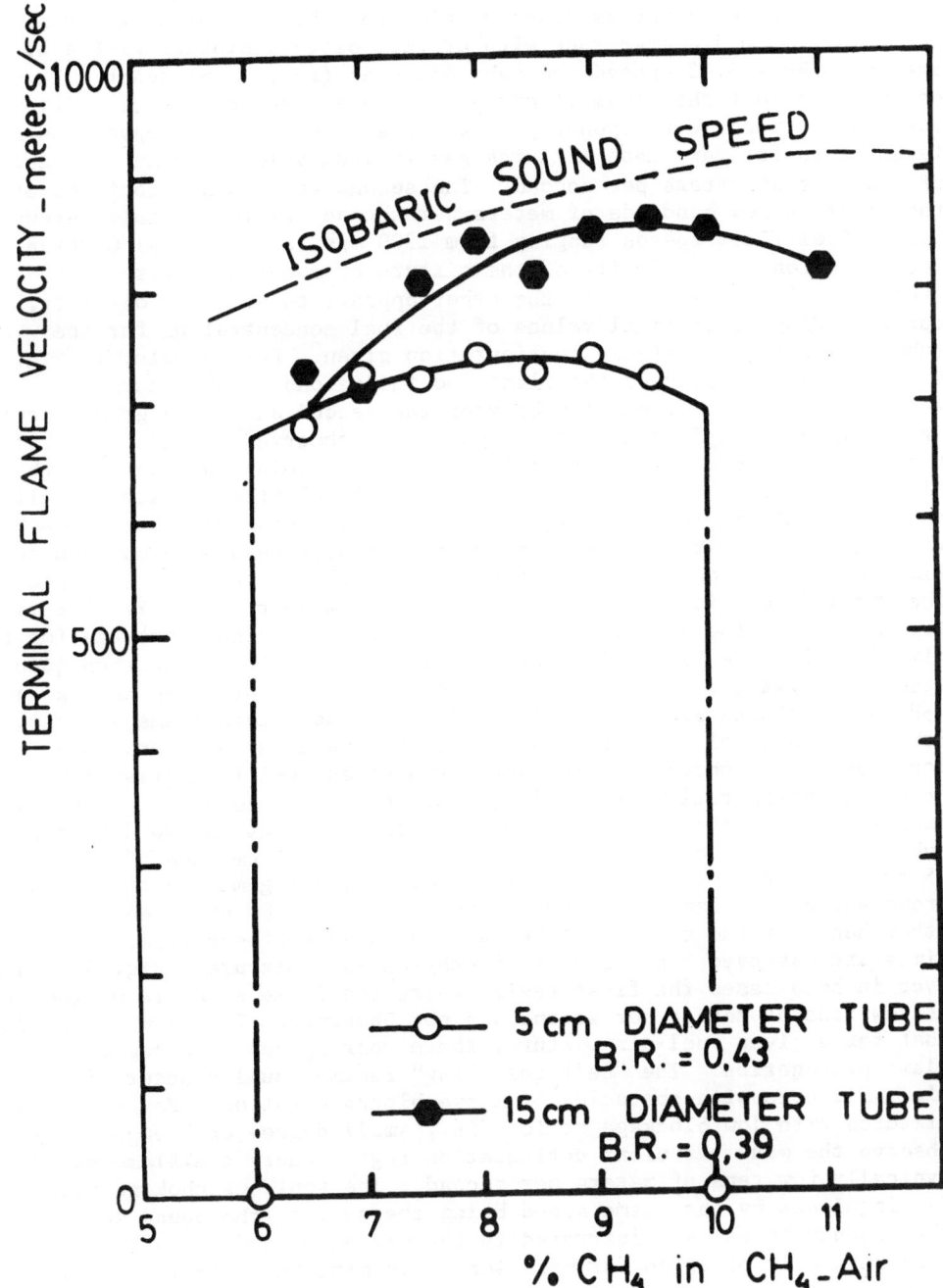

Fig. 5   Flame velocity versus fuel concentration for CH₄-air mixtures in the 5 cm and 15 cm tubes.

not meaningful to speak of a local flame velocity. Figure 4 shows a typical flame speed versus distance plot for the case of C_2H_2-air mixtures in a 5 cm diameter tube with orifice plate obstacles with a blockage ratio BR = 0.43 spaced one tube diameter (ie., 5 cm) apart. It can be observed that the final steady state flame speed can be classified into three groups corresponding to different regimes of propagation. The first group for very lean mixtures has typical velocities in the range of few tens of meters per second. The second group has velocities of the order of few hundreds of meters per second, while the third group has typical flame speeds ranging from 1200 m/s to the normal Chapman-Jouguet detonation velocity of the mixture of about 2000 m/s. The transition from one group to the other appears to be quite distinct, corresponding to critical values of the fuel concentration for the tube diameter and obstacle configuration given. For example, in Fig. 4 the transition between the first and the second group occurs at about 3.25% C_2H_2. Transition between the second and third group occurs precisely at 5% C_2H_2 since both regimes are observed at this particular fuel concentration. It is also interesting to point out that for the first two groups, the flame approaches the final steady state velocity asymptotically as shown in Fig. 2. However for group 3, the approach to the final steady state is typical of the transition phenomena from deflagration to detonation where the wave overshoots and then approaches the final detonation speed in an oscillatory manner (Fig. 1). For mixtures close to the lean and rich limits, there is also a propagation regime in which the flame first accelerates upon ignition and then quenches itself afterwards. This self-quenching regime is not shown in Fig. 4 for C_2H_2-air mixtures since there is no final steady state flame velocity. For different fuels in different tubes and obstacle configurations, similar results are observed. However for a given fuel in a given tube and obstacle configuration, not all the regimes can be observed. For example, in Fig. 5 where the results for methane-air mixtures in two different tube and obstacle configurations are shown (ie., 5 cm tube, BR = 0.43 and 15 cm tube, BR = 0.39), only the self-quenching regime and the second group where the flame speed is around 800 m/s can be obtained. On the other hand for the same two tubes and obstacle configuration, three regimes are observed for the case of ethylene-air mixtures (Fig. 6). However in both cases the first regime where the flame speed is of the order of few tens of meters per second are not observed. It may be concluded that for a given fuel-air mixture, there corresponds five regimes of flame propagation. The "self-quenching" regime usually occurs for near limit mixtures with obstacles of large blockage ratios. For near limit mixtures with low blockage ratios (ie., small degree of "roughness") we observe the weak turbulent deflagration regime where the flame speed is typically few tens of meters per second. The sonic or choking regime is distinguished by the flame speed being the same as the sound speed in the product gases as illustrated in the streak record shown in Fig. 2. This regime is observed for more sensitive mixtures with larger blockage ratios. The sonic or choking regime transits to the "quasi-detonation" or the "low velocity detonation" regime when the fuel concentration is increased further towards stoichiometric composition. The quasi-detonation speeds range from 50% to the full normal Chapman-Jouguet value of

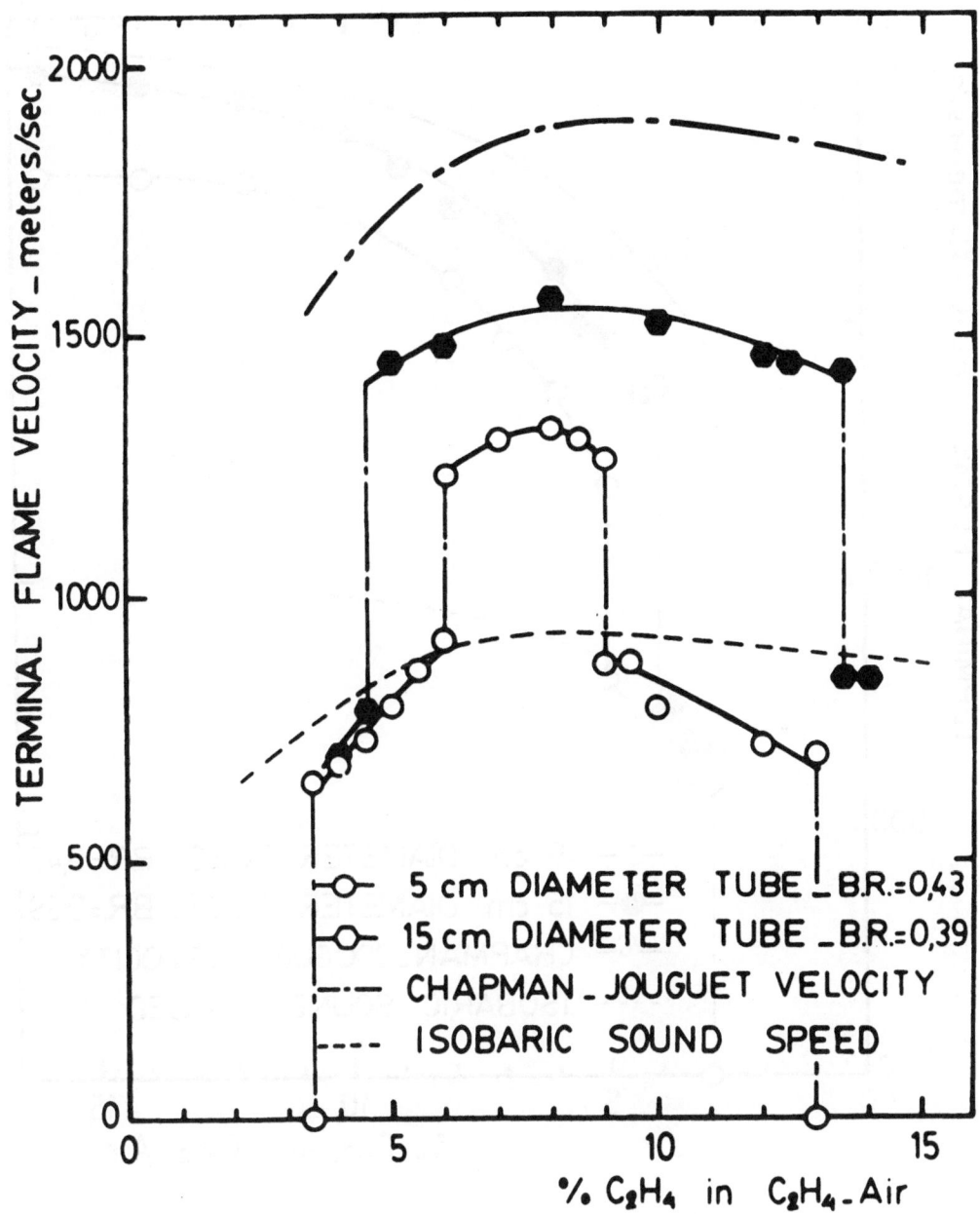

Fig. 6   Flame velocity versus fuel concentration for C$_2$H$_4$-air mixtures
in the 5 cm and 15 cm tubes.

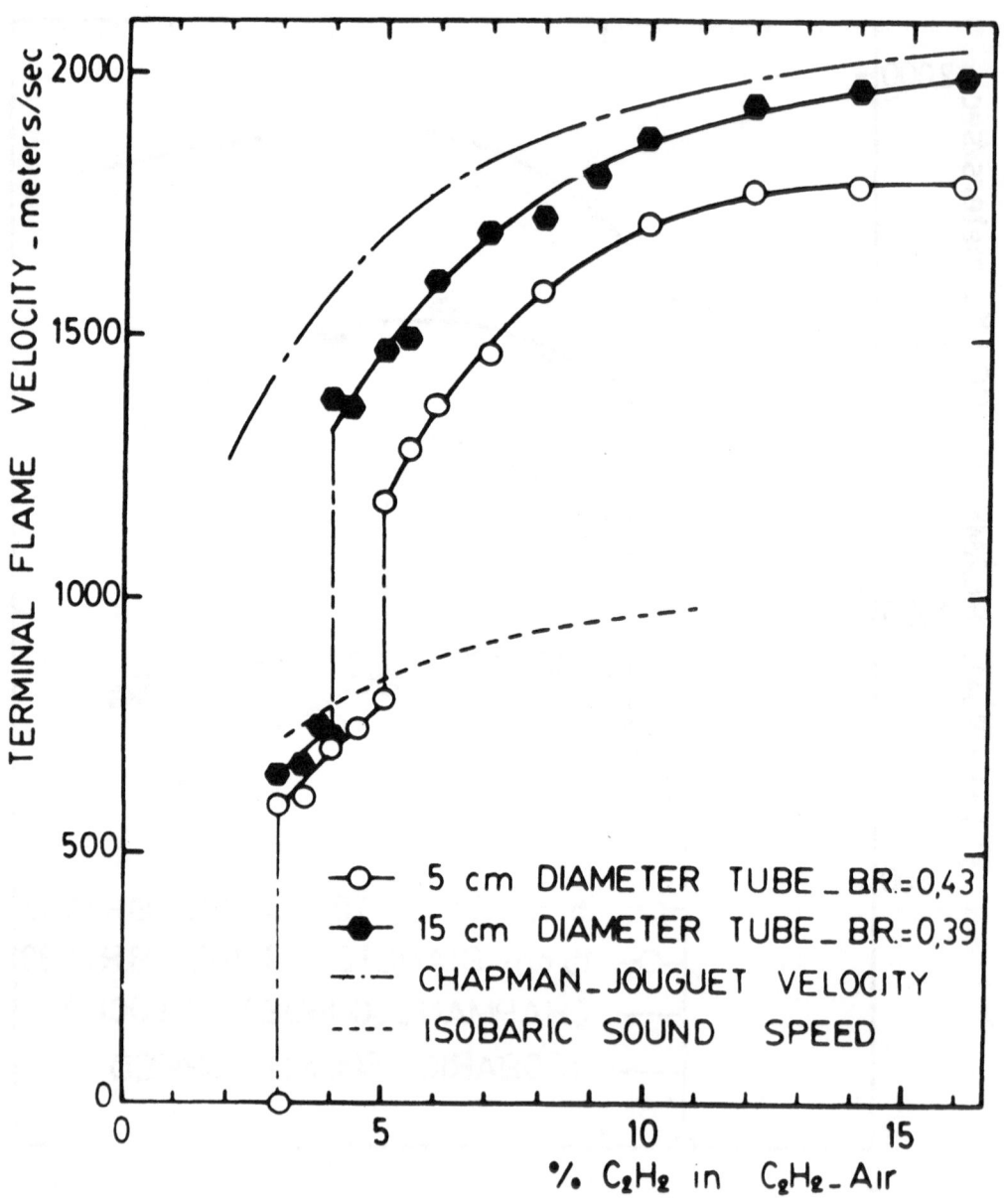

Fig. 7   Flame velocity versus fuel concentration for C_2H_2-air mixtures
in the 5 cm and 15 cm tubes.

the mixture. The large velocity deficit is due to severe momentum and heat losses to the wall caused by the obstacles. For very sensitive mixtures in large diameter tubes, the quasi-detonation regime appoaches asymptotically the normal Chapman–Jouguet detonation regime as is observed in large diameter smooth tubes. Although laminar deflagration with a flame speed of the order of a few meters per second corresponds to the lowest limit for a given mixture, it is not observed for flame propagation in tubes. This is due to the fact that laminar deflagrations are unstable and in tubes, they take on a global parabolic shape due to the interaction with the flow field generated in the unburned gases. The surface itself also takes on a cellular structure due to the presence of the various flame instability mechanisms. The "effective" burning rate or flame speed rate for such flames in tubes can exceed that of the non-existent planar laminar flame by many folds and are grouped into weak turbulent deflagration regime already described.

Figure 7 shows the case for C_2H_2-air mixtures in the 5 cm and 15 cm diameter tubes. Again the weak deflagration regime is not observed. In the 15 cm tube, the asymptotic approach to the normal Chapman-Jouguet regime can be observed as the fuel concentration is increased. However, in the smaller tube the normal Chapman-Jouguet regime is not approached due to the more severe losses associated with the smaller diameter tube. Similar results for propane-air mixtures are given in Fig. 8. In the smaller tube the quasi-detonation regime is not observed and only the choking regime is possible. For the 15 cm diameter tube however, both the choking and quasi-detonation regimes are observed, but the normal Chapman-Jouguet detonation regime is not attained. Figure 9 gives the results for H_2-air mixtures. The self-quenching regime is not observed due to the blockage ratios in all the tubes used are not sufficiently high. The weak turbulent deflagration regime is observed for the case of H_2-air mixtures and transition to the choking regime is quite distinct occuring at about 12.5% H_2. For the 30 cm tube where the orifice diameter is about 25.8 cm we see that normal Chapman-Jouguet detonations are observed quite readily even in the rough tube with orifice plate obstacles with BR = 0.28. The results for the various fuels in the 5 cm and 15 cm diameter tubes are summarized in Figs. 10 and 11. It is of interest to note that in the choking regime, the flame speeds for all the fuels in both tubes are of the order of 700 or 800 m/s. This is due to the fact that for all the various fuel-air mixtures shown, the sound speed in the combustion products over the range of fuel compositions do not change too much and is of the order of 700 or 800 m/s. The flame speed in the choking regime is also not too sensitive to the tube diameter and obstacle configuration. Thus in this choking regime, the flame speed being governed by the sound speed depends on the energetics of the mixture just like Chapman-Jouguet detonations. From Figs. 6, 7, 8 and 9, it can be observed that the range of fuel composition that defines the choking regime narrows with increasing tube diameter. This suggests that heat losses to the tube wall may in fact be the stabilizing mechanism for the existence of the choking regime. Without heat losses (as the tube diameter increases), it appears that only the quasidetonation regime is possible. This is in accord with gasdynamic theory of combustion waves in which if heat losses are negligible, the stable

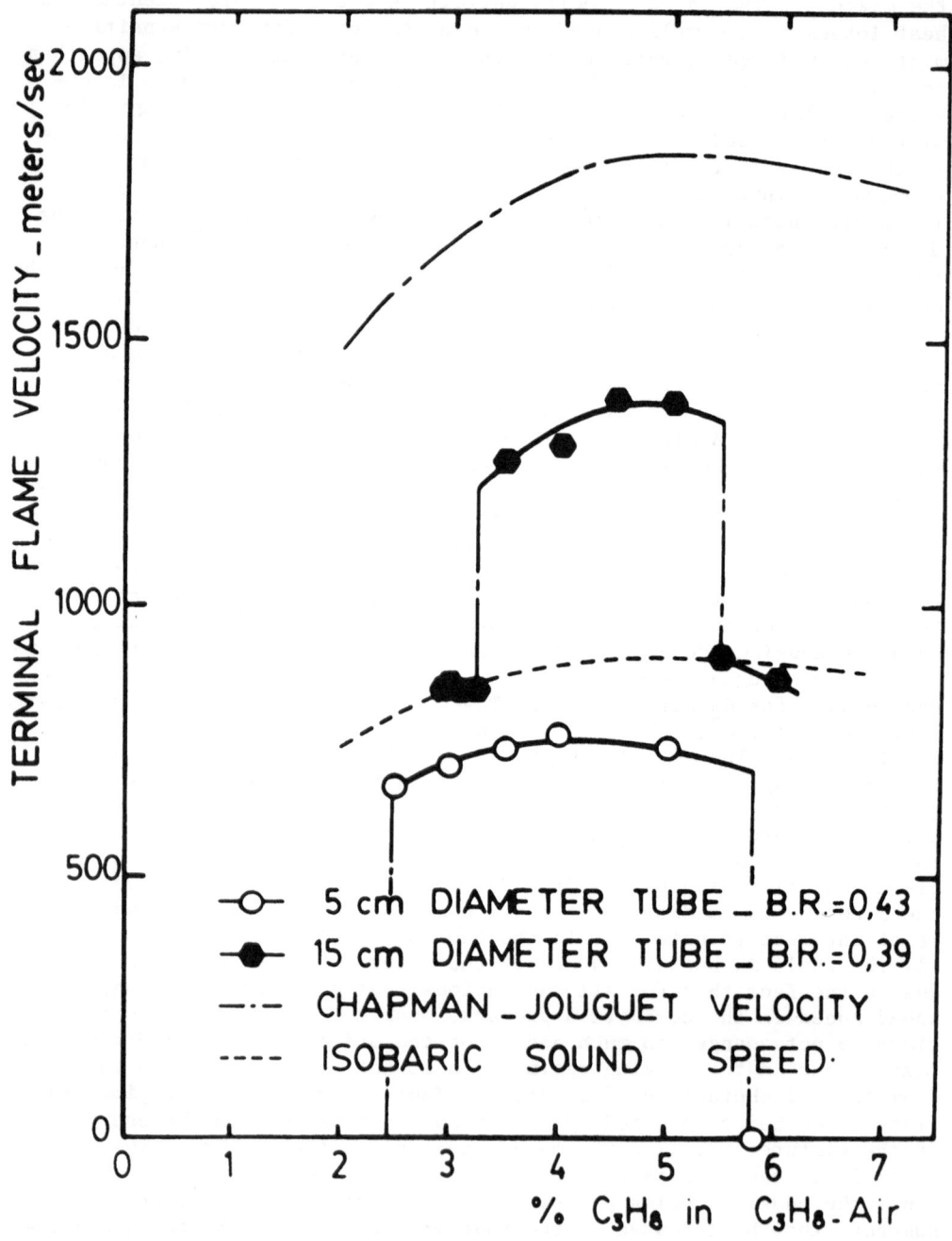

Fig. 8   Flame velocity versus fuel concentration for C_3H_8-air mixtures
          in the 5 cm and 15 cm tubes.

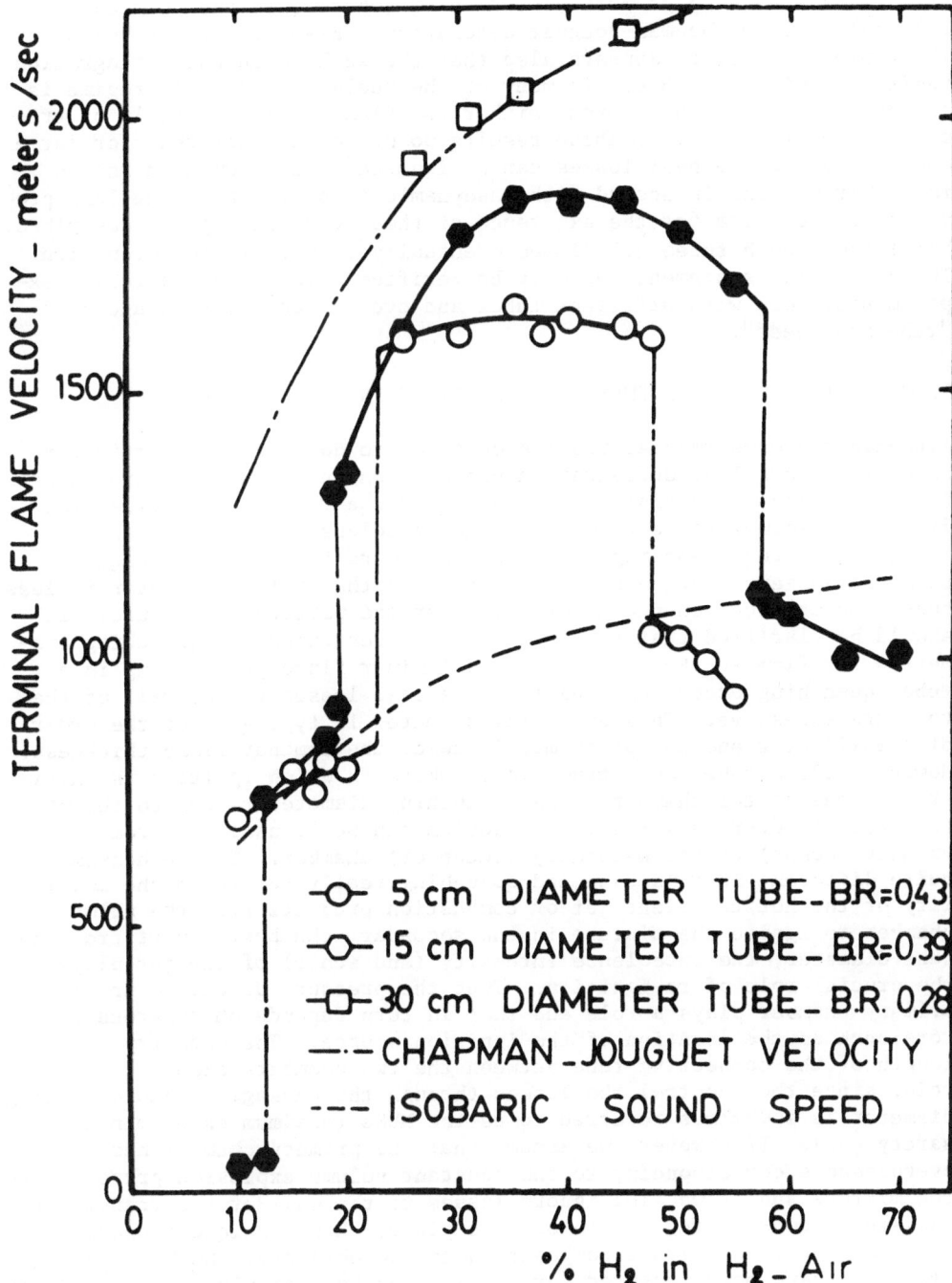

Fig. 9   Flame velocity versus fuel concentration for H₂-air mixtures
          in the 5 cm, 15 cm and 30 cm tubes.

steady state wave speed must depend on $\sqrt{Q}$ (where Q is the chemical energy release) as in a Chapman-Jouguet detonation. Examining the results for the various fuels, it appears also that the weak turbulent deflagration regime is quite unstable. In many of the fuels tested, this regime is not observed. In other words, either the flame quenches itself or transits to the choking regime. These results point to the fact that for large enough tubes where heat losses can be ignored, the detonation regime is the stable one in accord with gasdynamic theory. Thus heat loss provides the mechanism for the existence of these various regimes described. For large enough tubes all flames eventually accelerate to detonation! The preceeding statement can only be verified through more detailed experimental work with different tubes and over a much wider range of "tube roughness".

3. CRITERIA FOR THE REGIMES OF PROPAGATION IN ROUGH TUBES

Although the experimental results obtained to date on this particular problem of turbulent deflagration propagation in rough tubes are insufficient to provide definitive criteria for the transition between the different regimes to be formulated, nevertheless a pattern began to emerge. For the quenching regime which occurs for large blockage ratios, it appears that quenching occurs when the orifice diameter is less than some critical quenching diameter for the particular mixture. It should be clarified that there exists different kinds of quenching diameters for flames. For example for a laminar flame propagating in a tube, quenching occurs when heat and radical losses to the wall of the tube are excessive. This quenching diameter is typically of the order of a millimeter and is approximately twice the laminar flame thickness. However, for a tube connecting two chambers in which ignition is initiated in one of the chambers, then quenching diameter refers to the minimum tube diameter in which the explosion can be transmitted from the primary (donor) to the secondary (receptor) chamber. The mechanism is quite different in this case and quenching really refers to the inability of the hot turbulent jet of combustion products from the primary chamber to ignite the mixture in the secondary chamber. Apart from the tube diameter, the turbulence intensity (and scale) of the jet plays the crucial role of re-ignition. Thus the pressure developed in the primary chamber plays a role and that in turn depends on numerous factors such as the location of the ignition source. The geometry and length of the connecting tube between the two chambers also plays a role, since they control the losses through the passage. Thus quenching diameter is sometimes referred to as the MESG (maximum experimental safety gap). If however, we assume that the primary chamber has an overpressure corresponding to the constant volume explosion pressure of the mixture and if we also ignore losses to the wall of the transmitting passage by considering a thin orifice plate, then a unique quenching diameter for each mixture composition can be obtained. Such quenching diameters have been measured for a few gases by Thibault et al. [23] and in general for a given fuel, this quenching diameter is a minimum for stoichiometric composition and rises sharply near the lean and rich limits. For the present case of flame propagation with orifice plate

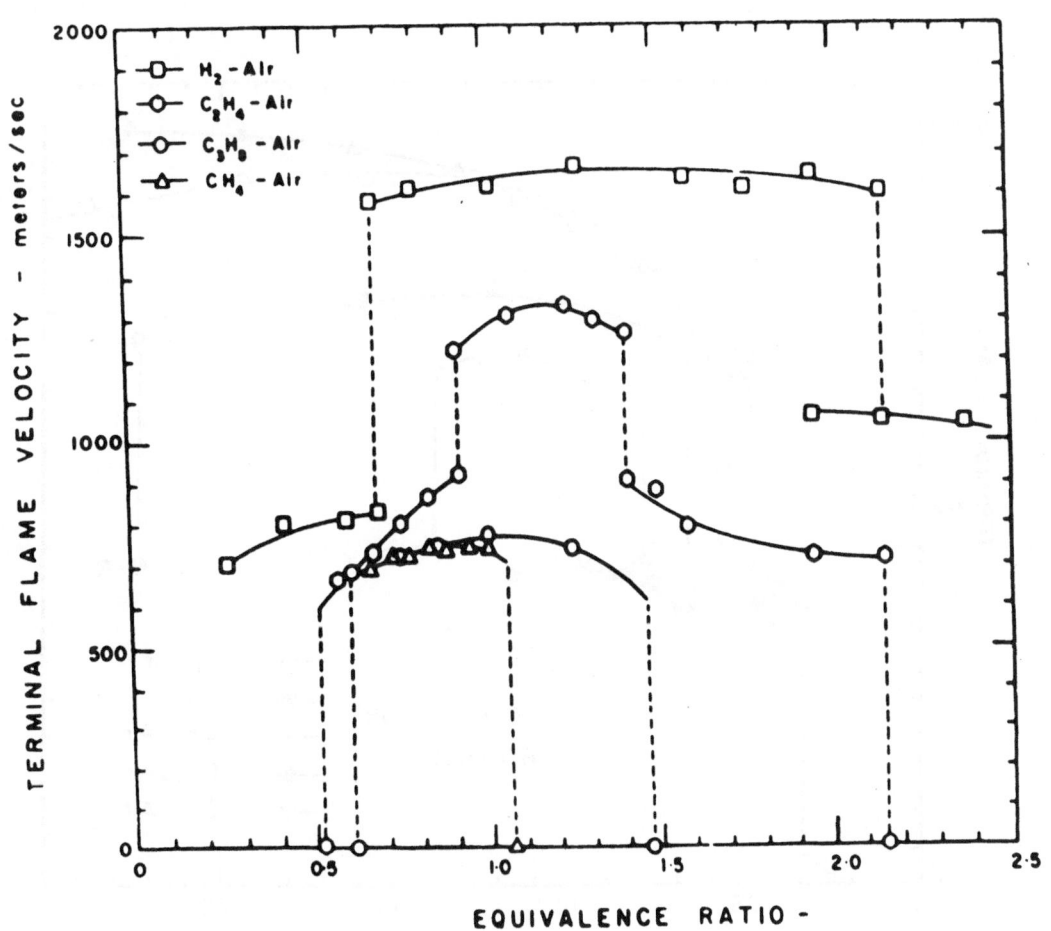

Fig. 10   Variation of the terminal flame velocity with mixture composition for the hydrogen and hydrocarbon fuel-air systems in the
5 cm tube.

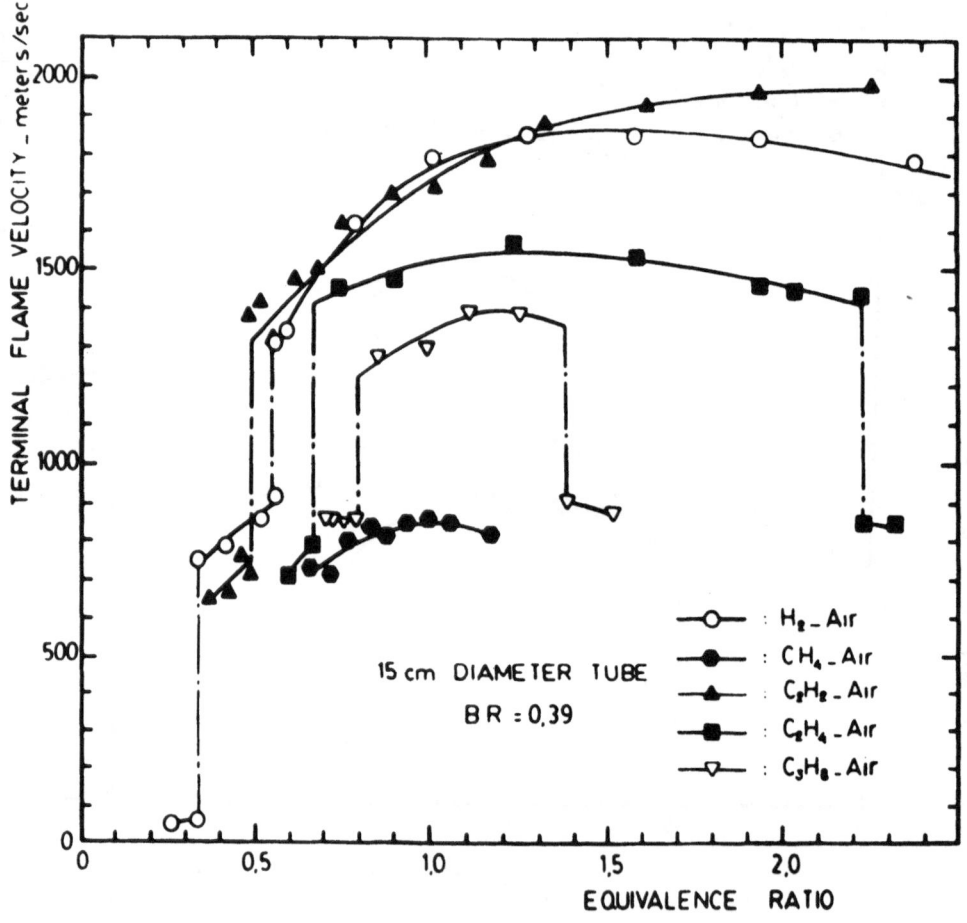

Fig. 11 Variation of the terminal flame velocity with mixture composition for the hydrogen and hydrocarbon fuel-air system in the 15 cm tube.

obstacles, we can consider the propagation process as the successive explosion of a continuous sequence of combustion chambers interconnected by the orifice plates. Ignition of the mixture in one chamber is achieved by the venting of the hot combustion products from the upstream to the downstream chamber through the orifice. Quenching results when the hot turbulent jet of product gases fails to cause ignition due to too rapid an entrainment and mixing rate between the cold unburned gases and the hot combustion products in the jet itself. Comparison between the orifice diameter for the quenching regime to the critical quenching diameter obtained by Thibault for the particular mixture composition shows good agreement. It establishes the fact that an accelerating turbulent flame can actually extinguish itself when the turbulent mixing rate in the flame zone is too high compared to the reaction rate. A criterion can be formulated on the basis of the turbulent mixing time "t_m" and the characteristic chemical reaction time "t_c" (ie., $t_c/t_m \leq C$, where C is some empirical constant depending on the definition of "t_c" and "t_m"). The existence of this quenching regime implies that turbulent mixing plays a dual role in combustion. On one hand it promotes heat and mass transfer resulting in a higher burning rate. On the other hand, rapid turbulent mixing puts a limit on the maximum transport rate permissible before extinction occurs. Thus turbulence is a self limiting mechanism in promoting the burning rate.

The weak turbulent deflagration regime is perhaps the most complicated one to describe quantitatively since it depends on the detailed structure of the turbulent flow generated in the unburned gases ahead of the subsonic flame zone. It has not been possible as yet to describe quantitatively the turbulent flow over complex geometry obstacles even for non-reacting incompressible fluids. Furthermore, the detailed mechanisms of interactions between turbulence and the flame front are not well understood. With the current state of the art computer codes, it has been possible to simulate the salient features of the turbulent flow structure ahead of the propagating flame and the behavior of the flame front as it advances into its self generated turbulent flow field. Figure 12 shows a typical numerical simulation of flame acceleration due to a sequence of orifice plate obstacles using the "vortex dynamics code" [24]. The flow is assumed two-dimensional and incompressible. No detailed mechanism of the interaction between the turbulence (vorticity) and the flame front has been considered and the flame is assumed to be just an interface advancing at a constant velocity normal to its surface. Such simulations do not yield any quantitative results nor reveal any "physics" that have not been already put into the code itself. However it does illustrate certain details of the flow structure which may not be evident. For example, the simulation shows the interesting fact that the vorticity generated upstream is convected downstream with the flow. Thus the turbulent flow structure at any one position depends not only on the local boundary conditions, but on the upstream conditions as well. Due to the difficulty in the experimental measurement of the transient turbulent flow structure, such simulation may serve as a useful tool to provide qualitative information regarding the flow field. There are other codes that attempt to provide quantitative results by putting in an appropriate turbulence model. However, there are a suffi-

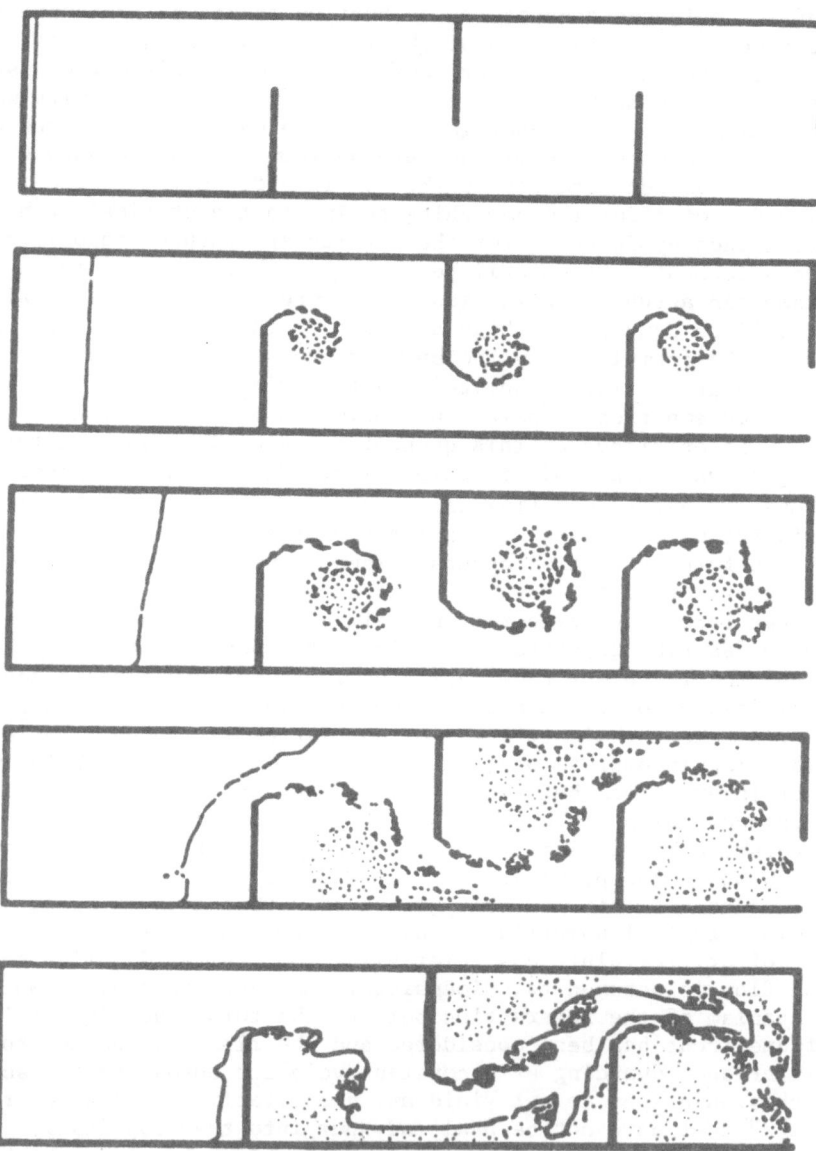

30 msec/frame

Fig. 12   Numerical simulation by the vortex dynamics code of flame ac-
           celeration in a channel. (Courtesy of P. Barr)

TRANSITION IN ROUGH-WALLED TUBE

	Mixture	Concentration (%)	λ (mm)	λ/d
	C_2H_2-Air	4.75	19.8	0.51
	H_2-Air	22	30.7	0.82
D = 5 cm		47.5	41.2	1.10
d ≃ 3.74 cm	C_2H_4-Air	6	37.8	1.01
(B.R. = 0.43)		9	30.1	0.81
	C_2H_2-Air	4	58.3	0.51
	H_2-Air	18	111	0.97
D = 15 cm		57	120	1.05
d = 11.4 cm	C_2H_4-Air	4.5	100	0.88
(B.R. = 0.39)		13.5	115	1.01
	C_3H_8-Air	3.25	112	0.98
		5.5	116	1.02

TABLE I

cient number of arbitrary constants and parameters that one can adjust
to yield the correct agreement with experiments. The general validity
of such numerical results is questionable. Thus for the weak turbulent
deflagration regime, very little can be said that is of general validity.
This regime constitutes the most difficult problem in the field of combustion.

Unlike the weak turbulent deflagration regime in which the burning
rate is governed by the details of the turbulent flow structure, the
sonic or the choking regime appears to be governed by the thermodynamics
of the mixture. Thus the flame speed in this regime becomes a fundamental property of the mixture, as in the case of Chapman-Jouguet detonation. The existence of the choking regime can be understood if we consider the flame propagation as a quasi-one dimensional compressible flow
in a pipe with friction (wall roughness and obstacles) and heat addition
(due to chemical reactions). It is well known in compressible flow
theory that both friction and heat addition will eventually drive the
flow towards the sonic condition (ie., M = 1) in which the flow speed
equals the sound speed. Thus if the mixture is sufficiently energetic
and the boundary conditions are such that quenching does not occur, the
turbulent flame will eventually accelerate until choking occurs. In the
weak turbulent deflagration regime, the final steady state flame speed
is due to a balance between the positive and negative effects of
turbulence. The positive effects being the enhancement of combustion
rate due to the more rapid turbulent heat and mass transport, while the
negative effects can be ascribed to quenching and flame stretch. However, as the mixture becomes more sensitive, the chemical time scale decreases, thus rendering the negative effects of turbulence due to quenching by mixing and flame stretch less effective. This allows the positive effect of turbulence to dominate and drive the flame speed eventually to choking or sonic condition. It is interesting to note that
the flame speed in the choking regime is supersonic with respect to the
unburned mixture whose sound speed is of the order of 330 m/s. Thus
unlike the weak deflagration regime in which the flame propagates into
its own displacement flow structure, the flame in the choking regime
propagates into an undisturbed medium (like a detonation wave). Since
the flame is supersonic, the shock waves generated do not decouple from
the reaction zone and propagate ahead of the flame, but move along with
it as a complex. Due to the finite spacing of the orifice plate obstacles, the flow fluctuates greatly in between the obstacles. We may
therefore speak of steady state conditions only in a global or averaged
sense. From Fig. 9 we see that the flame speed in the choking regime is
in general slightly below that of the isobaric sound speed (based on
constant pressure combustion). Hence, this provides a good approximation to the estimation of the flame speed in this choking regime which
should depend somewhat on the details of the flow processes generated by
the obstacles. From Figs. 10 and 11 we see that the flame speed in the
choking regime is only weakly dependent on the tube diameter and the
friction factor.

In the quasi-detonation regime, the speed of the wave front ranges
from about half the normal Chapman-Jouguet velocity of the mixture to
the full value depending on the obstacle configuration (hence frictional

losses). Since the quasi-detonation velocity can be only slightly higher
than the wave speed of the choking regime, it is difficult to make a
distinction between these two regimes on the basis of their speeds. However, for a given tube and obstacle configuration the transition from
the choking to the quasi-detonation regime is quite distinct. From the
experiments that have been carried out thus far [16], it appears that
transition occurs when the detonation cell size "λ" becomes less than
(or equal to) the orifice diameter "d" (ie., $\lambda \leq$ d). In other words
when λ/d > 1, the flame propagation is in the choking regime where the
flame speed corresponds to the sound speed of the combustion products.
However, when the mixture becomes more sensitive (ie., a reduction of
the cell size "λ"), then an abrupt transition to the quasi-detonation
regime occurs when λ/d $\leq$ 1. Table 1 gives the experimental values for
λ/d where transition takes place, except for the case of acetylene-air
mixtures. It appears that the detonation cell size measured for acetylene is about half the correct value. This discrepancy in the cell
size for C_2H_2-air mixtures has yet to be resolved. However, within experimental errors in cell size measurements for the other fuels, the
λ/d $\leq$ 1 criterion is confirmed. It should be noted that the value of
the quasi-detonation velocity may not differ much from that of the
choking regime in a given tube and orifice diameter. The transition
between the two regimes is identified by a distinct jump in the wave
speed indicating a change-over in the dominant propagation mechanism.
We identify this as a detonation regime on the basis of the transition
criteria λ/d $\leq$ 1. According to Schelkhin [25], the detonation limit in
tubes was postulated to be λ/d $\simeq$ 1. The logic behind this argument is
that the tube diameter must at least be big enough to accomodate one detonation cell. Schelkhin's criterion finds experimental support in
Vasiliev's [26] study where it was found that self-sustained detonation
can only propagate in a channel of width W $\geq$ λ. Thus the transition to
the quasi-detonation regime for λ/d $\leq$ 1 suggests that the wave is a detonation. However, the severe momentum losses reduce its velocity to a
value which can be significantly below that of the normal Chapman-
Jouguet value. The fact that the orifice diameter is of the order of
the cell size implies that the detonation mechanism dominates. A theory
has yet to be developed whereby the quasi-detonation velocity can be
determined for a given mixture when the momentum losses can be specified.

Finally, if the orifice diameter is sufficiently large, the influence of the momentum losses due to the rough walls on the quasi-detonation velocity should be negligible and the normal Chapman-Jouguet detonation regime is recovered. From the experimental results obtained, it
was found that the asymptotic approach to the normal Chapman-Jouguet
detonation regime occurs when the ratio d/λ $\geq$ 13. For example, in the
case of H_2-air mixtures shown in Fig. 9, the approach to the normal C-J
regime occurs at about 24% H_2 where the detonation cell size λ $\simeq$ 2 cm.
The orifice diameter is about 25.8 giving a d/λ $\simeq$ 12.9. This criterion
of d/λ $\geq$ 13 for the existence of the normal C-J detonation regime is
consistent with the fact that the critical tube diameter "d" in which
a confined planar detonation wave can continue to propagate as a spherical detonation without failure when the confinement is suddenly removed
is governed also by the criterion d/λ $\geq$ 13. The critical tube diameter

problem corresponds to the most severe form of a lateral perturbation to the forward motion of a detonation wave. Hence it seems reasonable that when $d/\lambda \gtrsim 13$, the detonation is not influenced by the weaker perturbations generated by the obstacles.

Summarizing the results, it appears that two of the five propagation regimes are detonation regimes since the detonation cell sizes provide the characteristic length scales for the scaling of the transition criteria for these regimes. The choking regime is essentially controlled mainly by the energetics of the mixture since the stagnation sound speed determines the flow velocity. The self-quenching regime is governed by the turbulent mixing and the chemical time scales. Perhaps the most difficult problem is to obtain understanding of the propagation mechanism of the weak turbulent deflagration regime since the details of turbulence flow structure and flame front interactions must be considered.

4. CRITERION FOR TRANSITION AND DETONATION LIMITS IN SMOOTH TUBES

For smooth tubes of relatively large diameters, there exists only two regimes of propagation, the weak turbulent deflagration and the normal Chapman-Jouguet detonation regimes. For very small diameter tubes or when the initial pressures of the mixtures are sufficiently low, the wall boundary layer effects may become dominant and the other regimes (e.g., choking, quasi-detonation) may exist also. The weak turbulent deflagration and the Chapman-Jouguet detonation regimes in smooth tubes are the same as that in rough tubes. It is, however, of interest to consider the criterion for transition from deflagration to detonation in smooth tubes. For transition to detonation to be possible, it must be established first that the explosive mixture is within the detonation limits. However, even if the conditions are such that the explosive mixture is within the detonability limits, transition may not necessarily occur The flame acceleration mechanisms must be sufficiently effective to bring the turbulent deflagration to a high enough velocity for the onset of detonation to occur. We shall consider in this section the current results regarding the detonability limits and the criteria for transition in smooth round tubes. We shall first discuss the problem of detonability limits. The detonability limits can be defined as the critical conditions that determine the possibility for the propagation of a self-sustained detonation wave. The critical conditions denote both the initial and boundary conditions of the explosive mixture. The initial conditions include the specification of the type of fuel, its concentration, initial thermodynamic state, the initial fluid mechanical state (ie., mean flow and turbulence characteristics), ignition source properties and all other relevant parameters that define the condition of the explosive prior to ignition. The boundary conditions represent size and geometry of the volume of explosive, the degree of the confinement, surface topology of the confining walls, as well as all other relevant parameters that are related to the boundary of the explosive. The current state of the art is that there exists no theory whereby the limits can be predicted from first principles. Experimental determination of the limits suffers from the lack of an operational definition of an experi-

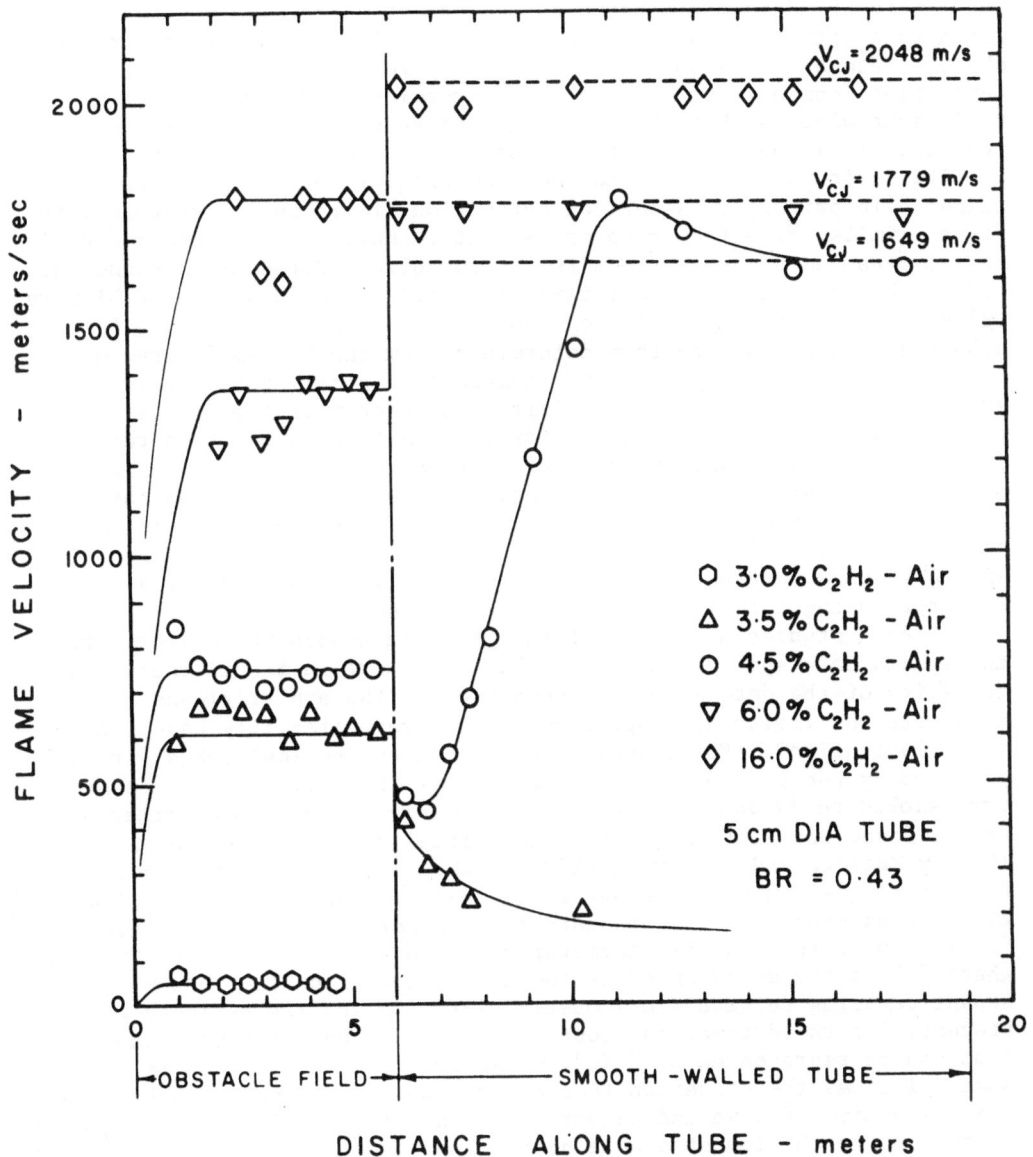

Fig. 13   The velocity history of a flame transmitted from the obstacle
field to the smooth-walled tube in the acetylene-air system.

mental criterion. Unlike the determination of flammability limits where
there exists a more or less generally accepted standard apparatus (ie.,
the flammability tube of Coward and Jones), no standard detonation tube
and procedures exist for the measurement of detonability limits. The
usual experiments were carried out in long tubes using powerful igniters.

The limits are determined based on failure to observe detonation in the given tube with the given igniter. In general, success or failure to observe detonation is quite distinct, to permit the critical fuel concentration (composition limit) to be determined. Although it is generally acknowledged that the limits are dependent on the apparatus (e.g., tube diameter) and ignition source strength, limits data are usually specified without reference to the conditions in which they were obtained. In principle, spherical detonation limits should represent the "true" limits, free from boundary effects. However, the actual experimental measurements of the limits for unconfined detonations cannot be rid of the influence of the initiation energy. Choosing some arbitrary value for the maximum initiation energy, the composition limits for spherical detonation are then determined from the U-shaped curve of initiation energy versus the fuel composition plot. Fortunately, the extremely rapid increase in the initiation energy as the limits are approached makes the actual values for the limiting composition rather insensitive to the maximum value of critical charge weight chosen. Nevertheless, a certain degree of arbitrariness is still associated with the actual value of the limits determined in this manner. It may be concluded that it is only meaningful to speak of the limits associated with a given apparatus (e.g., in a smooth cylindrical tube of a given diameter) using a certain reasonable criterion.

Recent studies attempt to formulate more meaningful criteria for the limits in a specific apparatus based on certain identifiable characteristics of the detonation as the degree of the arbitrariness associated with the limits are approached (e.g., wave stability based on velocity fluctuation). For example, for smooth cylindrical tubes, it was proposed by Lee [1] that $\lambda = \pi d$ (where λ is the cell size as measured from smoked foils and d is the tube diameter) be used as a criterion. It was found that this suggestion had actually been prposed as early as 1948 by Kogarko and Zeldovich [27]. Although they did not give the basis for arriving at this criterion, it can be reasoned as follows. Since the tube circumference "πd" represents the largest characteristic dimension for the tube, the longest characteristic acoustic time is thus "$\pi d/c$", where "c" is the sound speed of the detonation products. Assuming a resonant coupling between the acoustic vibration and the periodic chemical processes in the detonation front and that the characteristic chemical time can be represented by "λ/c", we arrive at $\lambda = \pi d$. This is in the same spirit as the criterion that $\lambda = d$ proposed by Schelkhin [25] and the suggestion of Dove and Wagner [28] that the single head spin should correspond to the lowest stable detonation mode (hence should represent limiting condition). It should be noted that if one formally adopts the criterion that the single head spin detonation is the limiting mode, then the acoustic theory of Manson [29] and Fay [30] gives a spin pitch to diameter ratio $P/d = \pi U/K_1 C$, where "P" is the spin pitch, "U" the C-J detonation velocity and $K_1 = 1.841$ is the first root of the derivative of the Bessel function of order one. In an earlier paper, Moen et al. [31] have assumed that the spin pitch "P" corresponds to the cell length (hence $P \simeq 1.6\ \lambda$) and this results in a limit criterion of $\lambda \simeq 1.7$ d (since $U/C \simeq 1.6$ for most detonating mixtures). It should be pointed out that there is no basis for the assumption that the single head spin pitch should correspond to the cell length. In fact, it is obvious that

the single head spin mode should occur in a mixture less sensitive than
that given by $\lambda = \pi d$. However, the assumption that the pitch equals the
cell length in the acoustic theory gives the contradictory result for
the limit criterion of $\lambda \simeq 1.7$ d which specifies a more sensitive mixture than that given by the $\lambda = \pi d$ criterion. Irrespective of the different criteria stated above, all indicate that the limit for round
tube occurs when the tube diameter is of the order of the cell size λ.
Recent experiments by Dupré [32] in tubes of difference diameters suggest that the $\lambda = \pi d$ criterion appears to be the most reasonable one.
For mixtures less sensitive than that specified by this criterion, ie.,
$\lambda > \pi d$, it is found that the detonation is highly unstable to finite
perturbations with velocity fluctuations in excess of 10% from the C-J
value. The study of detonation propagation in thin plastic wall tubes
by Murray [33] indicated that detonation failure occurs when the velocity deficit exceeds 10%. It should be noted that the $\lambda = \pi d$ criterion
applies to the condition where the detonation is initiated by a strong
source. For transition to occur, it is reasonable to expect that the
mixture should be more sensitive. The recent experiments of Knystautas
et al. [16] confirm that the transition crtierion in a smooth tube is
$\lambda \simeq d$ in accord with Schelkhin's criteria which requires a slightly more
sensitive mixture than that prescribed by the $\lambda = \pi d$ criteria.

   In Knystautas' experiment, a rough section of the tube is used to
bring the flame to a sufficiently high flame speed. Then upon transmission to the smooth section of the tube, the flame may either decay to
the weak turbulent deflagration regime or may transit to the normal
Chapman-Jouguet regime. By using the initial rough section of the tube,
it can be ensured that the maximum turbulent flame speeds have been attained for the transition to occur. Figure 13 shows the typical flame
speed versus distance along the tube for the 5 cm tube, with BR = 0.43
for the case of C_2H_2-air mixtures. When the propagation regime in the
obstacle section corresponds to the quasi-detonation regime, transition
to the Chapman-Jouguet detonation regime occurs immediately in the
smooth section. This is expected since the quasi-detonation regime is
in essence a detonation regime but with severe momentum losses. Thus
upon emergence into the smooth section where the losses are negligible,
immediate transition to the normal C-J detonation regime results. For
3.5% and 4.5% C_2H_2, the propagation regime in the rough section corresponds to the sonic regime. It can be observed in the case of 4.5% C_2H_2,
that the flame accelerates and transits to the normal C-J detonation regime. For 3.5% C_2H_2, the flame decelerates to the weak turbulent deflagration regime in the smooth section. Similar results have been obtained for other fuels in different tubes. The criterion for transition
to detonation has been found to correspond to the Schelkhin criterion
where $\lambda/d \lesssim 1$. Table II shows some typical values for the 5 cm tube for
C_2H_2, C_2H_4, C_3H_8, and H_2. Although much more experiments have to be
carried out before the criterion can be firmly established, the results
obtained to date seem reasonable.

   It should be pointed out that both detonability limits and the
transition criterion mentioned are for rigid circular tubes. For tubes
of other geometries, experiments have yet to be carried out to determine
both the limit and the transition criteria

TRANSITION IN SMOOTH-WALLED TUBE

4% C_2H_2 - Air   **D = 5cm** λ = 58.3 mm λ/D = 1.18

5% C_2H_4 - Air   **D = 5cm** λ = 65.1 mm λ/D = 1.32

10% C_2H_4 - Air   **D = 5cm** λ = 39.7 mm λ/D = 0.80

4% C_3H_8 - Air   **D = 5cm** λ = 52.2mm λ/D = 1.06

5% C_3H_8 - Air   **D = 5cm** λ = 59.0 mm λ/D = 1.19

20% H_2 - Air   **D = 5cm** λ = 55.4 mm λ/D = 1.12

51% H_2 - Air   **D = 5cm** λ = 52.5 mm λ/D = 1.06

TABLE II

5. GENERAL CONCLUDING REMARKS ON TURBULENCE AND COMBUSTION

We have seen that the propagation velocity of a combustion wave (hence the burning rate) can vary by three orders of magnitude as a result of turbulence and shock waves. In a confined tube where we can put obstacles to create any degree of turbulence intensity and scale, this range of propagation velocity can readily be achieved. In other geometries, the boundary conditions may not be as effective in producing turbulence and one observes only the weak turbulent flame and the Chapman–Jouguet detonation regimes. We have not thus far mentioned the precise nature as to how turbulence influences the combustion rate nor did we discuss the role of the shock waves when the flame is supersonic. It is of interest in concluding this lecture to give some physical picture of turbulent flames and detonations.

The propagation mechanism of a laminar flame is clear. It is in essence a diffusion wave . Thus its propagation velocity may be expressed as $S = \sqrt{\alpha/t_c}$, where α is the molecular diffusivity and t_c is the characteristic reaction time. Taking a typical value for $\alpha = 10^{-5}$ m^2/sec and $t_c \simeq 10^{-4}$ sec, the diffusion wave velocity is thus of the order of 0.3 m/s which corresponds to the value for atmospheric fuel-air flames. If one considers the influence of turbulence as an increase in the diffusivity, then we may consider the turbulent flame as a turbulent diffusion wave. Thus $S_T = \sqrt{\alpha_T/t_c}$, where the α_T denotes the turbulent diffusivity. The turbulent diffusivity α_T may be expressed as "u'ℓ", where "u'" is the turbulent fluctuation velocity and "ℓ" is the average eddy size. If we take typical values for u' = 1 m/s and $\ell = 10^{-2}$ m, the turbulent diffusivity $\alpha_T = 10^{-2}$ m^2/s. Thus the ratio of the turbulent diffusion wave velocity to the molecular counterpart is given by the square root of the ratio of the diffusivities, ie., $S_T/S = \sqrt{\alpha_T/\alpha}$. Therefore $S_T \simeq 30$ S for this case and is typical of turbulent burning velocities. In such a global consideration, the detailed physics of turbulent flames are missing.

Numerous models have been advanced. For example in the pioneering studies by Damköhler [34], it is assumed that the flame thickness "δ" is small as compared to the scale of turbulence "ℓ". Thus the turbulence simply wrinkles an otherwise laminar flame surface. The turbulent burning velocity could be calculated if the instantaneous surface area of the front and the laminar burning velocity are known. In this model turbulence serves to increase the burning surface area.

Damköhler also considered the case when the scale of the turbulence "ℓ" is less than the flame thickness "δ". In this case the role of turbulence is to increase the transport rate. Thus the turbulent burning velocity is as given previously, $S_T \simeq \sqrt{\alpha_T/t_c}$.

Significant progress has been seen since then in the understanding of the detailed structure of turbulent flows. In general there exists a spectrum of length scales. It is perhaps of interest to discuss a model where these scales are taken into account in the turbulent flame structure. We may represent the large scale turbulent motion by an averaged length scale (the integral scale or the average eddy size "ℓ"). If u' is the turbulent fluctuation velocity (ie., r.m.s. velocity), then the kinetic energy of the turbulent motion (per unit mass) is of the order of u'^2. The characteristic life time of the eddy is $t_\ell = \ell/u'$ and thus the energy dissipation rate $\varepsilon \simeq u'^2/(\ell/u') = u'^3/\ell$. Taking typical values for $u' \simeq 1$ m/s and $\ell = 10^{-2}$ m^2/s^3, if we compute the Reynolds number $Re_\ell = u'\ell/\nu$ and taking $\nu \simeq 10^{-5}$ m^2/s, we obtain $Re_\ell \simeq 10^3$. For such a high value of the Reynolds number, it is clear that viscous dissipation effects are negligible for the length scale of the average eddy size $\ell \simeq 10^{-2}$ m. For viscous dissipation to be important, the Reynolds number should be of the order of unity. Thus there arises the question as to how the turbulent kinetic energy of the eddy is dissipated. There must exist a fine structure within the eddy where the length scale "η" is sufficiently small to dissipate the energy via viscous shear. Consider the characteristic velocity of the small scale eddy to be "υ", then the Reynolds number $Re_\eta = \upsilon\eta/\gamma = 1$, we may replace υ by ν/η and thus the small scale eddy size $\eta = \varepsilon^{1/4}\nu^{3/4}$. The small scale eddy is generally known as the Kolmogorov scale and can be evaluated when the energy dissipation rate is known. Taking the typical values $\nu = 10^{-5}$ m^2/s and $\varepsilon = 10^2$ m^2/s^3, we can evaluate $\eta \simeq .1$ mm. Thus within the large eddy of the order of 1 cm, there exists the small eddies of the order of 0.1 mm. A physical model of the distribution of Kolmogorov eddies within the large eddy is given by Tennekes [35] who considered the Kolmogorov eddies to be vortex tubes of diameter η and spaced with an averaged distance "λ" within the large eddy of size "ℓ". The spacing of the Kolmogorov tubes "λ" is referred to as the Taylor microscale. The way energy is "pumped" into the small scale Kolmogorov vortex tubes where viscous dissipation takes place is via the mechanism of "vortex stretching". If we stretch a vortex tube, the work done in stretching goes to the spinning motion of the fluid inside the tube. Eventually, the kinetic energy of rotation dissipates into internal energy via viscous shear. Thus we see that the kinetic energy of the large scale motion goes into the small scale vortex tubes via stretching and deformation. Subsequently, viscous dissipation occurs in the small scale vortex tubes of scale "η". The Taylor microscale can be related to the large scale motion as follows: The characteristic diffusion time of vorticity over the distance λ is $t_\lambda \simeq \lambda^2/\gamma$ and this is proportional to the characteristic time of the large eddy $t_\ell \simeq \ell/u'$. Thus $\lambda^2/\nu \sim \ell/u'$ and $\lambda/\ell \sim Re_\ell^{-1/2}$. For isotropic turbulence the experiments give the relationship $(\lambda/\ell)^2 = 48 \, Re_\ell^{-1}$. Thus for the typical values considered, $\lambda \simeq 2$ mm. From the above picture we see that the model for the structure of turbulence consists of large eddies of size ℓ with a distribution of vortex tubes within it of size η and the average spacing between the vortex tubes is λ. This simple model is consistent with experiments although other physical pictures can also be constructed to account for the same observed effects.

We now ask the question "what happens when a flame propagates into a turbulent structure as described above?" The characteristic length scale of a flame is the thickness of the reaction zone "δ" which depending on the definition (reaction zone or preheat zone) ranges from 0.1 to 1 mm. Taking a typical value of say 0.5 mm, we see that $\delta < \eta$, $\lambda > \delta$. We may expect the large scale motion to wrinkle the flame front with characteristic dimension "ℓ" according to Damköhler's model. However, the small scale Kolmogorov vortex tubes play an important role. If the vortex tubes are severed by a flame front, then these vortex tubes act like "vacuum cleaner hoses" and pull the combustion gases and free radicals rapidly along their lengths. This is due to the low pressure generated inside the tubes from the centrifugal forces of the rotational motion. Experimental evidence of this effect can be found in the work of McCormack [36] who demonstrated that the propagation rate of a flame along a vortex ring can be very high. With a distribution of hot product gases and free radicals along these Kolmogorov tubes, combustion will occur at these sites and then spread over the distance "λ" between those burning tubular regions. The burn-out time of the regions between the tubes will then be of the order of λ/S, where S is the laminar burning velocity (or equivalently the burn-out time can be related to the diffusivity α, ie., λ^2/α since the propagation velocity of a laminar flame is essentially that of a diffusion wave). The above physical picture of a turbulent flame is largely due to Chomiak [37] who make some brilliant speculations regarding the detailed structure of turbulent flames. Whether such a model is the correct model is difficult to verify experimentally. As pointed out in this lecture, there is no universal model for turbulence and every special case has its own characteristics. There may be certain common features but no one model can be expected to provide the complete description of turbulence in general. The ideas of Chomiak contain many of the essential ingredients necessary for the interpretation of a wide class of high Reynolds number turbulent flames.

We now turn our attention to the role of the shock waves generated in the supersonic deflagration, quasi-detonation and the Chapman-Jouguet detonation regimes. In a subsonic deflagration, the expansion of the combustion gases generate a flow of the unburned gases ahead of the flame. In a supersonic deflagration, the expansion generates shock waves which propagate with the combustion zone. The adiabatic compression due to the shock waves increases the temperature of the unburned gases prior to combustion. The increase in the molecular diffusivities as a result of this temperature rise will no doubt play an important role in enhancing the local transport rates in the turbulent flame zone.

If the shock waves are sufficiently strong, then auto-ignition is possible. When this happens, an alternate mechanism for the propagation of the combustion wave is provided. According to the classical theory of detonation wave, this mechanism of auto-ignition by the shock wave is in fact considered to be the principle mechanism of propagation. However, the experimental studies carried out on the detailed structure of the detonation front over the past 25 years have shown the contrary. There is no concrete evidence that shock induced auto-ignition is the principle mechanism for the propagation of a real cellular detonation wave. In fact, the reactive blast calculations by Bach, etc. [38], Lundstrom and Oppenheim [39], and Thomas and Edwards [40] have shown that the amount of mixture that can be ignited by the blast wave in a detonation cell is an insignificant fraction of the total mixture defined by the cell boundary. In other words, the majority of the mixture is burned in the transverse waves or in the shear layers. This strongly indicates that it is the turbulent shear layer associated with the transverse wave of the triple shock configuration that provides the dominant mechanism of combustion in a real detonation front. Further support for the important role of turbulent shear in detonative combustion can be found in the observation of the onset of detonation. From the stroboscopic laser schlieren photographs of Urtiew and Oppenheim [4], it is evident that onset of detonation always occurs at the most intense shear region in the turbulent flame "brush" in the boundary layer near the wall of the tube. This indicates that auto-ignition is brough about by turbulent mixing between the hot product gases and the unburned mixture. In fact, it has been conclusively demonstrated by Meyer et al. [42] that the gasdynamic processes of compression ahead of the accelerating front are entirely insufficient to bring about the transition to detonation. They postulated that "the occurence of this event (ie., transition) must be due therefore to other phenomena of which the most influential should be those associated with heat or mass transfer from the flame". The conclusive demonstration that auto-ignition and subsequent onset of detonation can be brought about by rapid turbulent mixing alone without any shock pre-compression was later achieved by Knystautas et al. [43]. Thus it seems reasonable to assume that turbulence and shear play the dominant role not only in deflagration, but in detonations as well. For deflagration, intense turbulence must come from shear layers generated at the boundary layers or wakes of obstacles. In the detonation mode, the intense shear layers are generated at the triple shock configurations. Thus it may be said that triple shock configurations via shock collisions is nature's own way to generate shear and turbulence in the absence of boundaries. It appears that the mechanisms of propagation in both turbulent deflagration and detonation are in fact the same. Adopting this unified view enables us to understand the role of boundaries in controlling the combustion rate.

ACKNOWLEGEMENT

I would like to thank Dr. O. Peraldi for his assistance in preparing this manuscript.

REFERENCES

1. Lee, J.H.S. (1984) Annual Review of Fluid Mechanics 16, 311-36.

2. Lee, J.H.S. and Guirao, C.M. (1981) Gasdynamic Effects of Fast
    Exothermic Reactions, in Fast Reactions in Energetic Systems
    Eds. C. Capellos and R.F. Walker, D. Reidel Publ. Co.

3. Lee, J.H.S. and Moen, I.O. (1980 Progr. Energy Comb. Sci. 6,
    359-389.

4. Hjertager, B.H. (1981) 'Numerical Simulation of Turbulent Flame
    and Pressure Development in Gas Explosions'. Proc. Int. Mtg. on
    Fuel-Air Explosions, Eds. Lee, J.H. and Guirao, C.M., University of
    Waterloo Press.

5. Marx, K.D., Lee, J.H.S. and Cummings, J.C. (1985) 'Modelling of
    Flame Acceleration in Tubes with Obstacles'. Proc. 11th IMACS World
    Congress, Oslo, Norway, August 5-9.

6. Lee, J.H.S., Knystautas, R., Chan, C. Barr, P.K. Grcar, J.F. and
    Ashurst, W.T. (1983) 'Turbulent Flame Acceleration Mechanisms and
    Computer Modelling'. Proc. International Meeting on Light-Water
    Reactor Severe Accident Evaluation, Cambridge, Mass. August 28 -
    September 1. Also available as SANDIA Rept. 83-8655. Livermore,
    California.

7. Mallard, E. and Le Chatelier, H. (1881) C.R. Acad.Sci. Paris 9
    145-148.

8. Berthelot, M. and Vieille, P. (1882) C.R. Acad.Sci. Paris 94
    pp. 101-108, 16 Jan., pp. 882-823, 27 Mars, Vol. 95, pp. 151-157,
    24 Juillet.

9. Chapman, W.R. and Wheeler, R.V. (1926), (1927) J. Chem. Soc.
    (London), Vol. 37, p. 2139 and Vol. 38.

10. Schelkhin, K.I. (1940) J.E.P.T. (USSR), 10, 823-827.

11. Guenoche, H. and Manson, N. (1949) 'Influence des Conditions aux
    Limits Transversales sur la Propagation des Ondes de Shock et de
    Combustion. Revue de l'Institut Français du Petrole, No. 2, pp.
    53-69.

12. Brochet, C. (1966) 'Contribution à l'Etude des Détonations
    Instables dans les Mélanges Gaseux'. Thèses Presentées à la
    Fac. des Scie., Poitiers.

13. Wagner, H.G. (1982) 'Some Experiments about Flame Acceleration'
    First International Specialist Meeting on Fuel-Air Explosion,
    Montreal, 1981. University of Waterloo Press, SM Study No. 16,
    pp. 77-99.

14. Lee, J.H.S., Knystautas, R. and Freiman, A. (1984) <u>Combustion and Flame</u> <u>56</u>, 227–239.

15. Lee, J.H.S., Knystautas, R. and Chan, C. (1984) Proc. <u>20th Symp. (International) on Combustion.</u> The Combustion Institute, Pittsburgh, Pa.

16. Knystautas, R., Lee, J.H., Peraldi, P. and Chan, C. (1985) <u>10th ICDERS</u>, Berkeley, Calif., August 4–9.

17. Hjertager, B.H., Fuhre, K., Parker, S.J. and Bakke, J.R. (1983) 'Flame Acceleration of Propane-Air in a Large-Scale Obstructed Tube' <u>Presented at the 9th International Colloquium on Dynamics of Explosions and Reactive Systems</u>, Poitiers.

18. Urtiew, P.A. (1982) 'Recent Flame Propagation Experiments at LLNL within the Liquified Gaseous Fuels Spill Safety Program'. <u>First International Specialist Meeting on Fuel-Air Explosions</u>, Montreal 1981. University of Waterloo Press SM Study No. 16, 924–947.

19. Deshaies, B. and Leyer, J.C. (1981) 'Flow Field Induced by Unconfined Spherical Accelerating Flames', <u>Comb. and Flame</u> <u>40</u>, 141–153.

20. Zeeuwen, J.P. and Van Wingerden, C.J.M. (1983) 'On the Scaling of Vapor Cloud Explosion Experiments'. <u>Presented at the 9th International Colloquium on Dynamics of Explosions and Reactive Systems</u>, Poitier.

21. Yip, T.W.G., Strehlow, R.A. and Ormsbee, A.I. (1983) 'Theoretical and Experimental Studies in Acoustic Waves Generated by a Cylindrical Flame and 2-D Flame-Vortex Interactions'. <u>Presented at the 1983 Fall Technical Meeting, Eastern Section of The Combustion Institute</u>, November, Providence, R.I.

22. Sherman, M.P., Tieszen, S., Benedick, W., Fisk, J., Carcassi, M. (1985) 'The Effect of Transverse Venting on Flame Acceleration and Transition to Detonation in a Large Channel'. <u>10th ICDERS</u>, Berkeley, California, August 4–9.

23. Thibault, P., Liu, Y.K., Chan, C., Lee, J.H., Knystautas, R., Guirao, C., Hjertager, B. and Fuhre, K. (1982) 'Transmission of and Explosion Through an Orifice'. <u>Proc. 19th Symp. (International) on Combustion.</u> The Combustion Institute, Pittsburgh, Pa. pp. 599–606.

24. Ashurst, W. and Barr, P. (1982) 'Discrete Vortex Simulation of Flame Acceleration due to Obstacle Generated Flow'. <u>1982 Fall Meeting of Western States Section of The Combustion Institute</u>. Paper WSS/CI 82-84. Sandia National Lab. Rept. SAND 82-8724.

25. Schelkhin, K. (1966) 'Instability of Combustion and Detonation of Gases'. <u>Soviet Physics USPEKHI</u>, <u>8</u>, 5, 780.

26. Vasiliev, A.A. (1982) 'Geometric Limits of Gas Detonation Propagation', Fiz. Goreniya Vzryre 18, 132-136.

27. Kogarko, S.M. and Zeldovich, Y.B. (1948) Dokl. Akad. Nauk SSSR, 63, 553.

28. Dove, J.E. and Wagner, H.G. (1960) Proc. 8th Symp. (International) on Combustion, 589-600. The Combustion Institute, Pittsburgh, Pa.

29. Manson, N. (1946) Comp. Rendu 222, p. 46.

30. Fay, J.A. (1952) J. Chem. Phys. 20, p. 942.

31. Moen, I.O., Donato, M., Knystautas, R. and Lee, J.H.S. (1981) 18th Symp. (International) on Combustion, pp. 1615-1623, The Combustion Institute, Pittsburgh, Pa.

32. Dupré, G., Knystautas, R. and Lee, J.H. (1985) 10th ICDERS, Berkeley, California, August 4-9.

33. Murray, S. (1984) 'The Influence of Initial and Boundary Conditions on Gaseous Detonation Waves'. Ph.D. Thesis, McGill University.

34. Damköhler, G. (1940) Z. Elektrochemie Angewandte Phys. Chem. 46 601. (English Translation NACA TM 1112 (1947).

35. Tennekes, H. (1968) Phys. Fluids 11, 669.

36. McCormach, P., Scheller, K., Mueller, G., Tisher, R. (1972) Comb. and Flame 19, 297.

37. Chomiak, J. (1979) Progr. Energy Comb. Sci. 5, 207-221.

38. Bach, G., Knystautas, R., Lee, J.H.S. (1968) 12th Symp. (International) on Combustion, pp. 883-67. The Combustion Institute, Pittsburgh, Pa.

39. Lundstrom, E. and Oppenheim, A.K. (1969) Proc. Roy. Soc. London, Ser. A, 310, pp. 463-78.

40. Thomas, G.O. and Edwards, D.H. (1983) 'Simulation of Detonation Cell Kinematics Using Two-Dimensional Reactive Blast Waves'. J. Phys. D: Appl. Phys., 16, 1881-1892.

41. Urtiew, P. and Oppenheim, A.K. (1966) Proc. Roy. Soc., A295, 13-28.

42. Meyer, J.W., Urtiew, P.A. and Oppenheim, A.K. (1970) Comb. and Flame 14, No. 1, pp. 13-20.

43. Knystautas, R., Lee, J.H., Moen, I.O. and Wagner, H.Gg. (1979) Proc. 17th Symp. (International) on Combustion, pp. 1235-1245. The Combustion Institute, Pittsburgh, Pa.

KrF LASER INDUCED DECOMPOSITION OF TETRANITROMETHANE

C. Capellos, S. Iyer, Y. Liang*, and L.A. Gamss**
ARDC, Dover, New Jersey 07801-5001

ABSTRACT

Electronic excitation of tetranitromethane (TNM) in aerated or deaerated polar solvents generates the NO_2 radical and the nitroform anion, $(NO_2)_3C^-$, which has a maximum absorption at 350 nm. In nonpolar solvents electronic excitation of TNM leads to the formation of the NO_2 radical and the trinitromethyl radical, $(NO_2)_3C\cdot$, with an absorption maximum at 307 nm. Prolonged irradiation of TNM in nonpolar solvents results in the formation of nitroform, $(NO_2)_3C-H$, which is probably generated by the $(NO_2)_3C\cdot$ via hydrogen abstraction from the solvent. The nonpolar solvent radicals generated by the hydrogen abstraction reaction combine with NO_2 radicals in the solution to form nitroalkane molecules.

INTRODUCTION

Tetranitromethane has been used extensively for determining the yields of active reducing intermediates such as H atoms, HO_2 radicals, solvated electrons, and organic free radicals which are formed during γ-radiolysis or pulse radiolysis of various polar solvents (1-5). The reaction in any of these cases may be represented as a dissociative electron transfer from the reducing species to TNM.

$$e^- + (NO_2)_4C \rightarrow (NO_2)_3C^- + NO_2$$

The extent of this reaction was monitored in each case by determining the concentration of the stable nitroform anion $(NO_2)_3C^-$ from its maximum absorbance at 350 nm where the anion is known to have an extinction coefficient of $1.5 \times 10^4 M^{-1} cm^{-1}$ (2).

---

*Geo-Centers, Inc., Wharton, New Jersey 07885.
**National Research Council Research Associate at ARDC. Permanent
  address: Chemistry Dept., Bar-Ilan University, Ramat-Gan, Israel.

P. M. Rentzepis and C. Capellos (eds.), Advances in Chemical Reaction Dynamics, 379–394.

Previous investigators (6,7) have also detected the formation of nitroform anion during the electronic excitation of TNM in polar solvents. In nonpolar solvents, however, although the formation of the trinitromethyl radical has been postulated (7) it has not as yet been detected experimentally.

This paper presents nanosecond spectroscopic and kinetic data for the formation of $(NO_2)_3C^-$ as well as the previously undetected $(NO_2)_3C \cdot$ from electronically excited TNM in polar and nonpolar solvents, respectively.

EXPERIMENTAL

All solvents used in this work were spectrograde quality. TNM was purified as reported earlier (4). Deaeration, i.e., removal of oxygen, of the solutions was performed by flushing with nitrogen for 15 minutes in a flask attached to the optical cell. Spectroscopic and kinetic studies of polar and nonpolar solutions of TNM in the nanosecond range were performed with the excimer laser (Questek Model 2840) nanosecond kinetic spectrophotometer shown in Fig. 1.

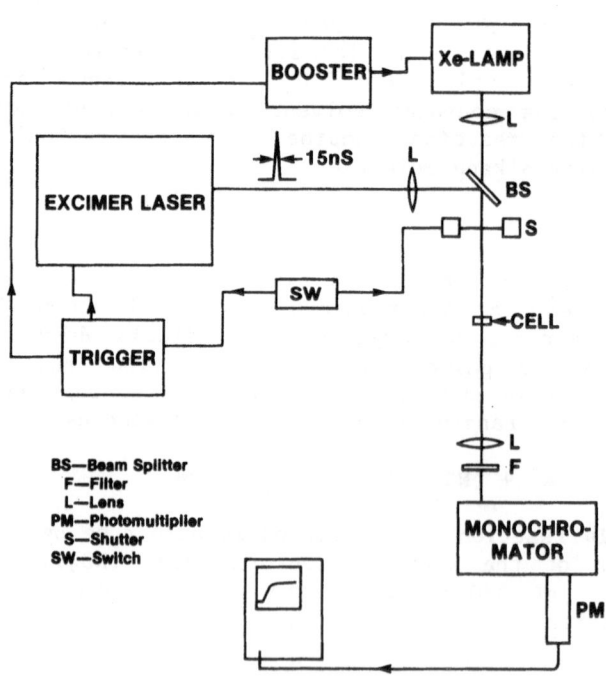

Figure 1. Schematic of nanosecond kinetic spectrophotometer.

The optical detection system consisting of the booster Xenon lamp, monochromator, photomultiplier tube and oscilloscope were described in an earlier publication (8). The kinetic data presented here were analyzed as described previously (9,10).

RESULTS AND DISCUSSION

Energetic Transient Species Formed From Electronically Excited TNM in Polar Solvents

Electronic excitation of TNM in aerated or deaerated polar solvents such as methanol results in the formation of a species with an absorption maximum at 350 nm as shown in Fig. 2.

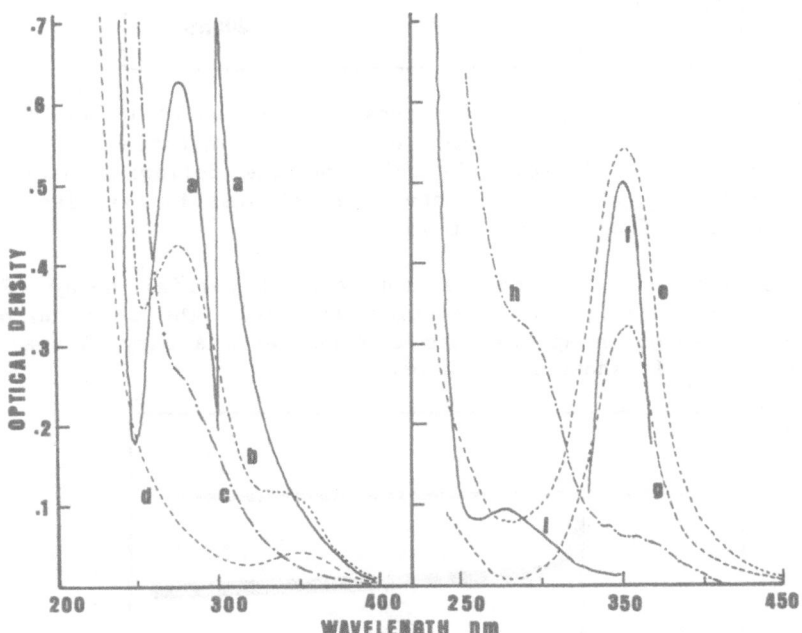

Figure 2. Absorption spectrum of TNM in a) gas phase; b) acetonitrile; c) hexane; d) methanol. Absorption spectrum of nitroform anion(NF^-); e) K^+NF^- in H_2O; f) TNM in EPA (77°K) after photolysis; g) TNM in methanol after after photolysis; h) TNM in n-hexane after photolysis; i) NF from NF^- on acidification.

As shown in the oscilloscope trace of Fig. 3, most of the transient absorption at 350 nm is formed within the first 40 nanoseconds followed by a slower build up which lasts for about 1.5 microseconds.

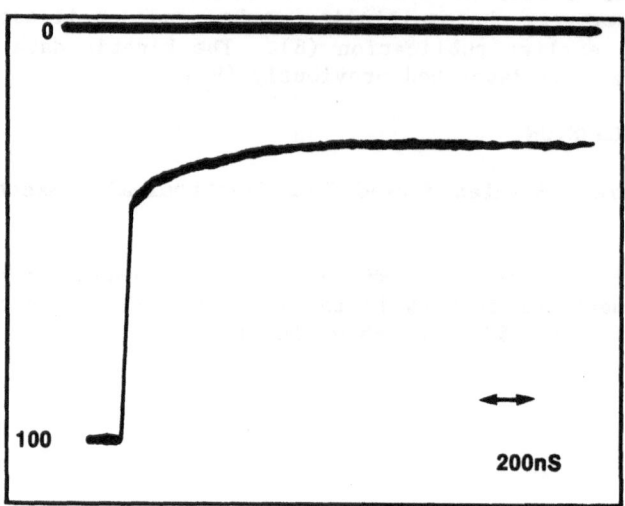

Figure 3. Oscilloscope trace showing the build-up of transient absorption at 350 nm, following electronic excitation of 2×10^{-3}M TNM in methanol. (Note: In this and the following oscilloscope traces of this paper the ordinate shows changes in transient optical absorption as a function of time.)

Subsequently, the species undergoes a partial decay (Fig. 4) resulting in a permanent absorption species. The spectrum of this permanent absorption species was recorded with a Cary 14 spectrophotometer and it is shown in Fig. 2g.

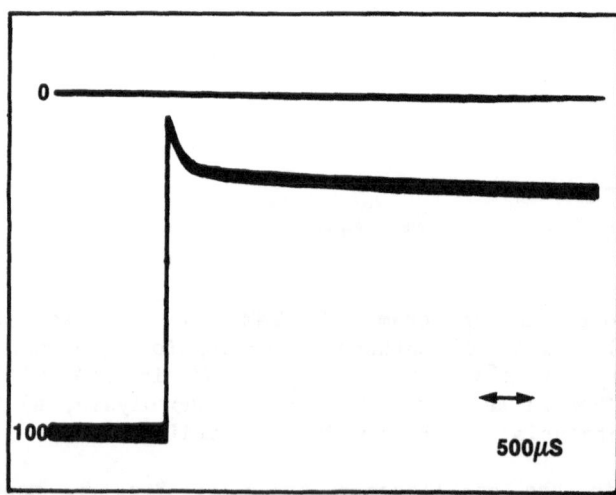

Figure 4. Oscilloscope trace showing the partial decay of the nitroform anion at 350 nm in deaerated methanol solution of TNM.

Assignment of the Transient Species and Mechanism

Oscilloscope traces like the one shown in Fig. 3 were obtained at different wavelengths in the spectral range 290 - 380 nm. The instantaneous absorptions (within 40 nanoseconds) as well as the build up absorptions (1.5 microseconds) from the beginning of the laser pulse were plotted as a function of wavelength as shown in Fig. 5.

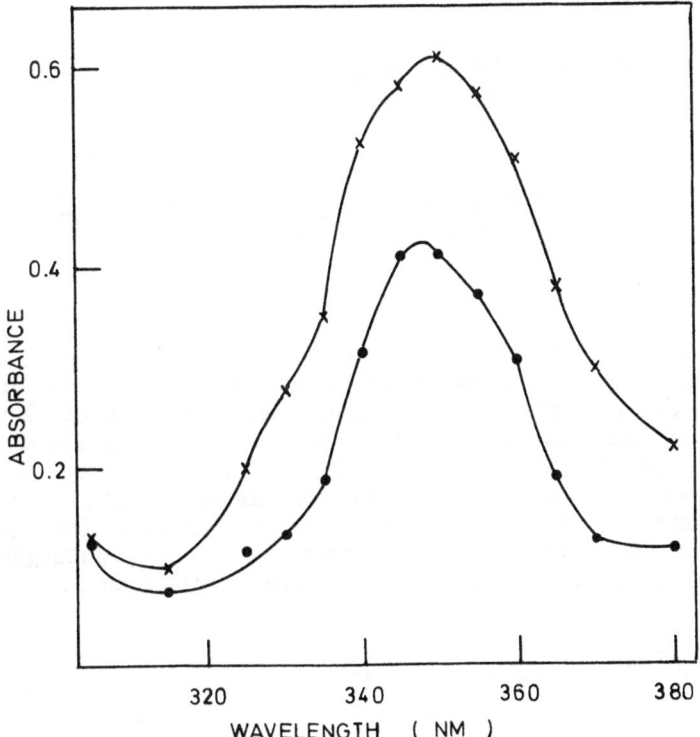

Figure 5. Transient absorption spectra of TNM in methanol solution on electronic excitation. ——•——•——•—— 40 ns after laser; ——x——x——x—— 1.5 μs after laser.

These transient spectra seem to be very similar to the permanent absorption spectrum of Fig. 2g and those spectra of photolyzed TNM solutions, in EPA at 77°K (Fig. 2f), in methanol at ambient temperature (Fig. 2g), and almost identical to the absorption spectrum of $(NO_2)_3C^-$ which results from the dissociation of $KC(NO_2)_3$ in water (Fig. 2e).

The experimental data presented above suggest that the transient absorption spectra of Fig. 5 are, probably due to the $(NO_2)_3C^-$ species which is formed through dissociative electron attachment as shown in the following reaction scheme:

$$\text{TNM} \xrightarrow{\text{248 nm}} \text{TNM*}(S_1) \tag{1}$$

$$\text{TNM*}(S_1) + \text{MeOH} \xrightarrow[\substack{\text{Electron} \\ \text{Transfer}}]{\text{Dissociative}} (NO_2)_3 C^- + NO_2 + \text{MeOH}^+ \tag{2}$$

$$\text{TNM*}(S_1) \xrightarrow[\text{Crossing}]{\text{Intersystem}} \text{TNM*}(T_1) \tag{3}$$

$$\text{TNM*}(T_1) + \text{MeOH} \xrightarrow[\substack{\text{Electron} \\ \text{Transfer}}]{\text{Dissociative}} (NO_2)_3 \bar{C} + NO_2 + \text{MeOH}^+ \tag{4}$$

The almost instantaneous build up (40 nsec) shown in the oscilloscope trace of Fig. 3 is probably due to the reaction of $\text{TNM*}(S_1)$ (singlet excited state) with methanol (MeOH) to form $(NO_2)_3C^-$ through dissociative electron attachment (reaction 2). The slower build up (1.5 μsec) of $(NO_2)_3C^-$ shown in Fig. 3 can probably be explained through reaction 4 which describes the dissociative electron attachment of $\text{TNM*}(T_1)$ (triplet excited state) to form $(NO_2)_3C^-$. This anion, which is formed through reactions 2 and 4, undergoes partial neutralization to form nitroform as shown in reaction 5.

$$(NO_2)_3C^- + H^+ \xrightleftharpoons{\qquad} (NO_2)_3C\text{-}H \tag{5}$$

This particular behavior of the nitroform anion has been observed in the present work (Fig. 4) and in pulse radiolysis studies of TNM in polar solvents (3). In these earlier studies the nitroform anion was monitored with time resolved absorption and conductivity measurements. Further confirmation for reaction 5 was obtained in dearated solution of TNM in MeOH containing 0.05 M H_2SO_4, where as it is shown in Fig. 6 the photochemically formed anion $(NO_2)_3C^-$ is neutralized with a rate constant $k_5 = 5.36 \times 10^6 M^{-1} sec^{-1}$, to form nitroform.

Energetic Transient Species Formed From Electronically Excited TNM in Nonpolar Solvents

Electronic excitation of TNM in deaerated nonpolar solvents such as n-hexane or cyclohexane results in the formation of a species with an absorption maximum at 307 nm as shown in Fig. 7.

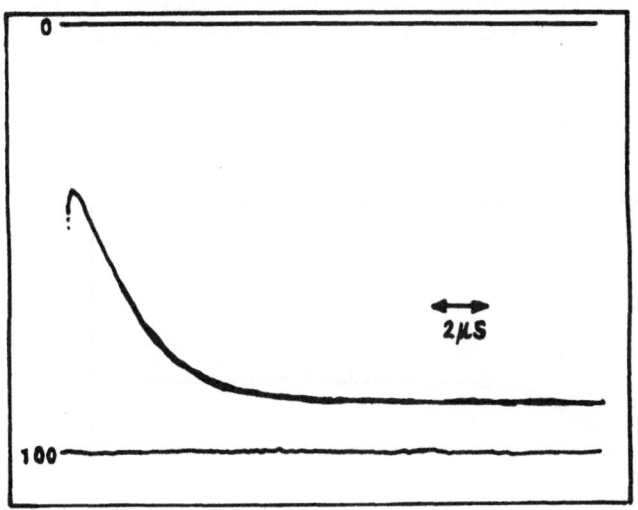

Figure 6. Decay at 350 nm of the nitroform anion in the presence of H_2SO_4. ([TNM] = 1 x 10^{-3}M, [H_2SO_4] = 5 x 10^{-2}M).

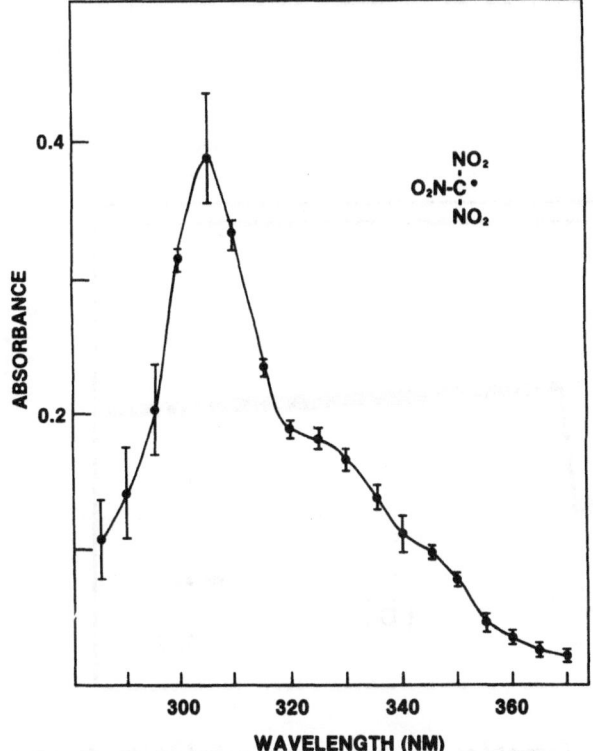

Figure 7. Transient absorption spectrum of TNM in cyclohexane solution after electronic excitation (~ 5μs after laser).

The oscilloscope traces of Fig. 8 demonstrate the build up (trace a) and decay (trace b) of this transient species.

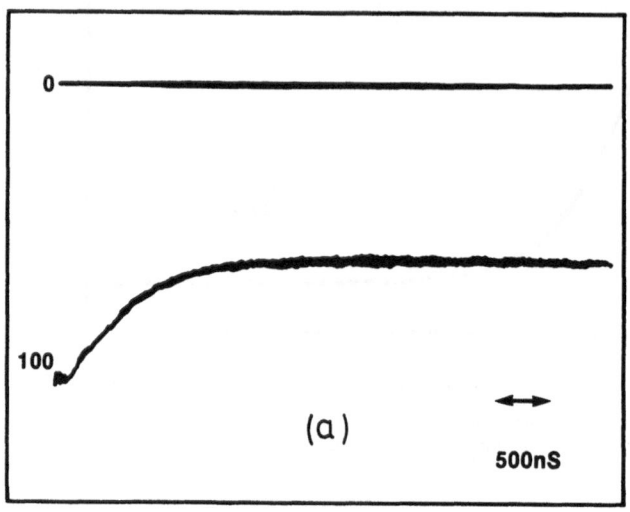

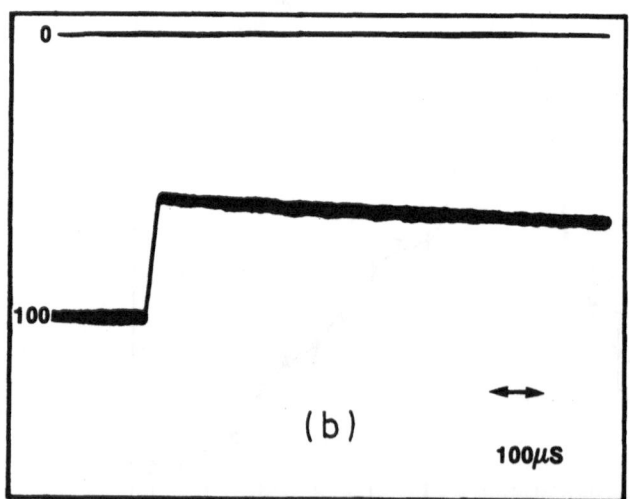

Figure 8. a) The formation of transient species in deaerated cyclo-hexane solution (325 nm); b) The decay of the same species in deaerated cyclohexane solution (320 nm)

Assignment of the Transient Species and Mechanism

The rise of the transient absorption at 325 nm, as shown in trace A of Fig. 8, was treated according to the kinetic expression 6.

$$A = A_\infty (1-e^{-k_6 t}) \tag{6}$$

where A_∞ is the maximum concentration of the transient species at the leveling off point which appears at approximately 5 μsec from the end of the laser pulse (Fig. 8a). $\underline{A}$ denotes the concentration of the building up species as a function of reaction time t, and k_6 is the rate constant. Analysis of the experimental data according to the kinetic expression 6 gave a rate constant $k_6 = 2.23 \times 10^6 \text{sec}^{-1}$.

The decay of this transient was measured from Fig. 8b and found to be pseudo-first order with a rate constant $k_d = 17 \text{ sec}^{-1}$. In air saturated solutions of TNM in cyclohexane this species reacts with oxygen, as shown in Fig. 9, in a pseudo first order fashion (much higher concentration of oxygen relative to that of the transient species) with a rate constant, $k_{ox} = 1.59 \times 10^4 \text{sec}^{-1}$.

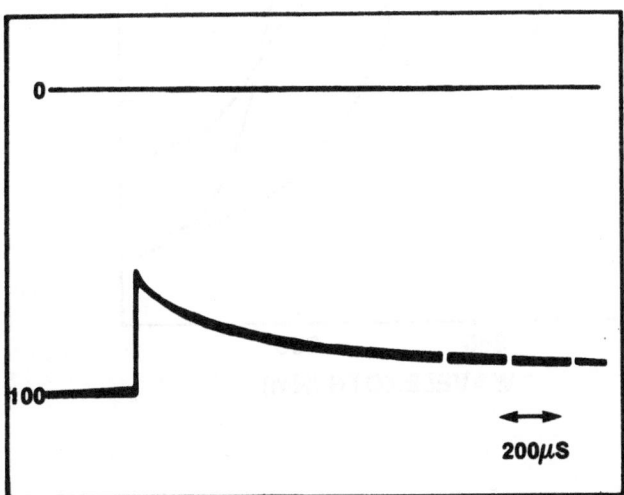

Figure 9. The effect of O_2 on the trinitromethyl radical decay monitored at 310 nm in cyclohexane solution.

In order to calculate the second order rate constant for the reaction of the 307 nm species with oxygen, the solubility of oxygen in aerated cyclohexane was assumed to be the same as the solubility of oxygen in aerated n-pentane which is known (11) to be 4.72×10^{-3}M. Using this solubility value the bimolecular rate constant for the reaction of the 307 nm transient with oxygen was found to be $k_{ox} = 3.37 \times 10^6 \text{M}^{-1}\text{sec}^{-1}$. The spectroscopic and kinetic data presented

above for the transient species with absorption maximum at 307 nm
(Fig. 7) coupled with the experimental evidence (Figs. 10 and 11)
showing that in nonpolar solvents electronic excitation of TNM leads
to an appreciable yield of $(NO_2)_3C-H$ (12) which is reduced in the
presence of oxygen (Fig. 10), suggests that the 307 nm transient
species is most likely the trinitromethyl radical.

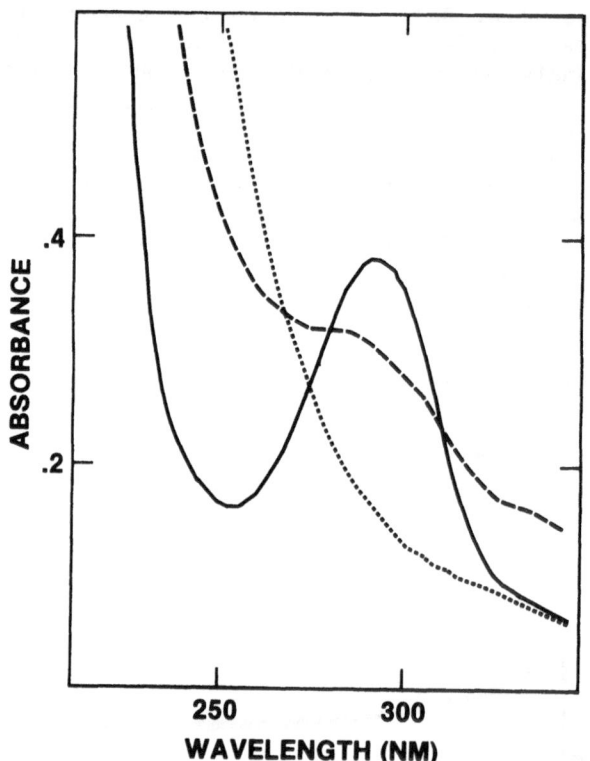

Figure 10. UV spectra of TNM in cyclohexane. (.....) nonirradiated
TNM, 4 x 10^{-4}M; (-----) irradiated TNM, 4 x 10^{-4}M with air; (————)
irradiated TNM, 1 x 10^{-4}M deaerated.

This assignment is further supported by the fact that electronic
excitation of TNM in gas phase or nonpolar solvents yields appreciable amounts of NO_2 as shown by the UV-visible and infrared spectra
of Figs. 12 and 13. This in turn indicates C-N bond rupture in the
TNM molecule which is bound to lead to simultaneous formation of the
trinitromethyl radical. It should be noted that C-N bond rupture
appears to be the initial step in the photochemistry of alkyl nitro
compounds (13-18).

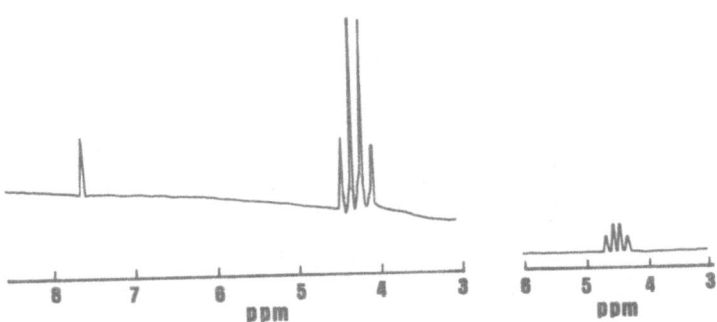

Figure 11. NMR spectrum of 2% (V/V) TNM in degassed hexane after irradiation with 2537Å Hg line. The single resonance peak at δ=7.6 ppm is due to trinitromethane protons (19). The four line spectrum in the ratio 1:2:2:1 in the δ = 4.1-4.4 region is very similar to the spectrum of 2-nitrobutane shown at right (20) and thus indicates 2-nitrohexane.

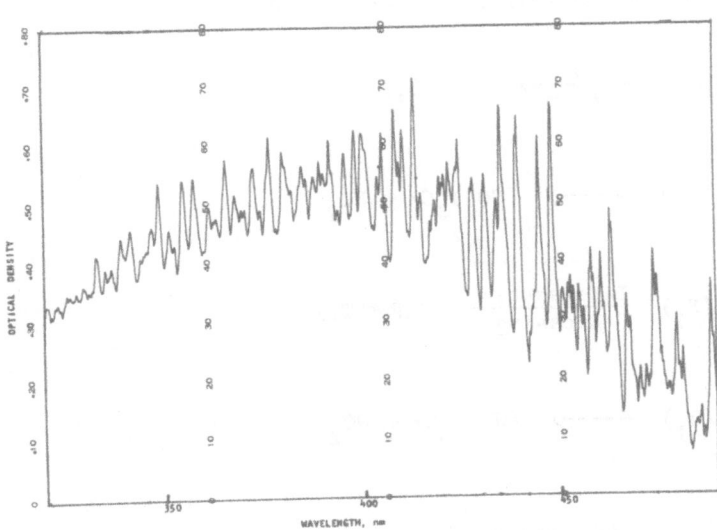

Figure 12. UV spectrum of NO_2 gaseous photolytic product of TNM.

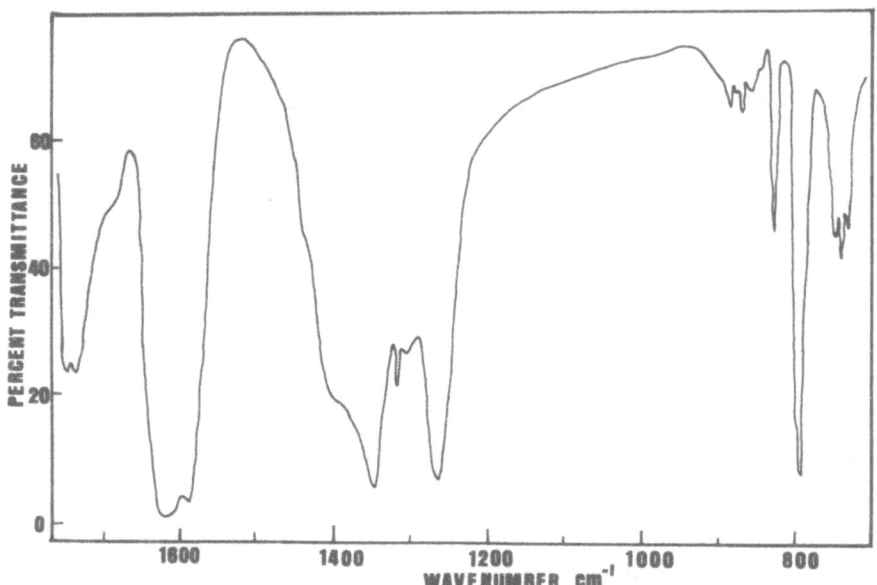

Figure 13.   IR spectrum of NO_2 gaseous photolytic product of TNM.

The formation of $(NO_2)_3C\cdot$, NO_2, $(NO_2)_3C-H$, and nitrohexane can be explained by the following reaction scheme:

$$TNM(S_o) \xrightarrow[\text{248 nm}]{h\nu} TNM^*(S_1) \tag{7}$$

$$TNM^*(S_1) \longrightarrow (NO_2)_3C\cdot + NO_2 \tag{8}$$

$$TNM^*(S_1) \xrightarrow[\text{Crossing}]{\text{Intersystem}} TNM^*(T_1) \tag{9}$$

$$TNM^*(T_1) \longrightarrow (NO_2)_3C\cdot + NO_2 \tag{10}$$

$$(NO_2)_3C\cdot + RH \longrightarrow (NO_2)_3C-H \tag{11}$$

$$R\cdot + NO_2 \longrightarrow R-NO_2 \tag{12}$$

where RH represents the molecule of the nonpolar solvent such as n-hexane or cyclohexane. The data indicate (Fig. 14) that the $(NO_2)_3C\cdot$ is formed to a smaller extent from the singlet excited state of TNM (instantaneous absorption of Fig. 14) and to a larger extent from the lowest triplet excited state of TNM (build up absorption of Fig. 14).

Experimental evidence for reactions 11 and 12 is presented in Figs. 10 and 11. Fig. 11 provides NMR spectroscopic evidence for both the formation of nitroform and nitrohexane (19, 20).

Effect of Oxygen on the Transient Species and Products Generated From Electronically Excited TNM

Figures 14 and 15 demonstrate quite clearly the effect of oxygen on the formation of $(NO_2)_3C\cdot$ and $(NO_2)_3C^-$, respectively. This effect is attributed to the quenching of the triplet excited state of TNM by oxygen (traces b and c of Figs. 14 and 15) which according to our reaction schemes (reactions 4 and 10) is bound to affect both the formation of $(NO_2)_3C^-$ and $(NO_2)_3C\cdot$.

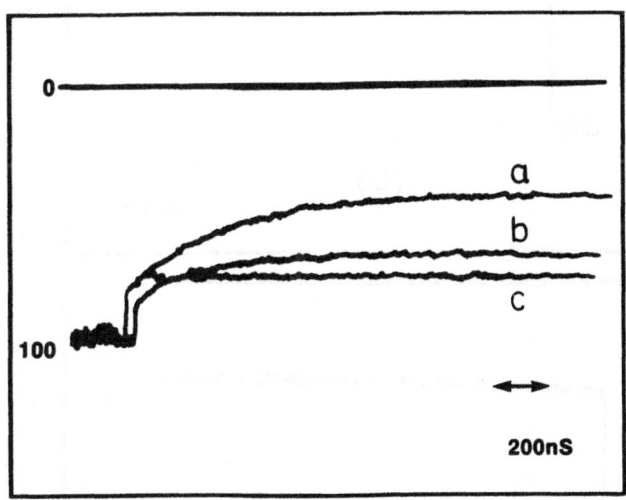

Figure 14. Effect of O_2 at 310 nm on the formation of the trinitromethyl radical in cyclohexane solution. a) deaerated solution; b) air saturated solution; and c) oxygen saturated solution, $P_{ox} = 1$ atm.

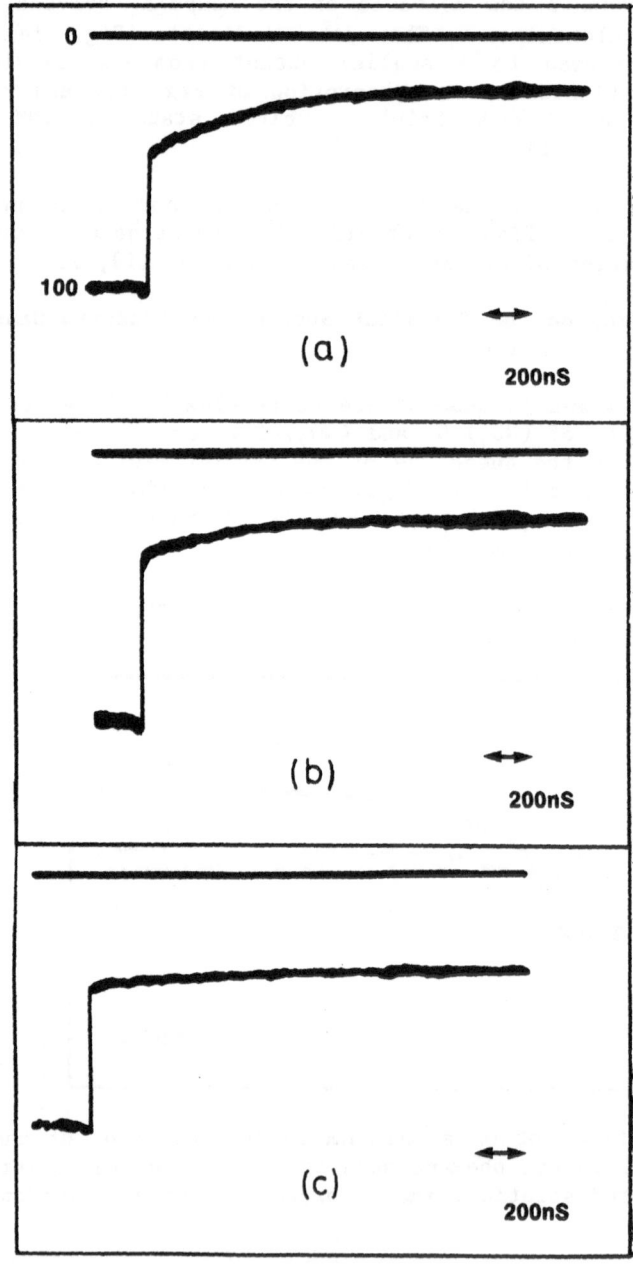

Figure 15. Effect of O_2 at 350 nm on the formation of the nitroform anion in methanol solution. a) deaerated solution; b) air saturated solution; and c) oxygen saturated solution P_{ox} = 1 atm.

The formation of $(NO_2)_3C\cdot$ will be further reduced in aerated nonpolar solutions of TNM due to the reaction of $(NO_2)_3C\cdot$ with oxygen with a rate constant $k_{ox} = 3.37 \times 10^6 M^{-1}sec^{-1}$ to most probably form alkoxy or peroxy radicals, in which case there will be a simultaneous reduction in the yield of $(NO_2)_3C-H$ as observed in the present work (Figs. 9 and 10).

It is interesting to note that the instantaneous formation of the $(NO_2)_3C^-$ and $(NO_2)_3C\cdot$ which is attributed to the reaction of the singlet excited state of TNM (reactions 2 and 8) is not affected by oxygen (Figs. 14 and 15). This is not surprising, however, since the relatively low concentration of oxygen used in these experiments $(4.72 \times 10^{-3}M)$ is not expected to quench the singlet excited state of TNM most of which appears to decay (Figs. 14 and 15) almost within the laser pulse via photophysical or photochemical processes to yield the $(NO_2)_3C\cdot$ in nonpolar solvents or the $(NO_2)_3C^-$ in polar solvents.

REFERENCES

1. Asmus, K.D., Henglein, A., Ebert, M., and Keene, J.P.: 1964, Ber. Bunsenges, Physik. Chem., 68, p. 657.

2. Bielski, B.H.J. and Allen, A.O.: 1967, J. Phys. Chem., 71, p. 4544.

3. Asmus, K.D., Chaudri, S.A., Nazhat, N.B., and Schmidt, W.F.: 1971, Trans. Farad. Soc., 67, p. 2607.

4. Chaudri, S.A. and Asmus, K.D.: 1972, J. Phys. Chem., 76, p. 26.

5. Johnson, D.W. and Salmon, G.A.: 1977, J. Chem. Soc., Faraday I, 73, p. 256.

6. Bielski, B.H.J. and Timmons, R.B.: 1964, J. Phys. Chem., 68, p. 347.

7. Frank, A.J., Grätzel, M., and Henglein, A.: 1976, Ber. Bunsenges., Physik. Chem., 80, p. 593.

8. Capellos, C. and Suryanarayanan, K.: 1976, Int. J. Chem. Kinet., 8, p. 529.

9. Capellos, C.: 1981, J. Photochem., 17, p. 213.

10. Capellos, C. and Iyer, S.: 1981, "Fast Reactions in Energetic Systems," edited by Capellos, C. and Walker, R.F.: p. 401, Reidel, D. Publishing Company, Dordrecht, Holland.

11. Stephen, H. and Stephen, T.: 1963, "Solubilities of Inorganic and Organic Compounds," p. 575, MacMillan, N.Y.

12. Slovetskii, V.N., Shlyapochnikov, V.A., Babievskii, K.K., and Novikov, S.S.: 1960, Izvestiya Akademii Nauk SSSR, Otdelenie Khimicheskikh Nauk, No. 9, p. 1709, English Translation p. 1589.

13. Rebbert, R.E. and Slagg, N.: 1962, Bull. Soc. Chim. Belg., 71, p. 709.

14. Napier, I.M. and Norrish, R.G.W.: 1967, Proc. Roy. Soc., (London), A299, p. 317.

15. Schoen, P.E., Marrone, M.J., Schnur, J.M., and Goldberg, L.S.: 1982, Chem. Phys. Lett., 90, p. 272.

16. Bulter, L.J., Krajnovich, D., Lee, Y.T., Ondrey, G., and Bersohn, R.: 1983, J. Chem. Phys., 79, p. 1708.

17. Blais, N.C.: 1983, J. Chem. Phys., 79, p. 1723.

18. Renlund, A.M. and Trott, W.M.: 1984, Chem. Phys. Lett., 107, p. 555.

19. Feuer, H.: 1969, "The Chemistry of Nitro and Nitroso Groups," p. 38, Interscience, NY.

20. Varian High Resolution NMR Spectra Catalog. Spectrum 84, Palo Alto, California, 1962.

INFRARED LASER MULTIPHOTON DECOMPOSITION OF 1,3,5-TRINITROHEXAHYDRO-S-TRIAZINE (RDX)

C. Capellos, S. Lee*, S. Bulusu, and L.A. Gamss**
ARDC, Dover, New Jersey 07801-5001

ABSTRACT

Powdered samples of 1,3,5-trinitrohexahydro-s-triazine (RDX), were exposed to nanosecond infrared laser pulses from a tuned CO_2 TEA laser. The laser pulses induced multiphoton vibrational excitation in RDX which decomposed very rapidly to form gaseous products. These products were detected and characterized with mass spectrometry, UV-visible, and infrared spectroscopy. Among the detected products were HCN, CO, N_2, NO, CH_2O, CO_2, N_2O, and NO_2.

INTRODUCTION

The initiation of exothermic decompositions of energetic materials very probably begins with the initial formation of energetic transient species, such as electronically or vibrationally excited states, ionic species, or free radicals caused by stimuli such as shock, thermal, laser pulse, or spark. Earlier papers presented data on the spectroscopy and chemical reactivity of ionic species (1, 2), free radicals (2), and electronically excited (3-8) states of nitro-aromatics formed via electronic excitation of the parent molecule.

The primary purpose of the research described here is to elucidate the role of vibrational excitation in the initiation of fast reactions in energetic materials. The data presented in this report show that infrared pulses from a CO_2 laser induce vibrational excitation in 1,3,5-trinitrohexahydro-s-triazine (RDX) which then decomposes very rapidly to form NO_2 and stable gaseous products essentially identical to those formed during the thermal decomposition of RDX.

---

*Geo-Centers, Inc., Wharton, New Jersey 07885. Presently with LeRon Associates, Inc., P.O. Box 84, Dover, NJ 07801.
**National Research Council Research Associate at ARDC. Permanent Address: Chemistry Dept, Bar-Ilan University, Ramat-Gan, Israel.

P. M. Rentzepis and C. Capellos (eds.), Advances in Chemical Reaction Dynamics, 395–404.
© *1986 by D. Reidel Publishing Company.*

EXPERIMENTAL

The laser used in the present studies is a Lumonics 601 TEA CO_2 laser, grating tuned with output in both the 10.6 μm and 9.6 μm bands. Its pulse repetition rate is maximally 0.166 Hz with an energy per pulse of approximately 7 J single line. It is operated with a mixture of CO_2-He-N_2 at ratios 11:8:3. The temporal pulse shape has the typical wave form of most TEA CO_2 lasers operating with a full component of nitrogen in that most of the energy is in an initial very intense spike of approximately 200 nsec and the rest of the pulse energy is in the weak tail of approximatley 1 μsec time duration.

The RDX was either sublimed at 165°C or recrystallized from acetone. The reagents methanol or acetonitrile used to clean the apparatus were spectroscopic grade (Gold Label) from Aldrich Chemical Company and used without further treatment.

The infrared spectra were obtained with a Perkin Elmer 621 or 580 infrared spectrometer. Spectra of the solid samples were taken using standard KBr pellet techniques. The spectra of the drawn off gaseous products were taken in a cell with NaCl windows and a light path length of 10 cm.

The UV-visible spectra of gaseous products were taken with a Cary 14 UV-visible spectrometer and a cell of light path length 10 cm and quartz windows. This cell was also used to test the UV-visible light transmission of NaCl windows. Mass spectroscopic data were obtained with a DuPont 21-492B instrument.

The RDX was irradiated on a stainless steel or NaCl plate in an evacuated cell. After irradiation, the gaseous decomposition products were drawn off with liquid nitrogen for analysis.

RESULTS AND DISCUSSION

Earlier work (9) discovered that the ring stretching vibrational mode (10) of RDX (fig. 1) can be excited by the 10.6 μm (944 cm^{-1}) laser pulses of the TEA CO_2 laser and, through multiphoton absorption, fast molecular decomposition of RDX was induced. Fully deuterated RDX, with the ring vibrational transition isotopically shifted by approximately 40 cm^{-1}, does not have an absorption overlapping the 10.6 μm laser line (fig. 1) and consequently did not undergo decomposition (9) when irradiated. It was also discovered (9) that RDX under suitable confinement, can be detonated with a single focused pulse of the CO_2 laser.

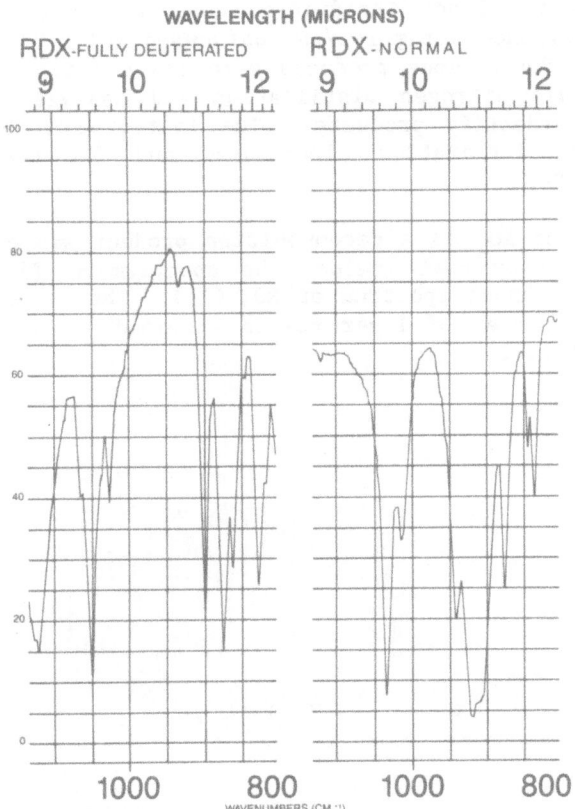

WAVELENGTH (MICRONS)

RDX-FULLY DEUTERATED   RDX-NORMAL

WAVENUMBERS (CM⁻¹)

Figure 1. Infrared spectra of fully deuterated and normal RDX.

In the present work it was found that RDX could be decomposed, under different conditions, by several of the CO_2 laser lines. Irradiation with an unfocused laser beam tuned to the 10.6 μm P(20) line resulted in the formation of a red-brown gas later shown to be NO_2. This corresponds to absorption by the RDX band centered at about 920 cm⁻¹ (fig. 1).

When the laser was tuned to the 9.6 μm P(20) line, corresponding to the RDX band centered at about 1025 cm⁻¹, no decomposition was found upon irradiation with an unfocused beam. However, irradiation with a softly focused beam did result in decomposition. Furthermore, irradiation with the 10.6 μm R(20) line, corresponding to the region between the above two RDX bands, where there is no apparent absorption as seen in the IR spectrum, also resulted in decomposition.

In all cases the laser irradiation dose was approximately 200 pulses, with the fluence being 1 J/cm^2 unfocused and 5 J/cm^2 focused. After irradiation the gaseous products were removed from the remaining solid by liquid nitrogen distillation and analyzed to confirm decomposition and identify products. The latter were the same for all the different irradiation lines used and for the different particle sizes of RDX.

The presence of NO_2 as a decomposition product was confirmed by visible/UV spectroscopy that yielded the spectrum of fig. 2, which matched exactly the known spectrum of NO_2 (11). The growth of NO_2 as a function of the number of laser pulses absorbed by the RDX sample is also shown in fig. 2.

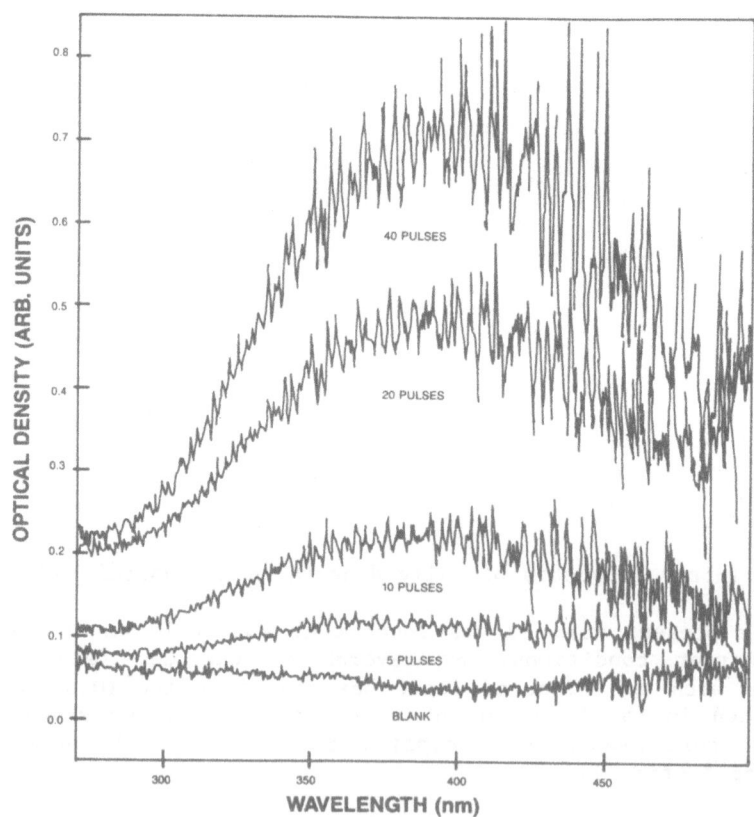

Figure 2. UV-visible spectra of gaseous products from CO_2 laser irradiation of RDX.

The UV-visible spectra did not indicate any other species with significant absorptions. We expect from the thermal decomposition studies (12-16) of solid RDX the formation of N_2O, CH_2O, N_2, CO_2, CO, NO, HCN, and H_2O as well. The infrared spectra of the gaseous products of a powdered RDX sample irradiated with 200 pulses of the CO_2 laser is shown in figs. 3 through 6. The spectra have many features but are dominated by the NO_2 peaks (17) at 1621 cm^{-1}, 1275 cm^{-1}, and 1375 cm^{-1}.

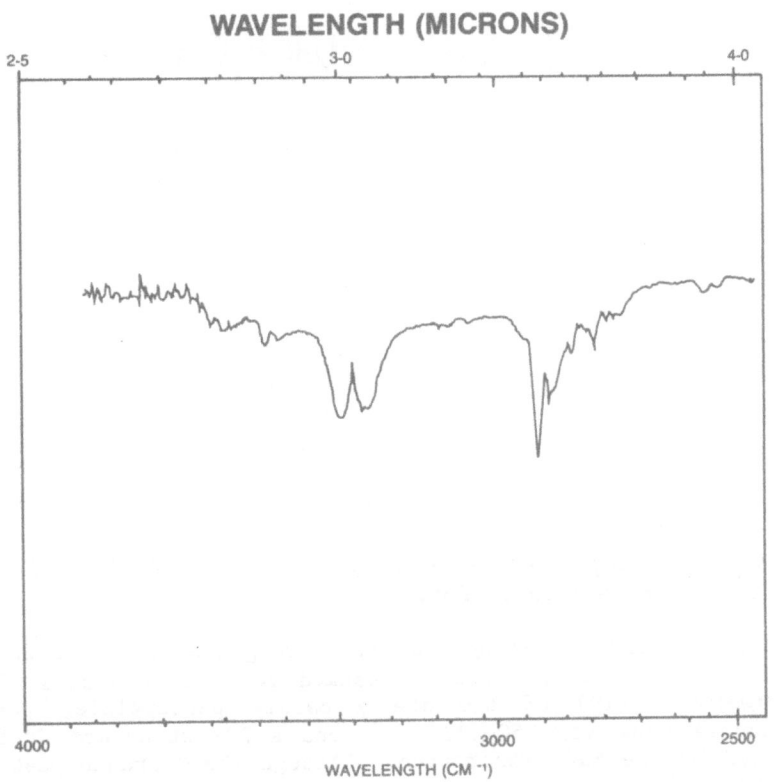

Figure 3. IR spectra of gaseous products formed from CO_2 laser multiphoton decomposition of RDX.

The very sharp peak at 717 cm^{-1} is characteristic of HCN. The P and R bands at either side of this sharp Q band (18) are evident. The R band is seen between 720 cm^{-1} and 770 cm^{-1} while the P band starts at 705 cm^{-1} and is convoluted into the NaCl absorption starting at approximately 650 cm^{-1}. Further evidence for the presence of HCN can be found as a broadening on the blue side of the NO_2 band in the 1300-1470 cm^{-1} region.

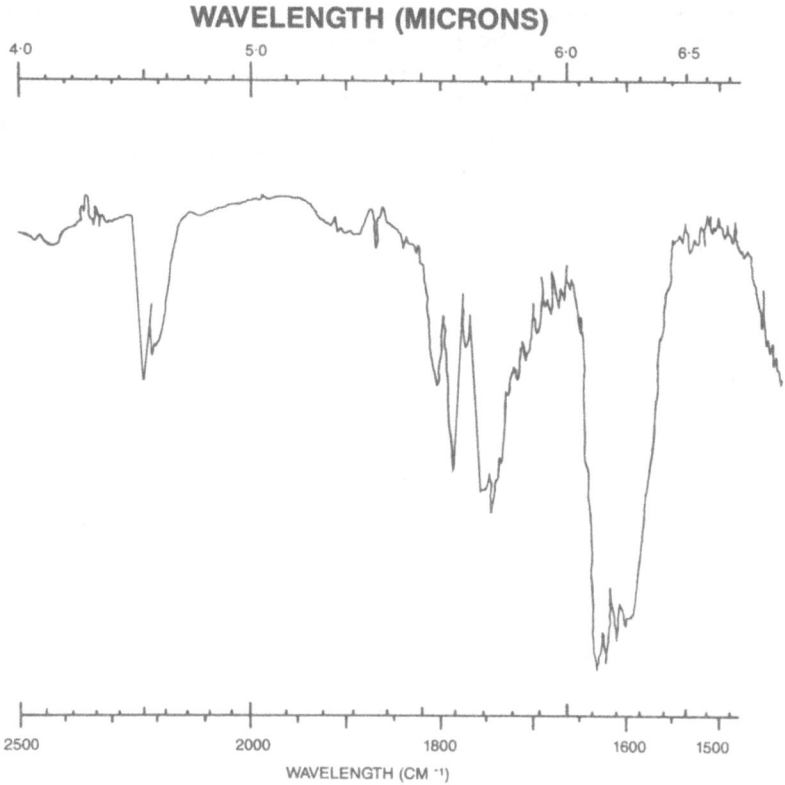

WAVELENGTH (MICRONS)

Figure 4. IR spectra of gaseous products formed from CO_2 laser multiphoton decomposition of RDX.

N_2O is indicated (19) by its very strong features at 2160 – 2260 cm^{-1} and a less strong, partially masked feature at 1250–1320 cm^{-1}. Weaker features (19) of N_2O are a barely perceptible, partially masked double peak at 1250–1320 cm^{-1} and a PQR structure at 550–650 cm^{-1} masked by the NaCl absorption, although the Q branch peak may be seen at 590 cm^{-1}.

The other products are in concentrations too low to appear strongly in this IR spectrum. There are, however, many features not attributable to any predicted product. One feature, the PQR structure at 1070 – 1130 cm^{-1}, has been identified (20) as probably being HCOOH. Other features suggesting this identification are a broad feature at 3570 – 3620 cm^{-1}, a sharp feature at 640 cm^{-1}, a weak double peak at 2180–2220 cm^{-1} masked by the N_2O, and a carbonyl peak 1775 cm^{-1}. The feature (20) at 2880 – 3020 cm^{-1} is probably masked by the many other unidentified features in that region.

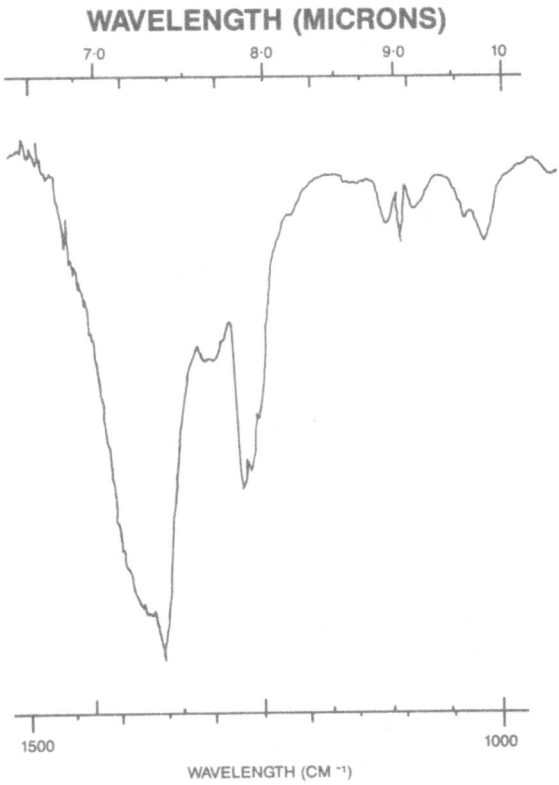

Figure 5. IR spectra of gaseous products formed from CO_2 laser multiphoton decomposition of RDX.

Mass spectrometry can often offer additional clues for analysis. The gaseous products formed from irradiated powdered RDX samples were analyzed, and typical results given in table 1 seem to be in good agreement with the known thermally induced reaction products (12–16).

Table 1. Mass spectra of gaseous products of irradiated RDX.
(100 Pulses – Unfocused)

M/E	Product
27	HCN
28	CO, N_2
30	NO, CH_2O
44	CO_2, N_2O
46	NO_2

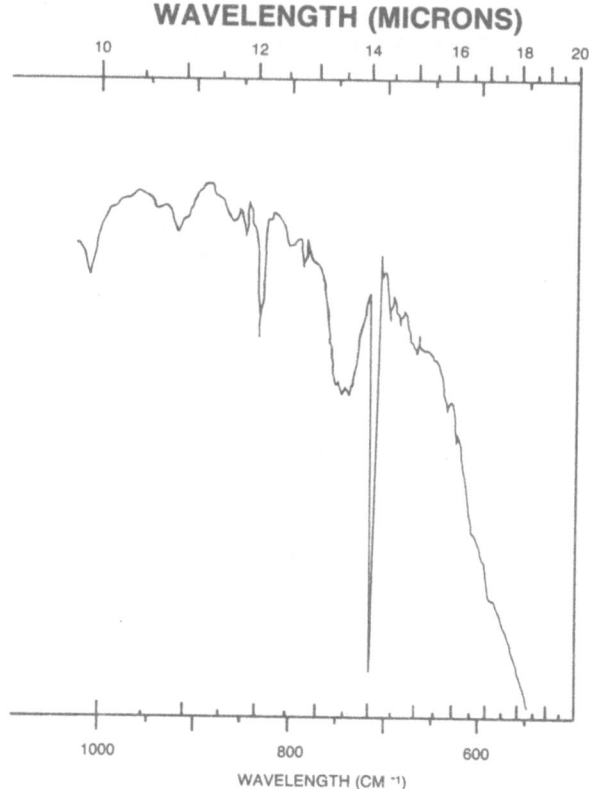

WAVELENGTH (MICRONS)

WAVELENGTH (CM ⁻¹)

Figure 6. IR spectra of gaseous products formed from CO_2 laser multiphoton decomposition of RDX.

Analysis of the solid products of the reaction is difficult because most of the reaction products are gaseous and the solid residue can be expected to be predominantly unreacted RDX, particularly for a laser driven reaction where only molecules near the surface interact with the laser light. This was clear when the mass spectra of the solid residues were taken and no significant differences from the original RDX mass spectra were evident.

The thermal decomposition of RDX has been studied extensively (12-16) in the gaseous and condensed phase. These earlier studies showed that all the gaseous products listed in table 1 are formed during the thermal decomposition of either gaseous or condensed RDX, with the only exception the formation of NO_2, which seems to form in appreciable amounts only during the thermal decomposition of gaseous RDX. This seems to be in contrast to the findings of the present study which indicate that irradiated powdered RDX yields relatively large amounts of NO_2 in addition to all the other gaseous products of thermally decomposed condensed RDX. This result suggests that the effects of the infrared irradiation of powdered RDX are distinct from those of conventional thermal degradation, or that the decomposition is actually taking place in gaseous RDX vaporized by the laser.

The formation of NO_2 from the laser irradiated RDX is not surprising since according to multiphoton dissociation theory (22), the thermodynamically directed dissociation should be at the N-N bond, the weakest bond, even though the vibrational excitation is at the ring stretching mode.

These results are interesting in their implication that the very fast interaction of the laser with a solid, perhaps by affecting surface molecules before intermolecular energy transfer occurs, yield results very similar to steady state gas phase thermal reactions.

CONCLUSIONS

This is the first study ever showing that vibrational excitation of an explosive material, (RDX), induced with infrared multiphoton processes, can lead to fast decomposition and under certain conditions to detonation.

The gaseous products formed from laser-induced vibrational excitation of powdered RDX are identical to those formed from thermally decomposed powdered RDX with the only exception the NO_2 radical, which is formed in appreciable amounts only during the fast decomposition of the laser induced vibrationally excited RDX.

ACKNOWLEDGMENT

L.A. Gamss wishes to acknowledge support from the National Research Council in the form of a Research Associateship at ARDC.

REFERENCES

1. Capellos, C. and *Suryanarayanan, K.: 1973, Intern. J. Chem. Kinetics, V, p. 305-320.

2. Suryanarayanan, K. and Capellos, C.: 1974, Intern. J. Chem. Kinetics, VI, p. 89.

3. Capellos, C. and Porter, Sir George: 1974, J. Chem. Soc. Farad. Trans, II, p. 1159.

4. Capellos, C. and Suryanarayanan, K.: 1976, Intern. J. Chem. Kinetics, VIII, p. 529-539.

5. Capellos, C. and Suryanarayanan, K.: 1976, Intern. J. Chem. Kinetics, VIII, p. 541-548.

_____

*Name changed, presently known as S. Iyer.

6.   Capellos, C. and Suryanarayanan, K: 1977, Intern. J. Chem.
     Kinetics, IX, p. 399.

7.   Capellos, C. and Lang, F.: 1977, Intern. J. Chem. Kinetics, IX,
     p. 409-415.

8.   Capellos, C. and Lang, F.: 1977, Intern. J. Chem. Kinetics, IX,
     p. 943-952.

9.   Capellos, C.: Unpublished work.

10. Iqbal, Z., Suryanarayanan, K., Bulusu, S., and Autera, J.R.:
     1972, Picatinny Arsenal Technical Report 4401, Dover, New
     Jersey.

11. Hsu, D.K., Monts, D.L., and Zare, R.N.: 1978, Spectral Atlas of
     Nitrogen Dioxide, Academic Press, New York.

12. Robertson, A.J.B.: 1949, Trans. Farad. Soc., 45, p. 85.

13. Cosgrove, J.D. and Owen, A.J.: 1968, Chem. Commun., p. 286.

14. Rauch, F.C. and Fanelli, A.J.: 1969, J. Phys. Chem., 73, p.
     1604.

15. Cosgrove, J.D. and Owen, J.D.: 1974, Comb. and Flame, 22, p.
     13.

16. Dubovitskii, F.I. and Korsunskii, B.L.: 1981, Russian Chemical
     Reviews, 50, p. 958.

17. Herzberg, G.: 1945, Molecular Spectra and Molecular Structure,
     D. Van Nostrand, New York, p. 284.

18. Choi, K.N., Barker, E.F.: 1932, Phys. Rev., 42, p. 777.

19. Plyler, E.K., Barker, E.F.: 1931, Phys. Rev., 38, p. 1828.

20. Williams, V.Z.: 1947, J. Chem. Phys., 15, p. 243.

21. Meeke, R., Langenbucher, F.: 1965, Infrared Spectra of Selected
     Compounds, Heyden and Son, London.

22. Ronn, A.M.: 1979, Sci. Amer., 240, p. 114.

TRANSIENT PHOTOCHEMISTRY OF NITROBENZENE AND NITRONAPHTHALENES

S. Iyer and C. Capellos
Energetics & Warheads Division, AED
US Army Armament Research and Development Center
Dover, New Jersey 07801-5001

ABSTRACT

Conventional flash photolysis of nitrobenzene and 1- and 2-nitronaphthalenes produces their respective radical anions in deaerated triethylamine-acetonitrile solutions. In deaerated triethylamine-nonpolar media, instead of the radical anion, their respective protonated neutral radicals are formed. 1-Nitronaphthalene yields in deaerated pure alcohols the neutral radical and the radical anion in deaerated alkaline alcohols. The transient absorption spectra of these species are presented. These transients, with the exception of neutral 1-nitronaphthalene radical in pure alcohols (where it decays via first order kinetics), decay via second order kinetics. The measured rate constants for their decay are given. Possible mechanism for the formation of these transient species is discussed.

INTRODUCTION

Light-induced reactions of aromatic nitrocompounds are of interest to this laboratory in that they have the potential to provide insights into the nature of transient species which might be formed in the fast decomposition of energetic materials like TNT or TNB when such materials are subjected to a variety of strong stimuli (e.g. shock, thermal pulse, or light pulse) (1). Fast decompositions of these materials could conceivably initiate involving ionic species, excited states or free-radicals which could cause buildup of uncontrolled chain reactions. Furthermore, looking from the scientific view point, nitroaromatics have very rich photochemistry in solution. For example, their triplets are very short-lived (i.e. lifetimes in the nanosecond time regime). Such reactive triplets of a variety of nitronaphthalenes have been researched and characterized in this laboratory (2). As will be seen from the discussion of this paper, these reactive excited states initially generated by light seem to undergo transformation, further on in the time scale, into transient basic radical anions or their protonated acid forms, depending on the acidity and/or polarity of the solvent medium.

P. M. Rentzepis and C. Capellos (eds.), Advances in Chemical Reaction Dynamics, 405–414.
© *1986 by D. Reidel Publishing Company.*

Henglein et al (3) have studied electron attachment reactions of nitrobenzene molecule in aqueous solutions via pulse radiolysis technique and have observed, as transients, the nitrobenzene radical anion and its protonated acid form, the latter at pH's below 3.2. During the course of this work, we also observed in flash photolysis of nitrobenzene, the nitrobenzene radical anion or its acid form as transients in nonaqueous solvent systems and have studied their reactivity. Furthermore, we observed from 1- and 2-nitronaphthalenes, their radical anions and their acid forms of the type $ArNO_2^{\cdot-}$ and $Ar-NO_2H$ as transients in flash photolysis of the two mononitronaphthalenes in different nonaqueous solvents and studied their reactivities. This paper describes the details of these results.

EXPERIMENTAL

Conventional microsecond (1a) and laser nanosecond flash photolysis (2b) apparatus used in this work have been previously described. Degassing of solutions was performed by inert gas bubbling through the experimental solutions (Ar or He gas). In conventional runs, 20 cm pathlength photolysis quartz cells were used which were jacketed for the purpose of providing appropriate filter solutions. Kinetic plots were obtained using NOVA minicomputer as described before (2d). All solvents used in these experiments were spectrograde quality.

RESULTS AND DISCUSSION

Nitrobenzene Transients

In this work, we observed, in nonaqueous media, the nitrobenzene anion radical and its acid form. The assignment of the observed spectra to these species is based on the similarity of the transient spectra of the nitrobenzene radical anion and its acid form reported in the literature (3, 4).

Figure 1, Spectrum A, shows the absorption spectrum of the radical anion of nitrobenzene formed in flash photolyzed deaerated solution of nitrobenzene in 20% V/V triethylamine in acetonitrile. This shows two maxima, viz. at 430 and 470 nm and is in excellent agreement with the absorption spectrum of the nitrobenzene anion radical observed by previous investigators (3, 4). This anion in aqueous alkaline solutions is reported (3) to be rather stable with a lifetime of about several seconds. The nitrobenzene anion observed in our laboratory in 20% Et_3N-CH_3CN solution is found to decay via second order kinetics (table 1) with $k/\varepsilon l = 18$ sec^{-1}. Assuming the molar extinction coefficient of the 430 nm band is the same in aqueous and in 20% Et_3N-CH_3CN media and using the value reported by Henglein et al (3) in aqueous media, the calculated rate constant for this decay amounts to $1.8 \times 10^5 M^{-1}sec^{-1}$. While this assumption is certainly not very accurate, nevertheless, such calculated rate constants will describe to us here and elsewhere the general photochemical picture and transient reactivities.

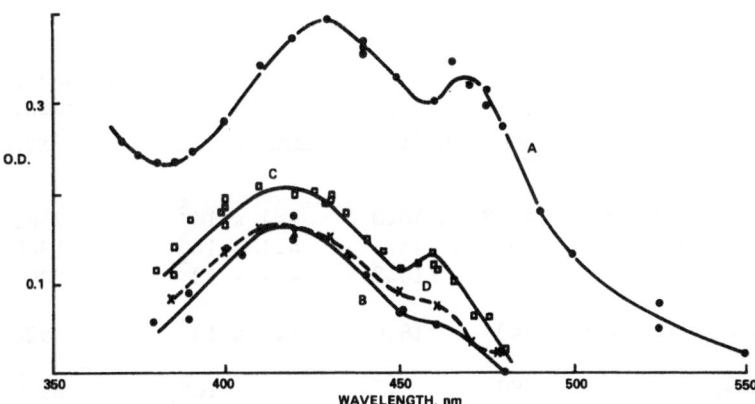

Figure 1. Transient absorption spectra of nitrobenzene species in deaerated:
A. 20% V/V Et$_3$N–CH$_3$CN medium (radical anion)
B. Cyclohexane – 0.253M Bu$_3$SnH (radical)
C. Hexane – 1.44M (20% V/V) Et$_3$N (radical)
D. Acetone – 0.190 M Bu$_3$SnH (radical)

The measured pK value (by Henglein et al (3)) of the equilibrium (in aqueous medium):

$$C_6H_5\overset{\bullet}{N}O_2^- + H^+ \rightleftarrows C_6H_5\overset{\bullet}{N}O_2H$$

is 3.2. Consequently, they observed the acid form, C_6H_5–$\overset{\bullet}{N}O_2H$, below pH = 3.2 in pulse radiolyzed aqueous solutions of nitrobenzene. We observed the same acid form in flashphotolyzed solutions of nitrobenzene in nonpolar solvent systems. Thus, it is formed in air-free cyclohexane containing 0.253 M Bu$_3$SnH and in hexane – 1.44 M Et$_3$N (20% V/V) solutions of nitrobenzene (fig. 1, Spectra B and C). The observed maxima are the same in both solutions, viz., 418, 460 nm, and the observed second order decay rates are (table 1) respectively:

$$5.4 \times 10^4 \text{ and } 3.1 \times 10^7 M^{-1}sec^{-1}$$

Table 1. $Ar\overset{\bullet}{N}O_2^-$ and $Ar\overset{\bullet}{N}O_2H$ Reactivities

| NN = nitronapthalene | 1 = 20 cm in all cases |
| NB = nitrobenzene | All systems deaerated |

A. Second order rate constants

System	λ_{nm}	$\dfrac{(2k/\varepsilon l)}{(sec^{-1})}$	$\dfrac{2k}{(M^{-1}sec^{-1})}$	Assumed ε for calculation (3,4)
Radical Anions				
1. 1-NN/CH$_3$CN/1.44 M	430*	10.0	4.4×10^5	2182
Et$_3$N (20% V/V)	600	15.6	4.1×10^5	1300
	660	10.9	4.4×10^5	2000
2. NB/CH$_3$CN/1.44M Et$_3$N	440	18.0	1.8×10^5	500
3. 1-NN/Acetone/0.09 M Bu$_3$SnH	390	1.74	2.4×10^5	6900
4. 2-NN/Acetone/0.19 M Bu$_3$SnH	400	2.39	1.3×10^5	2807
Protonated Neutral Radicals				
5. 1-NN/Hexane/1.44 M Et$_3$N	380	238	3.3×10^7	6900
6. NB/Hexane/1.44M Et$_3$N	420	3140	3.1×10^7	500
7. NB/Acetone/0.190 M Bu$_3$SnH	410	3.40	3.4×10^4	500
8. NB/Cyclohexane/ 0.253 M Bu$_3$SnH	420	5.4	5.4×10^4	500
9. 1-NN/Cyclohexane/ 0.001M Bu$_3$SnH	390*	0.55	4.8×10^4	4364
10. 2-NN/Cyclohexane/ 0.0095M Bu$_3$SnH	390*	0.62	2.1×10^4	1694

*not λ_{max}

B. First order rate constants

System	λ_{nm}	$\tau_{1/2}$, ms	k, $M^{-1}sec^{-1}$
1. 1-NN/MeOH	390	42	0.67
2. 1-NN/5% MeOH-Glycerol	390	50	1.01

The neutral nitrobenzene radical (i.e., the acid form) is also formed in deaerated acetone with 0.190 M Bu_3SnH (fig. 1, Spectrum D; maxima: 416 and 460 nm). Our observed spectra in these above-mentioned solvent systems are assigned to the neutral nitrobenzene radical by comparing them with its spectrum in aqueous solution (3) as follows. The absorption maximum for the radical in aqueous solution (3) is 430 nm. The above mentioned spectra show same absorption maxima in the vicinity of 430 nm. The small differences in λ_{max} and the shoulder fine structure are probably attributable to the different nature of the nonaqueous solvent systems.

The nitrobenzene radical decays in aqueous medium with a (bimolecular) disproportionation rate constant, $2k = 6 \times 10^8 M^{-1} sec^{-1}$ (3). The rate constant for the radical decay in deaerated hexane $-Et_3N$ (this work) is $3 \times 10^7 M^{-1} sec^{-1}$ which is comparable with that in aqueous solution. However, this value is higher (table 1), by about a factor of 200, than those in solvent systems containing Bu_3SnH as shown in table 1. The rate constants in systems containing Bu_3SnH seem to be anomalously low. Bu_3SnH, in view of its pronounced ability to donate H atoms, seems to stabilize the radical in these solvent systems.

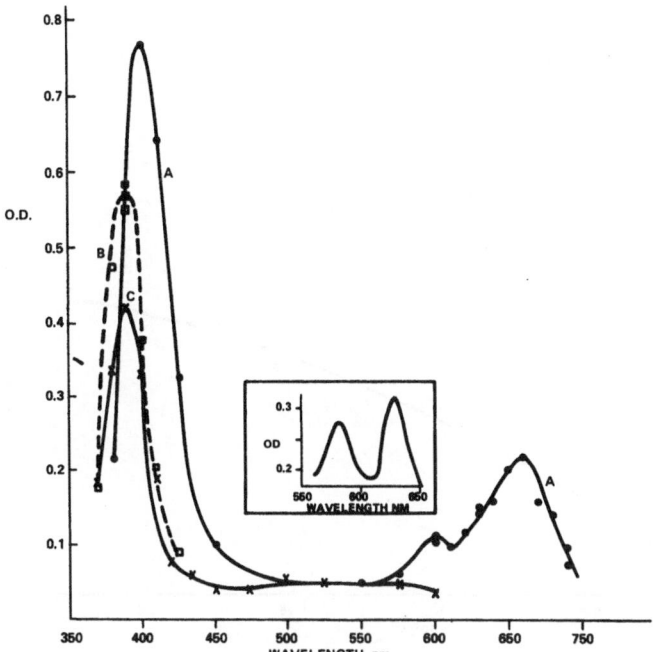

Figure 2. Transient absorption spectra of 1-nitronaphthalene in deaerated:
A. 20% V/V Et_3N - CH_3CN (radical anion); B. Methanol (radical); C. Ethanol (radical); (Inset) Radical anion as stable species in 4.67 mM NaOH in ethanol (generated by flash photolysis).

Nitronaphthalene Transients

In dearated solution of CH_3CN containing 20% V/V Et_3N, 1-nitro-
naphthalene produces a transient species which has an absorption
maximum at 400 nm in the UV and two maxima, 600 and 660 nm, in the
visible regions (fig. 2, Spectrum A). In the inset in fig. 2, a
spectrum of the stable product formed by flash photolyzing 1-nitro-
naphthalene in ethanol with $4.67 \times 10^{-3}M$ NaOH is shown. In the latter
case, the radical anion of 1-nitronaphthalene is formed and is
stabilized by the alkaline condition. The radical anion thus formed
has two absorption maxima, viz. 580 and 630 nm (see fig. 2, inset).
This compares quite well with the absorption maxima of the electro-
lytically formed radical anion of 1-nitronaphthalene (4) and of the
species produced in 20% Et_3N-CH_3CN solutions. On the basis of this
spectroscopic evidence and the kinetic data presented in table 1, the
above species in CH_3CN medium is assigned to the radical anion of 1-
nitronaphthalene. This radical anion is not formed in presence of
air or in pure CH_3CN. Triethylamine is necessary for its formation.
Figure 3 shows the yield of this anion as a function of triethylamine
concentration.

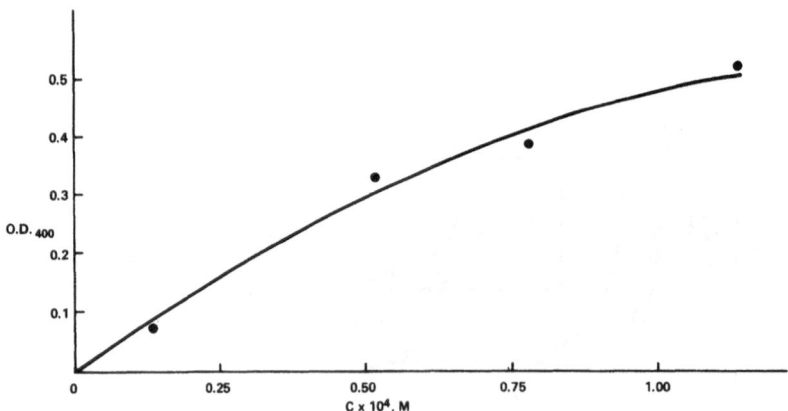

Figure 3.   Dependence of the yield of 1-nitronaphthalene radical
anion as a function of Et_3N concentration in CH_3CN (dearated).

This species is also formed in dearated solutions of 1-nitro-naphthalene in alcohols, isopropanol and ethanol, containing 0.1 to 0.5 mM NaOH. In these solutions the species behaves as a transient while in about 5 mM NaOH solution it is stabilized to a nontransient.

In dearated solutions of 1-nitronaphthalene in pure alcohols and in nonpolar media (i.e., hydrocarbon solvent systems with triethyla-mine) the radical anion is not formed. Instead, the protonated form of the anion, viz. the neutral radical, $1\text{-}C_{10}H_7\text{-}N\underset{O}{\overset{\overset{\displaystyle\cdot}{OH}}{\diagdown}}$, is formed. The absorption spectra of this species is shown in fig. 2 (Spectra B and C) and in fig. 4 (Spectrum A). In hydrocarbon media with tri-ethylamine, the neutral radical of the above type was also observed for 2-nitronaphthalene (fig. 4, Spectrum B).

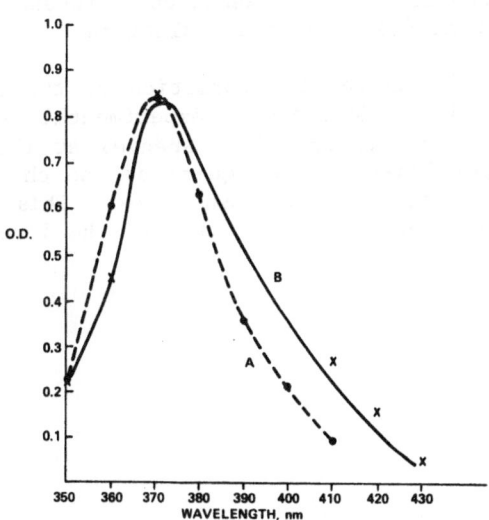

Figure 4. Transient absorption spectra from nitronaphthalenes
A. 1-Nitronaphthalene radical in deaerated 20% Et$_3$N-hexane
B. 2-Nitronaphthalene radical in deaerated 5% Et$_3$N-hexane

The 1-nitronaphthalene anion and the neutral radical in hydro-carbon solvent systems with the exception of pure alcohols decay via second order kinetics (table 1). In deaerated 20% Et$_3$N-CH$_3$CN solvent system, the radical anion decays with the same rate constant at wave-lengths of all the three bands which shows that all the three bands are due to the same species. The observed rate constant 4.1-4.4 x $10^5 M^{-1} sec^{-1}$ is similar to the reactivity of the nitrobenzene radical anion. Similarly, the observed second order rate constants for the radical decay in 20% Et_3N-hexane are same for the cases of both nitrobenzene and 1-nitronaphthalene.

In pure alcohols, the free radical decays via first order kinetics by reaction with the solvent. It is conceivable that the radical could abstract a hydrogen atom from the alcohol molecule to yield the species $1-C_{10}H_7-N\overset{OH}{\underset{OH}{\diagup}}$. The hydrogen abstraction rate constant is observed to be in the range 0.7 to $1M^{-1}sec^{-1}$ in methanol and glycerol (table 1). The reaction was studied in highly viscous glycerol to confirm that the radical does actually decay via first order kinetics. In flash photolysis of deaerated alcohol solutions with 0.005 mM H_2SO_4, we observed 1-nitronaphthalene to generate the free radical $1-C_{10}H_7-\overset{\cdot}{N}\overset{OH}{\underset{O}{\diagdown}}$. This indicates the acid-base nature of the free radical and the radical anion. Furthermore, this phenomenon is similar to the behavior of the analogous nitrobenzene species observed by Henglein et al (3) in aqueous solutions.

The biphotonic mechanism for the formation of the radical anion in all cases is excluded on the basis of experimental evidence showing the optical density of the transient species at the end of the flash is linearly proportional to the square of the charging voltage applied across the terminals of the flash lamps. This implies that the transient yield is linearly proportional to the intensity of the excitation light.

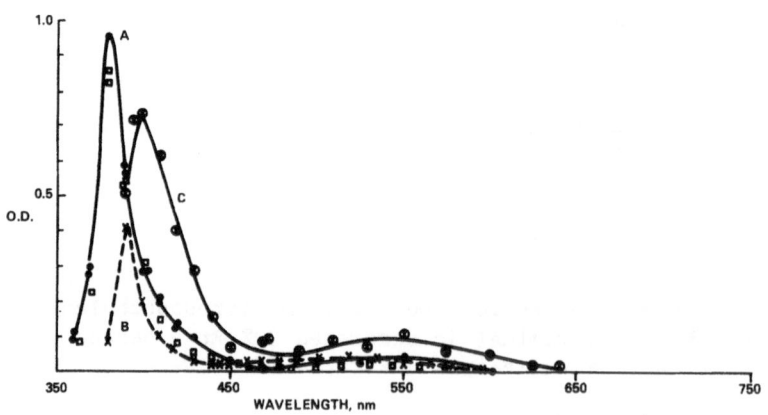

Figure 5. Transient absorption spectra from nitronaphthalenes.
A. 1- and 2-nitronaphthalenes show the same spectrum, in deaerated cyclohexane with 9.5 mM Bu_3SnH (radicals); B. 1-Nitronaphthalene radical anion in deaerated acetone with 90 mM Bu_3SnH; C. 2-Nitronaphthalene radical anion in deaerated acetone with 190 mM Bu_3SnH.

Nitronaphthalenes in Solvent Systems Containing Bu$_3$SnH

Figure 5 shows the spectra of transient species produced from 1- and 2-nitronaphthalenes in deaerated acetone and cyclohexane contain- ing Bu$_3$SnH. In contrast to nitrobenzene, 1- and 2-nitronaphthalenes yield the radical anion in acetone with Bu$_3$SnH. Nitrobenzene yields the neutral radical, $C_6H_5\overset{\bullet}{N}\diagup^{O}_{OH}$. (See discussion above.) In deaerated cyclohexane containing Bu$_3$SnH, nitrobenzene and nitronaphthalenes give the neutral radicals. The difference between nitrobenzene and nitronaphthalenes in acetone medium might be due to the difference in the pK value of the radical $Ar-\overset{\bullet}{N}\diagup^{O}_{OH}$. In acetone, in the case of nitronaphthalenes, when the anion is formed, it seems to decay with rate constants which are similar in value to those for the nitronaphthalene and nitrobenzene anion radicals formed in 20% Et$_3$N-CH$_3$CN media.

The rate constants for nitronaphthalenes' radical decay in cyclohexane-Bu$_3$SnH medium is reminiscent of the nitrobenzene case and indicates stabilization of the radicals by the strong H donating ability of Bu$_3$SnH.

Mechanism of Transient Generation

The transient species $Ar\overset{\bullet}{N}O_2H$ or $Ar\overset{\bullet}{N}O_2^{-}$ arising photochemically, depending on the conditions of the solvent system, probably comes from the first excited triplet state of the nitroaromatic molecule. This is consistent with ruby laser nanosecond flash photolysis exper- iments of alkaline solutions (4.67 x 10^{-3}M NaOH) of 1-nitronaphtha- lene in ethanol which showed the triplet excited state of 1-nitro- naphthalene being quenched via electron transfer by NaOH with a rate constant equal to 1.73 x 10^{8}M^{-1}sec^{-1} to form the 1-nitronaphthalene radical anion (2a). Other preliminary studies (5) in our laboratory on nanosecond quenching experiments provide as well some evidence for this suggestion. Thus, in deaerated polar solvents, an excited state complex between 1-nitronaphthalene and dimethylaniline was observed to form by light excitation (347 nm laser pulse). The complex was further observed to decay to ground state 1-C$_{10}$H$_7$NO$_2$ and the posi- tive ion of dimethylaniline, viz., Me$_2$-$\overset{+}{N}$-(C$_6$H$_5$). Both the complex and the 1-C$_{10}$H$_7$NO$_2$ have absorption maxima around 410 nm; however, the complex has higher extinction coefficient than the radical anion. They had different lifetimes (complex shorter-lived) and thus the absorptions were separated.

In acetonitrile solutions containing Et$_3$N the radical anion is presumably formed by electron transfer (probably to triplet) from Et$_3$N (6). In nonpolar media containing Et$_3$N, the anion radical formed initially picks up a proton from the hydrocarbon solvent yielding the neutral radical. In alcohols and other solvent systems, the triplet state could pick up a hydrogen atom thus yielding the neutral radical. This radical might dissociate into its basic form in acetone, when formed from the triplet picking up a H from Bu_3SnH.

The pK value of the equilibrium $C_6H_5\overset{\displaystyle\cdot}{N}\overset{\nearrow O}{\searrow_{O^-}} + H^+ \rightleftharpoons C_6H_5\overset{\displaystyle\cdot}{N}\overset{\nearrow O}{\searrow_{OH}}$ has been reported in aqueous solutions to be 3.2. Experiments are in progress to determine the pK values of the analogous equilibria in aqueous media for nitronaphthalenes.

REFERENCES:

1a. Capellos, C. and *Iyer, S.: 1973, Int. J. Chem. Kinet., 5, p. 305.

 b. Iyer, S. and Capellos, C.: 1974, Int. J. Chem. Kinet., 6, p. 89.

2a. Capellos, C. and Porter, G.: 1974, Faraday Trans. II, 70, p. 1159.

 b. Capellos, C. and Iyer, S.: 1976, Int. J. Chem. Kinet., 8, p. 529.

 c. Ibid., 8, p. 541.

 d. Ibid., 1977, 9, p. 399.

3. Asmus, K.D., Wigger, A., and Henglein, A.: 1966, Ber. Bunsunges Phys. Chem., 70, p. 862.

4. Kemula, W. and Sioda, R.: 1963, Bull. de L'Acad. Polon. des Sciences, 7, p. 395.

5. Work to be published from this laboratory.

6. Capellos, C. and Iyer, S.: 1981, Fast Reactions in Energtic Systems, NATO ASI, Ser. C (D. Reidel, Holland), p. 401.

---

*Present name; formerly known as K. Suryanarayanan.

DYNAMICS OF THE COLLISION FREE UNIMOLECULAR FRAGMENTATION OF PRIMARY ALKYL EPOXIDES

T. E. Adams, M. B. Knickelbein, D. A. Webb, and E. R. Grant
Department of Chemistry
Cornell University
Ithaca, New York 14853

INTRODUCTION

Virtually all reactive energetic processes begin with the elementary chemical step of unimolecular decomposition. The early kinetics of even complex detonation reactions depend critically on the nature of initiating and propogating dissociative events.[1] It follows that the construction of accurate rate models for chemically evolving energetic systems with widely and rapidly varying conditions of local temperature and pressure, demands a thorough theoretical understanding of key dissociative steps.

Our laboratory has made advances in the development of methods for the study of fundamental unimolecular reaction dynamics.[2-6] These methods initiate decomposition under collision-free conditions in a molecular beam by infrared laser induced multiphoton excitation. Products are detected in real-time by state resolved laser induced fluorescence and multiphoton ionization. A set of systems that perhaps best illustrates the power of this approach, as well as its limitations, is that of the primary alkyl epoxides. These are important strained ring heterolytic systems, which are widely used as monomers[7] and have recently been introduced as wide-overhead explosive agents effective in clearing mine fields.[8]

We have established that the principal channel for decomposition of the isolated molecule is methylene elimination:

$$R\text{-}\overset{\displaystyle O}{\overset{\displaystyle \triangle}{CH}}\text{-}CH_2 \longrightarrow RCHO + {}^1\text{:}CH_2$$

Both products are detected by laser spectroscopy. We use multiphoton ionization for the aldehyde and laser induced fluorescence for $^1CH_2$. From the data we obtain information on dynamics in the form of internal state and recoil velocity distributions, and kinetics in directly measured unimolecular lifetimes. Our results paint a revealing picture of the nature of the unimolecular bond breaking process, but leave in-

P. M. Rentzepis and C. Capellos (eds.), Advances in Chemical Reaction Dynamics, 415–424.

triguing questions about the dynamics of infrared photoexcitation
unanswered.

EXPERIMENTAL

Experiments are conducted in a laser-crossed pulsed molecular beam
apparatus which is diagrammed below:

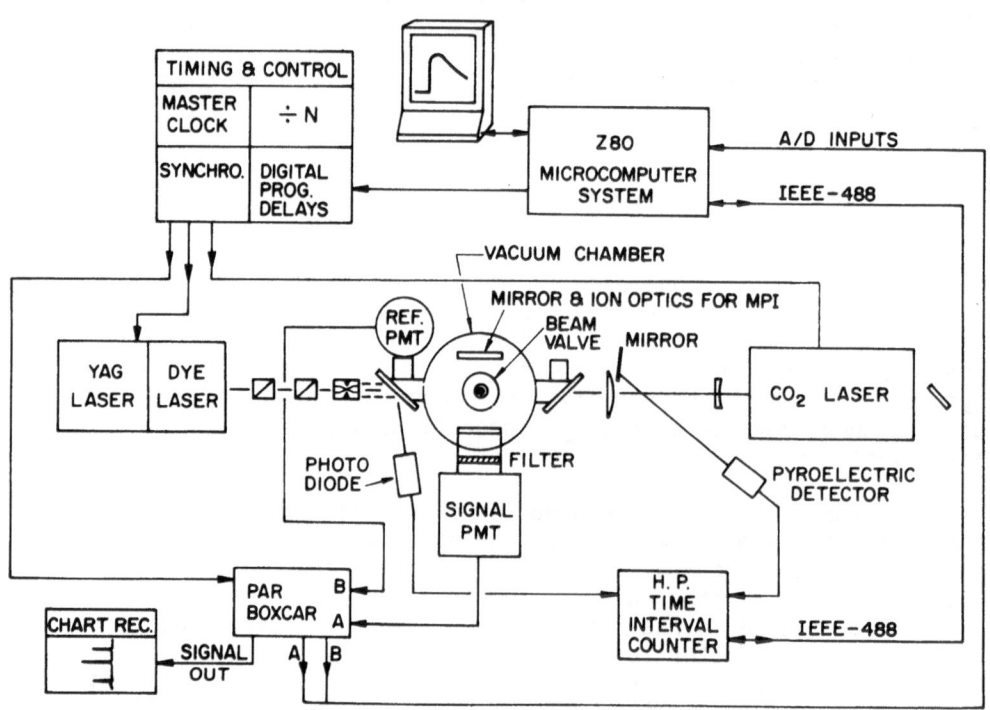

The experiment uses a pulsed supersonic jet of He, in which is
seeded alkyl epoxide ($R = CH_3$, C_2H_5 or C_6H_{13}) at a partial density of
less than 0.1 percent. This jet is crossed by the focussed output of a
specially designed fast-discharge CO_2 TEA laser, which produces up to
one Joule per shot at 30 Hz with more than 90 percent of the output
energy concentrated within a gain-switched pulse of 50 nsec fwhm. A
tunable pulsed dye laser probe is aligned to counterpropagate with the
IR pump. For LIF, fluorescent emission is collected by an F.1 lens-
mirror combination and dispersed by a 0.125 M monochromator for photomultiplier detection. For MPI, the probe laser is focussed. Ions are
collected by an electrostatic lens system (which is polished to serve
as the LIF-mirror noted above) and detected by a particle multiplier.
A laboratory microcomputer acquires the individual signal (LIF or MPI)
and measured pump-probe delay for each pair of laser pulses.

RESULTS AND DISCUSSION

That the reaction observed is really CH_2 elimination from the primary position on the heterocycle is confirmed by isotopic substitution; CH_3CHCD_2O yields only $^1CD_2$.[6] Isotope specificity, however, fails to answer the question of dynamical pathway. To help decide whether $^1CH_2$ elimination is concerted or the sequential product of ring opening followed by C-C (or C-O) scission, we have looked carefully at the spectrum of product states produced by this reaction.

Typical LIF excitation spectra for propylene oxide are shown below.

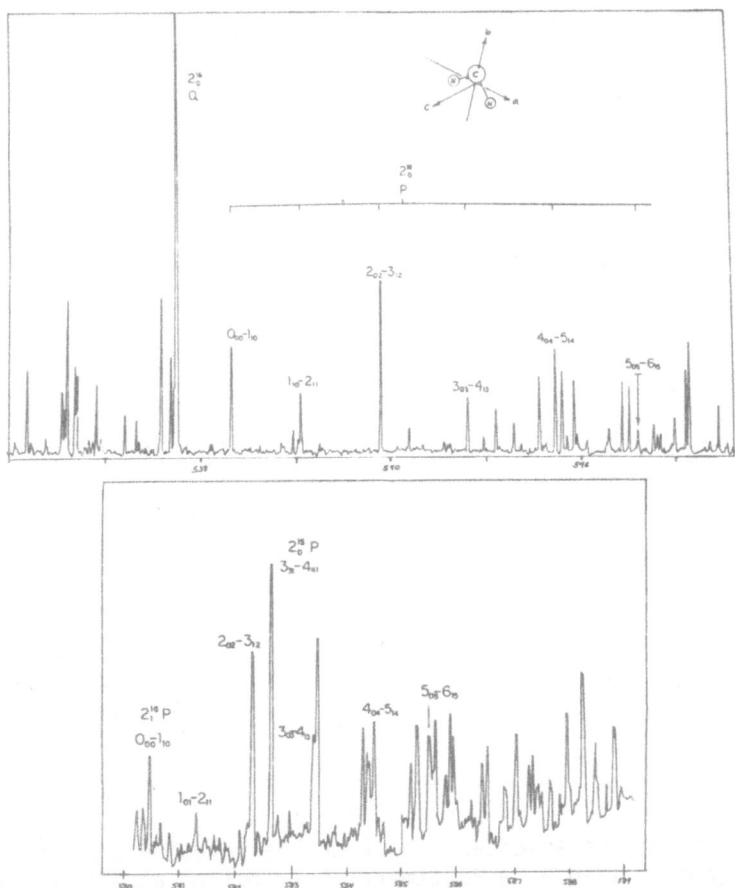

Those found for butyl and octyl epoxide are very similar. Rotational lines in the top frame are labeled with reference to a progression of initial state rotational quantum numbers $K_a = 0$; J, $K_c = 0$, 1, 2, 3, 4, 5, where J refers to total angular momentum and K_c its projection on the perpendicular axis. These transition intensities thus reflect the population distribution in what might be termed a boomerang mode of rotation.

At lower signal to noise, the bottom frame shows the same rotational sequence for a transition originating from the first excited bending vibration. In both cases spectra are completely unsaturated so that intensities are well related to populations by available Hönl-London factors. [9]

A plot of population per state versus rotational energy for the top spectrum above is shown below.

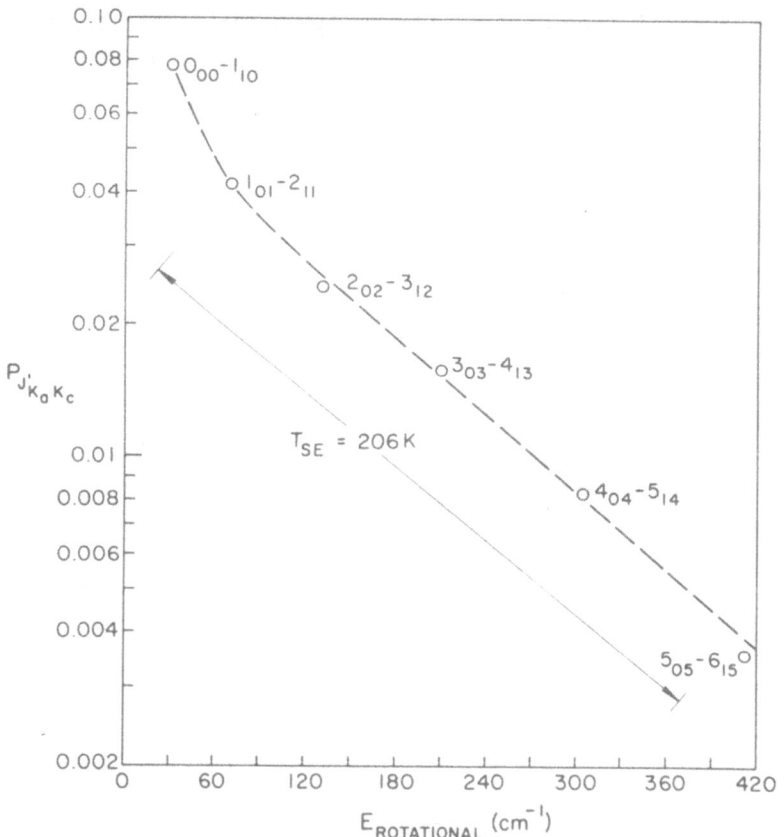

As a matter of convenience, note that the relative populations are approximately described by a Boltzmann distribution at approximately 200 K. This behavior, including both the approximate temperature and the deviation upward of the lowest rotational state, is common to all side-chain carbon numbers we have studied and virtually independent of CO_2 laser pulse energy. These results suggest that the decomposition dynamics are a local function of the properties of the heterocycle, and that in the exit channel, these local dynamics favor little CH_2 rotational motion perpendicular to the COC plane. This is most consistent with concerted molecular elimination.

Data reflecting the time-of-flight distribution of recoiling CH_2 and CH_3CHO fragments of propylene oxide decomposition are presented below.

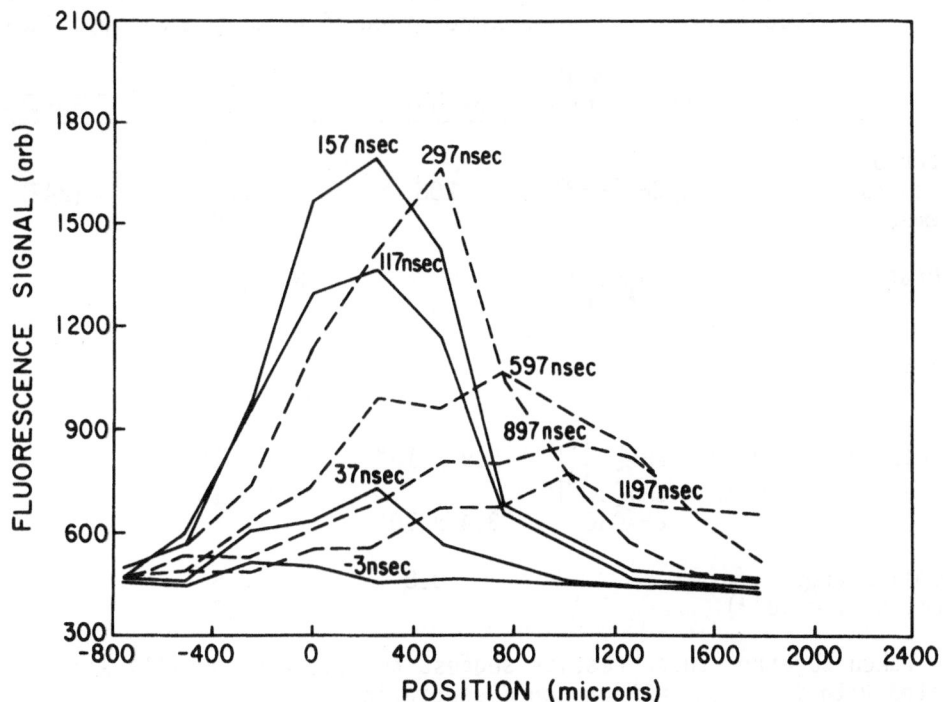

These curves show the spatial distribution of CH_2 along the axis of the beam at various delay times after the CO_2 laser pulse. The lowest curve, at 37 nsec, was taken while the CO_2 laser was still on. It gives a snapshot of the spatial distribution of CH_2 as a convolution of the CO_2 focal diameter (0.1 mm) and the finite width of the probe (0.5 mm). With increasing time, this instantaneous profile moves downstream with the beam velocity, and broadens due to the distribution of parent initial velocities combined with the vectorial distribution of recoil velocities. By knowing the initial velocity distribution for our expansion conditions we can estimate the kinetic energy added by fragment recoil.

The same technique can be applied to the CH_3CHO fragment using MPI for detection. This provides a confirmation which has even higher resolution because the probe diameter is smaller (focussed to 30 μm). We have done this, and the laboratory velocities measured for CH_2 and heavier CH_3CHO are mutually consistent with momentum conservation, yielding a confirmed estimate of the energy deposited in center-of-mass fragment recoil. The results obtained, as summarized below together with numbers characterizing the other sampled dynamical degrees of freedom, show an average recoil energy that is a mild function of IR laser fluence, but small compared with the energy of an IR photon.

TABLE I. Overview of energy disposal dynamics in the infrared laser induced unimolecular decomposition of primary alkyl epoxides

	Fragment [transition]	Parent propyl-	butyl-	octyl-
Rotational dynamics (temp, K)	$CH_2[2_0^{16}]$	206	209 229	(247)
Vibrational	$CH_2[\nu_2]$ (1424)		$N_1/N_0 = 0.25$	
Translational velocities (lab, cm sec^{-1})				
Fluence 630 mJ	$CH_2[2_0^{16}]$	5.9 x 10^4	5.2 x 10^4	
230 mJ	$CH_2[2_0^{16}]$		3.2 x 10^4	
	CH_3CHO	3.4 x 10^4		
Recoil Energy (com, kcal mole^{-1})		0.8	0.6 0.2	

Taken together these results suggest the picture normally associated with infrared multiphoton dissociation.[10] An ensemble of molecules is excited by sequential photon absorption to levels at some small energy above the lowest dissociation threshold. There decomposition takes place at a rate well determined by RRKM theory.[11] At high pumping intensities the process is characterized by a dynamic competition between dissociation and further up-pumping. Though the point in energy at which these processes become competitive depends on the cross section for photon absorption and molecular complexity (as it effects unimolecular decay rate), the natural assumption seems to have developed that IRMPD is a near-threshold process.

This question of the precise energy distribution of the reacting molecules presents difficulties for the interpretation of IRMPD results for unimolecular kinetics purpose: The confirmation of rate models and even the ordering of thresholds for observed reactions requires some knowledge of the distribution of energy in the field driven ensemble. The fullest understanding of the observed energy disposal dynamics also requires an idea of the average excess energy available in the excited parent. Experiments to date give only indirect information on this distribution. Purely spectroscopic pump-probe investigations measure internal populations in small-molecule fragment degrees of freedom.[12]

Molecular beam scattering experiments infer internal energy from recoil velocity distributions.[10]

By both of these conventional measures our alkyl epoxide systems appear to be excited little above threshold. However, this not-unconventional view of the excitation distribution differs substantially from that provided by another more direct measure of the vibrational energy content of our reacting molecules, the unimolecular decay lifetime. The figure below shows a waveform for $^1CH_2$ production from propylene oxide, which is measured by scanning the delay of the fluorescence probe relative to a t=0 mark furnished by the leading edge of the IR pump pulse.

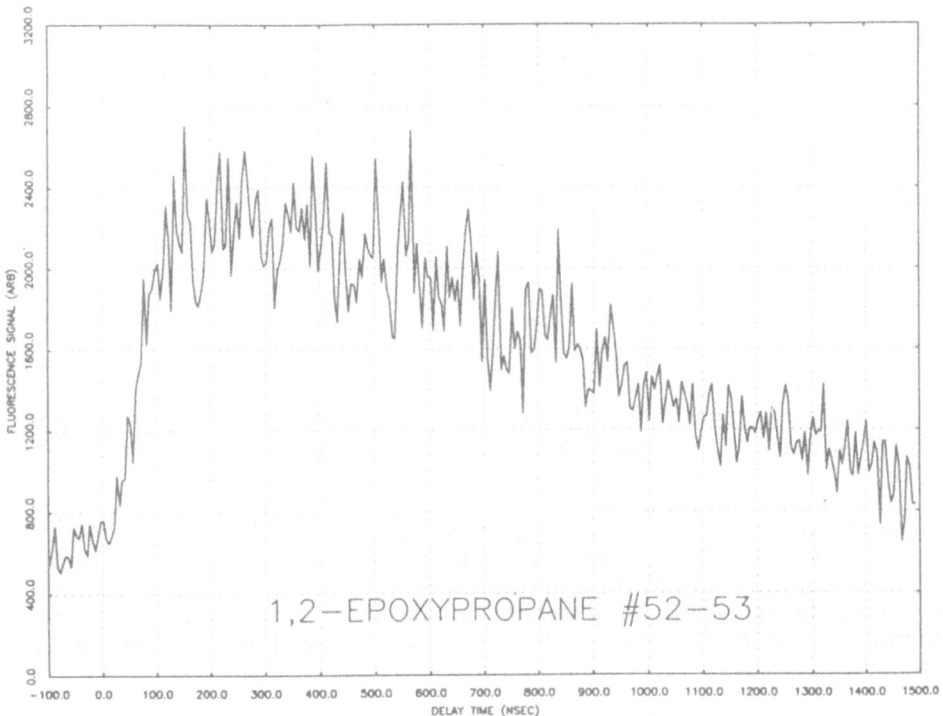

These measurements are jitter-free; a time interval counter measures the precise delay for every pair of pump/probe pulses. The pump pulsewidth is 50 nsec with a rise time of about 20 nsec. The formation rate of CH_2 product appears to fall in this same range. Thus, limited by our instrumental resolution, the lifetime of propylene oxide under these conditions is 20 nsec or less. On the basis of RRKM theory, for a molecule of this size to have a lifetime this short requires an excess energy of 90 kcal mole^{-1} or 30 IR photons.[11]

A similar risetime is shown below for butyl epoxide.

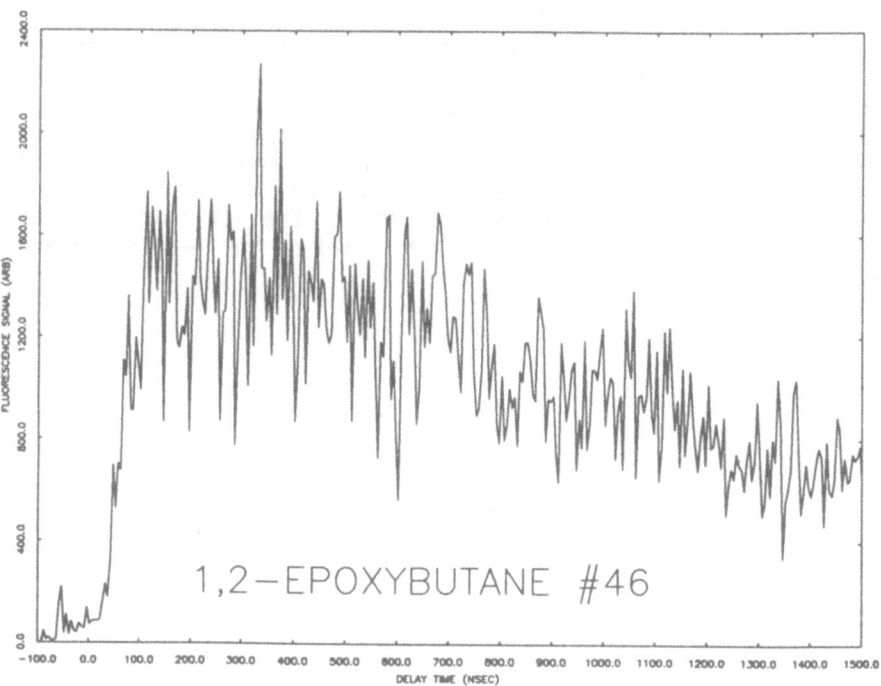

1,2—EPOXYBUTANE #46

In this case, by conventional application of statistical rate theory, the rise time observed requires at least 120 kcal mole[-1] above threshold.

These waveforms were recorded under laser power conditions identical to those which produced the spectra and velocity distributions described above. Apparently, either the lifetime of laser prepared alkyl epoxide is anomolously short, or the cold rotational and translational CH_2 product distributions observed issue from very hot molecules.

The standard theoretical models are not much help in resolving these two possibilities. The fragments are colder than would be expected from phase space theory applied to reactants with the excess energy required by RRKM theory.[5,13] These are large molecules, however, and such estimates depend precisely on how one counts states. This level of excitation can be made consistent with the usual rate-equations description of the pumping dynamics, but only by invoking a very large IR absorption cross section.

At this point the data appear to support either of two conclusions:

1) Infrared excitation and unimolecular decomposition proceed

statistically to produce CH_2, on the timescale observed, from a very hot reactant distribution. Dynamical effects in the exit channel contribute, perhaps, to cool the measured rotational and translational product state distributions. Cross sections for infrared absorption by vibrationally excited molecules must be as large as optical cross sections to support this high level of excitation. Or:

2) Infrared cross sections are more normal (like small-signal IR cross sections or smaller), producing a parent excitation distribution more consistent with observed small fragment rotational and translational energies, but parent molecules decompose much faster than statistically predicted.

To resolve these possibilities we must know directly the total energy absorbed per molecule. A more complete census of the energy in the products would give us this information. By far the greatest number of unsampled degrees of freedom lie in the aldehyde product. Resonant multiphoton ionization spectra give some information, and work to refine these diagnostics is continuing.

ACKNOWLEDGMENT

This work was supported by the U. S. Army Research Office. Acknowledgment is gratefully made to the Department of Defense for a DOD-University Research Instrumentation Grant.

REFERENCES

1. M. A. Schroeder, Critical Analysis of Nitramine Decomposition Data: Product Distributions from HMX and RDX Decomposition, Technical Report BRL-TN-2659, 1985; M. A. Schroeder, Critical Analysis of Nitramine Decomposition Data: Activation Energies and Frequency Factors for HMX and RDX Decomposition, Technical Report BRL-TR-2673, 1985.

2. B. H. Rockney and E. R. Grant, Chem. Phys. Lett. 79, 15 (1981).

3. B. H. Rockney and E. R. Grant, J. Chem. Phys. 77, 4257 (1982).

4. B. H. Rockney, G. E. Hall and E. R. Grant, J. Chem. Phys. 78, 7124 (1983).

5. B. H. Rockney and E. R. Grant, J. Chem. Phys. 79, 708 (1983).

6. T. E. Adams, Ph.D. Dissertation, Cornell University, 1986.

7. K. C. Frisch, ed., Cyclic Monomers, Wiley, New York, 1972.

8. M. H. Aley, J. A. Bowen and C. A. Glass, U. S. Naval Weapons Center, China Lake, CA, U. S. Patent 4,273,04

9. A. J. Grimley and J. C. Stephenson, J. Chem. Phys. <u>74</u>, 447 (1981).

10. P. A. Schulz, Aa. S. Sudbo, D. J. Krajnovich, H. S. Kwok, Y. R. Shen and Y. T. Lee, Ann. Rev. Phys. Chem. <u>30</u>, 379 (1979).

11. P. J. Robinson and K. A. Holbrook, <u>Unimolecular Reactions</u>, Wiley, New York, 1972.

12. F. F. Crim, Ann. Rev. Phys. Chem. <u>35</u>, 657 (1984).

13. I. Nadler, M. Noble, H. Reisler and C. Wittig, J. Chem. Phys. <u>82</u>, 2608 (1985).

COHERENT AND SPONTANEOUS RAMAN SPECTROSCOPY IN SHOCKED AND UNSHOCKED LIQUIDS*

S. C. Schmidt;   D. S. Moore;   D. Schiferl;   M. Chatelet;**
T. P. Turner; J. W. Shaner; D. L. Shampine; W. T. Holt
University of California
Los Alamos National Laboratory
PO Box 1663
Los Alamos, NM, USA

ABSTRACT. Coherent and non-coherent Raman spectroscopy is being used to study the structure and energy transfer in molecular liquids at high pressures. Stimulated Raman scattering, coherent anti-Stokes Raman scattering, and Raman induced Kerr effect scattering measurements have been performed in liquid benzene and liquid nitromethane shocked to pressures up to 11 GPa. Frequency shifts were observed for the 992 cm^{-1} ring stretching mode of benzene and the 920 cm^{-1} CN stretching mode of nitromethane. Results of these dynamic experiments are compared to spontaneous Raman scattering measurements made in a high temperature diamond anvil cell. Also, a picosecond infrared pump/spontaneous anti-Stokes Raman probe experiment is being used to measure CH stretch vibrational relaxation times in liquid halogenated methanes statically compressed to a few tenths GPa.

1.   INTRODUCTION AND OBJECTIVES

Presently most models of explosive and shock induced chemical behavior treat the medium as a continuum[1,2] that chemically reacts according to either a pressure dependent or Arrhenius kinetics rate law. One or more parameters are used to incorporate the global chemical behavior, hydrodynamic phenomenology and effects of material heterogeneity. In the past few years, several studies[3-10] have been started that attempt to improve the methodology by defining the continuum, not as a single component, but as one that incorporates ideas such as hot spots, voids or multicomponents. However, in all of these studies essentially no effort is made to incorporate any of the microscopic details of the shock-compression/energy transfer and release phenomenology that constitutes the detonation or reactive process.

* Work supported by the United States Department of Energy.
** Laboratory of Molecular Interactions and High Pressure, C.N.R.S., Villetaneuse, France.

P. M. Rentzepis and C. Capellos (eds.), Advances in Chemical Reaction Dynamics, 425–454.
© *1986 by D. Reidel Publishing Company.*

Ideally, for descriptions of reactive processes, we would like to treat the continuum as a mixture of pure components and incorporate changes in molecular structure resulting from shock compression, disequlibria due to shock compression, energy transfer from the hydrodynamic mode into the molecular internal degrees of freedom and the subsequent microscopic reaction history, energy release, and product formation. While such a goal may appear overly ambitious we feel that by using some of the diagnostics, particularly fast optical techniques, that have become available in the past few years, progress can be made toward understanding certain facets of this objective. For example, spontaneous Raman spectroscopy has already been used to make temperature estimates of shocked explosives[11-13] and examine the structure of shock-compressed materials.[14-15] In our own work, coherent Raman scattering techniques have been used to measure vibrational frequency shifts in benzene and nitromethane shock-compressed to pressures just below those where chemical reaction is expected.[16-22] Initial indications suggest the prospects for extending these measurements into the pressure regions where chemical reaction occurs are good.

Figure 1 depicts some of the consequences of the shock-compression

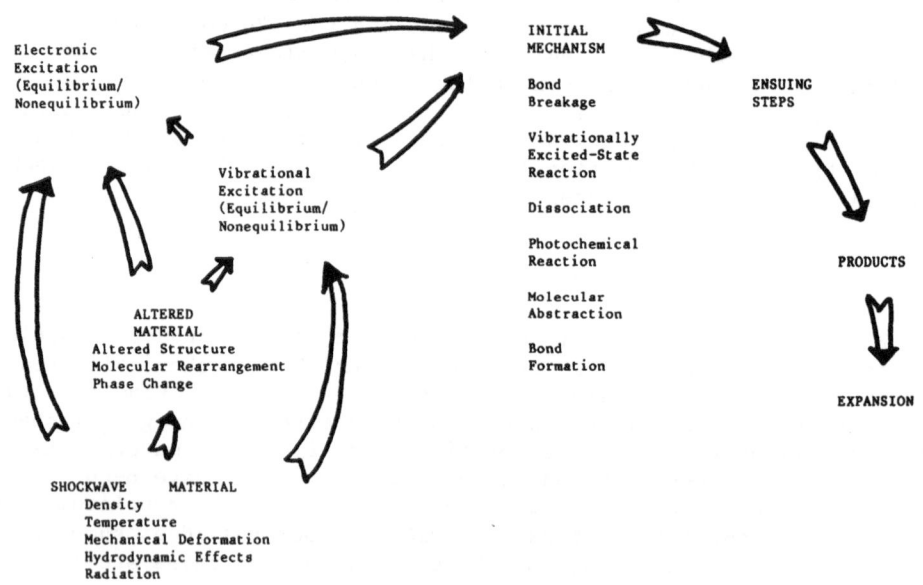

Fig. 1.   Condensed-phase molecular energy transfer.

of molecular materials.   In addition to the macroscopic continuum effects expected (e.g. hydrodynamic flow, density and temperature increases), molecular systems, because of vibrational and electronic energy levels that possibly lie close to the ground state, are expected to readily undergo a shock induced transfer of energy to these internal degrees of freedom. Under shock-compression, the molecular structure and hence the intramolecular and intermolecular forces will be altered considerably, consequently the energy transfer rates and mechanisms may be dramatically different from those expected on the basis of either extrapolation from ambient conditions or thermodynamic equilibrium. Depending on the vibrational and electronic relaxation rates and mechanisms in the high density/high temperature fluid, the excited states could have a nonequilibrium population density. Different authors[23-25] have proposed different initial steps for the chemical reaction schemes in detonating explosives. However, definitive supporting experiments have not been performed. The ensuing microscopic chemical reactions involving energy release and product formation also require experimental study.

The objective of our work has been two fold; (1) to determine the molecular structure and identify chemical species in unreacting and reacting shock-compressed molecular systems and (2) to study the effect of pressure and temperature on condensed phase energy transfer. Also, we would like to identify the unique features of a shock wave which contribute to the energy transfer processes. Achievement of these goals would contribute significantly to understanding the initial mechanisms governing shock-induced chemically reacting molecular systems and possibly to the steps controlling product formation. Two experiments are being employed in the pursuit of these objectives. A two-stage light gas gun is being used to dynamically shock-compress molecular liquids to pressures where chemical reaction occurs. The high density/high temperature fluid is then probed using coherent Raman scattering techniques. In the second effort which is still in the construction phase, a picosecond pump/spontaneous anti-Stokes Raman scattering probe experiment will be used to measure vibrational relaxation rates in liquids statically compressed using high pressure cells.

2. EXPERIMENTAL CONSIDERATIONS

Prior to discussing our experimental studies and results to date, several problems associated with conducting condensed-phase shock-wave experiments will be reviewed. These difficulties have historically limited the ability to conduct experiments in the adverse conditions through and immediately behind the shock-front and for our studies strongly governed the experimental techniques used.

For many materials shock waves are believed to be of the order of 1 μm or less in thickness.[26-28] The passage time through the front of a 1 μm-thick shock whose velocity is 5 km/s is thus of the order of 200 ps or less. Hence, if we desire to temporally and spatially resolve a measurement through a shock-front (5 data points), the diagnostic technique selected must be capable of spatial and temporal resolutions of 0.2 μm and 40 ps, respectively. Condensed-phase chemical reaction times could be of the order of 1 ps, thus necessitating

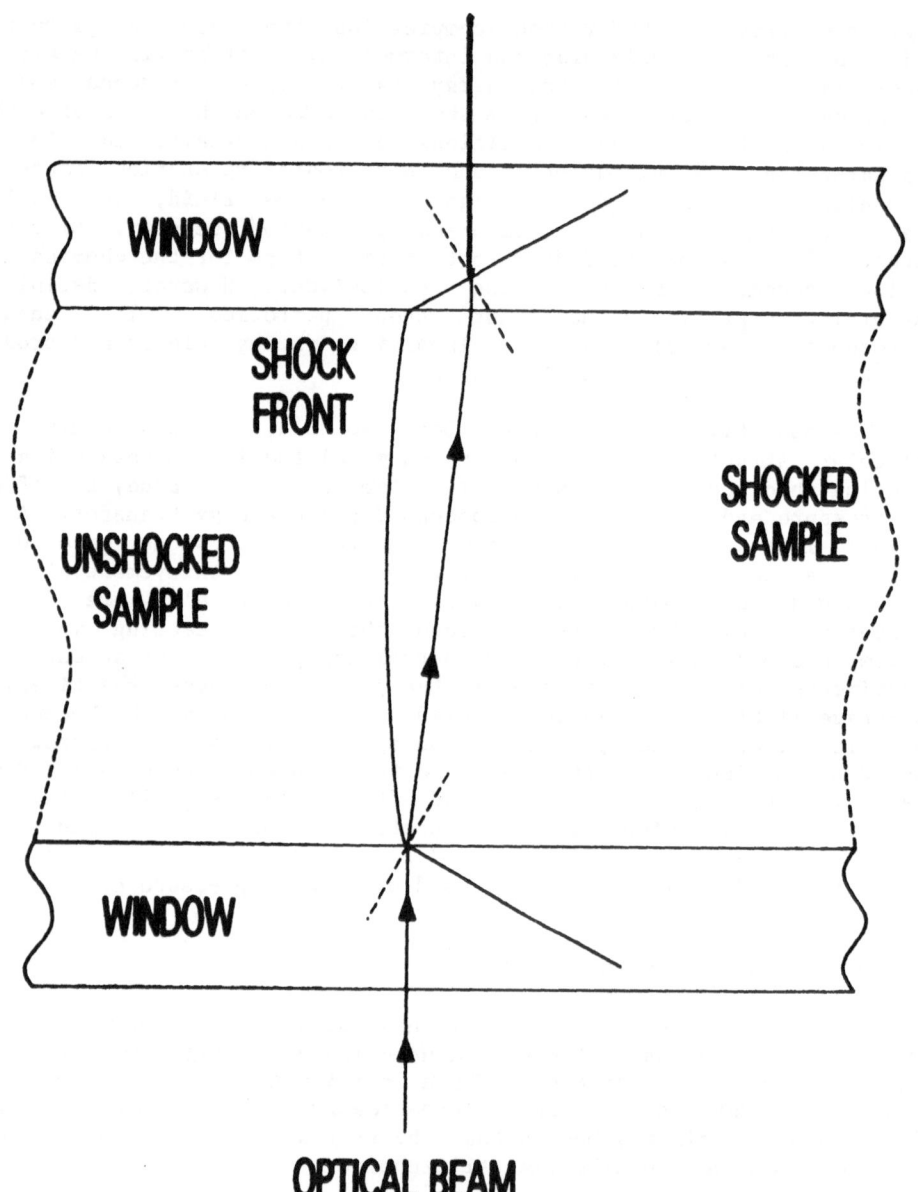

Fig. 2. Refractive effects of shock wave on optical beam.

even better temporal resolution. However, if all that is desired is to resolve features in the few mm long region behind the shock-front where relaxation and reaction processes may occur, then these requirements are drastically reduced.

Optical techniques offer some potential for achieving measurements within these stated limitations. However, with such methods some additional complications arise. Many materials are opaque or become opaque when shock-compressed. Consequently, the use of optical diagnostic techniques is limited to a few select materials primarily for phenomenological studies. Such studies may, however, have tremendous potential when used in conjunction with other techniques for determining phenomenology of shock-compressed materials. Two other difficulties inherent in optical shock-wave diagnostic techniques are the changes in material refractive index that accompany the density changes characteristic of shock waves and the possibility of photochemistry induced by the optical probes. Figure 2 shows the path deviation that occurs when an optical beam is passed through a hypothetical shock-compressed system. The trailing shock wave near the sample boundaries tends to bend the optical beam away from the shock-front thus making prediction of the expected optical path difficult. Any shock-front curvature will compound this difficulty. If the shock velocity in the

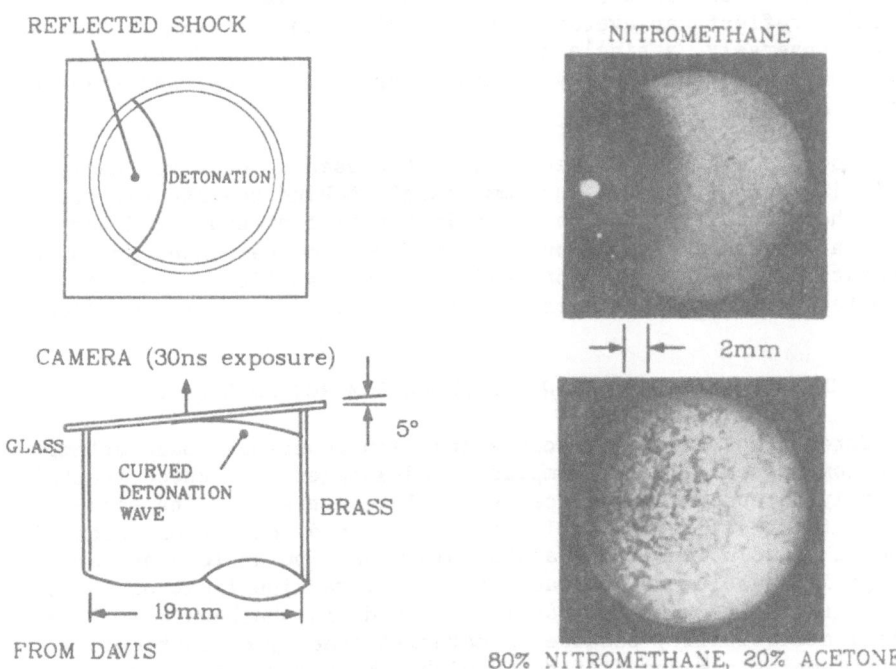

Fig. 3. Detonation wave microstructure for nitromethane and a nitromethane/acetone mixture.[29]

windows is greater than in the sample, additional complications could arise from the effect of the more complex wave structure on window transmission. Many molecules undergo photochemical reactions when exposed to light, particularly that in the ultraviolet region of the spectrum. If these reactions are fast compared to the characteristic time of the optical diagnostic, measurements could include the effects of both the shock stimulus and the photochemical reaction.

Measurements made using inhomogeneous samples often are averages over the nonuniformities and consequently do not reflect the details of the microstructure. For materials like granular explosives, the inhomogeneous nature is readily apparent and experiments are interpreted accordingly. For samples thought to be homogeneous, ambiguities can arise. For example, Fig. 3 depicts two image-intensifier-camera pictures[29] of the shock-front of detonating nitromethane and an 80% nitromethane/20% acetone mixture. These pictures show that microstructure exists in the vicinity of the shock front even in liquids, which are often thought to be homogeneous. Also, nothing is known about microstructure in the region immediately behind the shock front. When performing experiments on nitromethane or similar substances, especially experiments utilizing optical techniques where a spatial resolution of tens of microns is desired, one must be aware that results may actually reflect an average over a smaller characteristic microstructure. Conversely, a single measurement with spatial resolution smaller than the microstructure may be misinterpreted as representative of the average material.

Shock recovery experiments are often used to observe chemical and physical changes through and immediately behind the shock-front. However, these changes occur not only in the high pressure and temperature region at the shock front, but also in the somewhat lower pressures and temperatures of the expansion region. The inability to separate these two effects makes the interpretation of these experiments difficult.

3.   COHERENT RAMAN SCATTERING IN SHOCK-COMPRESSED LIQUIDS

Three coherent Raman scattering techniques have been attempted in shock compressed liquid samples. Advantages of these techniques, primarily because of large scattering intensities and the beam-like nature of the scattered signal, are increased detection sensitivity, temporal resolution limits approaching laser pulse lengths, and possible spatial resolution approaching the diffraction limit of the optical components. As with all optical methods in shock-wave applications, optical accessibility because of material opacity or particulate scattering remains a major difficulty with coherent Raman scattering.

Backward-stimulated Raman scattering (BSRS) has been observed in shock-compressed benzene up to pressures of 1.2 GPa.[17] Stimulated Raman scattering[30,31] (Fig. 4) occurs when the incident laser intensity in a medium exceeds a threshold level and generates a strong, stimulated Stokes beam. The threshold level is determined by the Raman cross-sec-

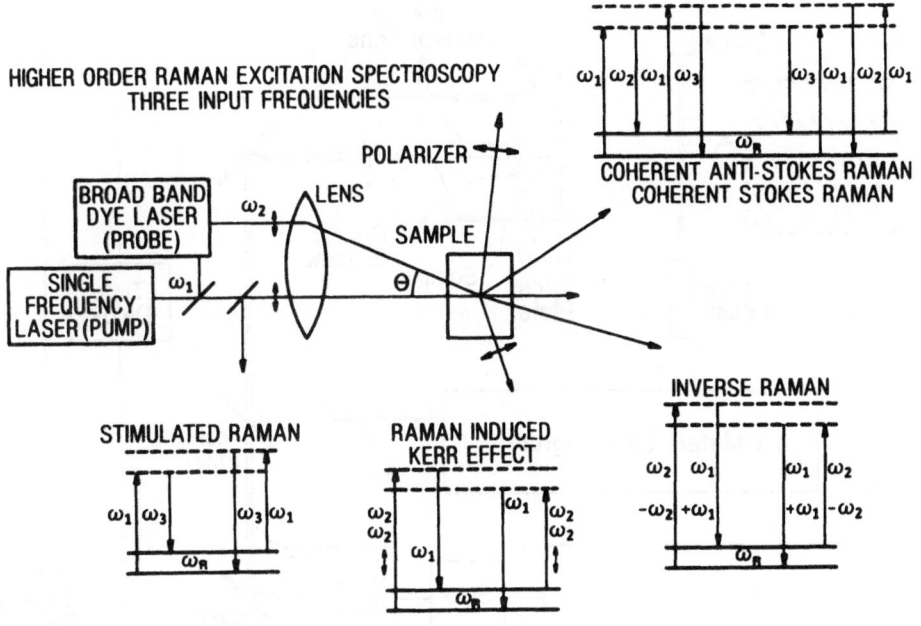

Fig. 4.   Coherent Raman scattering techniques.

tion and linewidth of the transition and by the focusing parameters of the incident beam. Typical threshold intensities are ~ 10-100 GW/cm^2. Figure 5 illustrates the arrangement used for the backward stimulated Raman scattering experiment.

An aluminum projectile of known velocity from a 51-mm-diam, 3.3-m-long gas gun impacted an aluminum target plate producing a shock wave which ran forward into a 7.5 to 8-mm-thick reagent grade benzene sample (Mallickrodt, Inc.).   Standard data reduction techniques[32] using published shock-velocity/particle-velocity data[33] were used to deter- mine the state of the shock-compressed benzene. Experiment design was greatly facilitated using the MACRAME one-dimensional-wave propagation computer code.[34]

A single 6-ns-long frequency-doubled Nd-doped yttrium aluminum garnet (Nd:YAG) laser pulse was focused using a 150-mm focal length lens through the quartz window to a point in the benzene 2 to 6 mm in front of the rear sample wall. The high intensity of the laser at the focus, coupled with the presence of a large cross-section Raman active vibrational mode in the sample, produces gain in the forward and back- ward directions along the beam at a frequency that is different from the Nd:YAG frequency by the frequency of the active mode. The timing sequence was determined by the incoming projectile. Interruption of a HeNe laser beam, in conjunction with an appropriate time delay,

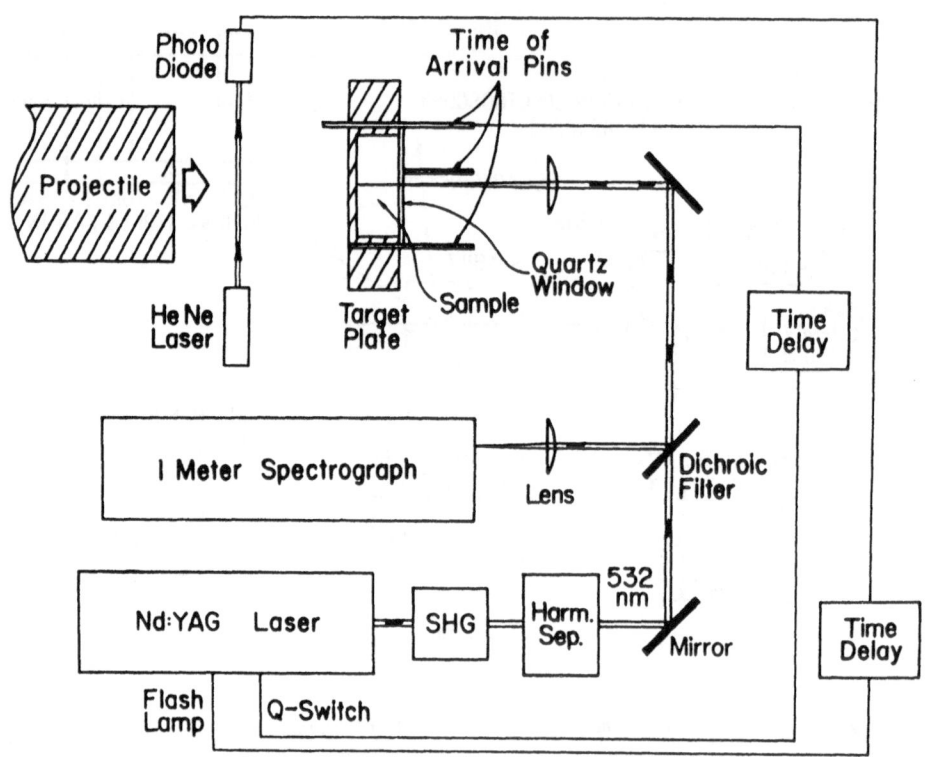

Fig. 5. Schematic representation of backward stimulated Raman-scattering experiment. SHG, second harmonic generator; Harm. Sep., harmonic separator. Sample, liquid benzene.

triggered the laser flash lamp approximately 300 µs prior to impact. A time-of-arrival pin activated just before impact and the appropriate time delay served to Q-switch the laser just prior to the shock wave striking the quartz window and after it was past the focal point of the incident laser light.

In liquid benzene, the ν_1 symmetric stretching mode[35] at 992 cm^{-1} has the lowest threshold for stimulated Raman scattering induced by 532-nm light, and was the transition observed in these experiments. As depicted in Fig. 5, the backward stimulated Raman beam was separated from the incident laser by means of a dichroic filter and was then focused onto the 10-µm-wide entrance slit of a 1-m Czerny-Turner spectrograph equipped with a 1200-grooves/mm grating blazed at 500 nm and used in first order. Figure 6 shows the resulting spectrogram for benzene shock-compressed to 0.92 GPa. The reflected incident laser line and the backward stimulated Brillouin-scattering line at 532 nm are observable, as are the backward stimulated Raman-scattering line from ambient benzene. The latter feature resulted as a consequence of

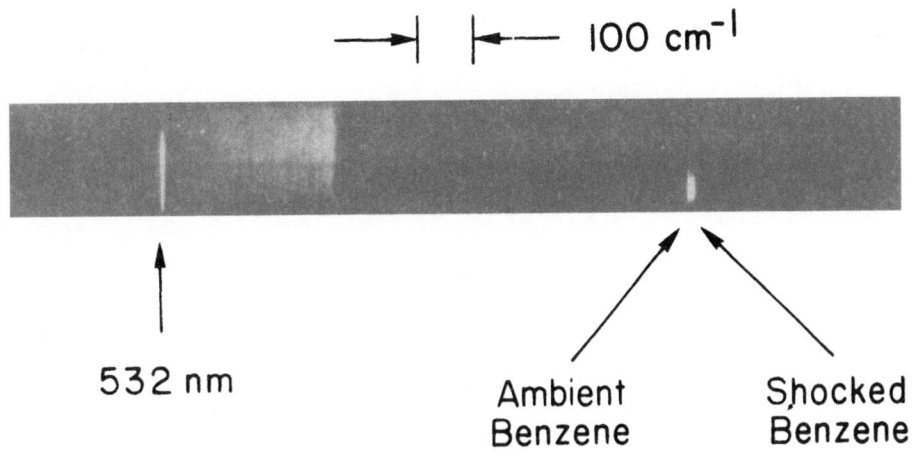

Fig. 6. Scattered light spectrogram for shock-compressed benzene.

the shock wave having passed only about two-thirds of the way through the sample, and hence a stimulated Raman signal was also obtained from the unshocked liquid.

The frequency shift of the Raman line has small contributions of approximately 0.1 cm^{-1} because the light crosses the moving interface between two media of different refractive indices and because of the material motion behind the shock wave.[36] Since these errors are considerably less than the experimental uncertainty of ± 0.5 cm^{-1} for the measured frequency shifts and are a small fraction of the shift due to compression, no attempt was made to correct the data for these effects.

Figure 7 gives the measured shift of the ν_1 ring-stretching mode vibrational wavenumber versus pressure of the shocked benzene. Observation of the ring-stretching mode at 1.2 GPa strongly suggests that benzene molecules still exist several millimeters behind the shock wave at this pressure, but does not, however, exclude some decomposition.[27,37,38]

Also depicted in Fig. 7 is the ring-stretching mode vibrational wavenumber shift measured for benzene isothermally compressed at temperatures between 24°C and 209°C with a diamond-anvil cell and techniques previously described.[39] Measurements of the phonon spectrum in the region 40-200 cm^{-1} were used to distinguish between benzene I, benzene II, and liquid benzene. The vibrational frequencies obtained from spontaneous-Raman-scattering measurements at 24°C with use of the 568.2-nm line of a krypton laser agree well with previous results for benzene.[35]

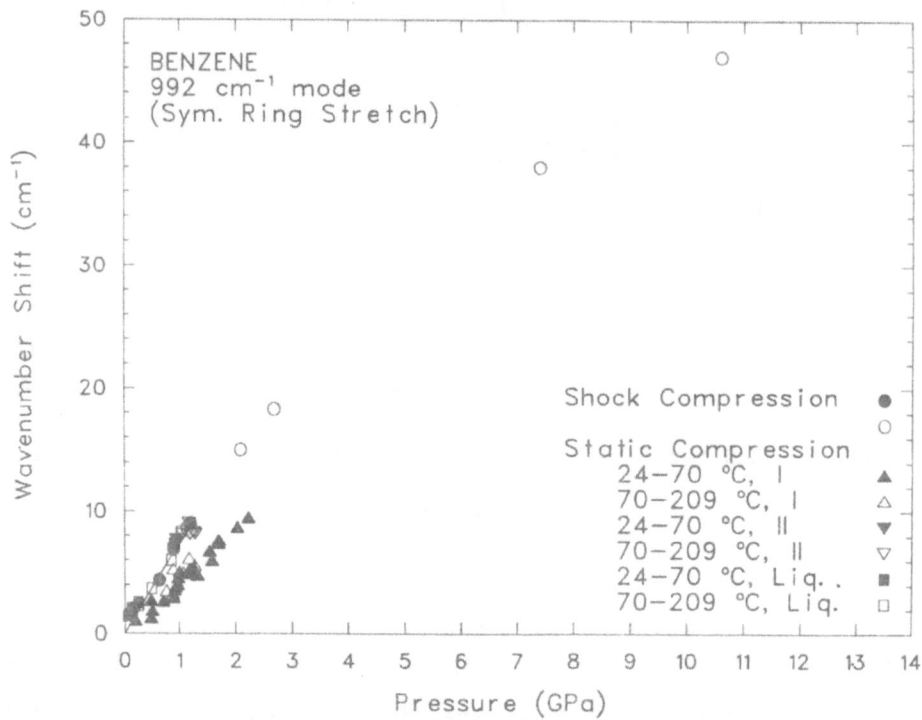

Fig. 7. Benzene ring-stretching mode vibrational frequency shifts (with respect to 992 cm⁻¹) vs pressure. The solid circles represent data obtained up to pressures of 1.2 GPa using the single stage gas gun and the open circles represent data obtained using the two stage light gas gun. The straight line is a fit of the shock-compressed data at pressures less than 1.2 GPa. The phase of the benzene during the diamond-anvil cell compression has been determined from phonon spectrum measurements. At 24 C benzene I was observed as a metastable phase above 1.2 GPa and benzene II was observed as a metastable phase below this pressure. Both spectrometers were calibrated with liquid benzene at room temperature.

At fixed pressure, no temperature shift was observed in these static measurements. The wavenumber shifts for the dynamic experiments agree well with the static data for either liquid benzene or benzene II, but differ substantially from those for benzene I. At pressures below the I-II-liquid triple point near 1.2 GPa,[40,41] the shocked benzene is therefore probably at temperatures high enough for it to be in the liquid state. At pressures near 1.2 GPa, the shock-compressed material could be either liquid or benzene II since both phases exhibit about the same magnitude of wavenumber shift for the ring-stretching mode, and the Hugoniot lies close to the phase boundary.

Beam intensities using BSRS are sufficiently large that film can be used as a detector. The large incident intensities required, however, can cause damage to optical components near focal points. Spatial and temporal resolution are determined by the confocal parameter of the focusing lens and the incident laser pulse duration. The BSRS technique also suffers because only certain molecules produce stimulated Raman scattering and of those molecules only the lowest threshold transition can be observed. Because of these limitations other coherent Raman scattering processes affording more experimental flexibility were attempted.

Coherent anti-Stokes Raman scattering (CARS)[42-45] (Fig. 4) occurs as four-wave parametric process in which three waves, two at a pump frequency, ω_p, and one at a Stokes frequency, ω_s, are mixed in a sample to produce a coherent beam at the anti-Stokes frequency, $\omega_{as} = 2\omega_p - \omega_s$. The efficiency of this mixing is greatly enhanced if the frequency difference $\omega_p - \omega_s$ coincides with the frequency of a Raman active mode of the sample. An advantage of CARS is that it can be generated at incident power levels considerably below those required for stimulated Raman scattering. However, since phase matching is required, possible geometrical arrangements are limited.

A schematic of the experimental apparatus used to perform reflected broadband coherent anti-Stokes Raman scattering (RBBCARS) in shock-compressed benzene and nitromethane is shown in Fig. 8. For pressures greater than 2 GPa, a two-stage light gas gun was used to accelerate a polycarbonate projectile with 4-mm-thick AZ31B magnesium or 2024 aluminum impactors to a desired velocity. The projectile struck an approximately 2.4-mm-thick 304-stainless-steel target plate producing a shock wave which ran forward into a 2.7-mm to 3.3-mm-thick benzene (or nitromethane) sample. Lower pressures were achieved using the previously described technique for backward stimulated Raman scattering. Stainless steel was chosen as the target plate because of previous experience and a series of reflectivity measurements which showed that polished steel would retain approximately 20 percent of its original reflectivity under shock compression at 11 GPa in the liquid sample (approximately 70 GPa in the stainless steel). This was necessary to reflect the CARS signal back out of the shock-compressed liquid. Reagent grade benzene (Mallinckrodt, Inc.) and commercial grade (Angus Chemical Co.) nitromethane were used. The state of the shock compressed material was determined as described above.

The timing sequence for the RBBCARS experiment was determined by the incoming projectile. A signal from three HeNe laser/photodiode detectors located in the barrel approximately 2.2, 1.2 and 0.7 m from the target, in conjunction with an appropriate time delay, triggered the laser flash lamps approximately 300 μs prior to impact. A time-of-ar-

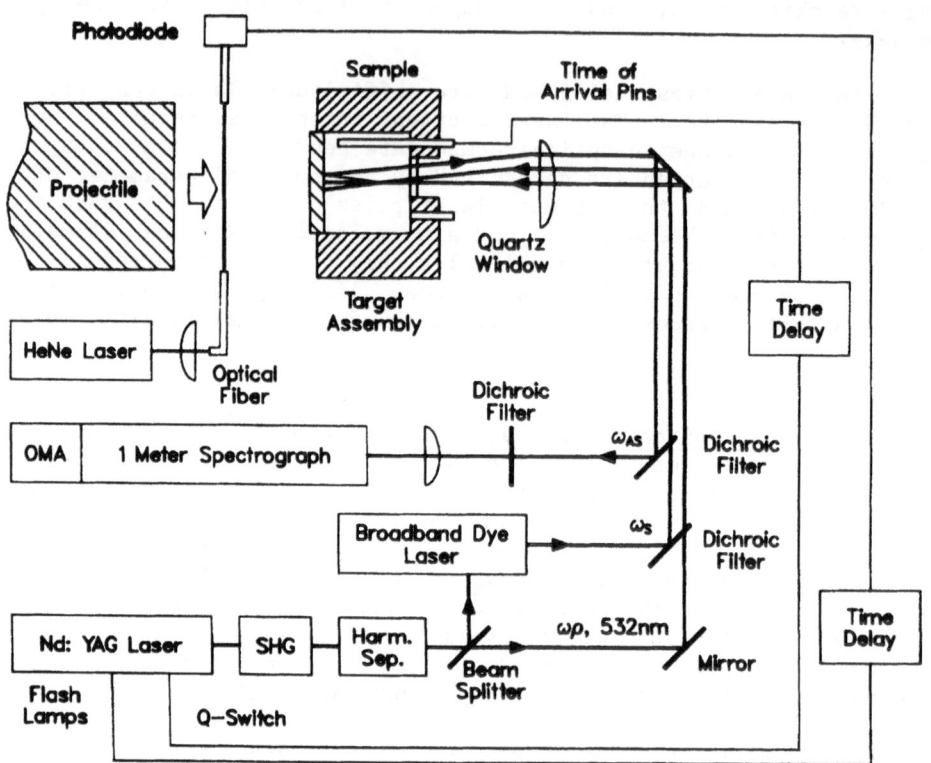

Fig. 8.   Schematic representation of the reflected broadband coherent anti-Stokes Raman scattering experiment.   SHG – second harmonic
generator; Harm. Sep. – harmonic separator; OMA – optical multichannel
analyzer; Sample – benzene and nitromethane.

rival pin activated just after the shock entered the liquid and another
time delay served to Q-switch the laser approximately when the shock
wave arrived at the quartz window.

Since the Raman frequencies of the shock-compressed materials are
not precisely known, and since we wish to produce CARS signals from
more than one mode or species, a broadband dye laser, with a bandwidth
equivalent to the gain profile of the dye, was used as the Stokes
beam.[46] A portion of the 6-ns-long frequency-doubled Nd:YAG laser
pulse was used to pump the dye laser. The dye laser beam was passed
through a Galilean telescope and sent along a path parallel to the
remaining pump laser towards the sample. The beams were focused and
crossed (with approximately 1-mm length of overlap) at a point
approximately 1 mm in front of the window using a previously described
technique.[47]   The beam crossing angle (phase-matching angle) was tuned
by adjusting the distance between the parallel beams using a turning
mirror on the pump laser beam. The CARS beam was reflected out of the
shocked sample by the highly polished front surface of the target plate

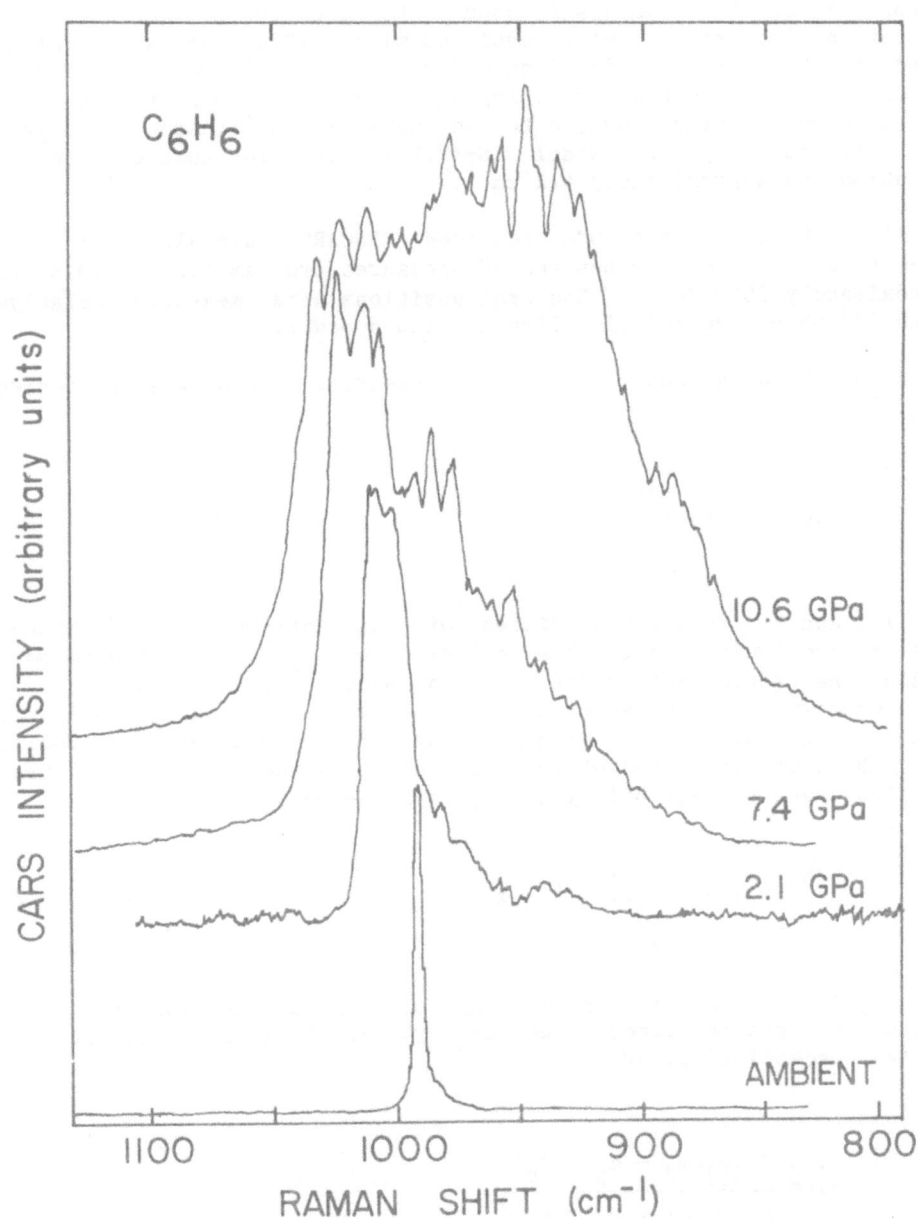

Fig. 9. RBBCARS spectra of ambient and shock-compressed benzene. The ambient peak position of the benzene is 992 cm^{-1}. Shock pressures are indicated. Wavelength calibration was done relative to the 253.652 nm Hg line in second order.

and along a path parallel to the two incoming beams. After being separated from the pump and Stokes beams using a long-wavelength-pass dichroic filter, the beam was focused onto the 100 μm-wide entrance slit of a 1-m spectrometer equipped with a 1200-grooves/mm grating blazed at 500 nm and used in first order. The signals were detected at the exit of the spectrometer using an intensified diode array (Tracor Northern Model TN-6133) coupled to an optical multichannel analyzer (OMA) (Tracor Northern Model TN-1710). The instrument spectral resolution was approximately 4.2 cm^{-1}.

Figure 9 shows the OMA recorded RBBCARS signals for the ring-stretching mode of benzene at pressures from ambient to 10.6 GPa (approximately 1000 °K).[48] Spectral positions were measured relative to the 253.652-nm Hg emission line in second order.

A preliminary analysis of the spectral lines was performed using:[44]

$$I_{as} \quad \alpha \quad \omega_{as}^2 |\chi^{(3)}|^2 \, I_s \tag{1}$$

where I_{as} and I_s are the intensities of the anti-Stokes and Stokes beams respectively. For this calculation, I_s was chosen to be a Gaussian that approximately fit the broad band dye laser profile. In future experiments, because of shot-to-shot variations and noise in the dye laser profile, a spectrographic record will be made of the profile for each shot and used directly to calculate the synthetic spectra. The third order susceptibility, $\chi^{(3)}$, is given by

$$\chi^{(3)} = \chi_{NR} + \sum_j \chi_j' + i \sum_j \chi_j'' \tag{2}$$

where χ_{NR} is the contribution from the nonresonant background and j is the sum over spectral lines. The real, χ_j', and imaginary, χ_j'', parts of the CARS susceptibility are

$$\chi_j' = \frac{\Gamma_j \chi_j (\omega_j - \omega_p + \omega_s)}{(\omega_j - \omega_p + \omega_s)^2 + \Gamma_j^2} \tag{3}$$

and

$$\chi_j'' = \frac{\Gamma_j^2 \chi_j}{(\omega_j - \omega_p + \omega_s)^2 + \Gamma_j^2} \tag{4}$$

respectively. ω_j, Γ_j and χ_j are the frequency, half amplitude half width (HWHM) and the peak amplitude of the corresponding spontaneous Raman scattered line. For this work no attempt was made to derive population densities (or temperatures) from χ_j using known Raman cross-sections. In fact, it may be necessary to re-determine the Raman cross-sections for the high densities characteristic of shock-compression.

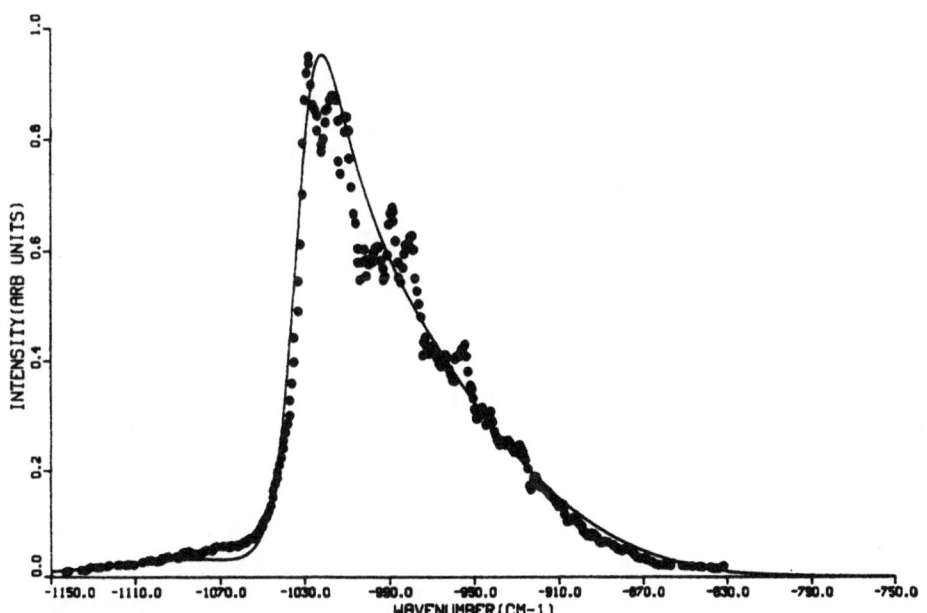

Fig. 10. Spectral fit of 7.4 GPa ring stretching mode of benzene. Wavelength calibration is with respect to the 253.652 nm Hg line in second order and the spectral slit width is 4.2 cm^{-1}.

Figure 10 depicts the experimentally measured and calculated spectra for the benzene ring stretching mode at 7.4 GPa. The intensity of both spectra has been normalized to a peak amplitude of 0.95. The structural features that appear in the measured spectra are thought to result from the noise in the broad band dye laser. An initial analysis

of the spectral shape at 10.6 GPa is dramatically different and requires two spectral lines to fit the measured profile. It is also
possible that a better fit may be obtained using an inhomogeneous
broadening component. These will be discussed in a future publication.

The frequency shifts estimated for the 7.4 GPa line and the lines

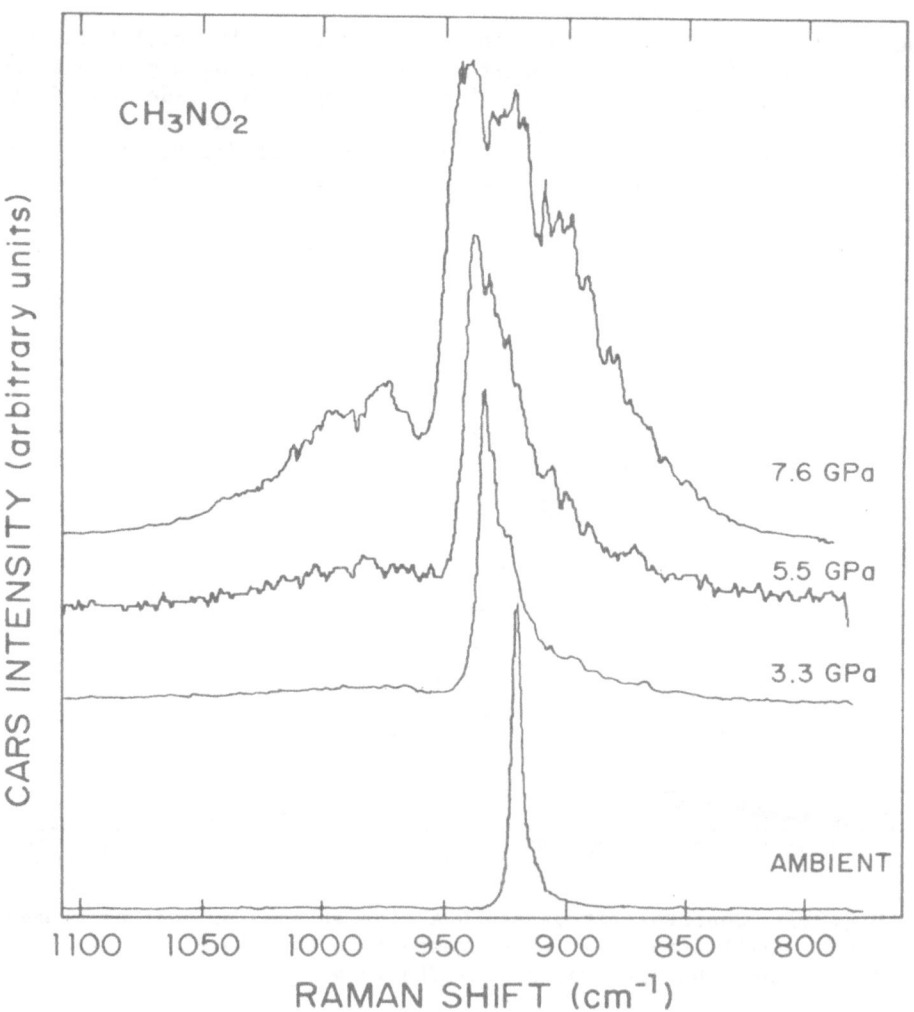

Fig. 11. RBBCARS spectra of ambient and shock-compressed
nitromethane. The ambient peak position is 920 cm^{-1}. Shock pressures
are indicated. Wavelength calibration was done with respect to the
253.652 nm Hg line in second order.

at other pressures are depicted in Fig. 7 along with the frequency shifts determined from BSRS measurements discussed previously. The results show an initial linear change of the frequency shift with pressure and then a weakening of this dependence as the region near 13 GPa is approached. Previous work[27,37,38,49,50] suggests chemical reaction occurs at these pressures. A plot of frequency shift versus volume change shows a nonlinear dependence at all pressures.

The RBBCARS spectra for the CN stretching mode of nitromethane at pressures from ambient to 7.6 GPa (approximately 950 °K)[51] are shown in Fig. 11. The existence of the CN mode at microsecond times after shocking implies that decomposition of the nitromethane has not oc-curred as has been observed for times of tens of microseconds[52] and in static high temperature/high pressure studies.[53] Measurements are presently being extended to higher pressures where nitromethane is thought to be reactive for very short shock run distances.[12,48,54-56]

Synthetic spectra were obtained for the nitromethane CN stretch mode using the procedure previously described for benzene. Figure 12 shows the experimentally measured and calculated spectra for 5.5 GPa normalized to a peak intensity of 0.95. At 5.5 GPa and for the lower pressure nitromethane spectra, this preliminary analysis suggests that the spectral signatures are much better matched using two spectral

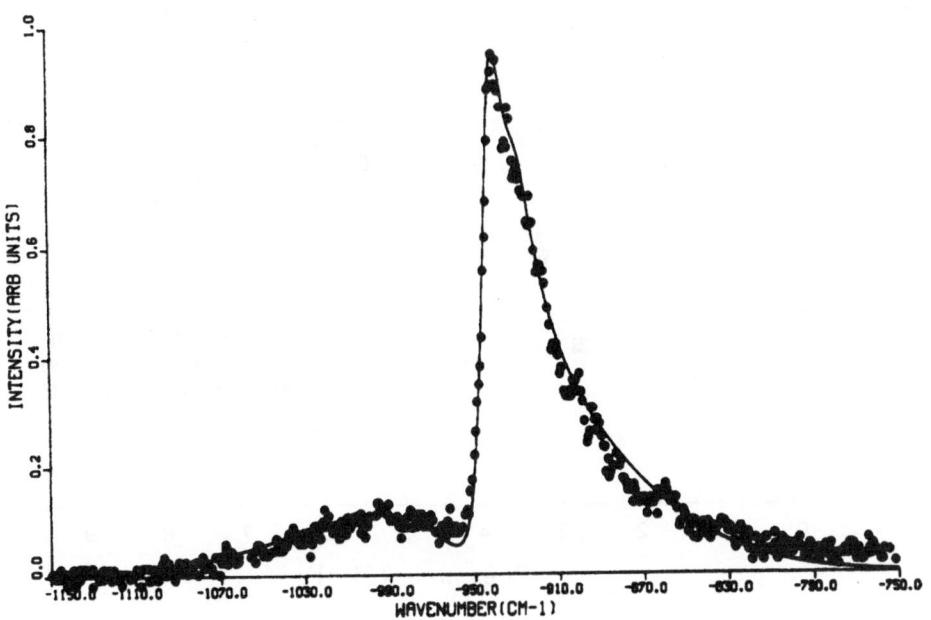

Fig. 12. Spectral fit of 5.5 GPa CN stretching mode of nitromethane. Wavelength calibration is with respect to the 253.652 nm Hg line in second order and the spectral slit width is 4.2 cm^{-1}.

lines separated by approximately 10 cm^{-1}. This second line has been observed previously in static high pressure Raman spectra.[57] The fitting at 7.6 GPa required a minimum of four or five spectral lines to represent the measured data. Discussion of these results will be withheld for a future publication when more accurate synthetic spectra can be calculated using measured dye laser profiles.

Figure 13 shows the estimates for the Raman frequency shifts of the more intense spectral feature versus pressure using the analysis indicated above. Also depicted are the frequency shifts measured for the CN stretch mode of solid nitromethane using a diamond anvil high pressure cell[58] and the spontaneous Raman measurements for nitromethane shocked to pressures of 5 GPa.[12] It is noted that our results obtained for the shock-compressed material do not differ significantly from the Raman shifts obtained for solid nitromethane. At present, we do not have an explanation for the difference between our results and those obtained by Delpeuch and Menil[12] using spontaneous Raman scattering. Earlier results[11] by these authors agreed more closely with our

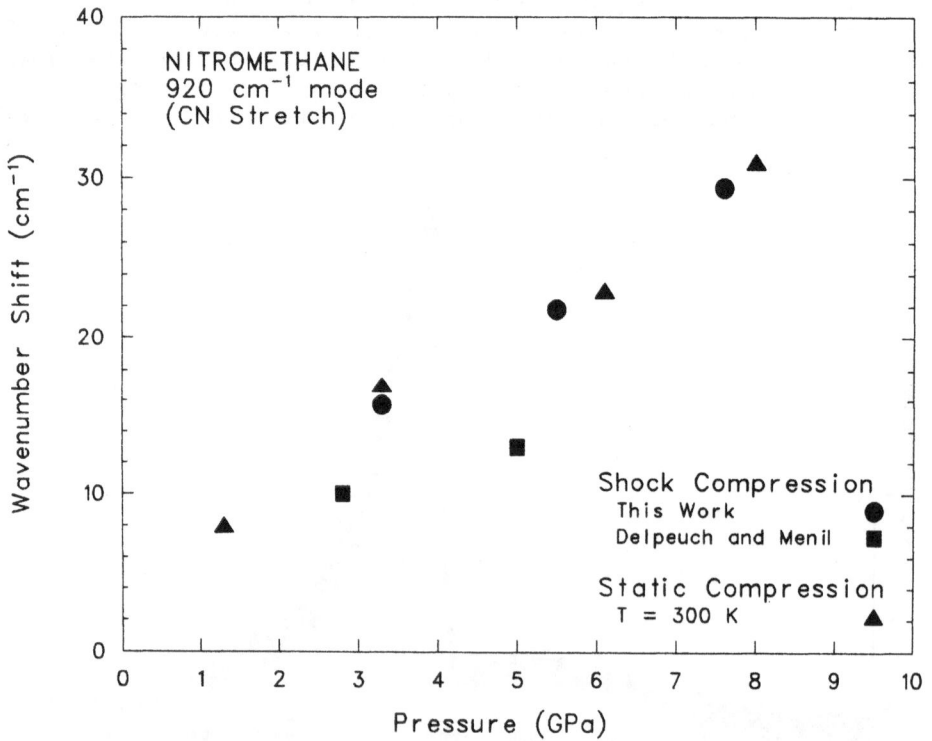

Fig. 13. Nitromethane CN stretching mode vibrational frequency shifts (with respect to 920 cm^{-1}) vs pressure.

results. The plot of frequency shift versus volume change was very similar to that observed for benzene.

Raman-induced Kerr effect spectroscopy (RIKES)[59] has been discussed as a diagnostic technique[19,20,60] for performing measurements in shock-compressed systems which may have a large non-resonant background. RIKES requires a single frequency pump beam, a broad-band probe source, no phase matching and lower incident power levels than stimulated Raman scattering (Fig. 4).

The effect can be described in terms similar to the above description of CARS. A linearly polarized probe laser beam is passed through the rotating electric field of a circularly polarized pump beam. The four-wave parametric process described above induces an ellipticity on the probe beam whenever the frequency difference between the two lasers equals that of a Raman active transition in the sample.[61,62] Since the RIKES involves the use of a single frequency pump laser and a broadband Stokes laser, it can be performed with an apparatus very similar to the above described RBBCARS apparatus. The modifications necessary are shown in Fig. 14 and are described below. The portion of the frequency-doubled Nd:YAG laser beam that does not pump the dye laser is

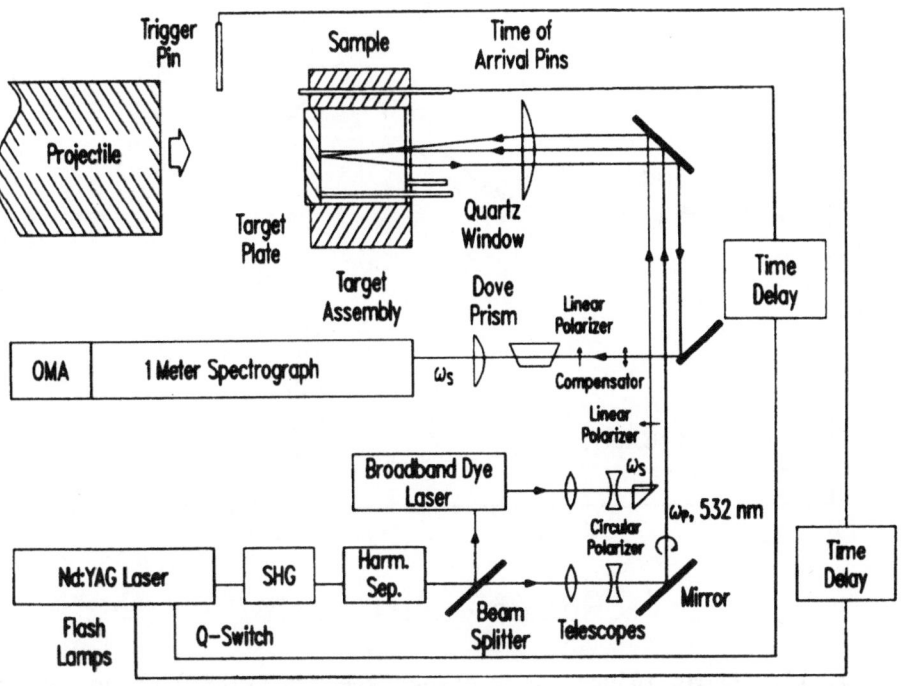

Fig. 14. Schematic representation of the Raman-induced Kerr effect spectroscopy experiment. SHG - second harmonic generator; Harm. Sep. - harmonic separator; OMA - optical multichannel analyzer; Sample - benzene.

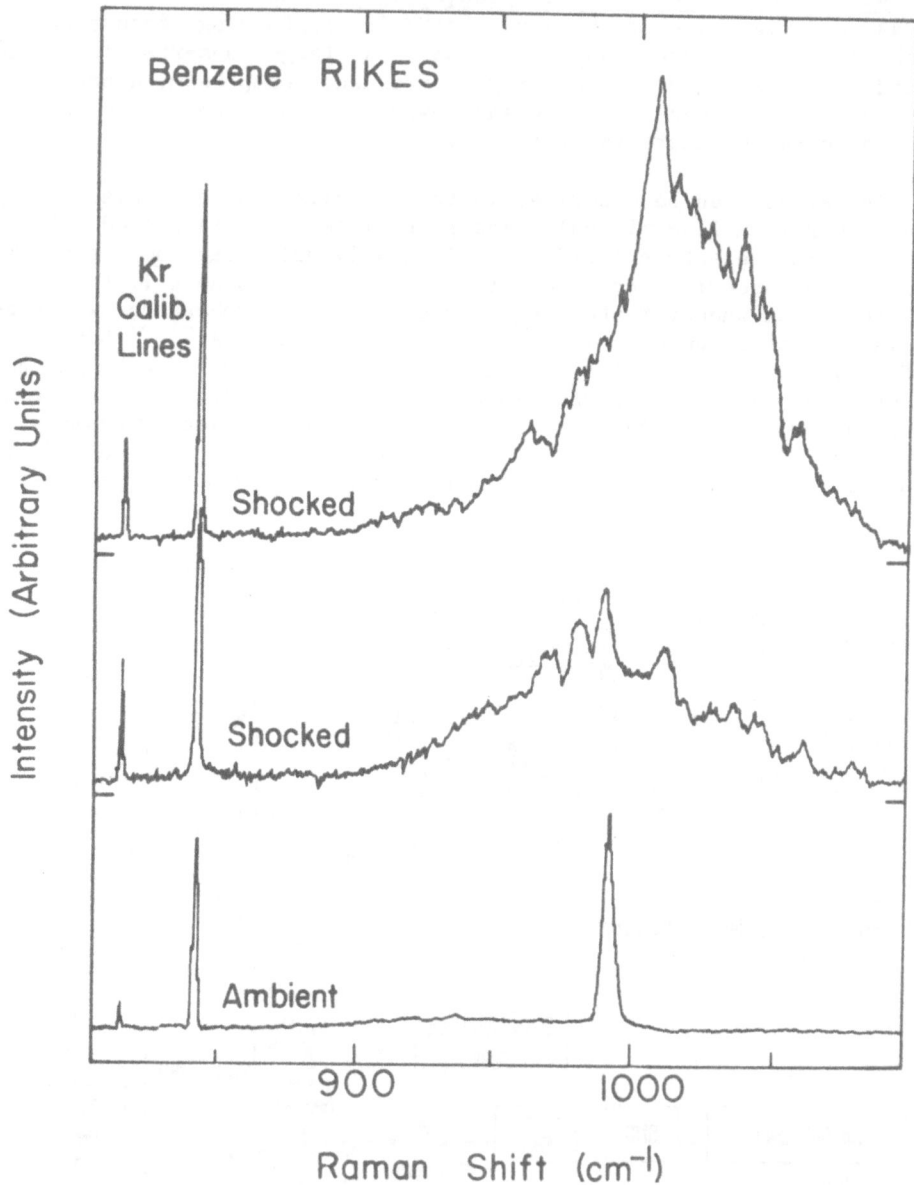

Fig. 15. Raman-induced Kerr effect spectra (RIKES) of an ambient and two shock-compressed liquid benzene samples. The shock pressure was 1.17 GPa and the 557.03 nm and 556.22 nm Kr calibration lines are shown. All spectra are obtained at the same power levels.

passed through a Fresnel rhomb to produce a beam of $>$ 99% circular polarization.   The dye laser beam (Stokes frequencies) is passed through
a high quality Glan-Taylor (air-gap) prism to produce a beam of $\sim$ 1
part in 10^6 linear polarization. The two beams are focused and crossed
in the sample using a 150 mm focal length, 50 mm diameter lens.   The
crossing angle is near 6 degrees, giving an overlap length of $\sim$ 150 μm
at the focus.   The Stokes beam is then reflected by the highly polished
front surface of the target plate back through the sample and along a
path parallel to the incoming beams. A mirror separates the reflected
dye laser beam from the other beams and directs it first through a
Babinet-Soliel polarization compensator and then through a Glan-Taylor
polarization analyzer.   The compensator was found to be necessary to
remove the ellipticity introduced into the linearly polarized Stokes
laser beam by the birefringence inherent in the optical components
located between the polarizers, including the ambient sample. When the
two Glan-Taylor prisms are crossed, the dye laser beam is blocked
except at frequencies corresponding to Raman resonances, where the
RIKES signals are passed. These signals are directed through a dove
prism, focused into the entrance slits of the 1 m spectrometer and
detected by the OMA system.

Figure 15 shows two RIKES 992 cm^{-1} region spectra of an ambient
and two shock-compressed liquid benzene samples.   The shock pressure
was 1.17 GPa and the 557.03 nm and 556.22 nm Kr calibration lines are
shown. All spectra are obtained at the same power levels. Both traces
have spectral features, however they are not consistent and do not
exhibit the pressure-induced frequency shift expected for the benzene
ring stretching mode based on previous BSRS and RBBCARS experiments.
In a polarization sensitive coherent Raman experiment, such as RIKES,
the possibility exists that shock-induced changes in a material would
perturb the probe laser polarization sufficiently to obscure the
desired signals. Therefore, the sensitivity of the RIKES apparatus to
minor rotations of the dye laser polarization was investigated.   The
figure of merit used was the polarization analyzer rotation angle necessary to saturate the detector with unblocked dye laser. It was found
that the detector could be driven from zero signal to saturation with a
polarization rotation angle of $\sim$ 20 arc minutes (20$'$) (using 50 μm
slits and 50 μJ dye laser energy). The RIKES signal found for the
ring-stretching mode of ambient liquid benzene nearly saturated the detector through 25 μm slits (using $\sim$ 200 μJ pump laser energy and 6°
beam crossing angle). These data suggest that, if the shock-compressed
sample induced a rotation of the probe laser polarization $>$ 20$'$, the
signal would be masked by the broad dye laser background passed by the
analyzer. The RIKES spectra (Fig. 15) obtained in shocked samples show
only broadband dye laser which has been passed by the polarization
analyzer.   These results indicate that the shock-compressed sample
induces a rotation of at least 20$'$ on the dye laser polarization. They
also lead to the conclusion that, while it may be possible to perform
RIKES experiments in shock-compressed materials in spite of our
failure, the experiment is considerably more difficult than techniques

not sensitive to the absolute polarization of the laser beam (such as BSRS and RBBCARS).

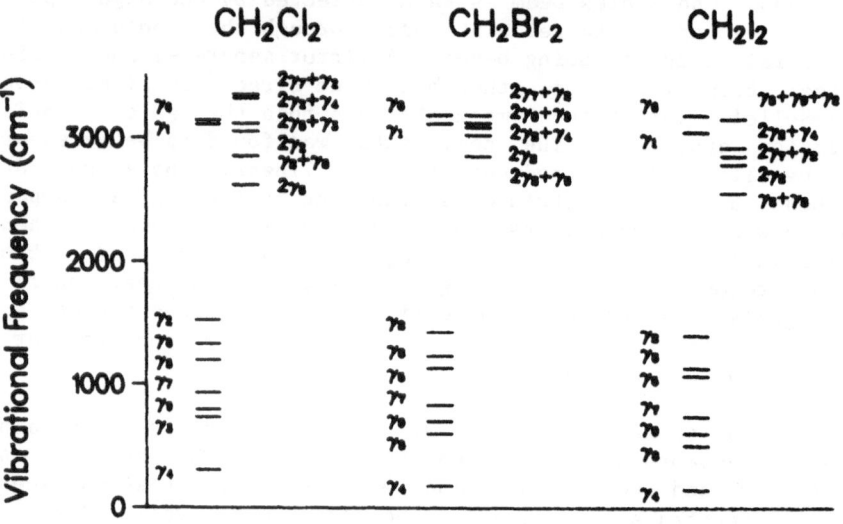

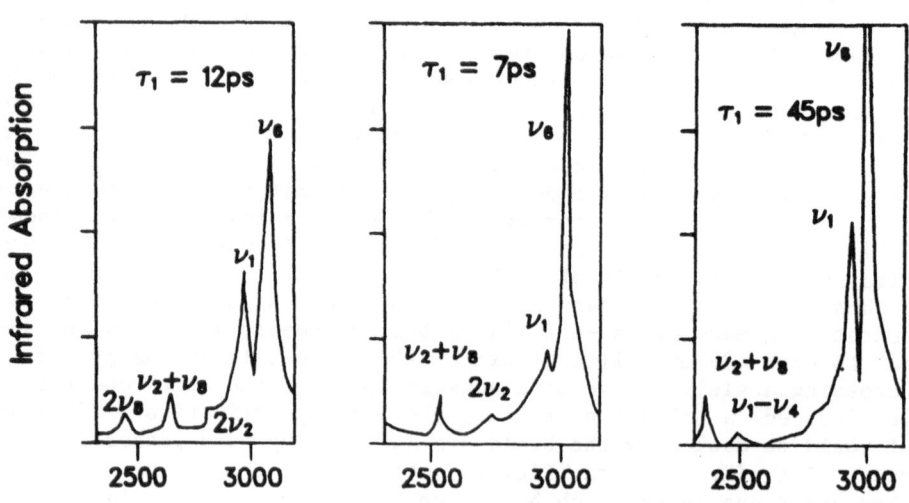

Fig. 16. Vibrational energy levels, some overtones and combinations and infrared spectra: CH_2Cl_2, CH_2Br_2 and CH_2I_2. The vibrational relaxation times, τ_1, shown with the infrared absorption spectra are the measured energy decay times of the CH stretching modes.

4.   ENERGY TRANSFER IN HIGH PRESSURE LIQUIDS

Although an abundance of literature[63-80] exists describing condensed phase energy transfer and relaxation phenomenology at ambient pressures and various temperatures, there is a dearth of studies showing behavior as a result of high pressure,[81-83] of large stress gradients, and of temperatures typical in shock-wave environments. Since understanding condensed phase molecular energy transfer is fundamental to understanding shock-induced chemical reactions and detonation, we have initiated an experiment to study the effects of pressure and temperature on condensed phase energy transfer. The experiment is based on the picosecond relaxation experiments of Laubereau et al.[84,85] and Fendt et al.[86] and ultrasonic studies of Takagi et al.[74,75] which study the vibrational energy transfer in substituted methanes. These materials were chosen because they have a simple molecular structure and have relaxation times comparable to those expected in shock compressed hydrocarbons. The lower vibrational energy levels and some overtone and combination levels in the vicinity of the CH stretch levels near 3000 cm^{-1} are shown in Fig. 16 for dichloromethane, dibromomethane and diiodomethane. The results of the studies for dichloromethane show that after populating the CH stretch modes using a picosecond infrared laser pulse (equilibration between the two modes is very rapid) these levels decay through a weak Fermi resonance to an overtone level of the bending modes. The presence of the Fermi resonance is deduced from a line in the infrared spectrum at 2832 cm^{-1} due to the first overtone of the ν_2 bending mode (Fig. 16). ν_1 and ν_6 are the two CH stretching modes and the peak at 2832 cm^{-1} is from $2\nu_2$. We believe that an important aspect of energy transfer during shock compression and shock-compression chemistry is how the energy flows through the vibrational degrees of freedom, i.e., how they are populated from the translational energy of the shock wave and if they are in equilibrium. In the case of the above system, compression either using shock-wave techniques or statically using a diamond-anvil cell, will induce a relative shift in the ν_1, ν_6 and ν_2 levels that should change the resonant coupling of the CH fundamental levels and the ν_2 overtone. The relaxation time should change accordingly.

Figure 17 schematically shows an experiment to measure the change in the CH stretch mode energy relaxation time at high pressure and temperature. Sub-picosecond pulses from a colliding-pulse-mode-locked ring dye laser[87] will be amplified using an excimer laser driven four stage amplifier and then used to generate picosecond infrared pulses which will vibrationally excite, by infrared absorption, the CH stretch levels of the substituted methanes. Part of the original amplified pulse will be optically delayed and used to probe the population density of the excited state by spontaneous anti-Stokes Raman scattering. Experiment repetition rate is 100 Hz.

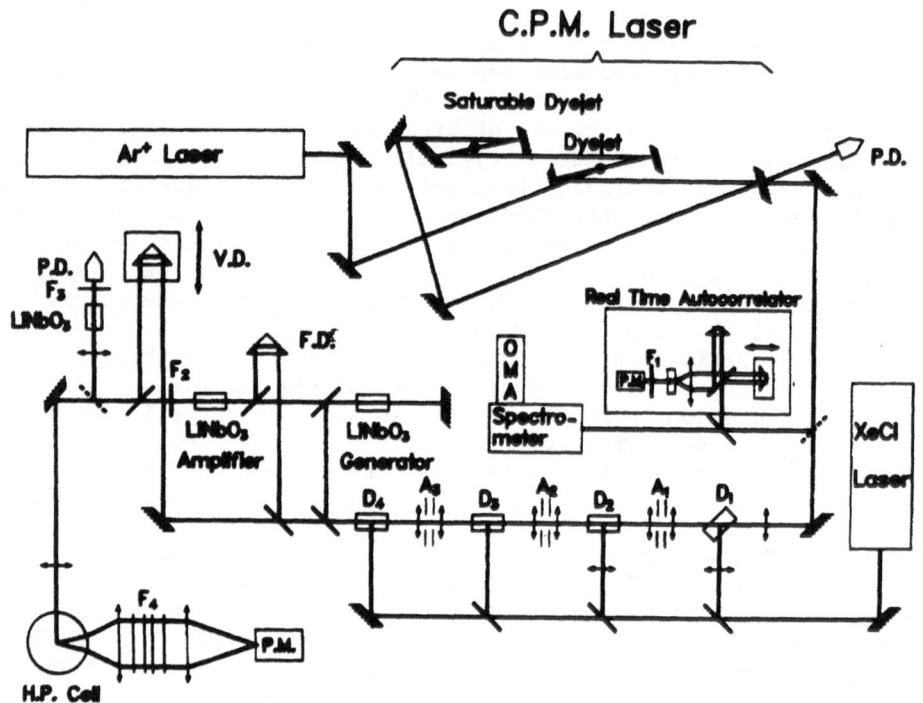

Fig. 17. Pressure dependent vibrational relaxation time experiment.

The Raman scattering signals will be detected by either a high quantum efficiency photo multiplier tube equipped with suitable filters or in a spectrograph using an optical multichannel analyzer. Compression of the sample to several GPa will be accomplished using a diamond anvil or other high pressure cell.

Experiment details and results will be described in a future publication. It is hoped that this experiment can be used to study the intra- or inter-molecular relaxation phenomenology at densities similar to those existing during shock compression. At pressures of several GPa, the molecules may not exist individually but as some other type of structure with radically shifted energy levels. Interpretation would require a theoretical approach which differs significantly from the frequently used bimolecular collision model.

5.   SUMMARY

Fundamental understanding of the detailed microscopic phenomenology of shock-induced chemical reaction and detonation waves is being sought by using pulsed coherent optical scattering experiments to determine the molecular structure, constituents and energy transfer mechanisms in both shock-compressed and static, high pressure/high temperature fluids. To date measurements of the ring stretching mode of benzene and the CN stretching mode of nitromethane up to shock-induced pressures just below those for which reaction is suspected to occur have shown both a shift in the vibrational frequencies and a definite change in the spectral profile. These results have confirmed that these molecules still exist on the microsecond time scale behind the shock front, but that some form of energy transfer is occurring from the hydrodynamic mode to the molecular internal degrees of freedom. Future experiments, both static high pressure picosecond vibrational relaxation and dynamic coherent Raman scattering at shock-compression pressures in the region where reaction is expected, should yield significant insight toward understanding the very complex and rapid processes that prevail in the shock environment.

6. ACKNOWLEDGEMENTS

The authors wish to thank C. W. Caldwell, R. L. Eavenson, and R. S. Medina for their assistance in performing the shock-compression experiments and V. A. Gurule, C. N. Gomez, and R. W. Livingston for machining and fabrication of the target assemblies. Special thanks is given to J. N. Fritz for use of the MACRAME computer code used to design target assemblies.

7. REFERENCES

1. C. A. Forest, "Burning and Detonation," LA-7245 (Los Alamos National Laboratory Report, Los Alamos, New Mexico 1978).

2. C. L. Mader, Numerical Modeling of Detonation (University of California Press, Berkeley, California 1979).

3. E. L. Lee and C. M. Tarver, Phys. Fluids 23, 2362 (1980).

4. J. Wackerle, R. L. Rabie, M. J. Ginsberg and A. B. Anderson in Proceedings of the Symposium on High Dynamic Pressures (Commissariat a l'Energie Atomique, Paris, France 1978) p. 127.

5. M. Cowperthwaite in Proceedings of the Symposium on High Dynamic Pressures (Commissariat a l'Energie Atomique, Paris, France 1978) p. 201.

6. J. W. Nunizato in Shock Waves in Condensed Matter - 1983, J. R. Asay, R. A. Graham, and G. K. Straub, eds. (Elsevier Science Publishers B. V., 1984) p. 293.

7. J. W. Nunizato and E. K. Walsh, Arch. Rational Mech. Anal. 73, 285 (1980).

8. J. N. Johnson, P. K. Tang and C. A. Forest, J. Appl. Phys. <u>57</u>, 4323 (1985).

9. P. K. Tang, J. N. Johnson and C. A. Forest in <u>Proc. 8th Symp. Detonation</u> (Albuquerque, New Mexico 1985), p. 375.

10. C. Mader and J. Kerschner in <u>Proc. 8th Symp. Detonation</u> (Albuquerque, New Mexico 1985) p. 366.

11. F. Boisard, C. Tombini and A. Menil in <u>Proc. 7th Symp. Detonation</u>, (Annapolis, Maryland 1981) p. 531.

12. A. Delpuech and A. Menil, in <u>Shock Waves in Condensed Matter - 1983</u>, J. R. Asay, R. A. Graham, and G. K. Straub, eds. (Elsevier Science Publishers B. V., 1984) p. 309.

13. S. Dufort and A. Delpuech in <u>Proc. 8th Symp. Detonation</u>, Albuquerque, New Mexico 1985) p. 221.

14. N. C. Holmes, A. C. Mitchell, W. J. Nellis, W. B. Graham and G. E. Walrafen, in <u>Shock Waves in Condensed Matter - 1983</u>, J. R. Asay, R. A. Graham and G. K. Straub, eds. (Elsevier Science Publisher B.V., 1984) p. 307.

15. W. M. Trott and A. M. Renlund in <u>Proc. 8th Symp. Detonation</u>, (Albuquerque, New Mexico 1985) p. 416.

16. D. S. Moore, S. C. Schmidt, D. Schiferl, and J. W. Shaner, in <u>Los Alamos Conference on Optics '83</u>, R. S. McDowell and S. C. Stotlar, eds. (Proceedings SPIE Volume 380, 1983) p. 208.

17. S. C. Schmidt, D. S. Moore, D. Schiferl, and J. W. Shaner, Phys. Rev. Lett. <u>50</u>, 661 (1983).

18. D. S. Moore, S. C. Schmidt, and J. W. Shaner, Phys. Rev. Lett. <u>50</u>, 1819, (1983).

19. S. C. Schmidt, D. S. Moore, and J. W. Shaner, in <u>Shock Waves in Condensed Matter - 1983</u>, J. R. Asay, R. A. Graham, and G. K. Straub, eds. (Elsevier Science Publishers B. V., 1984) p. 293.

20. D. S. Moore, S. C. Schmidt, D. Schiferl, and J. W. Shaner, in <u>High Pressure in Science and Technology, Part II</u>, C. Homan, R. K. MacCrone, and E. Whalley, eds. (North Holland Publishing, New York, 1984) p. 87.

21. S. C. Schmidt, D. S. Moore, J. W. Shaner, D. L. Shampine, and W. T. Holt in <u>Xth AIRAPT High Pressure Conference</u>, University of Amsterdam, Amsterdam, The Netherlands, 8-11 July 1985.

22. D. S. Moore, S. C. Schmidt, J. W. Shaner, D. L. Shampine, and W. T. Holt in Fourth APS Topical Conference on Shock Waves in Condensed Matter, Spokane, Washington, USA, 22-25 July, 1985.

23. A. Delpuech, J. Cherville, and C. Michaud in Proc. 7th Symp. Detonation, (Annapolis, Maryland 1981) p. 36.

24. J. Alster, N. Slagg, M. J. S. Dewar, J. P. Ritchie, and C. Wells in Fast-Reactions in Energetic Systems, C. Capellas and R. F. Walker, eds. (D. Reidel Publishing Co., 1981) p. 695.

25. R. Engelke, W. L. Earl and C. M. Rohlfing, J. of Chem. Phys., to be published.

26. S. B. Kormer, Sov. Phys.-Uspekhi 11, 229 (1968).

27. A. N. Dremin and V. Yu. Klimenko, "On the Role of the Shock Wave Front in Organic Substances Decomposition," Gas Dynamics of Explosions and Reactive Systems, Minsk, USSR 1981.

28. A. N. Dremin, V. Yu. Klimenko, K. M. Michaijuk and V. S. Trofimov in Proc. 7th Symp. Detonation (Annapolis, Maryland 1981) p. 789.

29. W. C. Davis in Proc. 7th Symp. Detonation (Annapolis, Maryland 1981) p. 531.

30. M. Maier, W. Kaiser and J. A. Giordmaine, Phys. Rev. 177, 580 (1969).

31. D. V. J. Linde, M. Maier and W. Kaiser, Phys. Rev. 178, 178 (1969).

32. M. H. Rice, R. G. McQueen and J. M. Walsh, Solid State Physics 6 (Academic Press, New York 1958) p. 1.

33. R. D. Dick, J. Chem. Phys. 57, 6021 (1970).

34. J. N. Fritz, in preparation for publication.

35. W. D. Ellenson and M. Nicol, J. Chem. Phys. 61, 1380 (1974), this mode is called ν_2 in G. Herzberg, Infrared and Raman Spectra (Van Nostrand Reinhold, New York 1968).

36. R. N. Keeler, G. H. Bloom and A. C. Mitchell, Phys. Rev. Lett. 17, 852 (1966).

37. A. N. Dremin and L. V. Barbare in Shock Waves in Condensed Matter - 1981, Am. Inst. Phys. Proc. 78, W. S. Nellis, L. Seaman, and R. A. Graham eds. (New York 1983), p. 270.

38. L. V. Barbare, A. N. Dremin, S. V. Pershin and V. V. Yakovlev, Fiz. Gor. i Var. 5, No. 4, 528 (1969).

39. R. LeSar, S. A. Ekberg, L. H. Jones, R. L. Mills, L. A. Schwalbe, and D. Schiferl, Solid State Comm. 32, 131 (1979).

40. S. Block, C. E. Weir, and G. J. Piermarini, Science 169, 586 (1970).

41. J. Akella and G. C. Kennedy, J. Chem. Phys. 55, 793 (1971).

42. P. D. Maker and R. W. Terhune, Phys. Rev. 137, A801 (1965).

43. W. M. Tolles, J. W. Nibler, J. R. McDonald and A. B. Harvey, Appl. Spectrosc. 31, 253 (1977).

44. J. W. Nibler and G. V. Knighten, in Raman Spectroscopy of Gases and Liquids, A. Weber, ed. (Springer-Verlag, Berlin Heidelberg, 1979) p. 253.

45. A. C. Eckbreth and P. W. Schreiber, in Chemical Applications of Nonlinear Raman Spectroscopy, A. B. Harvey, ed. (Academic, New York, 1981), p. 27.

46. W. B. Roh, P. W. Schreiber, and J. P. E. Taran, Appl. Phys. Lett. 29, 174 (1976).

47. J. J. Valentini, D. S. Moore, and D. S. Bomse, Chem. Phys. Lett. 83, 217 (1981).

48. W. J. Nellis, F. H. Ree, R. J. Trainor, A. C. Mitchell and M. B. Boslough, J. Chem. Phys. 80, 2784 (1984).

49. O. B. Yakusheva, V. V. Yakushev and A. N. Dremin, High Temp.-High Pres. 3, 261 (1971).

50. B. W. Dodson and R. A. Graham in Shock Waves in Condensed Matter - 1981, Am. Inst. Phys. Proc. 78, W. S. Nellis, L. Seaman, and R. A. Graham, eds. (New York 1981).

51. P. C. Lysne and D. R. Hardesty, J. Chem. Phys. 59, 6512 (1973).

52. F. E. Walker and R. J. Wasley, Comb. Flame 15, 233 (1970).

53. J. W. Brasch, J. Phys. Chem., 84, 2085 (1980).

54. D. R. Hardesty, Comb. Flame 27, 229 (1976).

55. A. A. Vorob'ev and V. S. Trofimov, Fiz. Gor. i Vzr. 18, Nov. 6, 74 (1982).

56. A. N. Dremin, V. Yu. Klimenko and I. Yu. Kosireva in Proc. 8th Symp. Detonation (Albuquerque, New Mexico 1985) p. 407.

57. D. Schiferl, private communication.

58. D. T. Cromer, R. R. Ryan and D. Schiferl, J. Phys. Chem. 89, 2315 (1985).

59. D. Heiman, R. W. Hellworth, M. D. Levenson and G. Martin, Phys. Rev. Lett. 36, 189 (1976).

60. W. G. VonHolle and R. A. McWilliams in Laser Probes for Combustion Chemistry (American Chemical Society Symposium Series 134), D. R. Crosley, ed. (American Chemical Society, Washington, D.C. 1983), p. 319.

61. G. L. Eesley, Coherent Raman Spectroscopy (Pergamon Press, Oxford 1981).

62. M. D. Levenson in: Chemical Applications of Nonlinear Raman Spectroscopy, A. B. Harvey, ed. (Academic Press, New York 1981) pp. 214-222.

63. W. F. Calaway and G. E. Ewing, Chem. Phys. Lett. 30, 485 (1975).

64. W. F. Calaway and G. E. Ewing, J. Chem. Phys. 63, 2842 (1975).

65. C. Manzanares and G. E. Ewing, J. Chem. Phys. 69, 1418 (1978).

66. C. Manzanares and G. E. Ewing, J. Chem. Phys. 69, 2803 (1978).

67. D. W. Chandler and G. E. Ewing, J. Chem. Phys. 73, 4904 (1980).

68. D. W. Chandler and G. E. Ewing, J. Phys. Chem. 85, 1994 (1981).

69. W. Kaiser and A. Laubereau in Nonlinear Spectroscopy (Proc. International School of Physics "Enrico Fermi," Course LXIV), N. Bloemberger, ed. (North-Holland Publishing Co., Amsterdam 1977), p. 404.

70. A. Laubereau and W. Kaiser, Rev. Mod. Phys. 50, 607 (1978).

71. D. Samios and Th. Dorfmüller, Mol. Phys. 41, 637 (1980).

72. Th. Dorfmüller and D. Samios, Mol. Phys. 43, 23 (1981).

73. K. Takagi, P.-K. Choi and K. Negishi, J. Acoust. Soc. Am. 62, 354 (1977).

74. K. Takagi and K. Negishi, J. Chem. Phys. 72, 1809 (1980).

75. K. Takagi, P.-K. Choi and K. Negishi, J. Chem. Phys. <u>74</u>, 1424 (1981).

76. P.-K. Choi, K. Takagi and K. Negishi, J. Chem. Phys. <u>74</u>, 1438 (1981).

77. J. T. Yardley, <u>Introduction to Molecular Energy Transfer</u>, (Academic Press, New York 1980).

78. C. Capellos and R. F. Walker, <u>Fast Reactions in Energetic Systems</u>, (D. Reidel Publishing Co., Dordrecht, Holland 1980).

79. P. M. Rentzepis, Science <u>218</u>, 1183 (1982).

80. J. Chesnoy and G. M. Gale, "Vibrational Energy Relaxation in Liquids," in preparation.

81. M. Chatelet, G. Widenlocher and B. Oksengorn in <u>High Pressure – Science and Technology</u>, Vol. 2, B. Vodar and Ph. Marteau, eds. (Pergamon Press, Oxford 1979), p. 628.

82. M. Chatelet, B. Oksengorn, G. Widenlocher and Ph. Marteau, J. Chem. Phys. <u>75</u>, 2374 (1981),

83. M. Chatelet, J. Kieffer and B. Oksengorn, to be published Chem. Phys.

84. A. Laubereau, S. F. Fisher, K. Spanner, and W. Kaiser, Chem. Phys. <u>31</u>, 335 (1978).

85. H. Graener and A. Laubereau, Appl. Phys. <u>B29</u>, 213 (1982).

86. A. Fendt, S. F. Fischer, and W. Kaiser, Chem. Phys. <u>57</u>, 55 (1981).

87. R. L. Fork, B. I. Greene, and C. V. Shank, Appl. Phys. Lett. <u>38</u>, 671 (1981).

REACTION PRODUCTS OF ENERGETIC MATERIALS HEATED BY SHORT LASER PULSES

Horst Krause and Achim Pfeil
Fraunhofer-Institut für Treib- und Explosivstoffe
D-7507 Pfinztal
FRG

ABSTRACT. Foccussed laser pulses were employed to pyrolyze nitrated cellulose and nitrocellulose containing propellants. The main products characterized by time-of-flight spectroscopy were oxohydrocarbons having C_3 to C_6 chains with masses up to 90 amu. Metal salts like Cu-ß-resorcylate strongly affect the pyrolysis process and products with markedly higher masses were observed.

1. INTRODUCTION

The use of laser pulses to pyrolyze energetic materials provides the following advantages: The heating rate is comparable to that present in the combustion process and the products obtained can be readily detected by TOF mass spectroscopy. Furthermore, these species are thermally highly excited and generated at a primary step of the pyrolysis process. Species of this kind, in contrast, are difficult to detect in the combustion process since the reaction path ways overlap and the primary, secondary and final products form a complex mixture.

We used laser pulses to pyrolyze nitrocellulose (NC) in different environments. The objectives were to get information on the pyrolysis products and on the effect of metal salts on this process.

2. EXPERIMENT

The samples consisted of pellets of neat NC (13% N) and a mixture of NC with 8% Cu-ß-Resorcylate (CußR). A CußR containing solid propellant having NC as the main component was also investigated. CußR catalyses the thermal decomposition of NC and strongly influences the burning rate of this propellant.

The pellets were irradiated by the foccussed 10 nsec pulse of a Nd/YAG laser (532 nm, 0.25 mJ). The power density was kept close to

P. M. Rentzepis and C. Capellos (eds.), Advances in Chemical Reaction Dynamics, 455–458.
© *1986 by D. Reidel Publishing Company.*

the ionisation threshold of the sample (10^8 W/cm^2). The focal
diameter was about 200 um. A typical crater produced by irridiation
on NC is shown in Fig. 1.

Fig. 1 Crater produced by pulse irridiation on NC,
enlarged 1000x .

The evolved species were detected at right angle by our spectrometer
(drift length 2,26 m, EMI 9643 detector). The mass detection limit
was 500 amu.

3. RESULTS AND DISCUSSION

The cation spectra were identical for all samples. In principal,
they showed plasma-type products like the atomic species H$^+$, C$^+$, N$^+$, O$^+$
and small molecular species like CH$^+$, CH$_2^+$ or NH$^+$, OH$^+$
and OH$_2^+$.

On high amplification, however, weak signals of products appeared
which, in the literature, are classified as typical primary species
of the combustion process. These were C$_2^+$,C$_4^+$, CO$^+$, HCO$^+$, H$_2$CO$^+$
and/or NO$^+$.

CußR did not effect the spectra.

A different situation was encountered in the anion spectra. Intensive signals of molecular species with masses between 48 and 90 amu could be detected.. A good resolution of these signals is shown in Fig. 2. We assigned these signals to $C_4H_n^-$, $C_3H_nO^-$, $C_5H_n^-$, $C_4H_nO^-$, $C_5H_nO^-$, $C_4H_nO_2^-$, $C_5H_nO_2^-$, with n = 0 to 4.

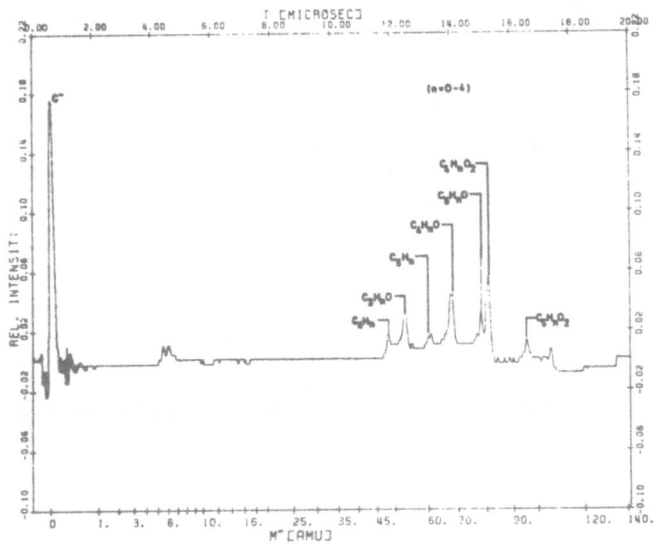

Fig. 2 Anion mass spectrum of NC.

As a result, the primary species of laser pyrolyzed NC appear to be oxohydrocarbons having C_3 to C_6 chains.

In contrast, the CuβR catalyzed samples clearly exhibit different spectra. Fig. 3 and 4 depict species which appear over the whole amu range up to the detection limit of our instrument. The signals above 90 amu must be attributed to segments of the NC chain. The most interesting signals are located at 130 amu (a denitrated NC segment) and at 292 amu (one NC segment).

It appears that the addition of CuβR causes the pyrolysis of the NC chain to be less severe so that larger fragments are generated. This is in accordance with the results obtained from comustion studies in visual bombs on catalyzed propellants (1). Carbonaceous residues (fibers) appear on the burning surface which is not observed for the uncatalyzed samples.

4. REFERENCES

1. N. Eisennreich, Propellants and Explosives 3 (1978) 141.

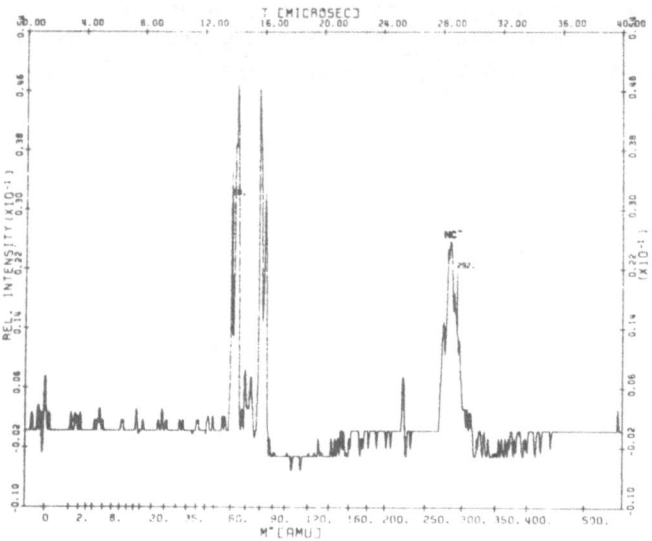

Fig. 3   Anion mass spectrum of CußR catalyzed NC.

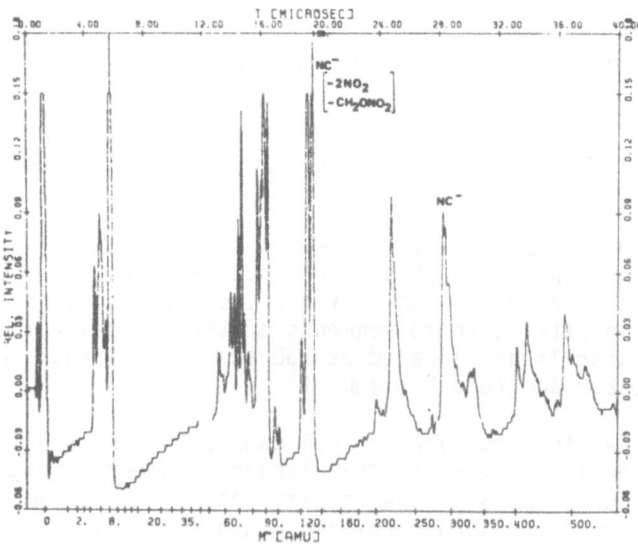

Fig. 4   Anion mass spectrum of CußR catalyzed propellant.

SPECTROSCOPY AS A TOOL TO PROBE LASER INITIATION OF PRIMARY EXPLOSIVES

M.W. Leeuw, A.J.Th. Rooijers and A.C. van der Steen
Prins Maurits Laboratorium TNO
P.O. Box 45
2280 AA Rijswijk
The Netherlands

ABSTRACT. Emission spectroscopy is used to monitor the emission of decomposition products of primary explosives after initiation with an excimer laser pulse. Using a gated optical multichannel analyser spectra can be obtained in the microsecond time domain.

These spectra provide detailed information about the decomposition reactions in the first stage of the initiation process. In this paper spectra obtained after initiation of mercury fulminate and lead styphnate are presented.

Among the reaction products of mercury fulminate atomic mercury, CN, C_2 and OH have been identified so far.

In the emission spectrum of lead styphnate atomic lead and lead oxide have been observed. The amount of lead oxide increases relative to the amount of lead as the emission farther away from initiation spot is monitored.

1. INTRODUCTION

For a better understanding of the explosive properties of explosive compounds it is of the utmost importance to know the kinetics of the decomposition reactions during and immediately after the initiation pulse. However, although extensive literature is available on the decomposition reactions of explosives, experimental data with respect to the first stages of the initiation process are scarce. In most experiments either the initiation techniques used, for example the drop hammer or the friction apparatus, or the detection techniques employed are too slow as compared to the rate at which explosives decompose in detonation- or deflagration-like reactions. In this paper it will be shown, that it is possible to obtain detailed information on the decomposition reactions in the microsecond time domain using a laser as an initiation source and using emission spectroscopy to monitor the reaction intermediates.

In recent years a number of techniques have been developed to initiate explosive compounds in a very short time interval. For instance, in our laboratory the Thermal Step Test (TST) is used to study

459

P. M. Rentzepis and C. Capellos (eds.), Advances in Chemical Reaction Dynamics, 459–467.
© *1986 by D. Reidel Publishing Company.*

the high temperature kinetics of explosive compounds [1]. With the TST explosives can be heated to temperatures up to 1400 K in about 30 µs. Subsequently, induction times which can be as short as 50 µs can be measured.

The laser is another energy source which can deposit a large amount of energy in an explosive in a very short time. Therefore lasers have been used since the mid-sixties to initiate explosieves. Brish [2] was the first to report on laser initiation in a study on the initiation of lead azide using a Nd glass laser. Since, a limited number of papers on this subject have been published [3,4,5].

The laser as an initiation source has two major advantages:
- Firstly, a well-defined amount of energy can be delivered to the explosive in a time interval which can be as short as 20 ns. In contrast, using most other initiation techniques the energy transfer to the explosive may not be clear at all and the interaction times are generally much longer.
- Secondly, the wavelength of the laser can be selected to coincide with an absorption band of an explosive, thus optimizing the interaction between the explosive and the radiation field. Furthermore, by changing the wavelength of the laser one can excite different electronic states of the explosive.

Most explosive compounds have absorption bands in the ultraviolet part of the spectrum. Therefore we have selected an excimer laser as an initiation source, since this kind of laser can emit high energy pulses in the ultraviolet.

With unconfined pellets of primary explosives induction times -the time lapse between the laser pulse and the moment the light intensity emitted by the sample is at a maximum - ranging from 500 ns (lead azide) up to 3.5 ms (mercury fulminate) have been observed [6].

Although these experiments provide valuable information on the kinetics of the initiation process, they do not provide information about what intermediates are formed. Therefore the experimental set-up has been extended with a spectrograph and an optical multichannel analyser to record the emission spectrum of the sample and to obtain this kind of information [7].

In this paper the emission spectra obtained the initiation of lead styphnate and mercury fulminate are presented and discussed.

2. EXPERIMENTAL

A block diagram of the experimental set-up is given in Figure 1.

About 40 mg of the explosive (with a purity better than 99%) is pressed to a pellet with a diameter of 5 mm and a thickness of about 0.5 mm. The density of the explosive can be varied by pressing the pellet with a force up to 0.15 MN. The pellet is placed on a hardened steel pin in a sample holder (S) and the laser beam is focussed directly on the sample. The pulse is delivered by a Lambda Physik EMG200 excimer laser. The laser can output up to 1.2 J at λ = 248 nm (KrF line); the pulse width is 15 ns. By means of a beam splitter (BS) part of the laser pulse is directed to a Joule meter (Gentech PRJ-A) to monitor the output of the laser system. The beam is focussed to a spot of about 1 x 2 mm^2. In that case the energy density is about 100 kJ/m^2.

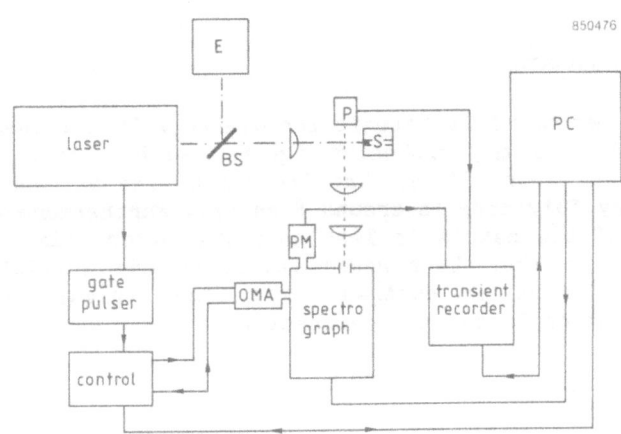

850476

Figure 1. Block diagram of the experimental set-up.

The light emitted in the course of the initiation process is detected in several ways. Firstly, to obtain induction times a photodiode (P) placed a few centimeters above the sample holder is used to monitor the total emission emitted by the decomposition products as a function of time. The output of the photodiode is fed into a 100 MHz transient recorder (LeCroy TR8818) or in a 20 MHz transient recorder (Nicolet Explorer).

Simultaneously, using two lenses a spot in front of the sample holder is focussed onto the entrance slit of a 0.6 m spectrograph (Jobin Yvon, HRS-2, grid 1200 lines/mm). The dispersed light is detected by a microchannel plate intensified optical multichannel analyser (Spectroscopy Instruments, IRY1024) connected to the side exit slit of the spectrograph. The spectral resolution of the detection set-up is better than 0.1 nm.

The optical multichannel analyser is activated by a gate pulser triggered by the laser. With the pulser an initial delay relative to the laser pulse from 270 ns up to 6 ms can be set. The gate width can be varied from 180 ns up to 6 ms.

Finally, a photomultiplier tube (EMI 9558 QB) is connected to the other exit slit of the spectrograph to monitor the spectrally-resolved light as a function of time. The output of the photomultiplier is fed into one of the transient recorders as well.

The control unit of the optical multichannel analyser, the transient recorders and the command unit of the spectrograph are all connected to a microcomputer (DEC Professional 350). The microcomputer is used to the control equipment, to store data, to perform data handling and to plot the data.

3. RESULTS AND DISCUSSION

When a pellet of mercury fulminate is irradiated with a focussed laser pulse of the KrF line the explosive ignites with a subdued report emitting a flame of about 30 cm. The induction time of unpressed and unconfined mercury fulminate is around 8 ms [7]. Furthermore the total light intensity of the sample is less than for other primary explosives. So it appears that in mercury fulminate the initiation only leads to a deflagration as contrasted to other primary explosives where laser initiation leads to a detonation.

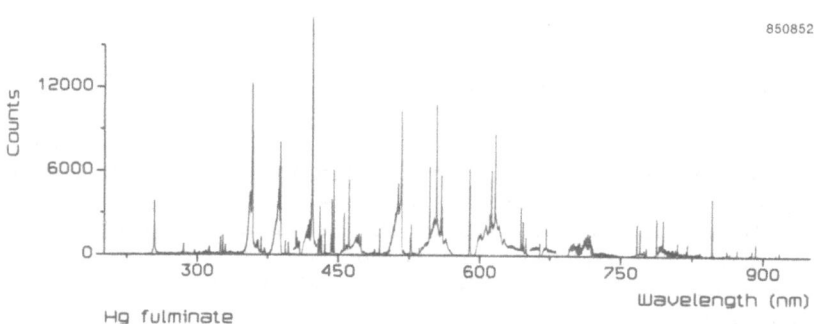

Figure 2. Emission spectrum obtained after the initiation of mercury fulminate.

The emission spectrum obtained after the laser initiation of mercury fulminate is shown in Figure 2; the parts of the spectrum between 370 and 400 nm and between 565 and 595 nm are attenuated 10 times with respect to the remainder of the spectrum. In Figures 3 and 4 more detailed parts of the emission spectra are presented. The spectrum has been recorded with the time window set between 5 and 11 ms relative to the laser pulse and by focussing the light originating from a spot around 1 cm in front of the sample holder. The spectral resolution is about 0.1 nm.

As can be seen the spectrum consists of a large number of more or less resolved band systems. Unfortunately, at the time of writing not all band systems had been assigned yet.

A number of lines in the emission spectrum can be assigned to atomic mercury. Rather prominent in the spectrum are the mercury lines [8,9] at 253.6, 435.8 and 546.0 nm. Less intense are the mercury lines at 296.7, 365.4 and 404.6 nm and the doublet at 313.1 nm.

A number of radicals well-known in flame spectroscopy have been identified as well.

Two very prominent band systems and one less intense band system are believed to be due to the presence of the CN radical as one of the reaction products of the initiation process.

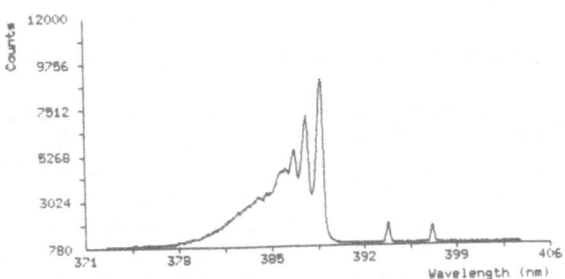

Figure 3. Electronic orign of the $B^2\Sigma$ - $X^2\pi$ transition of CN.

The band system between 375 and 390 nm shown in Figure 3 is assigned to the $B^2\Sigma$ - $X^2\pi$ transition of the CN radical [9,10]. The vibrationless transition is located at 388.3 nm. The other peaks in the band system are due transitions of higher vibrational levels with $\Delta\nu = 0$ (ν is the vibrational quantum number). Since the transitions tend to overlap as the vibrational quantum number increases, it is hard to make a precise assessment of the population distribution over the different virbrational levels. However, from the overall band shape it is estimated that levels with ν as high as 10 are populated.

Other band systems which can also be ascribed to the $B^2\Sigma$ - $X^2\pi$ transition of CN are located between 350 and 360 nm and between 415 and 422 nm. The first system is due to vibrational transitions with $\Delta\nu = -1$, whereas the second system is due to transitions with $\Delta\nu = +1$. The latter system is partly overlapped by a strong peak originating from Ca.

The above-mentioned band systems of the CN radical all belong to the so-called violet system. Another extensive band system of CN, the so-called red system extending into the near infrared, is not observed in the emission spectrum due to the decomposition of mercury fulminate.

Another radical which appears to be quite prominent in the decomposition of mercury fulminate is the C_2 radical.

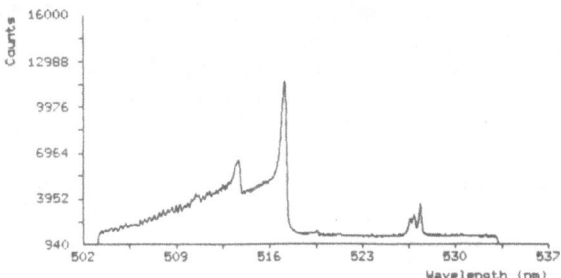

Figure 4. Electronic origin of the $A^3\pi_g$ - $X^3\pi_u$ transition of C_2.

Bands of the C_2 Swan ($A^3\pi_g$ - $X^3\pi_u$) system are clearly visible between 500 and 520 nm and between 460 and 475 nm. The electronic origin of this system is located at 516.0 nm which is in good agreement with our observations. The (1,1) hot band transition at 512.9 nm is clearly visible as well. The (2,2) transition is only observable as a weak feature upon the rotational contour of the other vibrational transitions. The band system between 460 and 475 nm is due to $\Delta\nu$ = -1 transitions and the peaks due to the levels with ν up to 4 are distinguishable. The band systems corresponding to $\Delta\nu$ = +1 and $\Delta\nu$ = +2 are present, but more or less masked by intense systems whose origins are not yet clear.

Since the vibrational transitions of the Swan system overlap each other, an exact vibrational temperature is hard to deduce, but from the intensity distribution of the vibrational $\Delta\nu$ = -1 transitions a vibrational temperature of about 11000 K is calculated. This result is in good agreement with the vibrational temperature of the C_2 radical created in low pressure oxyacetylene flames [11].

Finally, weak emission with a maximum around 306 nm is assigned to the $A^2\Sigma^+$ - $X^2\pi$ transition of the hydroxyl radical.

A number of lines are due to impurities present in the mercury fulminate. In all explosives studied so far sodium has been observed. The well-known Na doublet [8] at 589 nm is very dominantly present in the emission spectrum; a weak Na line appears at 331 nm.
In the red part of the emission spectrum two lines around 766 nm and 770 nm can be assigned to potassium.

Finally, calcium appears to be prominently present in the emission spectrum as well. The following lines are attributed to Ca: the strong line at 422.5 nm, the lines between 428 nm and 432 nm, the triplet of lines between 442 nm and 446 nm, the sharp lines between 558 and 561 nm superimposed on the broad band system emitting in that region and the unresolved lines between 525 and 527 nm [9]. The lines at 393.4 nm and 396.8 nm are attributed to Ca+.

Inspecting the molecular structure of mercury fulminate one would expect the NCO radical to be present as one of the decomposition products. The emission of this radical is well-documented [12,13] and the strongest emission occurs around 438 nm. As can be seen in Figure 2 in the emission spectrum of mercury fulminate no such emission occurs. So the absence of this radical and the low concentration of atomic mercury indicates, that the breaking of the bond between the mercury atom and the fulminate group is not the dominant decomposition process in the laser initiation of mercury fulminate.

After initiation of lead styphnate with a pulse of the KrF line of the excimer laser the sample ignites violently with a very loud report and emits a very intense radiation. So it appears that in lead styphnate the detonation regime is reached. The induction time of unconfined, unpressed lead styphnate is 120 μs.

In Figure 5 emission spectra recorded after the initiation of lead styphnate are displayed. The time window of the optical multi-channel analyser is set up to 200 μs. The spectral resolution is about 1 nm. The results presented in Figure 5 prove, that despite the short detection time it is possible to record spectra with a reasonable signal to noise ratio. The top spectrum in Figure 5 is obtained by focussing the light 1 cm in front of the sample holder; the bottom spectrum is recorded by focussing the emission 8 cm in front of the sample holder.

The emission spectra obtained after the initiation of lead styphnate consist of both sharp lines and broad bands. With the exception of the peaks around 357 nm and 373.6 nm all peaks below 410 nm can be ascribed to transitions of atomic lead [8]. Although the lines at 357.6 and 373.6 nm are not listed in [8] as being due to lead, it is our opinion that these lines can be attributed to lead as well. The shape of these two bands indicate that they are due to atomic transitions. Since at these two wavelengths emission of other elements are not known, it is assumed that the lines at 357.6 nm and 373.6 nm are due to atomic lead as well.

The broad ˎband emission starting around 400 nm and extending until 650 nm is assigned to emission due to the presence of the PbO molecule. PbO has a number of band systems which are more or less intense [9,14]. The so-called A band system extends from 475 nm up to 675 nm. In the bottom spectrum of Figure 5 emission bands which belong to this system can be clearly observed. Part of this emission is obscured by the strong emission of the Na doublet at 589 nm. The B system extends from 415 nm up to 580 nm. Bands of this system can also be observed in the emission of decomposition products of lead styphnate. The start of this emission is partly overlapped by the strong Pb emission at 405.8 nm. At the red side the B band system and A band system overlap. The C band system of PbO emits from 380 nm until 415 nm. The presence of this system is hard to establish, since again strong emission due to lead occurs in this region. However, the background under the Pb line at 405.8 nm is probably due to the C band system.

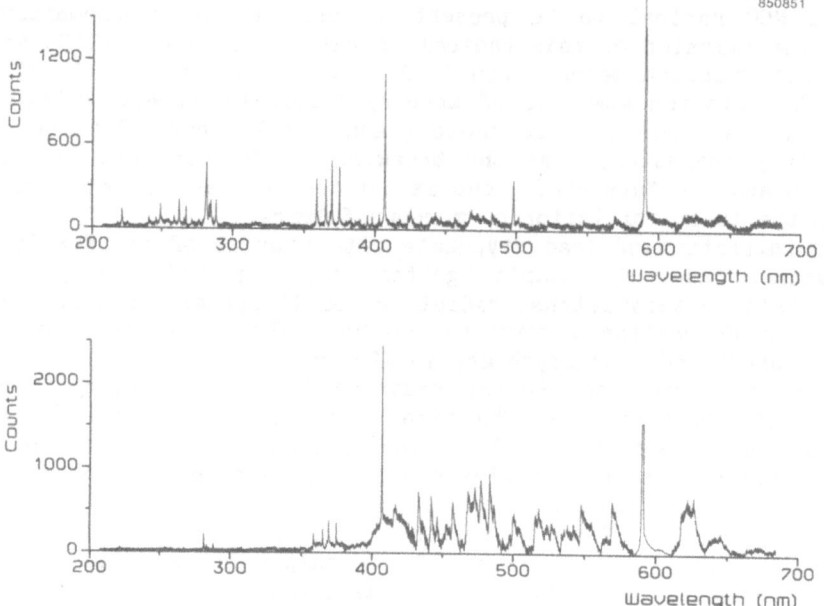

Figure 5. Emission spectra obtained after the initiation of lead styphnate.
top   : detection 1 cm in front of sample holder
bottom: detection 8 cm in front of sample holder

Two other band systems of PbO extending between 290 nm and 360 nm are known. These systems are not present in the emission spectra obtained after the laser initiation of lead styphnate.

The intensity distribution over the various band system changes considerably in two ways as the distance of the detection spot relative to the sample holder changes.

Firstly the intensity of the various lead lines relative to each other changes. Relative to the lead line at 405.8 nm all other Pb lines decrease in intensity. This decrease is most important for the lines below 300 nm and is due to collisional deactivation processes in Pb. At the start of the initiation Pb is dissociated from the styphnate group and free lead in various excited states is formed. Due to collisonal deactivation processes the higher electronic states of Pb are more easily depopulated resulting in a decrease in the intensity due to emission out of these states.

The emission due to PbO increases as the distance from the initiation spot increases. The atomic lead formed in the course of the initiation process reacts subsequently with oxygen to form PbO. The concentration of PbO in time increases, resulting in a decrease in the Pb emission and an increase in the PbO emission.

4. CONCLUSIONS

The results presented in this paper show that it is possible to obtain on a microsecond time scale detailed information about the decomposition products and the reaction kinetics in the first stages of the initiation process of primary explosives using a pulsed laser system as an initiation source and employing fast spectroscopic detection techniques.

Atomic mercury, CN, C_2 and OH have been identified among the reaction products of mercury filminate.

Atomic lead and PbO are the main decompositions products observed in the case of lead styphnate. The atomic lead is formed right at the start of the initiation process and reacts further to form lead oxide.

REFERENCES

1. M.A. Schrader, M.W. Leeuw and A.C. van der Steen, Proc. 9th international pyrotechnic seminar, Colorado Springs (1984)
2. A.A. Brish, I.A. Galeev, B.N. Zaitsev, E.A. Sbitnev and L.V. Tatarintsev, Comb., Expl. and Shock Waves 2, 81 (1966)
3. E.I. Aleksandrov and A.G. Voznyuk, Comb., Expl. and Shock Waves 14, 480 (1978)
4. J.T. Hagan and M.M. Chaudhri, J. Mat. Science 16, 2457 (1981)
5. L.C. Yang and V.J. Menichelli, Proc. 6th Symp. on Detonation, 612, Coronado (1984)
6. A.J.Th. Rooijers, M.W. Leeuw and A.C. van der Steen, Proc. 9th international pyrotechnic seminar, Colorado Springs (1984)
7. M.W. Leeuw, A.J.Th. Rooijers and A.C. van der Steen, Proc. 8th Symp. on Detonation, 724, Albuquerque (1985)
   A.J.Th. Rooijers, M.W. Leeuw and A.C. van der Steen, Proc. 16. internationale ICT-Jahrestagung, 10-1, Karlsruhe (1985)
8. P.W.J.M. Boumans, Coincidence tables for inductively coupled plasma atomic emission spectrometry, Pergamon Press, London (1980)
9. R.W.B. Pearse and A.G. Gaydon, The identification of molecular spectra, Chapman and Hall, London (1975)
10. W.M. Jackson and J.B. Halpern, J. Chem. Phys. 70, 2373 (1979)
11. R. Bleekrode, Thesis, University of Amsterdam (1966)
12. R.N. Dixon, Phil. Trans. Roy. Soc. 252A, 165 (1960)
13. D.D. Bell and R.D. Coombe, J. Chem. Phys. 82, 1317 (1985)
14. R.C. Oldenborg, C.R. Dickson and R.S. Zare, J. Mol. Spec. 58, 203 (1975)

DISLOCATION PILE-UP MECHANISM FOR INITIATION OF ENERGETIC CRYSTALS

R. W. Armstrong*, C. S. Coffey** and W. L. Elban***
*Department of Mechanical Engineering
University of Maryland
College Park, Maryland 20742
**Detonation Physics Branch
Naval Surface Weapons Center
Silver Spring, Maryland 2090 3-5000
***Department of Engineering Science
Loyola College
Baltimore, Maryland 2 12 10

ABSTRACT. The "hot spot" temperature produced during deformation by breakthrough of a dislocation pile-up is proportional to the ratio of the microstructural stress intensity and the thermal conductivity. On this basis, the calculated temperature rise for cyclotrimethylenetrinitramine (RDX) is 200 times greater than that for steel. Also, a log-log relationship is predicted for the drop-weight impact sensitivity and the inverse square root of RDX crystal size. Beyond the adiabatic heating effect, the pile-up model allows for the possibility of breaking molecules and separating their fragments.

1. ADIABATIC HEATING AT A PILE-UP AVALANCHE

Dislocation pile-ups provide an effective mechanism for locally concentrating energy within stressed crystals and polycrystals[1]. The pile-up energy is due essentially to the interaction of dislocations bunched up at the pile-up tip. A precipitate particle or chemical inclusion would be a suitable obstacle. The interaction energy can be built up isothermally during the very earliest stages of material deformation. Adiabatic dissipation of the interaction energy occurs if the internal obstacle blocking the pile-up collapses suddenly. The event should be accompanied by a discontinuous load drop in the stress-strain behavior. Heavens and Field[2] first reported that discontinuous load drops occurred in the transmitted stress pulses of compressively impacted energetic materials which were found to be susceptible to the generation of hot spots. Figure 1 gives a schematic description of the successive stages leading to pile-up collapse and sudden energy dissipation.

Brittle cracking at a pile-up tip should produce the greatest energy dissipation. At the theoretical fracture stress, the limiting temperature rise, ΔT, is given by[1]

$$\Delta T < \left[k_s \ell^{1/2} v \bar{\alpha} / 20 \pi^2 K\right] \ell n (2K/c^* v \Delta x_1), \tag{1}$$

469

P. M. Rentzepis and C. Capellos (eds.), Advances in Chemical Reaction Dynamics, 469–474.

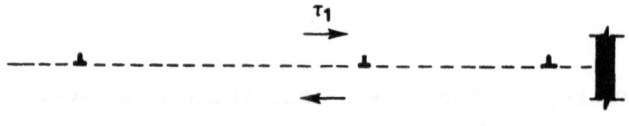

(a) isothermal stress build-up: n_1 dislocations

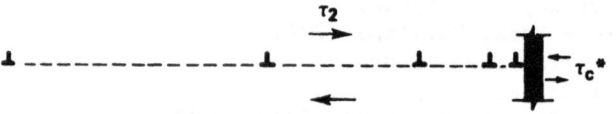

(b) critical stress concentration: $n_2 \tau_2 = \tau_c^*$

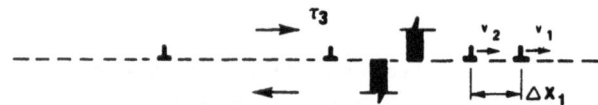

(c) adiabatic collapse-discontinuous load drop

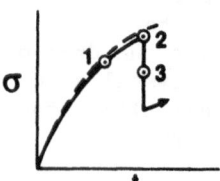

(d) pressure-time curve for τ_1, τ_2, and τ_3

Figure 1. Dislocation Pile-up Avalanche Model

where k_s is the pile-up stress intensity, ℓ is the slip plane diameter, v is the dislocation velocity, K is the thermal conductivity, c^* is the specific heat per unit volume, Δx_1 is the dislocation separation at the pile-up tip, and $\bar{\alpha} = 2(1-\nu)/(2-\nu)$ with ν being Poisson's ratio. For energetic crystals, ΔT is particularly large due to the ratio (k_s/K). A comparison of the stress intensity/thermal conductivity ratios for RDX (cyclotrimethylenetrinitramine), MgO, and steel shows that the temperature rise caused by pile-up induced cracking should be 40 times greater for RDX, relative to MgO, and 200 times greater for RDX as compared to steel[3].

2. CRYSTAL SIZE DEPENDENCE FOR INITIATION

The dislocation velocity, v, in Equation (1) is stress dependent. The relationship is described on a thermal activation-strain rate analysis (TASRA) basis as

$$v = v_o \exp[-\{U_o - \int bA^* d\tau_{Th}\}/RT] , \tag{2}$$

where v_o is the reference limiting dislocation velocity, U_o is the activation energy at $T = 0$, b is the dislocation Burgers vector, A^* is the dislocation activation area, τ_{Th} is the thermal component of shear stress, and R is the gas constant[4]. The TASRA relationship has been employed in combination with Equation (1) to describe the dependence on crystal size (ℓ) of the drop-height for 50% probability of initiation, H_{50}, of a loose pile of crystals in an impact test[5]. Figure 2 shows experimental results for RDX.

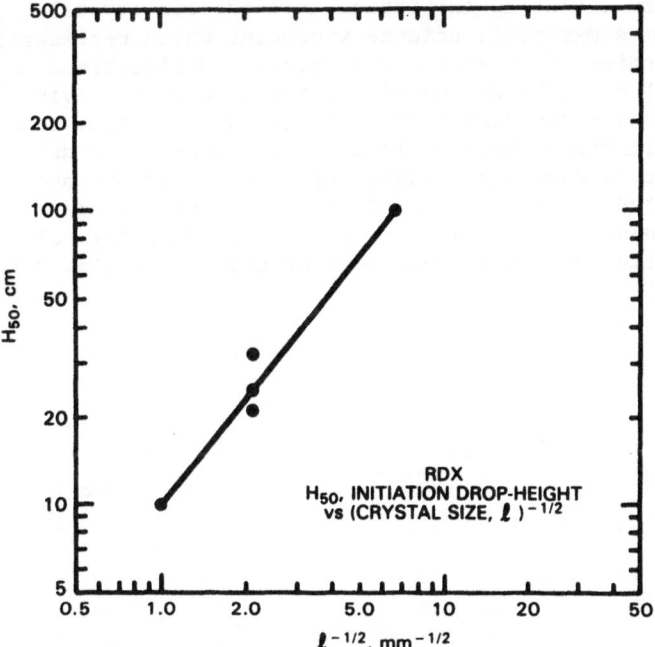

Figure 2. Crystal Size Dependence for Drop-Weight Impact Initiation

In accordance with other TASRA results[6], $(bA^*) = w_o/\tau_{Th}$. Based on the reasonable assumption that $\tau_{Th} \propto H_{50}^{1/n}$,

$$\log H_{50} \approx \log H_{50}^o + (n/m^*)\log[(20\pi^2 K\Delta T/B\bar{a}k_s v_o \exp\{-U_o/RT\})\ell^{-1/2}] , \tag{3}$$

where $m^* = (w_o/RT)$ is known as the velocity-stress exponent and $B = \ell n(2K/c^* v\Delta x_1)$. Supposing that H_{50} is obtained for an approximately constant value of ΔT, a log-log plot of H_{50} against $\ell^{-1/2}$ should give a straight line with slope (n/m^*). The predicted behavior is obeyed in Figure 2 with $(n/m^*) \approx 1.2$. The drop-height for initiation in Equation (3) increases with the following trends: increasing thermal conductivity (K); increasing temperature rise for initiation (ΔT); decreasing stress intensity for the strength of the obstacle (k_s); and decreasing temperature-modified reference dislocation velocity $\left(v_0 \exp\{-U_0/RT\}\right)$.

3. DISLOCATION PROPERTIES IN THE RDX LATTICE

The preceding model description gives emphasis to the microstructural parameters affecting initiation in energetic crystals, particularly, from a materials science viewpoint which necessarily deals with the properties of crystal dislocations. Dislocations have been observed in RDX crystals during pioneering studies involving chemical etch pitting[7] and x-ray diffraction topography[8]. Recent studies on RDX by Elban and Armstrong[9] have dealt with the deformation and cracking behavior at microindentation sites. Halfpenny, Roberts and Sherwood[10] have investigated the grown-in dislocation structures and the microindentation hardness properties of relatively perfect crystals. Extremely localized plastic deformation has been observed at the microindentation

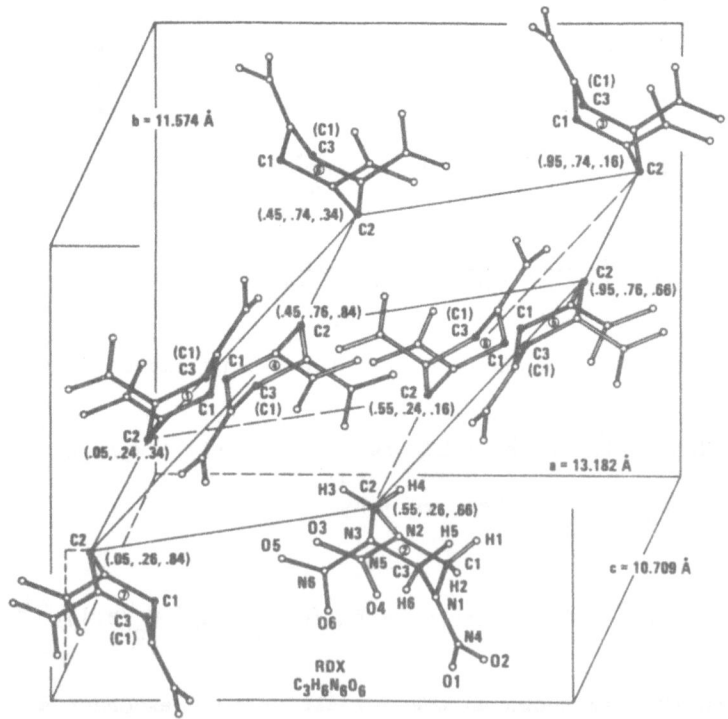

Figure 3. Orthorhombic RDX Unit Cell

sites in RDX. To some extent, the result must be explained by the relatively large self-energies of dislocations in RDX based on the large Burgers vectors characteristic of the orthorhombic unit cell containing 8 $C_3H_6N_6O_6$ molecules. The unit cell for RDX is shown in Figure 3 in accordance with the description given by Choi and Prince[11].

Using the RDX unit cell, Elban, Hoffsommer and Armstrong[12] have reasoned that the relatively large energy of the primary dislocation with Burgers vector $a[\bar{1}00]$ on an (021) slip plane can be reduced by a dislocation decomposition reaction. The result is 4 partial dislocations, involving molecules on interleaved planes, given in the reaction:

$$a[\bar{1}00] \rightarrow (a/6)[\bar{2}1\bar{2}] + (a/6)[\bar{1}00] + (a/6)[\bar{2}1\bar{2}] + (a/6)[\bar{1}00].$$

Dislocation movement by slip for the primary $[\bar{1}00](010)$ system[7] would seem to involve the interaction of specific oxygen and hydrogen atoms between certain neighboring molecules, for example, across the (040) plane in Figure 3. An interesting conjecture is that planar (or individual) molecular defects might produce out-of-place molecules which could serve as the obstacles to be broken by dislocation pile-ups. In this way, breaking and adiabatic heating of molecules (as per Figure 1) would occur sequentially at the same site during the deformation and fracturing of an energetic crystal. This conjecture is not out of line with the discussion given by Eyring[13] to the effect that initiation in the solid state must involve an effective space associated with holes, dislocations, or the separation of grains where the breaking up of a molecule can occur with respect to some strong bond in the gas phase.

4. ACKNOWLEDGMENT

This research effort has been supported at the University of Maryland (N00014-85-K-0264, NR 659-832) and at the Naval Surface Weapons Center (N00014-85-WR-24103, NR 659-797) by the Office of Naval Research, Dr. R. S. Miller, Scientific Officer. The research was accomplished while W.L. Elban was Materials Research Engineer at the Naval Surface Weapons Center.

REFERENCES

1. Armstrong, R.W., Coffey, C.S., and Elban, W.L.: 1982, Acta Metall. 30, 2111.
2. Heavens, S.N., and Field, J.E.: 1974, Proc. R. Soc. Lond. A338, 77.
3. Elban, W.L., Hoffsommer, J.C., Glover, D.J., Coffey, C.S., and Armstrong, R.W.: 1984, "Microstructural Origins of Hot Spots in RDX Explosive and Several Reference Inert Materials", NSWC MP 84-358.
4. "Dislocation Dynamics", 1968, edited by Rosenfield, A.R., Hahn, G.T., Bement, A.L., and Jaffee, R.I., McGraw-Hill Book Co., N.Y.
5. Armstrong, R.W., Coffey, C.S., DeVost, V.F., and Elban, W.L.: 1985, "Crystal Size Dependence for Impact Initiation of RDX Explosive",

submitted for publication.

6. Armstrong, R.W.: 1973, (Indian) J. Sci. Industr. Res. 32, 591.
7. Connick, W., and May, F.G.J.: 1969, J. Cryst. Growth 5, 65.
8. McDermott, I.T., and Phakey, P.P.: 1971, Phys. Stat. Sol. (a)8, 505.
9. Elban, W.L., and Armstrong, R.W.: 1982, "Proceedings of the Seventh Symposium (International) on Detonation", NSWC MP 82-334, p. 976; Armstrong, R.W. and Elban, W.L.: 1985, "Microindentation Techniques in Materials Science and Engineering", ASTM STP 889, edited by P.J. Blau and B.R. Lawn, ASTM, Phila., p.109.
10. Halfpenny, P.J., Roberts, K.J., and Sherwood, J.N.: 1984, J. Mat. Sci. 19, 1629.
11. Choi, C.S., and Prince, E.: 1972, Acta Cryst. B28, 2857.
12. Elban, W.L., Hoffsommer, J.C., and Armstrong, R.W.: 1984, J. Mat. Sci. 19, 552.
13. Eyring, H.: 1980, "Fast Reactions in Energetic Systems", edited by Capellos, C., and Walker, R.D., D. Reidel Publishing Company, Boston, p. 711.

LASER EMISSION AT 502nm INDUCED BY KrF
LASER MULTIPHOTON DISSOCIATION OF HgBr$_2$

P. Papagiannakopoulos* and D. Zevgolis**
Research Center of Crete
Institute of Electronic Structure and Laser
Heraklion, Crete, Greece

ABSTRACT: Emission resulting from KrF laser (248 nm) multiphoton dissociation of HgBr$_2$ was resolved in the 460-510 nm region. Vibrational analysis of the HgBr($B^2\Sigma - X^2\Sigma$) transition was also determined. The formation of electronically excited HgBr($B^2\Sigma$) radicals was induced by two-photon dissociation of HgBr$_2$ via a real intermediate state. The role of added inert gas on the observed emission was also examined, and the following order of efficiency was found N$_2$, Ar, Xe and Ne.

1. INTRODUCTION

The photophysics and photochemistry of molecular systems above 6eV to highly excited electronic levels (mainly of Rydberg nature), and below ionization has been examined in the past with the classical methods of vacuum ultraviolet photolysis and electron discharge. The first technique presents many experimental difficulties and also the problem of reaching certain states due to strong selection rules. The second technique provides a quite efficient method to achieve excitation of many states but the process is not selective due to the wide distribution of electron energies. The advent of high power excimer lasers enhanced the multiphoton excitation processes in molecules, and therefore the possibility of reaching selectively highly excited electronic levels, many of which

---

* Also Chemistry Department, University of Crete, Heraklion
  Crete, Greece
**Also Physics Department, University of Crete, Heraklion
  Crete, Greece

P. M. Rentzepis and C. Capellos (eds.), Advances in Chemical Reaction Dynamics, 475–482.

are prohibitted by single photon selection rules. Once the molecules acquire this level of excitation can undergo fragmentation, to electronically excited free radicals [1], or ionization with absorption of additional photons. The first process of photodissociation results to population inversion in one of the fragments and therefore to possible laser action. The first photodissociation laser was demonstrated in atomic Iodine at 1.315μm by dissociating CH_3I or CF_3I [2].

In this work we study the KrF laser (248nm) multiphoton excitation and dissociation of $HgBr_2$ vapors, which provides the possibility of achieving laser action on the $v'=0 \rightarrow v''=17-22$ transitions, of the resulting $B^2\Sigma \rightarrow X^2\Sigma$ emission band of HgBr radical. Laser emission from the above transitions in the green has been succesfully reported in the past either by electron-beam pumping [3] or by ArF laser pumping at 193 nm [4,5] of $HgBr_2$ vapors, and is very important in the development of blue-green lasers.

Moreover, we examine the vibrational analysis of the $B^2\Sigma \rightarrow X^2\Sigma$ transition of HgBr radical in the spectral region between 460-510nm. Finally we study the dependence of the above transition on the addition of an inert buffer gas, and in particular the dependence on the nature of added inert gas.

2. EXPERIMENTAL

The experimental setup has been described previously [6] and briefly consists of a KrF laser (Lambda Physik EMG 150) whose beam is collimated through a gross-shape metallic cell with 3 cm diameter and 18 cm long. The metallic cell was heated up to 100° C and the sample was effused into the centre of the cell through an orifice. A much higher pressure of inert buffer gas was initially introduced in the cell to prevent condensation on the windows. The fluorescence emission perpendicular to the laser beam was focused onto a 1m spectrometer (Jobin-Yvon HR1000) with a EMI (9683QR) photomultiplier. The photomultiplier signal was fed into a boxcar averager (PAR 162/165), where was sampled with a gate width 5-10 ns, and the output was then plotted into a chart recorder.

The mercury dibromide (Merck, 99,5% purity) was gently heated up to 100° C, resulting in a vapor pressure of approximately 100 mTorr. The pressure of added inert gases was measured with a capacitance manometer (MKS Baratron) before heating up the cell.

3. RESULTS AND DISCUSSION

The photolysis of $HgBr_2$ vapors with an unfocused KrF excimer laser beam ($I \cong 30MW/cm^2$) gave a strong fluorescence emission in the visible region below 510 nm, Fig.1. This emission band shows a clear vibrational structure and can be easily identified as the $B^2\Sigma \rightarrow X^2\Sigma$ transition of HgBr radicals [7]. A vibrational analysis of the band gave the following assignments: the lines between 485-502 nm correspond to the single transitions $v'=0 \rightarrow v''=17-22$, the lines between 503-510 nm correspond to a mixture of $v'=0,3$ $4,5 \rightarrow v''=23-29$ transitions, and the lines between 465-482

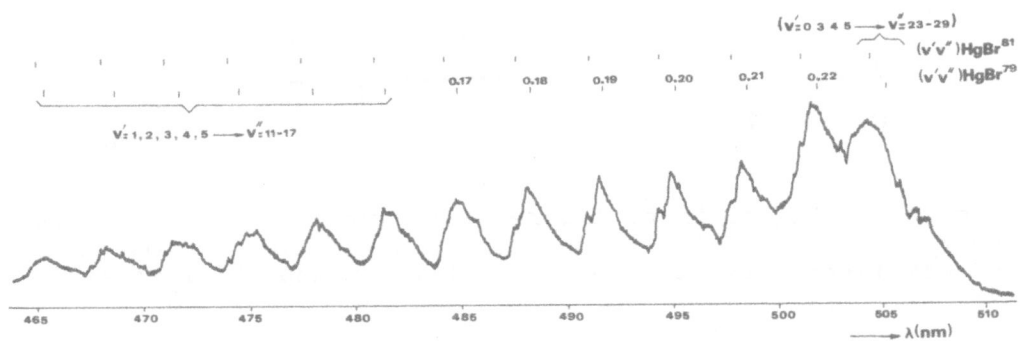

Figure 1. Emission band $B^2\Sigma \rightarrow X^2\Sigma$ of HgBr from the KrF laser photolysis of $HgBr_2$ vapor and 675 Torr Ar; cell temperature is 100° C.

nm correspond also to a mixture of $v'=1-5 \rightarrow v''=11-17$ transitions. The main molecular isotope was $HgBr^{79}$, with small traces of $HgBr^{81}$. The obtained molecular constants from this analysis are in good agreement with the literature [7,8].

Further, we studied the dependence of fluorescence
intensity of the 502nm HgBr ($v'=0 \rightarrow v''=22$) transition upon
the KrF laser intensity and was found to be two, Fig.2.
This indicates that the fluorescing HgBr radical has been
produced by a two-photon excitation process. It is known
that excitation of the first absorption band of

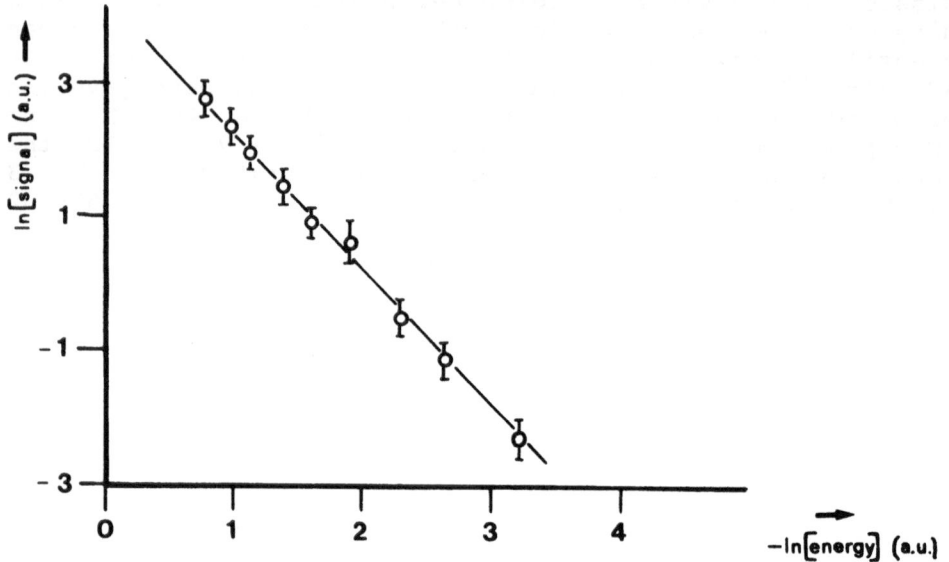

Figure 2. Logarithm of fluorescence intensity of 502 nm
HgBr$_2$ ($v'=0 \rightarrow v''=22$) transition as a function of logarithm
of KrF laser intensity for HgBr at 75° C and 675 Torr Ar.

HgBr$_2$ around 5eV ($\sigma=2 \times 10^{-18}$cm^2 [10]) does not produce any
fluorescence [9]. This band was later assigned as the $^1\Pi$u
excited state at HgBr [11], and correlates with the A
state of HgBr which is repulsive and leads to dissociation.
Therefore, the single photon KrF laser excitation of HgBr
will result to atomization, i.e.

$$HgBr_2(X^1\Sigma_g^+) \xrightarrow{h\nu} HgBr_2(1\,^1\Pi u) \rightarrow Hg(^1S) + 2Br(^2P)$$

without any fluorescence. Hence, the electronically
excited HgBr(B$^2\Sigma$) radical can only be formed by further
excitation of the parent molecule HgBr$_2$, with the
absorption of a second KrF photon to a highly excited
electronic state around 10eV. Indeed, the absorption

spectrum of $HgBr_2$ in the near UV region, between 7.5 eV and ionization (10.6 eV) is rich in Rydberg states, which are related to both the ground and excited ionic states [12]. The overall KrF laser photodissociation mechanism of $HgBr_2$

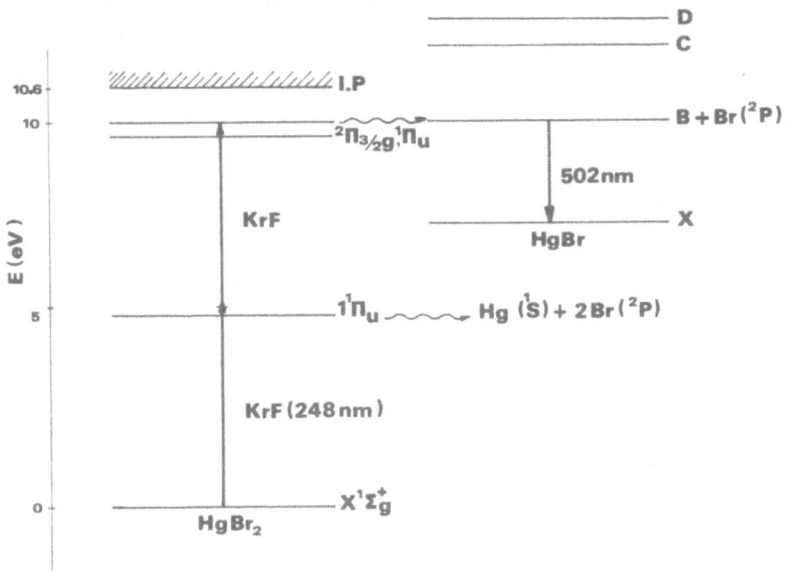

Figure 3. Diagram of possible channels of $HgBr_2$ photodissociation induced by KrF laser photolysis.

is presented in Fig.3, and includes the resonant two-photon excitation of $HgBr_2$ via a real repulsive intermediate state, and the subsequent dissociation to electronically excited $HgBr(B^2\Sigma)$ radicals and $Br(^2P)$ atoms. There is also same dissociation from the single excitation of $HgBr_2$ to $Hg(^1S)$ and $Br(^2P)$ atoms.

For higher laser intensities $(I \cong 300MW/cm^2)$ in a focused geometry, the $HgBr(B \to X)$ emission disappears and strong emission lines from highly excited Hg atoms are observed. Those emission lines are characterized as the $(6^3D \to 6^3P)$, $(7^3P_2 \to 7^3S_1)$ and $(7^3S_1 \to 6^3P)$ transitions of Hg atoms. All four excited Hg states 6^3D_3, 6^3D_2, 6^3D_1 and 7^3P_2 are in the same energy level of 8.8 eV above the ground $^1S_0$ state of Hg. Further, the dependence of the fluorescence intensity of 313.2nm $Hg(6^3D_1 \to 6^3P_1)$ transition upon the KrF laser intensity was determined and found equal to 2.5, which indicates that the formation of excited Hg atoms is induced by three or higher photon excitation processes.

The observed emission band B → X of HgBr has been found to increase with the addition of an inert gas. Furthermore, the fluorescence intensity has been found to depend on the nature of added inert gas, for the same experimental conditions, see Figure 4. For the four

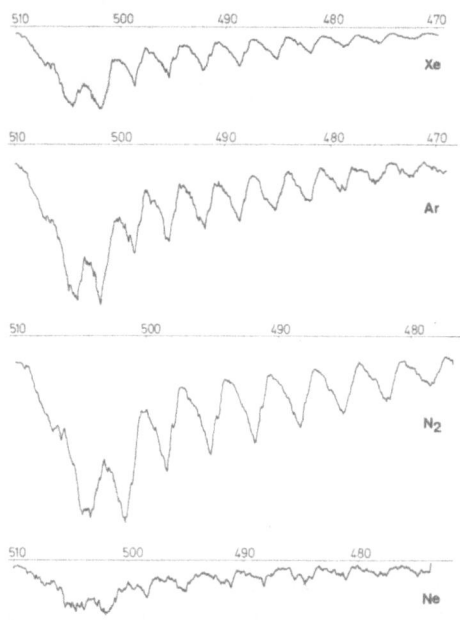

Figure 4. Emission band HgBr(B → X) from KrF laser photolysis of $HgBr_2$ vapor diluted in 375 Torr of different inert gases M=Xe, Ar, N_2, and Ne; cell temperature is 75°C.

different studied mixtures, the one with nitrogen showed the maximum emission intensity, while the neon mixture showed very little emission. The addition of inert gas in the sample results to an increase in the number of collisions, and according to the accepted photodissociation mechanism of $HgBr_2$, Fig.3, it can cause the following effects: a) decreases the up-pumping process of $HgBr_2$ which reduces HgBr photofragmentation and $HgBr(B^2\Sigma)$ formation, b) increases the population of bound excited $HgBr(B^2\Sigma)$ radicals by thermalization of initially formed vibrationally hot $HgBr(B^2\Sigma)$, and c) increases the electronic quenching of excited $HgBr(B^2\Sigma)$ radicals (the radiative lifetime of HgBr(B) state is 23.7 ns [13]). Further experiments are necessary to determine the relative importance of those processes.

4. CONCLUSIONS

The KrF laser two-photon excitation of HgBr vapor produces a strong emission band in the green around 502 nm, which is due to the $B \to X$ transition of HgBr radical. Vibrational analysis of the band showed a very good resolution for the $v'=0 \to v''=17=22$ transitions. Therefore, those transitions can be used to achieve lasing action in the green, with possible tunability over 20 nm range, by optical pumping HgBr₂ vapors with a KrF excimer laser.

The KrF laser induced emission at 502nm was also found to decrease with increasing laser intensity (300 MW/cm²). Therefore, the KrF laser intensity should not exceed a threshold value beyond which no laser action is possible.

The introduction of inert gas in the HgBr₂ sample increased the resulting emission HgBr($B \to X$), with different efficiency for the various inert gases. The observed order of efficiency for the used inert gases was, N₂, Ar, Xe and Ne. More experiments are necessary to determine the ultimate buffer gas conditions for maximum emission output.

REFERENCES

1. P. Papagiannakopoulos and C. Fotakis, in Proceedings of Conference "Photophysics and Photochemistry above 6 eV", F. Lahmani (Ed), Elsevier, Amsterdam, 1985.

2. J.V.V. Kasper and G.C. Pimentel, Appl. Phys. Lett. 5, 231 (1964).

3. E.J. Schimitschek and J.E. Celto, Opt. Lett. 2, 64 (1978).

4. E.J. Schimitschek and J.E. Celto, and J.A. Trias, Appl. Phys. Lett., 31, 608 (1977).

5. J. Husain, J.R. Wiesenfeld, and R.N. Zare, J. Chem. Phys. 72, 2479 (1980).

6. P. Papagiannakopoulos and C. Fotakis, J. Phys. Chem. 89, 3439 (1985).

7.   K.   Wieland,   Z.   Electrochem.,   64,   761   (1960).

8.   C.E.   Moore,   Atomic   Energy   Levels,   NSRDS-NBS35,
     Vol.3   (1971).

9.   K.   Wieland,   Z.   Phys.   76,   801   (1932);   77,   157
     (1932).

10. (a) J.   Maya,   J.   Chem.   Phys.,   67,   4976   (1977);
     (b)   B.E.   Willcomb,   R.   Burnham   and N.   Djen,
     Chem.   Phys.   Lett.   75,   239   (1980).

11. W.R. Wadt,   J.   Chem.   Phys.   72,   2469   (1980).

12. D. Spence, R.G. Wang, and M.A. Dillon,   J.   Chem.
     Phys.   82,   1883   (1980).

13. R.W. Waynant and J.G. Eden,   Appl.   Phys.   Lett.
     33,   708   (1978).

THE DYNAMICS OF ELECTRIC FIELD EFFECTS IN LOW POLAR SOLUTIONS: THE FIELD MODULATION METHOD

A. Persoons and L. Hellemans
Laboratory of Chemical and Biological Dynamics
Department of Chemistry, University of Leuven
B-3030 Leuven - Belgium

INTRODUCTION

The appearance of charged species and their subsequent interaction with the surrounding medium by long-range Coulombic forces are basic processes for many chemical phenomena, whether in the gas phase, in solution, or in the solid state. Traditionally ionic interactions are studied by conductance techniques and an overwhelming amount of data, as well as theoretical descriptions, exist on the conductive properties of chemical systems. Conductance measurements are, by their very nature, ideally suited to probe the distribution between conducting and nonconducting species although their exact chemical nature cannot be assessed by these techniques. The investigation of ionization mechanisms requires a knowledge of the distribution of the different species present at equilibrium, but the central problem remains the kinetic study of the interconversion processes between the conducting and nonconducting entities. A complete mechanistic study asks for a detailed investigation of number, nature, and interdependence of all elementary processes. Since these elementary processes are often fast, usually limited by diffusion, their kinetic investigation became only feasible with the advent of fast reaction techniques -photolysis, the method of choice in gaseous systems, electrochemical methods particularly suited for heterogeneous systems and chemical relaxation techniques for chemical processes in (aqueous) solution (1).

All fast reaction methods rely upon the analysis of the response of a system upon a fast appropriate perturbation. To attain a reasonable accuracy this perturbation should be large e.g., photolysis techniques or be applied to a system which is particularly sensitive to the perturbation parameter. The latter is mostly the case in chemical relaxation measurements where an intensive variable, determining the chemical equilibrium is rapidly changed. Usually an even distribution of products and reagents over the equilibrium state is required in this case for a measurement with an acceptable accuracy. This is a severe drawback of relaxation techniques when used in the

483

P. M. Rentzepis and C. Capellos (eds.), Advances in Chemical Reaction Dynamics, 483–502.

study of processes where charged species are present at very low concentration, but with a critical mechanistic function in the process.

In this paper we review the field modulation technique, a chemical relaxation technique used in the study of ionic phenomena. In this method the equilibrium between charged and uncharged species is perturbed by an electric field and the response of the systems is evaluated by a unique modulation method. A critical advantage of this modulation method is the sensitivity of the relative change in ionic concentration, even at exceedingly low concentrations of changed species (2, 3).

We will give first a critical discussion, stressing the physical aspects of electric field effects in ionic equilibria since some knowledge of this topic is a necessary prerequisite for a proper understanding of the field modulation method. In the subsequent description of the field modulation method and the results obtained we will stress the potential and generalilty of these modulation techniques for the investigation of fast processes.

THE FIELD DISSOCIATION EFFECT

The conductance of electrolyte solutions increases with electric field strength, limiting the validity of Ohms law to low field intensities. For solutions of strong electrolytes this conductance increase, known as the first Wien effect, is due to an increase in ionic mobility, itself a consequence of the impossibility to form an ionic atmosphere. For solutions of weak electrolytes the relative conductance increase with electric field strength is much more pronounced. This conductance enhancement effect is known as the second Wien effect, or the field dissociation effect, indicating that the effect is due to an increase in ionic dissociation (4, 5).

A purely thermodynamic treatment of the field dissociation effect is very difficult since an electrolyte solution is not an equilibrium state in the presence of an electric field. A first successful discription of the field dissociation effect was given by Onsager based on the laws of Brownian motion of the ions. Onsager's approach is based on the model for incomplete dissociation as put forward by Bjerrum; a configuration of two oppositely charged ions within a somewhat arbitrarily chosen distance is an ion-pair, an entity which does not take part in the conduction process. Onsager adopts the convention of Bjerrum taking for this distance q (= $e_1 e_2 / 2DkT$, symbols with their usual meaning), i.e., the distance at which the Coulombic interaction of the two ions equals their interaction with the surroundings (6).

Specifying to conditions of sufficiently small ionic concentration, hence a large Debye length, Onsager neglects shielding effects by the ionic atmosphere. Further simplifications arise from the assumption of point-like ions (their radius is negligible compared to q) and the neglect of hydrodynamic interactions.

In summary, Onsager's model is particularly simple: two oppositely charged ions are free, participate in the conduction process, when their separation is larger than q and are bound into an ion-pair at a distance smaller than q. The conceptual simplicity of this model is, unfortunately, in sharp contrast with the mathematical treatment. For the relative increase in dissociation constant with electric field strength, Onsager obtained a complex expression, the full derivation of which was not given explicitly; the significance of the assumptions made, and the approximations used in the mathematical treatment became research topics for itself.

The main result of Onsager's theory for the relative increase in dissociation constant is given as (6):

$$K(E)/K(E=0) = J_1(4(-\beta q)^{1/2})/2(-\beta q)^{1/2} = I_1(4(\beta q)^{1/2})/2(\beta q)^{1/2}$$

or, as a series expansion:

$$K(E)/K(E=0) = \sum_{n=0}^{\infty} \frac{(4\beta q)^n}{n!\ (n+1)!} \tag{1}$$

K(E=0) is the thermodynamic dissociation constant and K(E) the dissociation constant under field conditions. J_1 is the first order Bessel function and I_1 the first order modified Bessel function. q is the Bjerrum distance introduced before and 2β is a reciprocal distance defined as:

$$2\beta = |(e_1 u_1 - e_2 u_2)E|/(u_1 + u_2)kT \tag{2}$$

in which e_i, u_i are the charges, resp. the mechanical mobilities, of the ions and E the external field. The dependence upon the absolute value of the field is a noteworthy feature of Onsager's theory.

A simple picture of the physics involved in the dissociation enhancement in field conditions, and retaining the relevant features of Onsager's equation, may be derived from the very fact that 2β, the central parameter in Onsager's theory, becomes independent of the ionic mobilities for a symmetrical electrolyte. In this case the theory does not contain any parameter from transport theory and a thermodynamic approach becomes feasible. Specified for symmetrical electrolytes a physical meaning of 2β becomes clear: $(2\beta)^{-1}$ is the

distance at which the two ions form a dipole for which the electrostatic stabilization energy in an external field E equals $kT\cos\theta$, θ
being the angle between dipole and field.

Another dipole which is of some importance here is what we may
call the "Bjerrum-dipole" i.e., the configuration in which the two
ions are at the distance q. Compared with the configuration in which
the two ions are in close contact the relative probability of the
Bjerrum-dipole configuration is vanishingly small. Nevertheless, the
Bjerrum-Dipole is of critical importance for the dissociation process
since the probability for this state for collapsing into the ion-pair
equals exactly the probability for separation into free ions.

The enhanced dissociation under field conditions can be understood qualitatively from the reduction in electrostatic work needed
to separate the ion-pair into free ions. Alternatively this can also
be seen as a deformation of the Bjerrum-association sphere, radius q,
in the presence of an external field. In the following discussion,
we will show the probability for dissociation for a uni-univalent
ion-pair increases in an external field. The electric potential of
this ion-pair in the field E, oriented along the z-axis, is given
(Fig. 1) by:

$$U (r,\theta) = - \frac{e^2}{r D} - eEr \cos \theta \tag{3}$$

in which $(r,\theta$ are the coordinates of the negative ion, the positive
ion being at the center of the spherical coordinate system. This
potential has a maximum, viewed from the center, at a distance
$r_{max}(\theta)$ at which the potential is:

$$U (r_{max}(\theta),\theta) = -2 (\frac{e^3 E}{D} \cos \theta)^{1/2} \tag{4}$$

In the absence of a field the work to separate the ion-pair into free
ions is given by (per molecule):

$$W = \frac{e^2}{a D} = \Delta G^0(E=0) = -kT \ln K(E=0) \tag{5}$$

$\Delta G°(E=0)$ being the free energy change in the dissociation process and
a the distance of closest approach for the ions in the ion-pair.

In field conditions, the maximum work needed for ion-pair separation
is given by the height of the energy barrier, given by $U (r_{max}(\theta),\theta)$:

$$W (E) = \frac{e^2}{a D} + U_{max} = \frac{e^2}{a D} - 2 (\frac{e^3 E}{D} \cos \theta)^{1/2} = \Delta G^0(E=E) \tag{6}$$

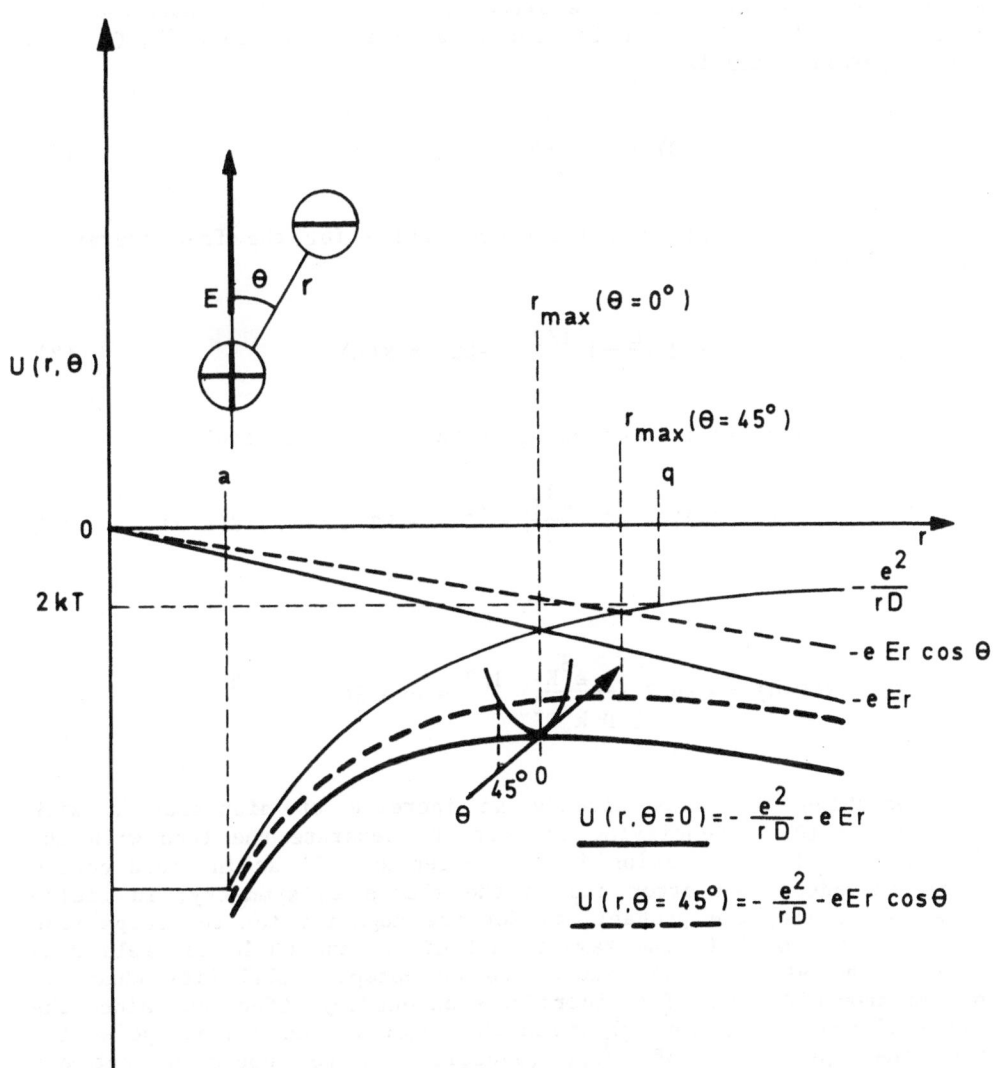

Figure 1. Dissociation of an ion-pair in the presence of an external field E. The energy barrier and relevant distance parameters: a, distance of closest approach (or contact distance of the ions); r_{max}, distance at which the electric potential is maximal in an external field E; q, Bjerrum distance, i.e., the distance at which the inter-action energy equals 2 kT. (See text for calculations.)

We now see that the work for ion-pair separation is lowered by an amount proportional to the electric field strength. At relatively low field strength we have a very small decrease in the separation work. As a result the direction at which the negative ion can escape from the potential well is relatively independent of θ. However, at increasingly higher fields the direction along $\Theta = 0$ becomes increasingly favorable. This can be clearly seen if we write U (E, Θ) to a first approximation is:

$$U (E, \Theta) = U (E,0) + \frac{1}{2} (\frac{e^3 E}{D})^{1/2} \Theta^2 \qquad (7)$$

At high field strength we, therefore, write for the free energy of dissociation:

$$\Delta G^0 (E) = \frac{e2}{a\ D} - 2 (\frac{e^3 E}{D})^{1/2} = -kT \ln K(E) \qquad (8)$$

and for the decrease in free energy with an external field:

$$\Delta G^0 (E) - \Delta G^0 (E=0) = -2 (\frac{e^3 E}{D})^{1/2} = - kT \ln K(E)/K(E=0) \qquad (9a)$$

or:

$$K(E)/K(E=0) = \exp 2 (\frac{e^3 E}{D\ k^2 T^2})^{1/2} = \exp 2(4\beta q)^{1/2} \qquad (9b)$$

This equation indicates already an increase of dissociation with electric field. Identifying the work to separate the ions with the free energy of dissociation in field-free as well as in field conditions introduces an error due to the change in symmetry; in field-free conditions, the probability for the negative ion to escape from the potential well is the same in all directions while in field conditions the symmetry is reduced to an escape probability which is peaked around $\Theta = 0$. This introduces an entropy effect and since the width of the "gate" through which the negative ion can escape scales with the square-root of $e^3 E/D$ (equation 7), we should correct our previous equation with this scale factor, properly adjusted dimensionally. We, therefore, should write for equation 9b:

$$K(E)/K(E=0) \sim (\frac{e^3 E}{D\ k^2 T^2})^{-n/2} \exp 2 (\frac{e^3 E}{D\ k^2 T^2})^{1/2}$$

$$= (4\beta q)^{-n/2} \exp 2 (4\beta q)^{1/2} \qquad (10)$$

This expression is to be compared with the asymptotic expansion of Onsager's equation, given as:

$$K(E)/K(E=0) = (\frac{1}{4\pi})^{1/2} (4\beta q)^{-3/4} \exp 2 (4\beta q)^{1/2} \qquad (11)$$

We obtain here the asymptotic expansion of Onsager's equation since the approximations we made become valid at high field strength. Also, the distance for the maximum of the electric potential becomes equal to the Bjerrum distance at a value of $4\beta q$ equal to 4.

Equation 10 has been derived very heuristically, but this derivation has the important advantage to show some of the physics involved in the field dissociation effect.

THE FIELD MODULATION TECHNIQUE

As a consequence of the field dissociation effect the concentration of free ions, hence the conductance of a weak electrolyte solution, will increase upon the application of an electric field of sufficient intensity. The relative increase of the dissociation constant is, according to equation 1, in the linear approximation for a uni-univalent ion-pair given by:

$$\frac{K(E) - K(E=0)}{K(E=0)} = \frac{\Delta K}{K} = 2\beta q = \frac{e^3 |E|}{2 D k^2 T^2} \qquad (12)$$

The relative change in conductance σ, equal to the relative change in degree of dissociation α, is therefore:

$$\frac{\overline{\Delta\sigma}}{\sigma} = \frac{\overline{\Delta\sigma}}{\alpha} = \frac{1 - \alpha}{2 - \alpha} \frac{\Delta K}{K} = \frac{1 - \alpha}{2 - \alpha} \frac{e^3}{2 k^2 T^2} \frac{1}{D} |E| \qquad (13)$$

From this equation some important features of the field dissociation effect emerge clearly. First, the effect is largest under conditions where dissociation in free ions is almost negligible, i.e., $\alpha \to 0$. This situation is predominantly encountered in low-polar media, where the effect is for that matter amplified by the very long range of Coulombic interactions, the reason the field dissociation effect is inversely proportional with dielectric constant. The dependence upon the absolute value of electric field strength is a shortcoming of the theoretical description; even a linear dependence, used as a verification of his theory by Onsager, is not always borne out by the experimental data. It is also important to note that equation 13 describes the effect of d.c. fields, for which the

Onsager theory was made, and neglects all possible frequency-dependent effects.

From the viewpoint of the experimental physicist a very important conclusion emerges from the foregoing conclusion: a solution of a weak electrolyte behaves as a nonlinear, field-dependent resistance at high field intensities. The resistance of such a field-dependent element, when subjected to a high a.c. voltage, can be written in a simplified form as:

$$R = \overline{R} + \Delta R \sin \omega_E T \qquad (14)$$

where $\overline{R}$ is the frequency-independent part of the resistance, ω_E describes the frequency dependence, and ΔR is a measure of the non-linearity of the resistance. If we now measure the resistance R with another a.c. voltage V, of frequency ω, applied at the same or other terminals of a properly designed cell, we see a current flowing in the measuring circuit which has frequency-components originally not present:

$$I = V.R. = V_M \sin \omega t\ (\overline{R} + \Delta R \sin \omega_E t)$$

$$= \overline{R}\, V_M \sin \omega t + 1/2.\ V_M\, R \cos\ (\underline{\omega_E - \omega})t - 1/2.\ V_M\, R \cos(\underline{\omega_E + \omega})t$$

fundamental   intermodulation products

These simple arguments are basic for all electronic modulation techniques and are at the core of the whole field of electronic communication. Another very important aspect of modulation techniques is the bandwidth of the nonlinear element. Indeed, if the nonlinear element (resistance) in a modulator circuit is subjected to (very) high frequencies, the modulation efficiency will decrease, according to some dispersion law describing the frequency dependency of the nonlinearity (ΔR).

Returning to the solution of a weak electrolyte as a field-dependent resistance, we immediately see a physical cause for the dispersion of the field dissociation effect with a.c. fields: the increase in ionic dissociation cannot be instantaneous, but is determined by the relaxation time of the equilibrium between free ions and ion-bound states. The time dependence of the conductance change can be obtained from the general equation:

$$\tau\, \frac{d\Delta\sigma}{dt} + \Delta\sigma = \overline{\Delta\sigma} \qquad (16)$$

τ is the relaxation time for the conductance increase and $\overline{\Delta\sigma}$ represents the equilibrium conductance change, the change which would be measured if ionization were instantaneous, and is given by equation 13. The application of equation 16 implicitly assumes that the relative conductance change is very small.

Since a weak electrolyte solution behaves as a field dependent resistance under high field conditions, a cell containing such a solution subjected to electric field pulses is a nonlinear resistance. Such a cell can be used as the nonlinear element in a modulator circuit. The efficiency and frequency-dependent properties of such a modulator circuit are related to the field dissociation effect in the electrolyte solution and to its dispersion, which depends upon the relaxation of the ionization equilibrium.

The schematic circuit for the modulation measurements is given in Fig. 2a together with the general specifications for the experimental set-up. In this experimental set-up high-frequency, high-voltage pulses are switched periodically on and off between the electrodes of a (three-electrode) cell containing a weak electrolyte solution. As a consequence the conductance of this solution is modulated with the switching repetition frequency. The degree of modulation is measured by an a.c. voltage signal, of small amplitude, applied to the high-voltage electrodes. The modulation of the cell resistance introduces intermodulation signals at frequencies equal to sum and difference of the pulse repetition frequency and the measuring signal frequency. These frequencies are chosen equal to that we obtain, as the difference term, a d.c. signal. The d.c. current flowing in the circuit can be used to charge a capacitor. Since the d.c. signal appears in a circuit in which only a.c. signals are present originally from the generators this d.c. signal can be measured very accurately, after suitable filtering of a.c. components. The d.c. intermodulation signal is a measure of the magnitude of the field dissociation effect at low modulation frequencies where we have an in-phase shift of the ionic equilibrium with the field pulses. At higher frequencies the d.c. signal will show a dispersion in its amplitude, the frequency behavior of which is linked to the relaxation of the ionization process.

In Fig. 2b the equivalent measuring circuit is shown from which the d.c. intermodulaton product and its dispersion can be derived. The voltage over capacitor C, at modulation frequency ω, is given by (2):

$$\frac{dV}{dt} + \frac{\sigma(\omega,t)}{C} V = \frac{C_c}{C} \omega V_M \cos \omega t + \frac{\sigma(\omega,t)}{C} V_M \sin \omega t \qquad (17)$$

subject to the conditions that $C_c \ll C$ and that we have no phase-shift between the envelope of the high-voltage, high-frequency pulses and the measuring voltage signal.

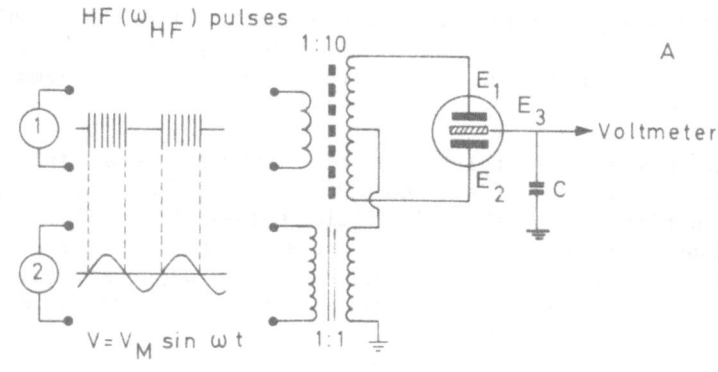

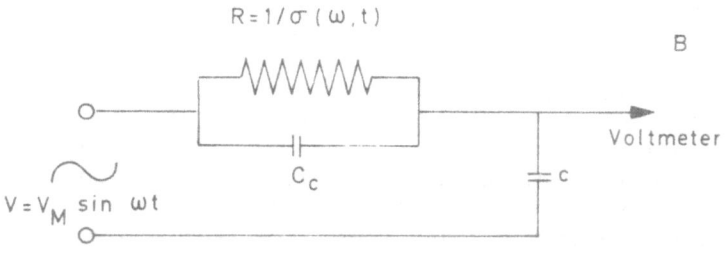

Figure 2a. Schematic circuit for the measurements of the field dissociation effect and its dispersion. Generator 1 is a square wave (0.01-30 kHz) modulated high frequency (0.1-1 MHz) high voltage (0.1-1 kV_{pp}) sinusoidal oscillator driving a step-up (1:10) transformer with a ferrite core. Generator 2 is a low voltage (1-20 V_{pp}) oscillator synchronously locked on the square wave modulating the high voltage oscillator. The sample cell contains, in a very symmetric arrangement, three electrodes, two of which (E_1 and E_2) are connected to the high voltage transformer output terminals. Electrode E_3 is, over capacitor C (10pF-22nF), grounded for a.c. signals. Electrode area is relatively uncritical while the spacing between the HV electrodes and the center electrode E_3 is of the order of 0.2-1 mm, depending on particular cell design. The center tap of the high voltage transformer is connected, over an isolation transformer, to oscillator 2.

Figure 2b. Equivalent measuring circuit used in the calculations of the intermodulation products.

Particularly difficult is the determination of $\sigma(\omega,t)$. Indeed we use in this experimental set-up high-frequency fields to induce the field dissociation effect while Onsager's theory, and all other theoretical approaches, are based upon d.c. fields. We, therefore, must define an effective field strength E_o which is equivalent to the d.c. field used in the theory. Here we have taken the envelope of the high-frequency, high-voltage signal as the electric field with an effective strength given by the linear average of the absolute value of the high field:

$$E_o = \frac{1}{T_{HF}} \int_0^{T_{HF}} | E_M \sin \omega_{HF}t| \ dt = E_{pp}/\pi \qquad (18)$$

where E_M is the amplitude of the high-voltage, high-frequency field, E_{pp} its peak-to-peak value, and T_{HF} the high frequency period.

Although the use of high-frequency, high-voltage pulses is experimentally obvious, we must be prepared for specific frequency effects which are neglected altogether by considering only the envelope of these high-frequency pulses as the effective electric field. Indeed we are aware that at very high frequencies in the pulses the external field will no longer perturb the Brownian motion of the ions and the field dissociation effect may vanish completely.

With the previous discussion in mind we can write for $\sigma(\omega,t)$:

$$\sigma(\omega,t) = \sigma_o[1 + \frac{\Delta\sigma(\omega,t)}{\sigma_o}] \qquad (19)$$

where σ_o is the equilibrium conductance measured at low field strength and $\Delta\sigma(\omega,t)$ is the shift in conductance at the modulation frequency. This shift may be calculated from equation 16 with $\overline{\Delta\sigma}$ given as:

$$\overline{\Delta\sigma} = \begin{cases} g\overline{\Delta\sigma} \text{ for } nT < t < (n + 1/2)T \\ \\ 0 \quad \text{ for } (n + 1/2)T < t < (n + 1)T \end{cases} \qquad (20)$$

where T is the modulation period ($= 2\pi/\omega$) and g is a correction factor taking mainly into account the deviations from the theoretical predictions, deviations which are due to the use of alternating high fields. g is of the order of one, but the exact value, its frequency dependence, the possible effects of the field intensity, the effect of the ionic concentration...are unknown.

Solving equation 16 with $\overline{\Delta\sigma}$ as given by equation 20 we obtain for $\sigma(\omega,t)$, neglecting transient terms which disappear after sufficient $(t \gg \tau)$ time:

$$\sigma(\omega,t) = \overline{\sigma}\left[1 + m\sum_{k=0}^{\infty}\frac{\sin(2k-1)\omega t - (2k-1)\omega\tau\cos(2k-1)\omega t}{(2k-1)(1+(2k-1)^2\omega^2\tau^2)}\right] \quad (21)$$

where $\overline{\sigma}$ is the mean equilibrium conductance under field conditions $(\overline{\sigma} = \sigma_o + \overline{\Delta\sigma}/2)$ and m equals $2g\overline{\Delta\sigma}/\pi\overline{\sigma}$. Equation 17 can now be written as:

$$\frac{d}{dt}V + \frac{\overline{\sigma}}{C}(1+m\Sigma)V = \frac{C_c\omega V_M}{C}\cos\omega t + \frac{\overline{\sigma}V_M}{C}(1+m\Sigma)\sin\omega t \quad (22)$$

Σ is the summation given in Equation 20. We hence obtain a nonhomogeneous linear differential equation with time dependent coefficients. This equation can be solved by a perturbation method since the parameter m, measuring the modulation properties of the nonlinear cell resistance, is small. Writing the voltage over capacitor C as:

$$V = V_o + m V_1 \quad (23)$$

we obtain a differential equation for V_1 which gives the harmonic components due to the modulation:

$$\frac{D V_1}{dt} + \frac{\overline{\sigma}}{C}V_1 = (\frac{\overline{\sigma}}{C}V_M\Sigma)\sin\omega t - \frac{\overline{\sigma}}{C}V_o\Sigma \quad (24)$$

The solution of this equation is straightforward and we have as a solution a series in harmonic components containing a frequency-independent term which is the d.c. voltage measured over capacitor C:

$$V_{dc} = g\frac{\overline{\Delta\sigma}}{\overline{\sigma}}\frac{V_M}{\pi}\frac{1}{1+\omega^2\tau^2}\frac{(\overline{R}C\omega)^2}{1+(\overline{R}C\omega)^2}(1-\frac{C_c}{C})(1-\frac{\omega\tau}{\overline{R}C\omega}) \quad (25)$$

$\overline{R}$ is the mean equilibrium cell resistance under field conditions ($=1/\overline{\sigma}$). In low polar media $\overline{R}$ is rather large, experimentally a necessary condition for the application of high electric fields, and with a reasonable choice for C, we see that $\overline{R}C\omega$ is much larger than 1, even at very low values of ω; normally, $\tau < \overline{R}C$ and we are free to choose $C_c < C$. For reasonable experimental conditions we therefore have:

$$V_{dc} = (g \frac{\overline{\Delta\sigma}}{\overline{\sigma}} \frac{V_M}{\pi}) \frac{1}{1 + \omega^2\tau^2} = \frac{V_{dc\ max}}{1 + \omega^2\tau^2} \qquad (26)$$

The amplitude of the d.c. voltage measured over capacitor C shows, therefore, a simple dispersion from which the relaxation time of the ionization equilibrium is readily determined. The d.c. signal measured at frequencies much lower than the reciprocal relaxation time is a direct measure of the field dissociation effect in the experimental conditions used. The difference between this maximum value and the value calculated from Onsager's theory is contained in the correction factor g. This allows to investigate high frequency effects in the field dissociation effect.

RESULTS AND DISCUSSION

In Fig. 3 some typical measurements of the dispersion of the field dissociation effect in low polar solutions of a weak electrolyte (tetrabutylammonium picrate, TBAP) are presented. The excellent agreement between experimental data and equation 26 is obvious for all cases and the relaxation time for the relative conductance increase can easily be determined with very good accuracy. To obtain information on the ionization mechanism(s) in the systems investigated, this relaxation time must be linked to (the) chemical relaxation time(s) of the equilibria present in solution.

A particularly rewarding feature of the field modulation technique is the relatively simple and uncritical design of the measuring cell. Field modulation experiments can be carried out easily at a wide range of temperature and hydrostatic pressure in order to obtain information on the activation parameters for the processes involved (7, 8).

At not too low polarity of the medium and relatively low salt concentration, conductance data indicate a sole equilibrium between ion-pairs and free ions in solutions of tetraalkylammonium salts. Chemical relaxation theory predicts for such an equilibrium a relaxation time τ:

$$\tau^{-1} = k_d + 2k_r K_d^{1/2} C_o^{1/2} \sim 2k_r K_d^{1/2} C_o^{1/2} = 2(k_r k_d)^{1/2} C_o^{1/2} \qquad (27)$$

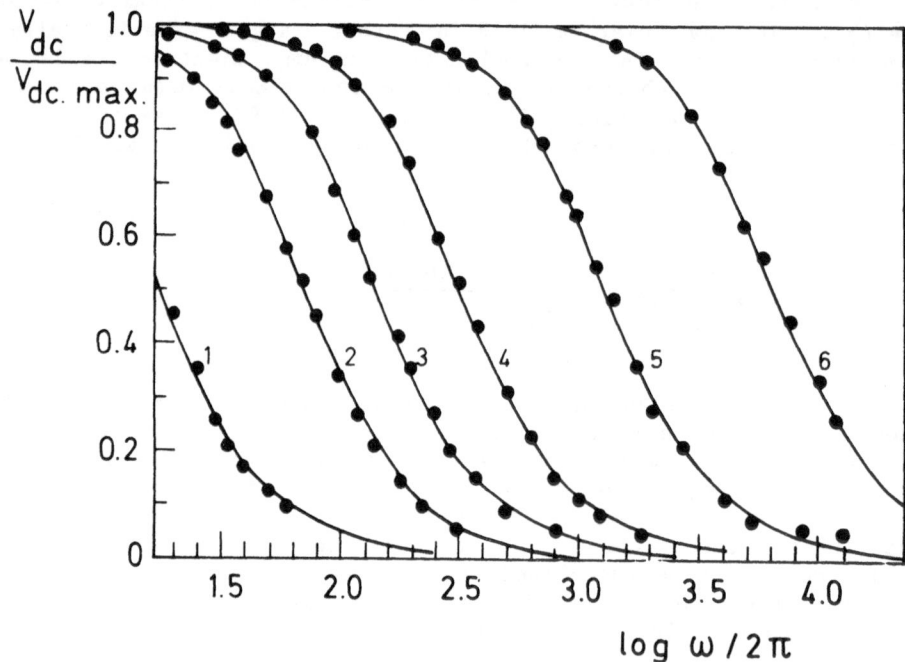

Figure 3. Dispersion of the field dissociation effect with the modulation frequency for solutions of TBAP in different solvents at 298 K, experimental points, theoretical curves (Equation 26).

1. Benzene, $8.74 \; 10^{-5}$M, $\tau = 8.78 \; 10^{-3}$ s
2. Benzene, $1.52 \; 10^{-4}$M, $\tau = 2.20 \; 10^{-3}$ s
3. Benzene-chlorobenzene (70:30 v%), $6.4 \quad 10^{-6}$M, $\tau = 1.16 \; 10^{-3}$ s
4. Benzene-chlorobenzene (70:30 v%), $1.95 \; 10^{-5}$M, $\tau = 5.15 \; 10^{-3}$ s
       P = 650 atm
5. Benzene-chlorobenzene (70:30 v%), $8.8 \quad 10^{-5}$M, $\tau = 1.27 \; 10^{-4}$ s
6. Benzene, $3.60 \; 10^{-3}$M, $\tau = 2.28 \; 10^{-5}$ s

where C_o is the total salt concentration and we used the fact that K_d, the ionization equilibrium constant $K_d(=k_d/k_r)$, is very small in these media. k_d and k_r are the dissociation, respectively recombination, rate constants for the ionization reaction. Equation 27 predicts a linear relation between the reciprocal relaxation time and the square-root of total salt concentration. This simple relation is always borne out by all experiments on systems where the conductance data are also to be described by a unique equilibrium between ion-pairs and free ions. A knowledge of the ionization equilibrium constant K_d, obtained from conductance data, and the reciprocal relaxation time, Equation 27, allow the determination of the ion-pair dissociation and ionic recombination rate constants.

The unique conclusion of all experimental observations in these simple systems is that the dissociation-recombination processes can be described entirely, and very successfully, by a sphere-in-continuum model where association and dissociation rates are given by the laws of electrodiffusion and the Coulombic interaction with the polarisable medium determines completely the energetics and volume-effects of these processes (8). Some experimental data and the corresponding calculated values are given in Table 1.

Table 1. Rate and Activation Parameters for the Dissociation of TBAP in Benzene-Chlorobenzene (C) Mixtures at 298 K.

Experimental

D (vol% C)	$k_r/10^{-11}$ $M^{-1}s^{-1}$	k_d/s^{-1}	$(E_a)_r$	$(E_a)_d/kJ \cdot mol^{-1}$	$\Delta V_r^{\ddagger}$	$\Delta V_d^{\ddagger}/$ $cm^3 mol^{-1}$
3.22 (30)	3.01	0.109	9.7	41.2	19.7	-32.3
3.55 (40)	2.61	1.00	9.3	36.8	18.0	-30.8
3.87 (50)	2.64	6.45	7.0	34.4	20.5	-30.6

Calculated*

D (vol% C)	$k_r/10^{-11}$ $M^{-1}s^{-1}$	k_d/s^{-1}	$(E_a)_r$	$(E_a)_d/kJ \cdot mol^{-1}$	$\Delta V_r^{\ddagger}$	$\Delta V_d^{\ddagger}/$ $cm^3 mol^{-1}$
3.22 (30)	2.56	0.089	11.3	43.3	18.9	-33.0
3.55 (40)	2.26	0.83	11.1	40.1	18.4	-31.1
3.87 (50)	2.02	4.75	11.0	36.3	18.0	-29.8

*Calculated from a sphere-in-continuum approach; for details of the calculation and accuracy of data, see J. Everaert and A. Persoons, J. Phys. Chem., 85 3930 (1981), 86 546 (1982).

The importance of a kinetic investigation and especially the knowledge of the activation parameters for the elucidation of reaction mechanisms is clearly shown in field modulation measurements of solutions of fluorenyl-lithium ion-pairs in diethylether as a solvent (9). Although the rate constants fitted reasonably well with diffusion-controlled association-dissociation processes at one temperature, the activation energy for the dissociation process was negative. This is definite proof for the intervention of a solvent-separated ion-pair in the dissociation pathway; such a species was postulated as àn intermediate on analogy with other systems, but was too low in concentration to be detected spectrophotometrically in this system.

At increasing concentration of salt, or at decreasing polarity of the solvent, the conductance of tetra-alkylammonium salt-solutions indicates the emergence of new species such as triple ions, dimers (quadrupoles), and even higher aggregates, in equilibrium with free ions and ion-pairs. Normally one expects, therefore, the emergence of more relaxation processes under these experimental conditions. The data on the dispersion of the field dissociation effect for TBAP in benzene solution, presented in Fig. 3, show convincingly this is not the case: a simple dispersion, corresponding to a single relaxation process, is observed. This is always the case and no single example is known of a system showing several relaxation processes in its dispersion. This odd behavior is intimately linked to the mechanisms underlying charge-separation processes in solution. The application of an electric field does not perturb equilibria between ion pairs and free ions, but rather the equilibria between conducting and nonconducting states, the ionization mode proper. This ionization mode is slow compared to all other processes since charge-separation is energetically very favorable compared to ion-dipole and dipole-dipole interactions. The perturbation of the latter interactions by an electric field is, on the level of the field dissociation effect, a second order effect.

If higher aggregates are present in solution, e.g., triple ions and quadrupoles, the reaction scheme for the formation of charged species becomes exceedingly complex, as seen from Fig. 4. For such a complex scheme a single relaxation time can be calculated if all equilibria without charge separation (the "horizontal" equilibria in the scheme) are assumed to be fast and that the concentration of all species are negligible compared to the ion-pair concentration. The expression for this relaxation time, equation 4, is a very involved equation, with pleasing symmetrical appearances, but useless for experimental tests. Fortunately, depending upon the relative stability of the species present in solution, according to scheme IV, this expression has limiting forms which clearly explains many experimental observations.

Gratifyingly, equation 4 reduces to equation 27 at low concentrations of ion-pairs and/or low stability of the aggregates species. At higher ion-pair concentration and/or high stability of both kinds of triple ions, and assuming that all ionic recombination rate constants are diffusion controlled and, therefore, of comparable magnitude, we obtain for the recombination term in equation 4:

$$\text{recombination term} = 2\,k_T\,\left(\frac{K_d}{K_3^- K_3^+}\right)^{1/2}\,c_o^{3/2} \tag{28}$$

This 3/2 power law relation is easily understood since at high stability of triple ions the predominant charged species under the assumed conditions, the main recombination process will be between cationic and anionic triple ions, the concentration of which depends on the 3/2 power of the total ion-pair concentration.

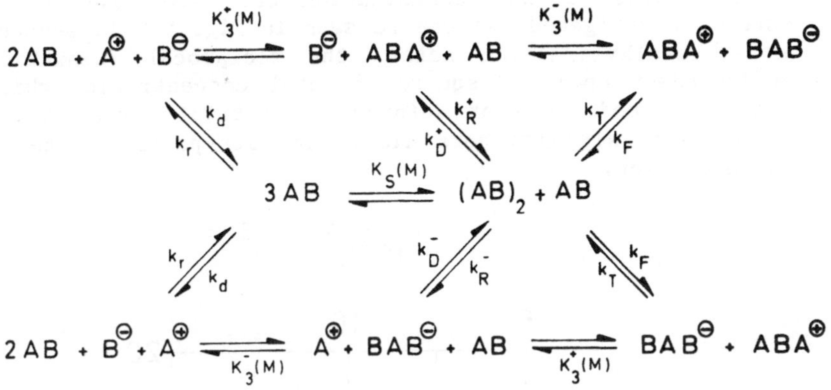

$$2AB + A^{\oplus} + B^{\ominus} \underset{}{\overset{K_3^+(M)}{\rightleftharpoons}} B^{\ominus} + ABA^{\oplus} + AB \underset{}{\overset{K_3^-(M)}{\rightleftharpoons}} ABA^{\ominus} + BAB^{\ominus}$$

$$\underset{k_r}{\overset{k_d}{\searrow}} \qquad \underset{k_D^+}{\overset{k_R^+}{\searrow}} \qquad \underset{k_F}{\overset{k_T}{\searrow}}$$

$$3AB \underset{}{\overset{K_S(M)}{\rightleftharpoons}} (AB)_2 + AB$$

$$\underset{k_d}{\overset{k_r}{\nearrow}} \qquad \underset{k_R^-}{\overset{k_D^-}{\nearrow}} \qquad \underset{k_T}{\overset{k_F}{\nwarrow}}$$

$$2AB + B^{\ominus} + A^{\oplus} \underset{K_3^-(M)}{\rightleftharpoons} A^{\oplus} + BAB^{\ominus} + AB \underset{K_3^+(M)}{\rightleftharpoons} BAB^{\ominus} + ABA^{\oplus}$$

dissociation term

$$\tau^{-1} = k_d + (k_D^- + k_D^+)\, 4C_0/K_S + 9/16 \cdot k_F K_S (4C_0/K_S)^2 \Big/ (1 + 4C_0/K_S)$$

$$+ 2 \; \frac{K_d^{1/2} C_0^{1/2} (1 + C_0/K_3^+)^{1/2} (1 + C_0/K_3^-)^{1/2}}{1 + 2C_0/K_3^+ + 2C_0/K_3^- + 3C_0^2/K_3^- K_3^+} \left[k_r + 2\left(\frac{k_R^+}{K_3^+} + \frac{k_R^-}{K_3^-}\right) C_0 + \frac{k_T}{K_3^- K_3^+} C_0^2 \right]$$

recombination term

Figure 4. Ionization scheme of an ion-pair (AB) in the presence of dimerization and triple ion formation. Expression (Equation 4) for the reciprocal relaxation time for the equilibrium ionphores – ions.

Assuming only the anionic triple ion as stable, which is equivalent with the condition $K_3^+ < C_0 < K_3^-$, we find for the recombination term:

$$\text{recombination term} = 2\, k_R^- \left(\frac{K_d}{K_3^-}\right)^{1/2} C_0^1 \tag{29}$$

We see, therefore, that a linear dependence upon total ion-pair concentration is an indication of unilateral triple ion formation and the recombination process is mainly between a simple and a triple ion.

Tentatively we may conclude from the foregoing discussion that the concentration dependence of the reciprocal relaxation time is a direct measure of the main ionic recombination process in solution.

All limiting forms of the recombination term are observed in different systems investigated, as can be seen in Fig. 5. In benzene solutions of tetra-alkylammonium salts, the reciprocal relaxation time is even dependent upon the square of total concentration which would indicate a preponderance of quintuple and triple ions in the recombination process, although here dimers and ion-pairs may be in comparable concentration.

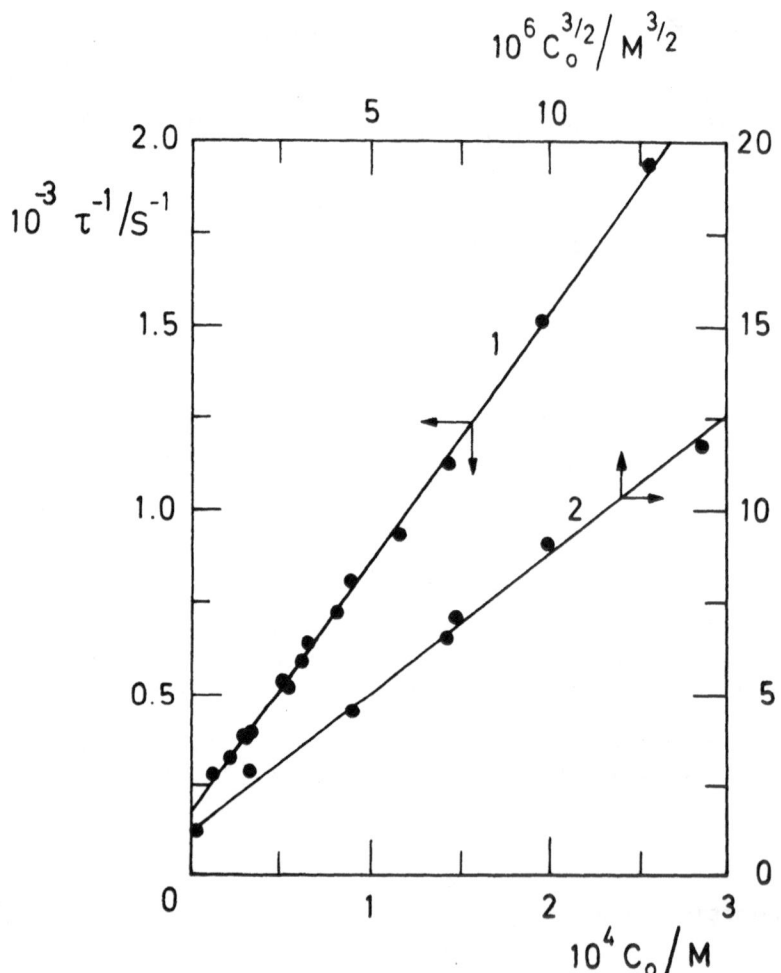

Figure 5. Dependence of the reciprocal relaxation time upon concentration for solutions of TBA-salts in the presence of higher aggregates (as determined from conductance data).

1. TBA-bromide in benzene-nitrobenzene (3.22 v%, D = 2.90) at 298K:
$\tau^{-1} = 1.06 \ 10^3 + 7.68 \ 10^8 \ c_o^{3/2}$
2. TBA-picrate in benzene-chlorobenzene (16 v%, D = 2.78) at 298 K:
$\tau^{-1} = 1.69 \ 10^2 + 6.88 \ 10^6 \ c_o$

Until now we have mainly focussed attention on the recombination term. The dissociation term in equation 4 poses some difficult problems when we compare with experimental data. At low concentration of ion-pairs the dissociation term is equal to k_d, hence small enough to be neglected. At higher concentration the dissociation term, as calculated, is concentration-dependent. This is in contradiction with most experimental observations: when the concentration-dependence of the reciprocal relaxation time is higher than square-root there is <u>always</u> an important intercept. This intercept indicates the presence of a relatively important monomolecular process involved in the relaxation. A satisfactory explanation of this behavior would be the uncoupling of the aggregation processes from the charge-generating processes. Such an uncoupling would lead to a dissociation term which is simply the sum of all dissociation rate constants for the dissociation processes present in solution. Such an uncoupling is very plausible since the concentration of ions is very small in these media and their concentration change upon the application of the field is therefore negligible compared to the concentration of nonconducting species. This is equivalent to stating that the aggregation equilibria are not perturbed by the field and are hence not coupled to the relaxation of charge-generating reactions. This explanation is corroborated by the activation parameters for the intercept which identify this monomolecular process as linked to ionic dissociation phenomena.

CONCLUSIONS

The field modulation method is a very versatile tool for the study of the dynamics of ionization processes in solution, over a wide range of experimental conditions. Being a stationary technique, using high electric fields, the method is limited to systems of relatively low conductance, although a large variety of systems can still be investigated. The field modulation method is moreover not restricted to solutions and was successfully applied to the investigation of autoprotolysis phenomena in ice and solid acetic acid. In general, due to the nature of the perturbation and the response detection used in this method, the relaxation mode(s) of processes generating charge carriers, perhaps coupled to slower processes, is specifically detected.

The measuring principle, using a chemical system as a modulator-element in a well-chosen circuit and taking advantage of the extreme sensitivity of frequency-conversion methods, can generally be exploited for the investigation of the dynamics of processes underlying nonlinear behavior in (chemical) systems. An exciting possibility is the study of photo-conductive processes in liquid or solid phase, where a conductance modulation can be impressed by intensity modulated light.

The use of (chemical) systems where an external perturbation modulates their optical properties in electro-optical circuits opens also novel ways to study the dynamics of many important phenomena.

REFERENCES

1. Weissberger, A. and Hammes, G. Ed.: 1973, "Techniques of Chemis-
    try," Interscience, New York, NY.
    Bernasconi, C.F.: 1976, "Relaxation Kinetics," Academic Press,
    New York, NY.
    Strehlow, H. and Knoche, W.: 1977, "Fundamental of Chemical
    Relaxation," Verlag Chemie, Weinhein.

2. Persoons, A.: 1974, J. Physics. Chem., 78, 1210.

3. Persoons, A., and Hellemans, L.: 1978, Biophys. J. 24, 119.

4. Wien, M.: 1927: Ann. Phys. 83, 327; ibid.: 1928, 85, 795,
    Physik. Z., 28, 834 (1927); ibid, 29 751 (1928).

    Eckstrom, H.C. and Schmelter, C.: 1939, Chem. Rev., 24, 367;
    Harned, H.S. and Owen, B.B.: 1958, "Physical Chemistry of
    Electrolyte Solutions," 3rd Ed, Reinhold, New York, NY;
    Falkenhagen, H.: 1971, "Theorie der Elecktrolyte," Hirzel
    Verlag, Stuttgart.

5. Nackaerts, R., De Maeyer, M., and Hellemans, L.: 1979, J. of
    Electrostatics, 7, 169.

6. Onsager, L.: 1934, J. Chem. Phys.,2, 599.

7. Nauwelaers, F., Hellemans, L., and Persoons, A.: 1976, J. Phys.
    Chem., 80, 767.

8. Everaert, J. and Persoons, A.: 1981, J. Phys. Chem., 85, 3930;
    ibid, 1982: 86, 546.

9. Persoons, A. and Van Beylen, M.: 1979, Pure Appl. Chem., 51,
    887.

HIGH ELECTRIC FIELD PERTURBATION AND RELAXATION OF DIPOLE EQUILIBRIA. A HAZARDOUS UNDERTAKING?

L. Hellemans and A. Persoons
Department of Chemistry,
University of Leuven,
B-3030 Heverlee, Belgium

In physics there exists a long tradition of perturbing objects at rest for the purpose of learning about their dynamic properties. This information is disclosed by observing the rate of response to the perturbation. When it was realized that chemical systems in equilibrium or in steady state conditions could be perturbed and studied in much the same way, the expansion of research work in the field of chemical relaxation studies became explosive.

Significant progress has been made in the elucidation of chemical reaction dynamics, mainly by using pressure and temperature changes as external parameters. Whatever happened to the application of electric fields as perturbing factors?

In this contribution we shall examine the conditions necessary for an electric field to disturb the equilibrium composition of fluid systems involving dipolar species. (The interaction of the field with ions and ionizable particles in solution is being dealt with in another section). We shall concentrate our attention on the case where high field strengths are required to achieve measurable equilibrium shifts; next, an original detection method will be described and its effectiveness will be illustrated.

THEORETICAL

A Condensed Introduction To Dielectrics

The microscopic effect of a uniform electric field on a dielectric is to distort the distribution of the charges carried by electrons and nuclei. The charge density in the molecules is displaced over distances that are usually not larger than the molecular size. Particles with a permanent or field-induced asymmetry in their charge distribution experience a reorienting couple by the field as well.

503

P. M. Rentzepis and C. Capellos (eds.), Advances in Chemical Reaction Dynamics, 503–523.
© *1986 by D. Reidel Publishing Company.*

In macroscopic terms, the excess surface charge density appearing at the boundary of the dielectric in a plane perpendicular to the field as a result of the charge movements just described, is known as the polarization P. The electric moment of the sample of volume V subjected to the field is given by M = PV. The change of the polarization with the field E defines a material constant, the permittivity $\varepsilon \cdot \varepsilon$.

$$\partial P/\partial E = \varepsilon \cdot (\varepsilon - 1) \tag{1}$$

with $\varepsilon \cdot$ the permittivity of vacuum ($\varepsilon \cdot = 8.86 \times 10^{-12} C/Vm$) and ε the relative permittivity. The quantity $\varepsilon \cdot (\varepsilon - 1)$ is referred to as the susceptibility of the material. Its variation with the change of external parameters allows the inspection of all kinds of charge displacements at the molecular level.

In general, ε is a tensor, which becomes a scalar for isotropic substances. It is complex to account for eventual phase lags between the oscillating field and the induced polarization. The relaxation rates are related to the microscopic processes by which charges are transported in the material on a molecular scale. The Debye dipolar relaxation and the resonant relaxation of bound electrons are classic examples of such events.

At this point, we are interested in the case where chemical change lies at the root of contributions to the overall polarization. The dielectric relaxation then provides clues as to the extent and the rate of the chemical transformation.

Chemical Perturbation And Relaxation In A Field

If we want to use the language of chemical perturbation and relaxation methodology (1), we shall first have to find out how the polarizing field can affect the chemical equilibrium.

The change of the Gibbs free energy G is given as

$$dG = -SdT + Vdp - dw - Ad\xi \tag{2}$$

with S, V, and ξ the extensive properties entropy, volume, and extent of reaction, while T, p, and A are the corresponding conjugate intensive variables temperature, pressure, and affinity; dw represents all flows of work other than expansion work. At present, our interest is focused on the interchange of electromagnetic work (dw = -EdM), which through polarization enriches the free energy of the system. The change at constant temperature and pressure of a new state property G$\cdot$defined by G$\cdot$ = G - EM can now be written as

$$dG\cdot = - Ad\xi - MdE \tag{3}$$

from which one immediately concludes that

$$(\partial A/\partial E)_\xi = (\partial M/\partial \xi)_E \qquad (4)$$

The affinity is defined as follows

$$A = - \sum v_i g_i \qquad (5)$$

where v_i is the stoichiometric coefficient of the chemical i taking part in the equilibrium (with $v<0$ for reactants, $v>0$ for products), and g_i is the individual chemical potential approximated by

$$g_i = g^\circ_i + RT \ln c_i \qquad (6)$$

when the activity coefficients are assumed to be unity. R is the gas constant and g°_i is the standard value, while c_i represents the actual concentration.

Taking the variation of A with E at constant composition one finds

$$(\partial A/\partial E)_\xi = \partial(-\Sigma v_i g^\circ_i)/\partial E = RT(\partial \ln K/\partial E) \qquad (7)$$

where K is the equilibrium constant. Using the identity of Equation 4 we find an expression giving the change in composition to be expected when a chemically reactive system is subjected to a field.

$$(\partial \ln K/\partial E)_{T,p} = \Delta M/RT \qquad (8)$$

The driving force $\partial M/\partial \xi = \Delta M$ is related to the difference between the electric moment of products and reactants. The equation reflects the increased stability of properly oriented polar (or polarizable) particles in the field over nonpolar or nonaligned molecules.

The relation was established 25 years ago by Bergmann, Eigen, and DeMaeyer (2), and by Schwarz (3), although inferred earlier by Frank (4). The interaction of the field with the reactive system can now be used to perturb its composition. The rate of adaptation to the new external conditions will depend on the dynamics of the system which we are trying to explore.

Nonlinear Dielectric Effects

In the following we shall try to evaluate the molar change of the electric moment ΔM. A closer look, however, at the origin of nonlinear behavior of the polarization is in order. For the sake of clarity, we choose to limit this discussion to the case of dilute gases, where the interaction between the particles is negligible, so that we can take the external field equal to the field that is actually experienced by the molecules.

The molecular moment **m** is given by

$$m = \mu + \alpha \cdot E \tag{9}$$

where μ is the permanent dipole moment and α the polarizability tensor.

The polarization in the direction of the field is the average projection of the molecular moment on the field direction times the particle density N

$$P = N \langle \mathbf{m} \cdot \mathbf{e} \rangle = N \left(\langle \mu \mathbf{e} \rangle + \langle \mathbf{E} \cdot \mathbf{a} \cdot \mathbf{e} \rangle \right) \tag{10}$$

with **e** the unit vector along the field.

The averages can be computed by weighting all contributions **m**·e according to their probability expressed by the appropriate Boltzmann factor. Thus,

$$\langle \mathbf{m} \cdot \mathbf{e} \rangle = Q^{-1} \int (\mathbf{m} \cdot \mathbf{e}) \exp(-W/kT) d\tau \tag{11}$$

where the integration covers all possible orientations and Q is the normalization factor $Q = \int \exp(-W/kT) d\tau$.

The change of the potential energy of each molecule in the field is given by

$$W = - \mu \cdot \mathbf{E} - (\mathbf{E} \cdot \mathbf{a} \cdot \mathbf{E})/2 \tag{12}$$

In the limit of weak fields one recovers the Debye expression for the molar polarization

$$P(E \rightarrow 0) = N \left(\mu^2/3kT + \bar{\alpha} \right) E \tag{13}$$

where $\bar{\alpha}$ is the average (isotropic) polarizability of the particle $\bar{\alpha} = \Sigma \alpha_i /3$ with α_i one of the three principal polarizabilities defining the polarizability ellipsoid. We note that to first order the polarization varies linearly with the field. At stronger fields one is bound to retain more terms in the calculation so that

$$P = P(E \rightarrow 0) + N[(\Sigma \alpha_i^2 - 3\bar{\alpha}^2)/15kT +$$
$$+ 2(\Sigma \mu_i^2 \alpha_i - \bar{\alpha}\mu^2)/15(kT)^2 - \mu^4/45(kT)^3]E^3 + \dots \tag{14}$$

where μ_i are the three components of the permanent moment along the principal axis of the polarizability.

At this point, it appears that the complete polarization will be given by a series expression in odd powers of the field. This, of course, is necessary in view of the directional character of the polarization which requires that $P(E) = - P(-E)$. We recognize in equation 14 nonlinear contributions that are connected with (i) the anisotropy of polarizability, and the reorientation of the polarizability ellipsoid by the field, (ii) the competition for interaction of the field with the polarizability ellipsoid and the permanent moment, and (iii) the saturation of dipole reorientation. The first two contributions are related to the Kerr effect (5); the last term reflects the onset of the deviation from a linear dependence of the dipole orientation on the field.

We have in equations 13 and 14 the first and second terms of the series representing the Langevin function L, which determines the average $\langle \cos \theta \rangle$ of the angle between the permanent moment and the field direction (6)

$$L(\mu E/kT) = \langle \cos \theta \rangle = \mu E/3kT - (\mu E/kT)^3/45 + \ldots \qquad (15)$$

A similar saturation behavior exists for the other modes of reorientation mentioned (7).

We have found expressions for the polarization to sufficient accuracy; P must be written as a series $P = \alpha E + bE^3 + O(E^5)$. It is evident now that the value of the derivative $\partial P/\partial E$, which defines the permittivity, depends on the field strength at which it is determined. This leads to the nonlinear dielectric change $\varepsilon \cdot \Delta\varepsilon$. It is a series in even powers of E.

$$\varepsilon \cdot \Delta\varepsilon(E) = (\partial P/\partial E)_E - (\partial P/\partial E)_{E->0} = 3bE^2 + \ldots \qquad (16)$$

In particular, for the Langevin saturation we have a contribution

$$\varepsilon \cdot \Delta\varepsilon_L = - N[\mu^4/15(kT)^3]E^2 + \ldots \qquad (17)$$

We need to be vigilant now for extra contributions to $\Delta\varepsilon$, which originate under certain conditions by interaction of the field with a reactive system in equilibrium.

Nonlinear Effects and Fluctuations

Before attempting to write ΔM explicitly, we shall return to the statistical-mechanical determination of $\Delta\varepsilon$. The calculation was performed in the fifties by Buckingham (8), Schellman (9), and Kielich (10). Assuming a Taylor series expansion for P, one may write that

$$b = (\partial^3 P/\partial E^3)_{E->0}/3! \qquad (18)$$

with $P = N\langle \blacksquare^{\bullet}e\rangle$. Part of the result obtained by these authors, corresponding to $\varepsilon \cdot \Delta\varepsilon$ derived in equation 17, but containing interesting new detail is

$$\varepsilon \cdot \Delta\varepsilon = N\{5[\langle\mu^4\rangle\cdot - \langle\mu^2\rangle\cdot^2] - 2\langle\mu^4\rangle\cdot\}E^2/30(kT)^3 + \dots \quad (19)$$

where the averages had to be taken in the limit of zero field. (The description of polarizability fluctuations and their coupling with the permanent moment is more involved. The effect is probably quite small in practice. Therefore we shall omit it).

The expression tells us that, over and above normal saturation, fluctuations of μ^2 contribute to $\Delta\varepsilon$. The fluctuation is nonzero (in so far as we are referring to dielectric detection) for as long as the field recognizes each state individually by its own permanent moment. This is accomplished when the particular state survives for a time longer than its average reorientation time, so that it can contribute to the polarization in a unique way. Otherwise, the effective dipole moment would be

$$\mu_{eff} = \Sigma \ X_i\mu_i \quad (20)$$

with X_i the probability (or time fraction) of finding the particle in a short-lived state i, for which a charge distribution may be anticipated (by some other method than a dielectric experiment) leading to a moment μ_i.

On the other hand, fluctuations and their time course may be observed by following the positive contributions to the nonlinear part of the polarization in case the net moment is composed of the mole fractions X of individual moments squared, or

$$\mu^2_{eff} = \Sigma \ X_i\mu_i^2 \quad (21)$$

Without the fluctuation term equation 19 is identical with equation 17.

The competition between two rate processes with characteristic time contants τ, first that of chemical change (τ_{chem}) and second, that of dipolar reorientation (τ_{or}), determines how the kinetics of the chemical equilibrium can be studied dielectrically.

If $\tau_{or} \gg \tau_{chem}$ the orientational distribution of the reaction partners is mainly adjusted to the pull of the electric field by chemical transformation (3). The molar change ΔM is to be defined for every angle θ, as is $\partial \ln K/\partial E$. For the simple case of an isomerization accompanied by a change of the dipole moment magnitude without change of direction, we have simply that $\Delta M(\theta)=N_A(\Sigma v_i\mu_i)\cos\theta$, with N_A as Avogadro's number.

If originally, in the absence of the field, the distribution was purely random, there exists for every direction in space another one diametric to it, that it equally populated. An increase induced by the field in the number of large dipoles pointing in the direction θ at the expense of species with smaller moments, will be canceled by $\pi + r$. The net change ΔM integrated over all space is zero.

While there is no change in the overall number of dipoles (which is what determines the chemical composition of the sample) the alignment of the dipoles in the field, of course, remains assured. At field frequencies below $1/\tau_{chem}$ the net polarization follows the oscillating field in phase; dielectric relaxation at higher frequencies reflects the rate of the chemical process.

In the limit of $\tau_{or}\rightarrow\infty$, neither in weak, nor in strong fields there would be a shift of the chemical composition. Only when this last conditions is relaxed, and reorientation by Brownian movement is allowed, the depletion of the population in the unfavorable direction will succeed at high fields. Both chemical and rotation diffusion rates define the relaxation spectrum. There is no particular advantage in studying this type of situation at large fields. Anyway, the departure from the linear polarization law described by the Langevin saturation expresses how far at high field strength the random orientation distribution is modified. Eventually, a positive contribution to the polarization may ensue as a consequence of chemical perturbation, but then only as a second order effect to the already higher order field effect of saturation.

The present situation is of relevance in the study of very fast processes such as proton (11,12) and electron transfer reactions, charge transfer complex formation (13), conformational change in general, but particularly in polymers (14). It is likely to be of importance in the case of solids.

When $\tau_{or} \ll \tau_{chem}$ orientational equilibrium is maintained for all reaction partners, irrespective of perturbation of the chemical equilibrium. On inserting the Debye expression for polarization and remembering that $d\xi = VdN_i/v_i$, we find

$$\Delta M = V \, \partial \, \Sigma \, N_i(\mu_i^2/3kT + \alpha_i)E/\partial\xi$$
$$= N_A\Sigma v_i(\mu_i^2/3kT + \alpha_i) \, E$$

(22)

If we assume that there is practically no change in polarizability with the reaction advancing ($\Sigma v_i\alpha_i \approx 0$), we have ultimately

$$\partial \ln K/\partial E = [\Sigma v_i\mu_i^2/3(kT)^2] \, E$$

(23)

In this instance the degree of perturbation of the equilibrium depends on $\Sigma v_i \mu_i^2$ and on E. If E->0, there is no change in composition, as it is the case in classic dielectric measurements. At high field strength, however, the degree of perturbation increases with E, contributing to the growth of the observed permittivity. This reveals à shift in the equilibrium in favor of the species with the highest dipole moments.

The Chemical Contribution To The Nonlinear Effect

From now on we shall restrict our discussion to systems which are in rotational equilibrium under the conditions of the experiment. The permittivity in a strong field is given by

$$\varepsilon \cdot (\varepsilon - 1)_E = (\partial P/\partial E)_\xi + (\partial P/\partial \xi)(\partial \xi/\partial E) \tag{24}$$

where we separated the term including all sorts of nonlinear phenomena from the one induced chemically by the probing field, which is superposed on the perturbing field. Elaborating on this equation provides

$$\varepsilon \cdot (\varepsilon - 1)_E = (\partial P/\partial E)_\xi + (\partial P/\partial \xi)(\partial \xi/\partial \overline{\xi})(\partial \overline{\xi}/\partial \ln K)(\partial \ln K/\partial E) \tag{25}$$

with the second term on the right hand side being the dielectric increment due to the equilibrium shift; it is complex in view of the relaxation function, which for a single step mechanism may be represented by $(\partial \xi/\partial \overline{\xi}) = (1 + j\omega\tau_{chem})^{-1}$ with $\overline{\partial \xi}$ the shift at zero angular frequency ω and j the imaginary unit. Thus,

$$\varepsilon \cdot \Delta \varepsilon_{chem} = V\Gamma (\partial P/\partial \xi)(\partial \ln K/\partial E) / (1 + j\omega\tau_{chem}) \tag{26}$$

where we have used $\partial \overline{\xi}/\partial \ln K = V\Gamma$ as definition of relaxation strength (1). Here Γ is given by $\Gamma = (\Sigma v_i^2/\overline{c}_i)^{-1}$ with $\overline{c}_i$ the equilibrium concentrations. Finally, we have

$$\varepsilon \cdot \Delta \varepsilon_{chem} = \Gamma (\Delta M)^2 / RT (1 + j\omega\tau_{chem}) \tag{27}$$

It is important to note that $\Delta \varepsilon$ is proportional to $(\Sigma v_i \mu_i^2)^2$ and to E^2 by virtue of $(\Delta M)^2$. The effect will be particularly sensitive to changes involving large dipole moments. The chemical relaxation rate should become apparent from the dispersion of $\Delta \varepsilon_{chem}$ with increasing ω. For an isomerization it is easy to show that equation 27 corresponds to the fluctuation term of equation 19. A more general equivalence has been attempted by Malecki (15).

We would like to emphasize here that a fortunate experimental aspect has been implied in the foregoing; there is no need to use high frequency, high voltages to study the relaxation. It will be sufficient to observe the permittivity with a weak high frequency signal superposed on a high (static) voltage.

Up to this stage we have failed to discuss in detail the contents of the first term on the right hand side in equation 25. This particular term measures the permittivity of the system at high field strength without allowing the probing high frequency field to alter the chemical composition. It reflects the change of the permittivity as compared to its zero field value. Two important contributions to the overall $\Delta\varepsilon$ are to be included: first, a part connected with the saturation of the dipole orientation according to equation 17, and second, a part as consequence of the change in composition on rising the static field to its maximum value. In order to compute this last contribution we multiply the shift in composition induced by the field (expressed as a certain advancement of reaction $\Delta\xi$) by $\partial\varepsilon/\partial\xi$, the parameter translating the concentration changes into a permittivity increment. It is important to note at what frequency this incremental permittivity is probed. In order to include the effect of the permanent dipoles explicitly, the frequency must remain well below the inverse orientation relaxation time of the dipolar molecules. We represent the contribution as follows

$$\varepsilon \cdot \Delta\varepsilon_{shift} = \varepsilon \cdot (\partial\varepsilon/\partial\xi) \, (1 + j\omega\tau_{or})^{-1} \int_0^E (\partial\xi/\partial E) \, dE \qquad (28)$$

In this equation we have neglected to take into account the weak dependence of $\partial\varepsilon/\partial\xi$ on the field magnitude. From the definition in equation 1 we know that $\varepsilon \cdot (\partial\varepsilon/\partial\xi)_E = \Delta M/VE$. Since the equilibrium shift is effected by the static field, it was not necessary to introduce the chemical relaxation function once more. With the help of a procedure similar to the one used above, it is found that

$$\varepsilon \cdot \Delta\varepsilon_{shift} = \Gamma(\Delta M)^2 \, / \, 2 \, RT \, (1 + j\omega\tau_{or}) \qquad (29)$$

The amplitude of this contribution is just one half the contribution $\varepsilon \cdot \Delta\varepsilon_{chem}$ in equation 27. These terms, however, show a different dependence on the frequency of the probing field as emphasized by the following equation, which collects the nonlinear effects of dipolar origin, as distinguished by different time constants particular to each mode (16).

$$\Delta\varepsilon = \Delta\varepsilon_L(\tau_{or}) + \Delta\varepsilon_{shift}(\tau_{or}) + \Delta\varepsilon_{chem}(\tau_{chem}) \qquad (30)$$

The present state of affairs is the consequence of the current experimental practice of superposing a high frequency field on a high static field.

It is time now to indicate a correction for the relation of the internal field in dense media to the externally applied field E. The simplest and quite satisfactory refinement consists of the Lorentz field factor (8), so that every term $\Delta\varepsilon$ is proportional to $[(\varepsilon+2)/3]^4 E^2$.

It will become clear from a numerical example that the effects to be expected are quite small, even at a field strength of 100 kV/cm, which is about one half the breakdown strength of good insulators.

Consider the equilibrium A ⇌ B with K = 1 and μ_A = 4D (a horrendous 1.3 x 10^{-29} Cm) and μ_B= 0D . This set of data is typical for the case of internal rotation in some molecules (17). At a field of 100 kV/cm and for a formal concentration of 0.1 mole/l of the flexible compound in a solution with ε = 2, one obtains at room temperature $\Delta\varepsilon_{chem}/\varepsilon$ = 13.4 ppm according to equation 27. The relative change of the equilibrium constant $\Delta K/K$ is about 3 x 10^{-4}.

The change $\Delta\varepsilon/\varepsilon$ is complex in view of relaxation. The real part $\Delta\varepsilon'/\varepsilon$ of the overall effect in equation 30 decreases from +4.1 ppm, measured a low frequency, to -9.3 ppm at frequencies above the chemical relaxation region. All contributions disappear at still higher frequencies corresponding to orientational relaxation. The imaginary part $\Delta\varepsilon''/\varepsilon$ is positive and shows a maximum for $\omega=1/\tau_{chem}$, then changes sign and presents a minimum at $\omega=1/\tau_{or}$. A diagram representing this behavior is given in Figure 1.

We shall continue to forego the effects due to electrostriction, polarizability anisotropy, and hyperpolarizability, although in some cases state-of-the-art measurements allow these to be determined (18,19); also, the interference from electrode movements under the influence of the large fields has to be considered.

EXPERIMENTAL

A sensitive method is required to detect the minute changes $\Delta\varepsilon'$ and $\Delta\varepsilon''$ that may appear in systems under high electric stress. We shall describe here the "high field modulated resonance" technique, an original product of our laboratory (16). Instead of using a static high field, as suggested until now, a slowly alternating field is applied to the measuring cell, which is incorporated in a parallel resonant circuit. A properly tuned high frequency signal of low amplitude excites the LC-circuit. The high field causes the resonance properties to vary periodically, which results in amplitude modulation of the resonant voltage. The modulation depth leads to the field-induced changes of ε' and ε", as we shall see in the following paragraphs.

One practical merit of the present technique resides in the fact that the frequency of the information carrying signal is brought down from the radio frequency range to the audio domain, where the electronics of signal handling are much simpler. Since the measuring circuit is composed of "passive" elements only, the overall stability is limited essentially by the quality of the signal generator (synthesizer) used. In this respect, we are in a position to gain directly from the technological developments in the area of telecommunications.

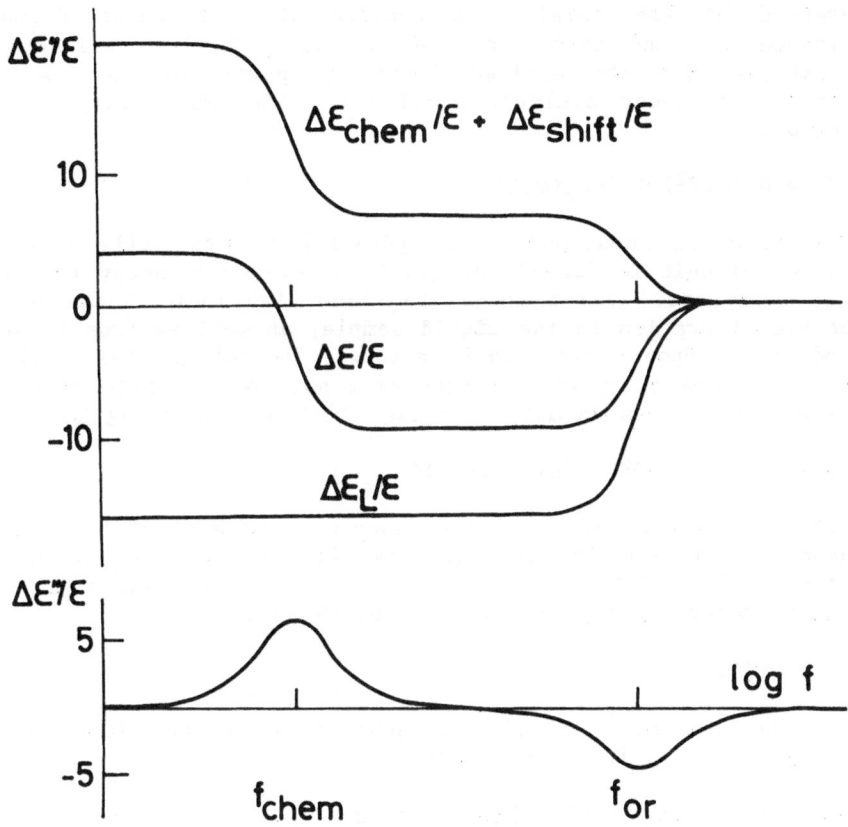

Figure 1.   The real and imaginary components of the nonlinear
dielectric effect calculated in ppm as function of frequency for the
case of an isomerization A $\rightleftharpoons$ B with μ_A = 4D, μ_B = 0 D at E = 100
kV/cm, T = 300 K and ε = 2. The equilibrium constant K = 1 and the
formal concentration is 0.1 mole/l. The chemical relaxation occurs
at lower frequencies than the orientational relaxation. The upper
figure contains the net effect $\Delta\varepsilon/\varepsilon$ and its components
$(\Delta\varepsilon_{chem} + \Delta\varepsilon_{shift})/\varepsilon$ and $\Delta\varepsilon_L/\varepsilon$ separately, while the lower figure
represents the overall effect only.

L-C Resonance

With our method we take advantage of the great sensitivity of
high Q circuits to variations in one of the circuit elements.
Network theory may be applied to an assembly of (linear!) lumped
circuit components of reasonable size in the frequency range up to
10^8 Hz.

A two electrode cell filled with a dielectric liquid can be represented by the parallel connection of a frequency-dependent capacitance $C(\omega)$ and conductance $G(\omega)$. Both of these elements may show changes with the applied field, in particular here with E^2 because of nonlinear dielectric effects. The admittance Y of the cell equals

$$Y_x = G_x(\omega,E^2) + j\omega C_x(\omega,E^2) \tag{31}$$

By connecting an inductance L in parallel to the cell, a sharply tunable LC-circuit is formed, provided the electric energy losses are not too important. These could stem from high ionic conductance or dielectric absorption in the liquid sample, as well as from losses in the inductor. Upon excitation by a source loosely coupled to the LC-circuit, and operating at a frequency ω not too far from resonance, the network generates an output voltage $U(\omega)$ corresponding to

$$(U\cdot/U)^2 = 1 + Q^2 \ (\omega/\omega\cdot - \omega\cdot/\omega)^2 \tag{32}$$

U reaches a maximum value $U\cdot$, the resonance voltage, at the resonant frequency $\omega\cdot$, and Q is the quality factor of the circuit at resonance. This last quantity is related to the bandwidth $\Delta\omega$ of the resonance characteristic $U(\omega)$ in the following way

$$Q = \omega\cdot/\Delta\omega \tag{33}$$

with $\Delta\omega$ the difference of the frequencies for which $(U/U\cdot)^2 = 0.5$, also known as the half-power points.

Figure 2 shows a simplified diagram of a parallel resonant circuit. It includes an ideal voltage source U_g (with zero output impedance), which is coupled to the circuit by a small capacitor C_k. DeMaeyer has discussed in great detail the relation of the circuit components of interest G_x and C_x to the equivalent conductance G', capacitance C', and inductance L' taking into account the parasitic and distributed components as well as the losses of the inductor (20). It is necessary to make this type of analysis for each practical situation in order to have a precise idea of the sensitivity of the resonance parameters to the changes of interest ΔC_x and ΔG_x. It will suffice to state that if certain precautions are taken, one may assume $\Delta G_x = \Delta G'$ and $\Delta C_x = \Delta C'$.

By applying Thévenin's theorem to the circuit in Figure 2, we find that the output voltage is given by

$$U = U_g \ j\omega C_k/ \ \{G' + j \ [\ \omega(C'+ C_k) - 1/\omega L'] \ \} \tag{34}$$

The resonance condition may be defined as the frequency for which the admittance of the circuit becomes real (21). Consequently, one has

$$\omega\cdot^2 = 1/ \ (C' + C_k) \ L' \tag{35}$$

and for the rms voltage at resonance

$$|U \cdot| = U_g \omega \cdot C_k / G'$$ (36)

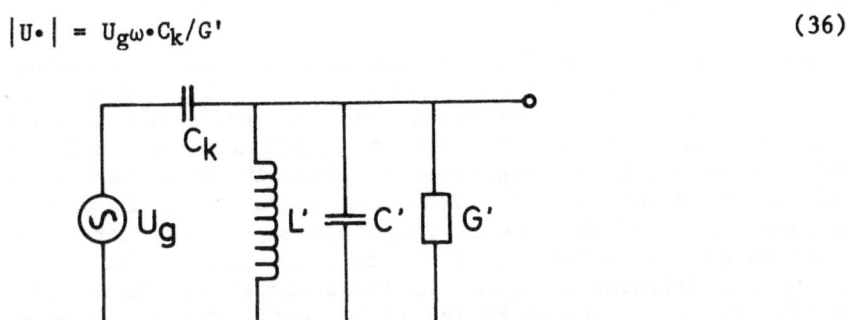

Figure 2. Block diagram of the equivalent circuit.

The quality factor of the circuit is determined by the ratio of the out-of-phase current component to the in-phase component.

$$Q = \omega \cdot (C' + C_k) / G'$$ (37)

In order to define the quality of a capacitor with the ideal capacitor as a standard, the concept of loss tangent has been introduced. The angle δ is the complement of the phase angle between the current through the component and the voltage applied over it. An ideal capacitor presents no losses so that $\delta = 0$. For the measuring cell we have

$$\tan \delta_x = G_x / \omega C_x = (\sigma + \omega \varepsilon \cdot \varepsilon'') / \omega \varepsilon \cdot \varepsilon'$$ (38)

The specific admittance of the sample is characterized by its conductivity σ and permittivity $\varepsilon = \varepsilon' - j\varepsilon''$.

Modulated Resonance

The objective of the measurement is to determine the complex change $\Delta \varepsilon$ induced by the high field. The nonlinear behavior of the sample entails changes of G_x and C_x, which affect the resonance characteristics $U(\omega)$. If the high alternating field is chosen as $E = E \cdot \cos wt$ with $w \ll \omega$, the resonance voltage amplitude becomes modulated in the following manner, when the field is switched on

$$U_{mod}(\omega) = U(\omega) \{ 1 + [\Delta U(\omega, E \cdot^2) / U(\omega)] (1 + \cos 2wt) \}$$ (39)

where it is assumed that the chemical and physical effects responsible for the change ΔU develop in phase with the high field. The harmonics generation is inherent in the occurrence of nonlinearity; higher harmonic components may arise eventually, though they will be limited to even multiples of w. In practice, we have chosen w = 85 Hz, a frequency whose harmonics are different enough from the line voltage frequency of 50 Hz.

The useful information is concealed in the modulation depth $\Delta U/U$, which can be measured accurately by filtering out all components with frequencies different from 2w. While the resonance band is slowly swept, the envelope of the amplitude-modulated resonance signal is detected and amplified selectively by a narrow bandpass amplifier tuned at 2w. The resulting signal $\Delta U(\omega)$ is a difference resonance curve, whose general shape approaches the differential of the well-known characteristic $U(\omega)$ given in equation 32. The amplitude of the difference signal is proportional to the shift of the resonant frequency imposed by the field, while the asymmetry reflects changes of U• and Q accompanying the field effects.

The sign of the difference can be deduced from the phase of ΔU with respect to the field; the shape of a Lissajous figure formed by ΔU and E solves this question. A schematic representation is given in Figure 3 for different situations.

With the objective of extracting values of $\Delta\epsilon'$ and $\Delta\epsilon''$ from the demodulation measurement, we have equated the experimental $\Delta U(\omega)$ with the differential of $U(\omega)$

$$dU(\omega) = (\partial U/\partial\omega\cdot)d\omega\cdot + (\partial U/\partial U\cdot)dU\cdot + (\partial U/\partial Q)dQ \qquad (40)$$

wherein one may substitute Q $(dU\cdot/U\cdot - 2d\omega\cdot/\omega\cdot)$ for dQ as seen from the differentiation of equations 35, 36, and 37. It is of interest to retain here

$$2\ d\omega\cdot/\omega\cdot = -\ dC_x/(C' + C_k) = -\ [C_x/(C'+ C_k)]\ (d\epsilon'/\epsilon') \qquad (41)$$

where we have introduced C_x, $= \epsilon'C\cdot$ with C• the vacuum capacitance of the cell. Approximating differentials with finite changes, we obtain

$$\Delta\epsilon'/\epsilon' = -2\ \gamma\ \Delta\omega\cdot/\omega\cdot \qquad (42)$$

The factor $1/\gamma = C_x/(C' + C_k)$ represents the proportion of the ϵ dependent part of the capacitance to the total capacitance on which resonance conditions depend. This ratio can be determined according to equation 35 with calibration measurements of $\omega\cdot^{-2}$ as a function of the known ϵ of reference compounds. Ideally, the ϵ-dependent capacitance should coincide with that part of the cell where the high and homogeneous field E will be generated.

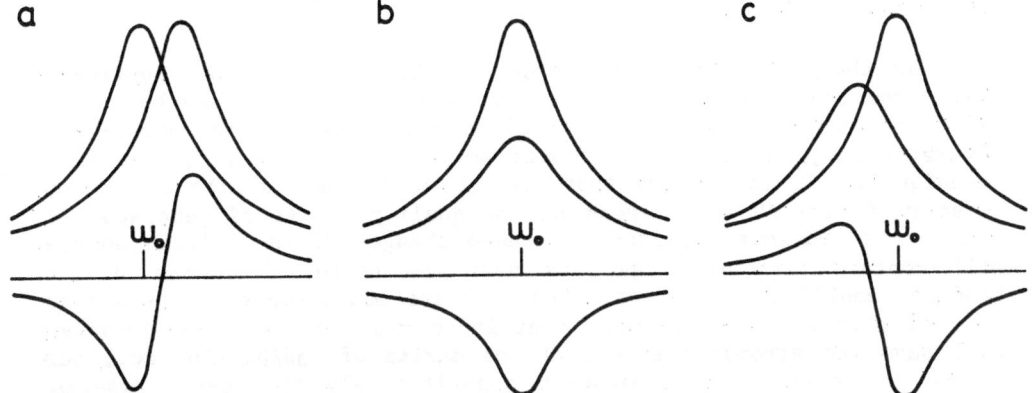

Figure 3. Resonance cuves $U(\omega)$ at high field and at zero field. The values $\omega\cdot$ for resonance in the field free state are indicated. The lower curves represent the shape of the difference signals for

(a) $\Delta\omega\cdot > 0$, $\Delta U\cdot = 0$ and $\Delta Q = 0$
(b) $\Delta\omega\cdot = 0$, $\Delta U\cdot < 0$ and $\Delta Q < 0$
(c) $\Delta\omega\cdot < 0$, $\Delta U\cdot < 0$ and $\Delta Q < 0$.

On further rearranging terms in equation 40, a linear relation of the form $y(\omega) = (\Delta U\cdot/U\cdot) - (\Delta\omega\cdot/\omega\cdot) x(\omega)$ is obtained, where y and x are relatively complicated functions involving ω and the resonance parameters, and where y includes also the experimentally recorded values $\Delta U(\omega)$ (16). $\Delta\varepsilon'/\varepsilon'$ can be obtained immediately from the slope of this relation on the basis of equation 31 for each frequency $\omega\cdot$ where one was able to set up resonance. This is accomplished experimentally by means of a set of exchangeable inductance coils of different size, which are connected in parallel to the cell to provide resonance at conveniently spaced frequencies.

Differentiating equation 36, on the other hand, provides a relation between the intercept $\Delta U\cdot/U\cdot$ and the desired $\Delta\varepsilon''/\varepsilon'$ value, by way of the definition of the loss tangent in equation 38. We find, to a high degree of accuracy, that

$$\Delta\tan\delta_x = (\Delta\sigma + \omega\cdot\varepsilon\cdot\Delta\varepsilon'')/\omega\cdot\varepsilon\cdot\varepsilon' = (\gamma/Q) (\Delta\omega\cdot/\omega\cdot - \Delta U\cdot/U\cdot) \quad (43)$$

We note that nonlinearities of the Ohmic conductance σ induced by the field (as in the Wien effects) will appear as contributions to $\Delta\tan\delta_x$ in the form of $\Delta\sigma/\omega\cdot\varepsilon\cdot\varepsilon'$. In nonpolar media where the ionization is very weak, it cannot be expected a priori that the changes $\Delta\sigma$ be in phase with the high field, even when this is only oscillating at 85 Hz. The analysis of compound signals involving nonlinear changes in both the conductance and the capacitance has been performed (22). This particular aspect is important when investigating the behavior of ion-pairs and other ionizable species with high field methods. If $\Delta\sigma = 0$, we have simply that $\Delta\tan\delta_x = \Delta\varepsilon''/\varepsilon'$.

The Technique Applied

   Now that we understand how $\Delta\epsilon'$ and $\Delta\epsilon''$ are obtained experimentally from the amplitude modulated resonant voltage, it may be worth-
while to return to some experimental detail. Because the measurement
is stationary, its precision should be good as compared to transient
techniques. It is obvious that the ultimate limit of the signal to
noise ratio for ΔU will depend on the quality of the rf generator at
hand. Any spurious amplitude or phase changes of the voltage source
will contain harmonic components near the center frequency of the
bandpass amplifier, that spoil the accurate measurement of the degree
of modulation. AM-noise will contribute most at ω•, while FM-noise
will have its strongest effect at the maxima of $\partial U/\partial\omega$ (In fact, our
method is ideally suited to test quantitatively the phase noise of
signal generators).

   The requirement of minimal phase noise is particularly stringent
when slowly scanning the resonance band. Usually a width of 2 $\Delta\omega$ is
swept in about 40 s. In this way, a large number of individual difference measurements are made (at a rate of 2 x 85 per second) and
averaged at each point of the sweep. The main filtering, of course,
is effected by the bandpass amplifier. Additional signal averaging
is accomplished when 1024 digital data points representing the signal
$\Delta U(\omega)$ are pooled to a manageable number of $\pm$ 40. A least-squares
line through the computed $y(\omega•+d\omega)$ vs. $x(\omega•+d\omega)$ values, where the
precise value of ω• in the computations is defined from the minimum
X^2 value found by iteration for such lines (16), further enhances the
precision of the slope $\Delta\omega/\omega$• and intercept $\Delta U•/U•$. This extensive
filtering action explains also why during the experiment a certain
amount of drift (by heating of the cell, say) can be tolerated.

   The apparatus and the measuring cell in different stages of
development are described elsewhere (16,18,22). Two capacitors in
series hold the sample liquid; they are constructed in a symmetrical
manner with the purpose of minimizing deformations by the electrostatic forces. The high voltage is coupled to the circuit at the
series connection between the identical measuring capacitors. The
rest of the circuitry is also laid out in a symmetric fashion: by
connecting the center of the inductors to ground, the point, where
the high voltage is injected into the resonant circuit, remains at
virtual ground as far as the high frequency signal is concerned.
This procedure optimizes the sensitivity of the circuit by reducing γ.

   In a typical experiment one of the induction coils is plugged
into place. Then the resonance parameters are determined. Q values
range between 200 and 500 for frequencies between 0.1 and 100 MHz.
With the high field on, the amplitude modulation depth is detected,
amplified, and recorded. After calculating the $\Delta\epsilon'$ and $\Delta\epsilon''$ values
for the resonant frequency ω• chosen, the experiment is repeated with
the next coil in place, until a wide enough frequency range is

covered to be able to recognize relaxation. Ultimately, the relaxation parameters are determined.

A valuable aspect of the present technique consists in the possibility to test the kinetic results for coherence with thermodynamic parameters derived independently from the same data. Clearly, the dispersion data do not offer any more information on the dynamics of the process than do the absorption data. In the present context, however, it should be realized that once the chemical relaxation strength $\Delta\varepsilon_{chem}$ is determined experimentally from the dispersion of the effect, the other contributions $\Delta\varepsilon$ in equation 30 can be computed immediately. It is important to notice that both $\Delta\varepsilon_L$ and $\Delta\varepsilon_{chem}$ depend on the formal concentration of the reaction partners, but each one according to a different function. This function involves in both cases the values of μ_i and K as parameters. We have seen in equation 27 how $\Delta\varepsilon_{chem}$ is related to the equilibrium concentrations; for $\Delta\varepsilon_L$ it is obvious that the net saturation effect corresponds to the sum of the individual contributions related to all dipolar species present, taking into account their equilibrium concentration. (In the case of the perturbation of multistep equilibria similar arguments can be made).

In earlier dynamic studies, performed mainly by Hopmann (23,24), only the changes $\Delta\varepsilon''$ became available. Without recourse to other experimental resources, it was not possible to ascertain both parameters μ_i and K.

As an illustration of the capability of our method we present preliminary data on the relaxation of the nonlinear dielectric effect in solutions of $AgClO_4$ in toluene. Experimental values of $\Delta\varepsilon'/\varepsilon'$ and $\Delta\tan\delta$ as function of the frequency are presented for different concentrations in Figures 4 and 5. The field strength is 71 kV/cm, the relative permittivity of the solutions is about 2.45. Under the experimental conditions the degree of ionization is negligible: there is hardly a contribution $\Delta\sigma/\omega\varepsilon\cdot\varepsilon'$ noticeable in the $\Delta\tan\delta$ curves. Theoretical lines corresponding to a single step (Debye) relaxation are drawn through the points.

The effects show relaxation in the MHz region; the relaxation frequency varies with concentration. The disturbed equilibrium is identified as the association equilibrium of the polar ion-pairs (M) into nonpolar aggregates (quadrupoles Q) according to $2M \rightleftharpoons Q$. First, from the variation of the amplitudes of $\Delta\varepsilon_{chem}$ and $\Delta\varepsilon_L$ with the concentration it is deduced that the ion-pairs have $\mu = 10.0 \pm$ .2D, while the moment of the quadrupoles is $\mu = 0 \pm 2D$. The association equilibrium constant is $K(-5.5°C) = 2500M^{-1}$. Second, from the concentration dependence of τ^{-1} and the value of K it may be concluded that the rate constant for association is nearly diffusion-controlled with $k_{as}(-5.5°C) = 5.4 \times 10^9 M^{-1}s^{-1}$. It was not possible here to obtain from the relaxation data both rate constants k_{as} and k_{dis} independently: on plotting τ^{-2} as function of the formal concentration c• according to the following relation pertinent to self-association (1)

$$\tau^{-2} = k_{dis}^2 + 8 \, k_{as} k_{dis} c \bullet \tag{44}$$

we found the intercept too small to be reliable. In this case, it was of great advantage to dispose of the equilibrium information obtained from the relaxation amplitudes in order to complete the kinetic analysis. In favorable cases the value of K would have served to confirm the ratio of the rate constants obtained separately.

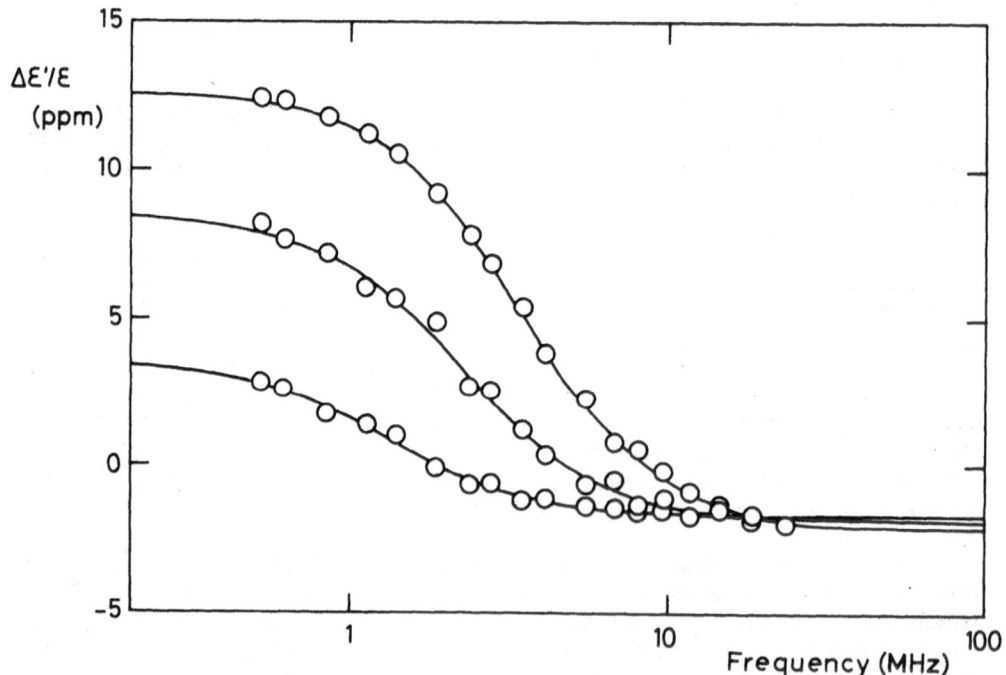

Figure 4. Dispersion of the nonlinear dielectric effect in solutions of AgClO₄ in toluene at -5.5°C and 71 kV/cm. The concentration decreases from top to bottom: 4.94, 2.73, and 0.91 x 10^{-3} mole/l. The data have been corrected for the solvent contribution $\Delta\varepsilon/\varepsilon$ =5.01 ppm. The full line is a best fit to a Debye-type relaxation.

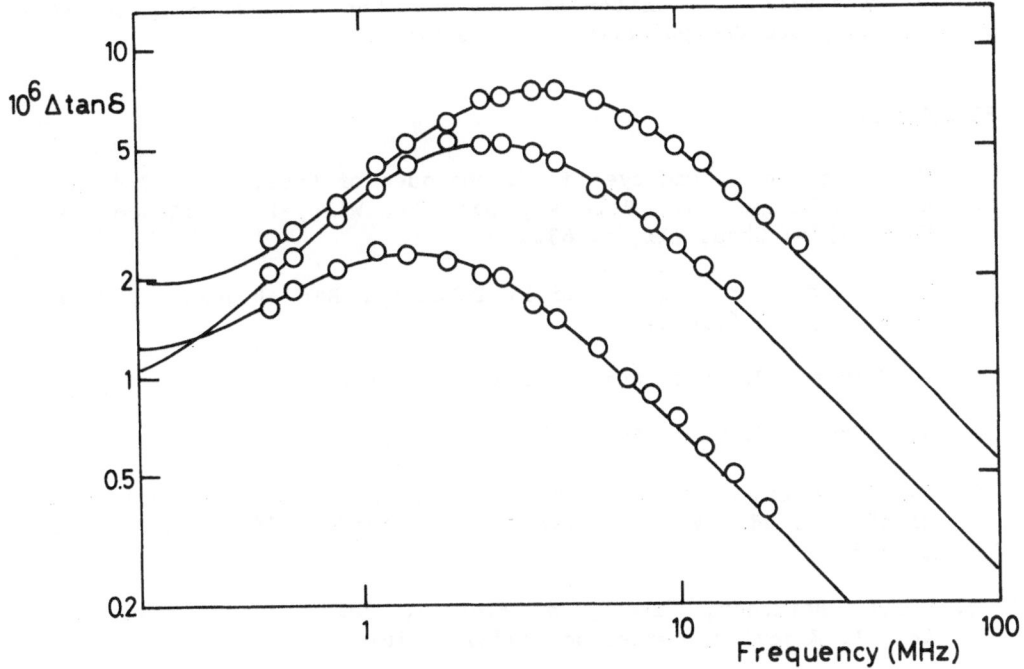

Figure 5. Double logarithmic plot of the field-induced increment of the loss tangent against frequency for solutions of AgClO$_4$ in toluene at -5.5°C and 71 kV/cm. The concentration decreases from top to bottom: 4.94, 2.73, and 0.91 x 10^{-3}mole/l. The full line is a best fit to a Debye-type absorption including field-induced conductance change.

In this paper we have reviewed the conditions for the perturbation of chemical equilibria in high electric fields. The functioning of a novel and uncommon experimental technique has been examined, which allows to detect the complex nonlinear changes of the permittivity. With the complete dispersion behavior available now for the first time, any interpretation of the data must explain a larger amount of observations and therefore is strengthened.

The chemical relaxation of the nonlinear dielectric effect allows the investigation of the rates and mechanism of the perturbed system. It appears that even in the strongest fields the equilibrium shifts involving small molecules remain quite small. One may assume on this basis that the same will be true for the individual rate

constants. The measurement of the variation of the rate constant
with the field strength is claimed in only one instance for a case
involving biopolymers (25). Obviously, a new generation of experimental techniques will be necessary to observe such changes in
smaller molecules as a matter of routine. The results could satisfy
the need to probe the polarity of the activated state.

REFERENCES

1. M. Eigen and L. DeMaeyer in "Techniques of Chemistry", 3rd ed.,
    A. Weissberger ed., Vol. VI, part II, Wiley-Interscience, New
    York (1974) Chap. III, p. 63.

2. K. Bergmann, M. Eigen, and L. DeMaeyer, Ber. Bunsenges. Phys.
    Chem. 67, 819 (1963).

3. G. Schwarz, J. Phys. Chem. 71, 4021 (1967).

4. H.S. Frank, J. Chem. Phys. 23, 2023 (1955).

5. C.J.F. Böttcher and P. Bordewijk, "Theory of Electric Polariza-
    tion", 2nd ed., Vol. II, Elsevier, Amsterdam (1978) Chap. XIII,
    p. 315.

6. C.J.F. Böttcher, "Theory of Electric Polarization", 2nd ed.,
    Vol. 1, Elsevier, Amsterdam (1973) p. 161.

7. S. Kielich in "Dielectric and Related Molecular Processes", M.
    Davies, ed., Vol. 1, The chemical Society, London (1972) p. 192.

8. A.D. Buckingham, J. Chem. Phys. 25, 428 (1956).

9. J.A. Schellman, J. Chem. Phys. 26, 1225 (1957).

10. S. Kielich, Acta Phys. Polonica 17, 239 (1958).

11, G. Schwarz, J. Phys. Chem. 74, 654 (1970).

12. M. DeMaeyer, P. Wolschann, and L. Hellemans in "Techniques and
    Applications of Fast Reactions in Solution", W.J. Gettins and E.
    Wyn-Jones eds., Reidel Publ., Dordrecht (1979) p. 501.

13. J.E. Anderson and C.P. Smyth, J. Am. Chem. Soc. 85, 2904 (1963).

14. G. Schwarz and J. Seelig, Biopolymers 6, 1263 (1968).

15. J. Malecki, J.C.S. Faraday II 72, 104 (1976).

16. L. Hellemans and L. DeMaeyer, J. Chem. Phys. 63, 3490 (1975).

17. A. Persoons and L. Hellemans, Biophys. J. $\underline{24}$, 119 (1978).

18. L. Hellemans and M. DeMaeyer, J.C.S. Faraday II $\underline{78}$, 401 (1982).

19. J. van der Elsken, P. van Zoonen, and J.C.F. Michielsen, Chem. Phys. Letters $\underline{106}$, 252 (1984).

20. L. DeMaeyer in "Methods in Enzymology", S.P. Colowick and N.O. Kaplan eds., Vol. XVI, Academic, New York (1969) p. 80.

21. G. Lancaster, "DC and AC Circuits", 2nd ed., Oxford University Press, Oxford (1980) p. 85.

22. R. Nackaerts, M. DeMaeyer, and L. Hellemans, J. Electrostat. $\underline{7}$, 169 (1979).

23. R.F.W. Hopman, J. Phys. Chem. $\underline{78}$, 2341 (1974).

24. R.F.W. Hopman, J.C.S. Faraday II $\underline{71}$, 1844 (1975).

25. D. Porschke, H. Meier, and J. Ronnenberg, Biophys. Chem. $\underline{20}$, 225 (1984).

THE MECHANISM OF ENERGY STORAGE IN THE REACTION CENTER IN PHOTOSYNTHETIC BACTERIA AND ATTEMPTS TO SIMULATE PROCESSES USING MONOLAYER ASSEMBLY TECHNIQUE

Hans Kuhn
Max-Planck-Institute for Biophysical Chemistry
(Karl-Friedrich-Bonhoeffer-Institut)
D 3400 Göttingen-Nikolausberg
Germany

ABSTRACT. Organized assemblies of monolayers are useful tools in studying energy and electron transfer in planned molecular arrangements of well-defined geometry. Systems of dye aggregates and energy or electron acceptors designed for studying exciton motion and trapping are investigated and the results are found to be in good agreement with a simple model based on a classical description of interactions. Monolayer assemblies of particular architecture are used to find the role of energy transfer and electron injection in the spectral sensitization of the photographic process. Arrangements are investigated where a π-electron system is positioned between a dye molecule and an electron acceptor. Such arrangements allow photoinduced charge separation at high energetic level. They illustrate design principles for energy storing systems. The structure of the photosynthetic reaction center recently resolved by Deisenhofer et al. is considered in connection with these design principles. Each chromophore component in the reaction center appears to be positioned optimally for the purpose of energy storage. This can be shown by calculating the rate of each electron transfer step for the arrangement given by Deisenhofer et al. and for other similar arrangements. The rates are deduced by taking the electronic structure of the chromophores and their geometrical arrangement explicitly into account, while the protein portion is considered as a dielectric continuum. The values thus obtained are in good agreement with the experimental data.

A) EXCITON MOTION AND TRAPPING IN MONOLAYER ASSEMBLIES

Exciton motion and trapping are of interest in simulating photosynthesis and for technological process such as photography. The processes can be studied in organized monolayer assemblies in great detail. Depending on the conditions, the dye chromophores can be disordered in the monolayer or they can be highly ordered.

In the first case the excitons produced by illumination are incoherent, they hop from dye molecule to dye molecule and are trapped if at least one trap per about 500 dye molecules is present [1]. Such

P. M. Rentzepis and C. Capellos (eds.), Advances in Chemical Reaction Dynamics, 525–550.
© *1986 by D. Reidel Publishing Company.*

an arrangement mimics the situation in plant and bacterial photosynthesis.

In the second case, after illumination, coherent excitations extending over a certain number of dye molecules move over the monolayer
and the energy is trapped even if one trap is among about 10 000 dye
molecules. This has been observed in the case indicated in Fig. 1 [2].

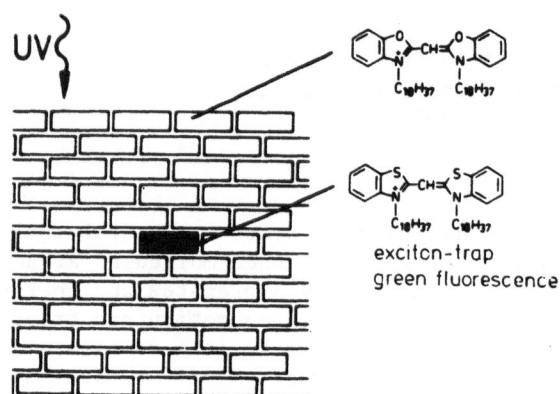

Fig. 1. Aggregate of oxacyanine with incorporated thia-
cyanine (trap). Fluorescence of oxacyanine quenched by
50 % at molar ratio N = 10 000.

The interaction of this excitation with the trap can be treated by
representing the excited dye chromophores as oscillating dipoles. In
the present case the dipoles oscillate in phase. The electric field
of these dipoles at the location of the trap determines the perturbation energy ε. It can be easily shown for the case of Fig. 1 that

$$\varepsilon = \frac{0.087 \text{ eV}}{\sqrt{\nu}} \tag{1}$$

if the exciton extends over ν molecules and if the exciton is in the
range of a trap. Then the rate of energy transfer is

$$k = \frac{2\pi}{\hbar} \varepsilon^2 S \tag{2}$$

where

$$S = \int f_D a_A dE = 1.6 \text{ eV}^{-1}$$

(f_D fluorescence quantum distribution of donor, a_A absorption of acceptor, normalized according to $\int_{\text{abs.band}} a_A dE = 1$). To obtain the rate

of energy transfer this rate k must be multiplied by the probability
that the exciton is in the range of a trap. This probability is ν/N if N
is the number of dye molecules per trap. The ratio (rate of energy transfer $(k\nu/N)$ to rate of deactivation in the absence of the trap
$(k_0 = 10^{10}\ s^{-1})$ gives the amount of quenching of the fluorescence of
the dye aggregate by the trap:

$$\frac{I_o}{I} - 1 = \frac{k\frac{\nu}{N}}{k_o} \tag{3}$$

(I and I_o fluorescence intensity of the dye aggregate with and without
the trap). If (1) and (2) are introduced in (3) ν cancels and we obtain

$$\frac{I_o}{I} - 1 = const \cdot \frac{1}{N} \text{ with const } = 1.2 \cdot 10^4 \tag{4}$$

-The result is in good agreement with the observation that the fluorescence is quenched by 50 % if a trap is among N = 10 000 aggregating molecules. The calculated dependence on N is observed.
    In the case considered above the trap is an energy acceptor incor-
porated in the aggregate (Fig. 1 and arrangement on the right of
Fig. 2). The trap is less effective if it is incorporated in a layer

$$\frac{I_o}{I} -1 = \frac{const}{N}$$

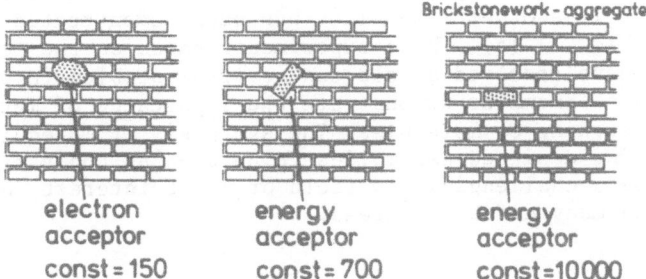

electron   energy   energy
acceptor   acceptor   acceptor
const = 150   const = 700   const = 10000

Fig. 2.   Aggregate of oxacyanine, energy or electron
acceptor in adjacent monolayer (left and middle) and
incorporated in the aggregate monolayer (right)

deposited on top of the aggregate layer [3] (case on the left for an
electron acceptor, in the middle for an energy acceptor), (const = 150
and 700 in these cases respectively instead of 10 000 in the first
case). Again the trap is at direct contact with the aggregating dye,
but in a smaller perturbing field of the dipoles representing the chromophores. The difference in the three cases can be quantitatively interpreted by the formalism indicated.

B) PHOTOGRAPHIC SENSITIZATION PROCESS

The probability of tunneling from an excited dye molecule to an acceptor
at 5 nm is negligibly small (see section C). However, energy transfer is
easily possible if an appropriate acceptor is present. This allows to
discriminate between the two processes. This was demonstrated in the
case of the photographic process which is sensitized by dyes at the silver bromide surface. Using the monolayer assembly technique, the dye molecules can be fixed at defined distances from the silver bromide surface. It is important in evaluating such experiments to make sure that
the sensitization of the photographic process measured in such arrangements is due to dye molecules present in the intended architecture and
not due to molecules that have reached the silver bromide surface by
imperfections of the assembly or by undesired rearrangement processes.
      This possibility can be excluded in different ways [4]. One way is
to add an energy acceptor to the organized assembly at a distance of
5 nm from the sensitizing dye. It acts as competitor of well defined
strength depending on its surface density. If the sensitization would
be due to imperfactions or layer reorganization, the energy acceptor
would be ineffective as a competitor.
      The contributions of energy transfer and electron-injection to
the spectral sensitization of AgBr vary strongly with the energetic position of the excited dye level and with the amount of neutral silver
atoms at the silver bromide surface acting as acceptors for the excitation energy of the dye. With increasing surface density of these acceptors energy transfer becomes increasingly favourable as compared to
electron injection, going from pure electron injection in one case to
pure energy transfer in another case. This has been demonstrated recently by Steiger (Fig. 3) [5]. These experiments based on the monolayer
assembly technique have resolved an old controversal problem in photographic science. These examples demonstrate the importance of methods
allowing exact positioning of molecules. The synthesis of molecules
that interlock forming designed assemblies to be used as tools of molecular size should be a challenging new field of great interest in future technologies in many different areas.

C) ELECTRON TRANSFER IN MONOLAYER ASSEMBLIES

The development of organized systems that mimic photosynthesis is of
interest in studying possibilities of solar energy conversion and for
that purpose it is important to investigate in some detail the photoinduced electron transfer in monolayer assemblies. An arrangement in
Fig. 4 can be built up on a glass slide: A layer of a cyanine dye D,
a spacer layer and a layer of an acceptor A (viologen). If the acceptor
is separated from the donor by only 2 nm using a fatty acid interlayer
with 14 carbon atoms in the hydrocarbon chain, the fluorescence is
strongly quenched [6]. The interlayer thickness d can be varied by
using fatty acids of various chain lengths, and for each system the
amount of quenching is measured.
      One difficulty arises in realizing such systems. The hydrophilic

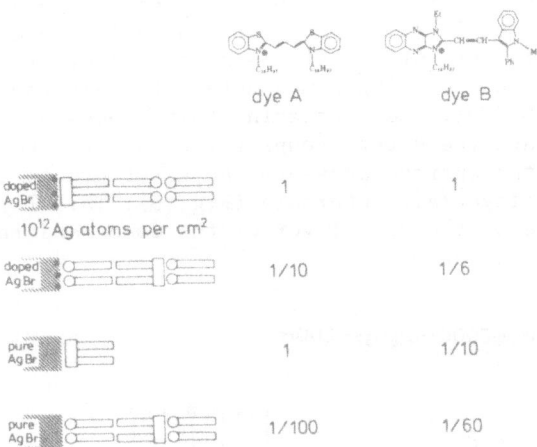

Fig. 3. Spectral sensitization of photographic process.
Effect of doping AgBr by neutral Ag atoms (according to
Steiger).
Dye A (excited dye level within conduction band of AgBr).
At contact: No effect by doping: dye acts by electron injec-
tion. At 5 nm: Strongly enhanced spectral sensitization by
doping; dye acts by energy transfer to neutral Ag atoms.
Dye B (excited dye level below conduction band of AgBr). At
contact and at 5 nm: Strong enhancement of spectral sensiti-
zation by doping; dye acts by energy transfer in both cases.

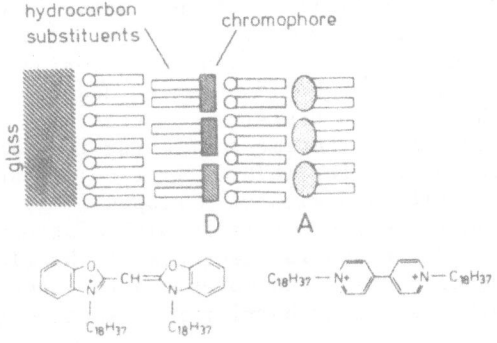

Fig. 4. Photoinduced electron transfer from cyanine dye (D)
to viologen (A) by tunneling through fatty acid spacer layer.
D and A are incorporated in a mixed methyl-arachidate and
Cd-arachidate matrix omitted in the symbolic representation
of the assembly architecture.

groups in the acceptor layer do not easily bind to the hydrophobic surface of the fatty acid spacer layer. The deposition of the acceptor
layer is easily possible if the surface of the spacer layer is made
partially hydrophilic by incorporating a monoester of a dicarboxylic
acid, such as $HOOC(CH_2)_{16}COOCH_3$. When spreading this monoester and fatty acid at the water surface the ester groups first turn to the water
surface; when increasing the surface pressure the ester groups are
forced to move to the monolayer/air interface (Fig. 5). This layer is
deposited on a glass slide on top of a layer of the donor and it is covered by a layer of the acceptor.

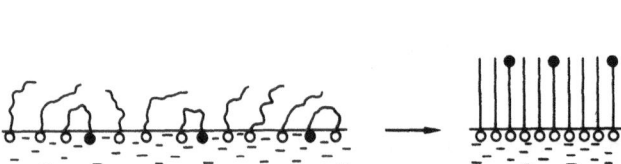

Fig. 5. Production of mixed layer of fatty acid and octa-
decane dicarboxylic acid monoester. Layer used as spacer in
the arrangement of Fig. 4 according to Möbius and Vogel [7].

The amount of quenching is strongly dependent on the thickness of
the spacer layer. For 2.0 nm the fluorescence is quenched by 70 %, for
2.2 nm by 50 %, for 2.7 nm by 10 %. From the amount of fluorescence
quenching the rate of electron transfer k, relative to the decay rate
without acceptor, k_O, can be given ($k/k_O = I_O/I - 1$). The rate k thus
obtained is found to decrease exponentially with increasing interlayer
thickness. An independent way to demonstrate the exponential dependence
of the rate constant k on the distance d is given by measuring the fluorescence lifetime of the donor, which should be shortened due to the
electron transfer, because the electron transfer competes with the fluorescence emission. The fluorescence lifetime shortening is a measure
of the rate constant of electron transfer ($k/k_O = \tau_O/\tau - 1$ (τ and τ_O
are the fluorescence lifetimes with and without A respectively)), and
again this rate is found to decrease exponentially with distance d of
donor and acceptor. The observed exponential decrease of the electron
transfer rate with increasing distance d indicates quantum mechanical
tunneling of electrons through the energy barrier representing the fatty acid ester interlayer.

For energy storage by photoinduced vectorial charge separation the
electron should be removed from the excited dye D fast enough to avoid
deactivation and this electron must be kept in an acceptor A at highest
possible energy level for sufficiently long time. In the arrangement
discussed above where the trap is separated from the dye by a saturated
hydrocarbon the electron cannot be removed far enough with the necessary speed to avoid back transfer.

To reach that goal the excited dye molecule should be in contact

with a π-electron system acting as a molecular wire W leading from excited dye D to a trap A.

$$D\ W\ A\ \xrightarrow{h\nu}\ D^*W\ A\ \rightarrow\ \overset{+}{D}\ W\ \overset{-}{A}$$

The trap must be at a distance far enough to avoid back reaction by
tunneling.

    To demonstrate the action of a molecular wire in a model system
molecules were synthesized that were constructed in such a way that the
π-electron portion is interlocked in the gap between the hydrocarbon
substituents of a cyanine dye. By spreading the two components and fatty acid on the surface of water and pushing the molecules together, the
molecular assembly is assumed to self-organize as indicated in Fig. 6a.

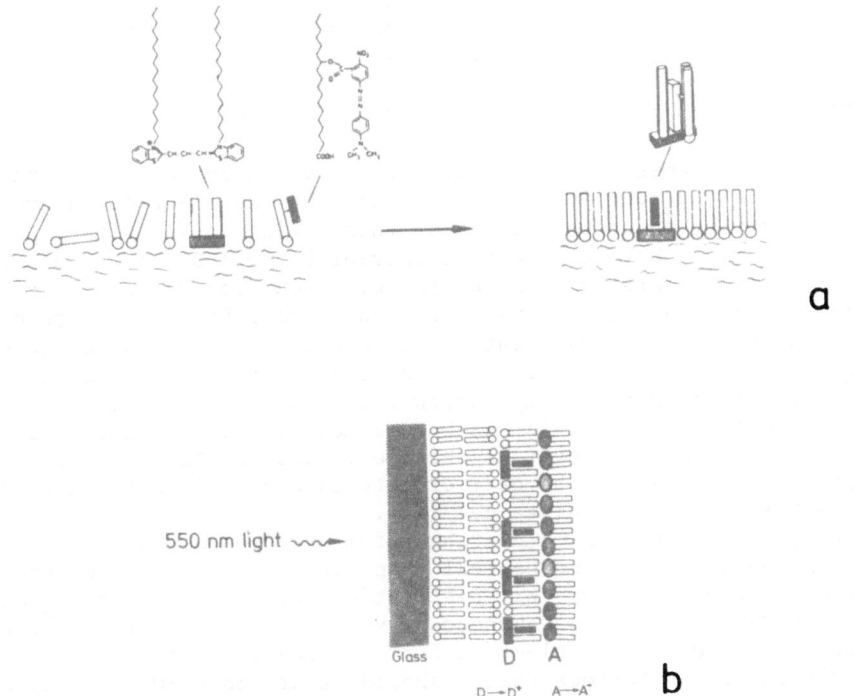

Fig. 6. Experimental realization of the idea of a molecular
         wire
         a) formation of organized assembly of cyanine and azo dye
            at water surface
         b) monolayer of cyanine and azo dye transferred to glass
            slide and superimposed by layer with viologen derivative.
            Cyanine acts as photoinduced electron donor, azo dye as mo-
            lecular wire, viologen as acceptor.

This follows from absorption measurements with polarized light in the
band of both molecules and from the surface pressure area isotherm which
is in a characteristic way different from the isotherm of the components,
indicating the interlocking of the two dyes.

Several effects can be demonstrated that support this view. The
monolayer can be incorporated in an assembly between metal electrodes
[8]. The system is illuminated with light absorbed by the cyanine dye
and the photocurrent can be measured [9]. It is by an order of magnitude
larger than without the azo dye demonstrating that the azo dye acts as a
molecular wire. The photoelectrical properties of such a system can be
analyzed in detail. In another experiment, the layer is deposited on a
glass slide and a layer of an electron acceptor is deposited on top it
[10](Fig. 6b). The cyanine dye is excited by illumination and the rate
of electron transfer from excited dye D to acceptor A is measured by
the absorption change indicating the formation of A⁻; the absorption
band of A⁻ appears. Compared with a corresponding arrangement without
molecular wire (an arrangement where the dye is separated from the acceptor by hydrocarbon chains) the electron transfer rate is enhanced.

In a device for energy storage a donor D' should be present besides the functional unit D W A to regenerate photocatalyst D [11]

$$D' \quad D W A \xrightarrow{h\nu} D' \quad D{*}W A \rightarrow D' \quad D^+W A^- \rightarrow D'^+ \quad D W A^-$$

This solid arrangement is regenerated to D' D W A by reaction with mobile oxidizing and reducing agents. To prevent the electron in excited
dye D from moving to another acceptor than moving to W and then to A
it is necessary to have a potential barrier between D and D'. But then
the difficulty appears to transfer the electron from D' to D⁺. It may
seem impossible that the barrier acting as a wall for the photoexcited
electron in D can be transparent for an electron of D', but by quantum
mechanical tunneling an electron can be transferred through a barrier
if the barrier is sufficiently narrow and if sufficient time is available. As mentioned above this has been demonstrated by constructing a
number of different monolayer organizates. Based on these experiments
the optimal thickness of the barrier between D and D' is found to be
about 14 Å (edge-to-edge distance) [12].

Acceptor A must be at a distance from D large enough to prevent
back reaction by quantum mechanical tunneling within a time of the order of milliseconds. Under this restriction this distance should be as
small as possible to prevent unnecessary loss of free energy in the
process of conducting the electron from excited D to acceptor A. From
such considerations follows that A should be at about 30 Å distance.
In a reasonable device two molecules with π-electron system (W_1 and
W_2) are required to bridge that distance [12]. D, W_1 and W_2 must be
close in order to remove the electron from D* within some picoseconds.
The contact of W_2 with A can be looser since the time to transfer the
electron from W_2 to A must be short only in comparison with the time
for returning the electron from W_2 to D⁺ (by tunneling or by thermal
activation and tunneling via W_1). The need for the transient trapping
of the electron in W_2 requires that its energy level in W_2 is by 0.1 to
0.2 eV below its level in W_1. On the other hand, in an optimized device

W_1^- should be at the energy level of D*. If W_1^- would be at a higher level
the required fast electron transfer would not be possible; being at a
lower level would result in an unnecessary loss of free energy.

The energy level of A must be as high as possible, but low enough
to prevent the back transfer of the electron from A^- to D^+. For trapping the electron in A for milliseconds its level in A must be by 0.6 to
0.7 eV below the level in excited D.

D) MECHANISM OF ENERGY STORAGE IN REACTION CENTER OF PHOTOSYNTHETHIC BACTERIA

In the following we consider the reaction center in the light of
these design principles. According to a recent X-ray analysis in the
case of bacterium Rhodopseudomonas viridis [13] the main portion of the
reaction center consists of heme group (D') with a pair of two bacteriochlorophyll molecules at edge-to-edge distance 15 Å (the special pair
(D) (Fig. 7). In contact with the special pair is a bacteriochlorophyll
molecule (W_1) and in contact with W_1 a bacteriopheophytine molecule (W_2)
and at the end a quinone, the electron acceptor A. The chromophores are
fixed in their position by a protein matrix. The proposed solid components D' D W_1 W_2 A and the energy barrier between D' and D are indeed
present. The distances between D' and D (15 Å edge-to-edge distance) and
between D and A (22 Å edge-to-edge distance) are in the predicted range.
The molecular wire connecting donor D and acceptor A is realized.

From spectroscopic evidence [14] it is known that the time of electron transfer from the special pair to the bacteriopheophytine is ≤ 4 ps.
This time is indeed short in comparison with the time for deactivating
the excited special pair which seems to be about 20 ps. The time of
electron transfer from bacteriopheophytine to quinone is 230 ps. This
is indeed short compared with the time to transfer the electron from
reduced bacteriopheophytine to oxidized special pair (15 ns). The time
of electron transfer from cytochrome to oxidized special pair is 270 ns.
This is short compared to the time of electron transfer from reduced
quinone to oxidized special pair (9 ms) (Fig. 7a). The mid-point potential (oxidized special pair)/(photoexcited special pair) is about
$E = -0.83$ V, the mid-point potential of quinone $E \approx -0.165$ V, the level
difference then about 0.7 eV. The level of W_2^- is about 0.2 eV below the
level of photo-excited special pair D*. The level difference of W_2^- to
quinone then is about 0.5 eV. All values are consistent with the design
principles considered above.

For a more detailed study [15] we have calculated the rates of all
electron transfer steps for the given arrangement (Fig. 7b) and for
other arrangements (variing distances d_1 to d_6) by taking the electronic structure of the chromophores and their geometrical arrangement ex-
plicitely into account, while the protein portion was considered as a
dielectric continuum. The distances d_1 to d_6 calculated from the measured rate constants (Fig. 7a) are compared with the values given by
the X-ray analysis (Fig. 7b):

	d_1	d_2	d_3	d_4	d_5	d_6	(in Å)
calculated	20	13	10	15	10	23.5	
observed	21±3	13±3	11±3	17±3	13±3	27±3	

The small values calculated for d_5 and d_6 may indicate some flexibility of the quinone (the shorter distances reached by fluctuation determine

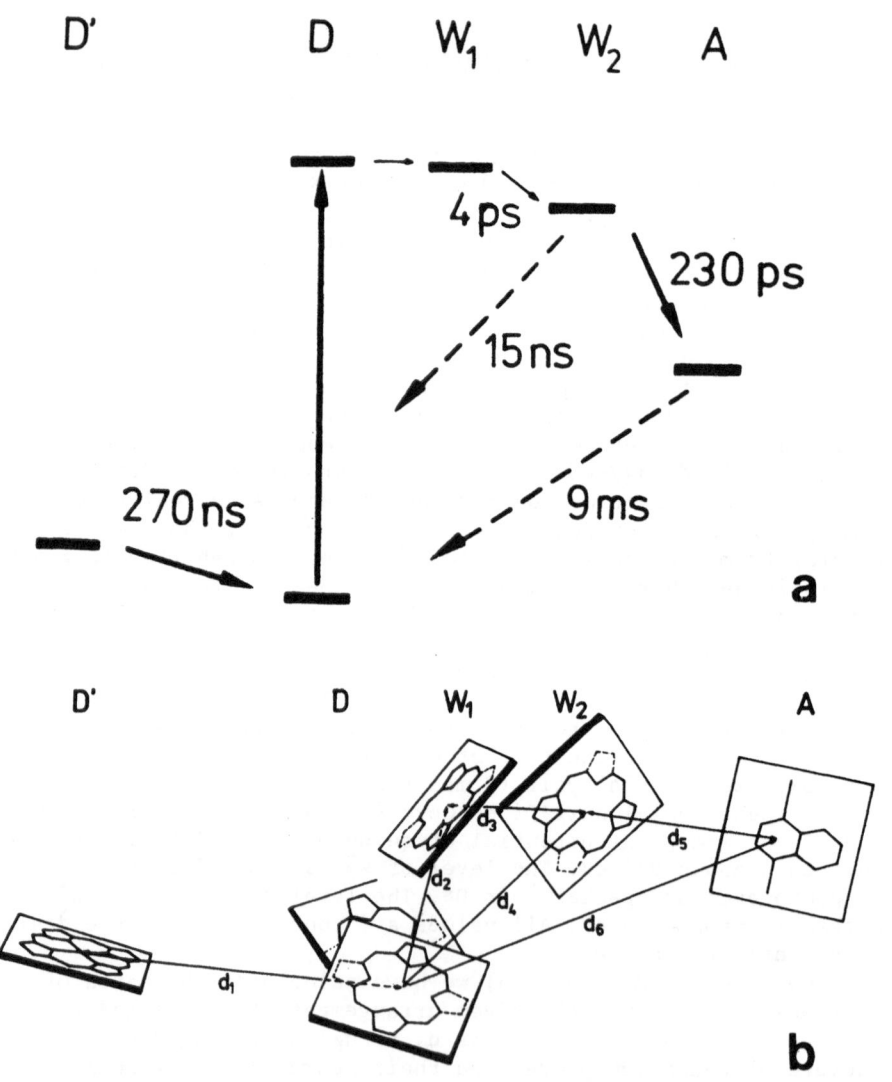

Fig. 7. Main portion of photosynthetic reaction center in Rps. viridis
a) energy levels
b) arrangement of chromophores according to Deisenhofer et al.

the tunneling rate).

From considering the various arrangements of the chromophores it can be concluded that the actual arrangement (Fig. 7b) is a well adjusted device for energy storage. Some results are discussed below.

Arrangement of Molecules in Special Pair D

The special pair captures excitons from the antenna system made of bacteriochlorophyll molecules. This is only possible if the strong absorption band of D is at sufficiently long waves. This can be realized by arranging two molecules of bacteriochlorophyll in an appropriate manner. A calculation based on the free electron model has shown that the arrangement of the two molecules in the special pair given by the X-ray analysis constitutes the arrangement with almost the maximum possible bathochromic shift.

Arrangement of $D \ W_1 \ W_2$

To avoid deactivation the electron must be transferred from D* to W_1 and from W_1 to W_2 within some picoseconds. This limits distances d_2 and d_3 (Fig. 7b). Within this limitation these distances must be as large as possible. The optimal arrangement of D, W_1 and W_2 is the arrangement with the smallest possible rate of tunneling from W_2^- to D^+, and this is a bent arrangement. In this case the electron must tunnel through the high potential barrier of the protein portion. The electron transfer in the backward direction is thus efficiently suppressed by the particular arrangement observed in the reaction center. The observed time for this process (15 ns) corresponds roughly to what is expected in this limit.

Arrangement of W_2 and A

The time for electron transfer from W_2^- to A must be short compared to 15 ns (time for transfer to D^+). With this restriction the time should be as long as possible since d_5 (and thus d_6) should be as great as possible. The observed time (230 ps) and the observed distance d_5 correspond roughly to what is expected for the optimum.

Arrangement $D \ W_1 \ W_2 \ A$

The time for tunneling from A^- to D^+ should be as long as possible. Due to the banana-shaped arrangement of $D \ W_1 \ W_2 \ A$ the electron must tunnel through the high potential barrier of the protein portion. This bending leads to maximum suppression of the electron transfer in the backward direction. The tunneling from A^- to D^+ then needs milliseconds as required for the proposed device.

Energy Level of W_2^-

We consider again the electron in W_2. Besides moving to A and tunneling back to D^+ it can be thermally activated to the level of W_1^-, tunnel to W_1 and from there back to D^+. The level difference between W_1^- and W_2^- should be optimized for minimum energy loss. This is found to be the case for a level difference of 0.15 eV. If the energy difference is made smaller the energy loss is increased due to the increasing probability of back transfer instead of transfer to A. If it is made larger the energy loss is increased due to the lowering of the energy level. This optimum condition is roughly fulfilled in the reaction center.

Energy Level of A^-

Besides tunneling to D^+ the electron in A^- can be thermally activated to the level of W_2^- , tunnel to W_2 and from there be transferred back to D^+ (directly or via W_1). This process should not occur with appreciable probability within some milliseconds and this means that the energy difference between W_2^- and A^- cannot be smaller than about 0.34 eV (corresponding to an electron transfer time of 7 ms) and in the optimum it should not be much larger. The value 0.46 eV is actually observed.

What Is the Action of The Second Branch?
Why Evolved a Branched Arrangement?

The reaction center has another equivalent molecular wire W_1 W_2 contacting the special pair, but the quinone is missing at the end. A quinone seems to bind occasionally to the second branch, to accept the electron from the quinone bound to the first branch, to diffuse away and to transfer the electron to some reactant, thus acting as an electron shuttle. Such an arrangement allows removal of the negative charge from A^- within a fraction of a millisecond. The electron transport system then has recovered by the time the next exciton arrives in the reaction center under natural conditions. The inactivity of the second branch can be due to the fact that W_2^- in the second branch is at a slightly higher energy level due to the changed environment. Then it does not act as a trap; the excited electron is trapped by W_2 in the first branch, even if it has been transferred primarily to the second branch.

It can be imagined that the second branch took an active part in electron transfer at an earlier evolutionary stage. Assuming for such a stage that both quinones are loosely bound and only part of the time present, photoreduction takes place if at least one of the two quinones is bound. The electron is transferred to the quinone in either case, if it is bound to the first or the second branch. Therefore, the quantum yield is increased by the action of the second branch as conducting element.

A further increase in yield is given by tightly binding quinone Q_A and loosely binding quinone Q_B. In this case quinone Q_A is always present and ready to accept the electron (except for the short time of 6 μs needed to transfer the electron from Q_A to Q_B). Then the development of such an asymmetrical arrangement (with only one branch acting as electron transport system) has a selectional advantage and the present system can have developed in this way in the course of evolution. It permits the separation of the device for photoinduced electron transfer and the device for carrying the electron into the pool where it is delivered at some acceptor. The proposed separation of an originally symmetric arrangement into two cooperating devices corresponds to a general pattern of evolutionary processes.

The originally symmetric arrangement is assumed to have evolved from a simple electron transfer system obtained by binding two bacteriochlorophylls and a bacteriopheophytine or some ancestors and a quinone to a protein. It is assumed that two such systems joined forming the special pair. The arrangement then had a strong selectional advantage, since a better exciton trap now was present. This hypothesis predicts the involvement of two proteins in the fixation of the chromophores forming the two branches. These proteins having evolved from the same ancestor must be genetically related (Fig. 8).

The results indicate that the arrangement of the chromophores in the reaction center is optimized in an astonishing fashion. Surprisingly the protein seems to be acting essentially as a spacer allowing an extremely specific arrangement of the chromophores. We attempted to understand the mechanism on the basis of more general design principles indicating the functional relations behind the given structure. This approach was found to be fruitful in the past in predicting the arrangement $D' \, D \, W_1 \, W_2 \, A$ and it was useful in the present paper in rationalizing structural details in the reaction center.

APPENDIX

Evaluation of Rate Constants of Electron Transfer Reactions

Donor and acceptor molecules are assumed to be in a dielectric with D = 4 (protein) in the geometrical arrangement given by Deisenhofer et al. and in alternative arrangements. For finding rate constant k the wave functions of the electron before and after transfer from donor (D) to acceptor (A) must be given.

Calculation of Rate Constant k

Considering downhill electron transfer (from electronic wave function $\psi_{el,D}$ to $\psi_{el,A}$) the system is first in the vibronic ground state (vibronic wave functions of donor and acceptor φ_{Do} and φ_{Ao}) and at the end in the vibronic state with quantum numbers v and w respectively (φ'_{Dv} and φ'_{Aw}). (For simplicity it is assumed that only one normal vibration is excited in each molecule). The transition (from state with wave function $\psi_{el,D} \, \varphi_{Do} \, \varphi_{Ao}$ to state with wave function $\psi_{el,A} \, \varphi'_{Dv} \, \varphi'_{Aw}$) takes place if the two states energetically match. This requires some thermal activation energy Δ_{vw}. The rate of electron transfer then is given by

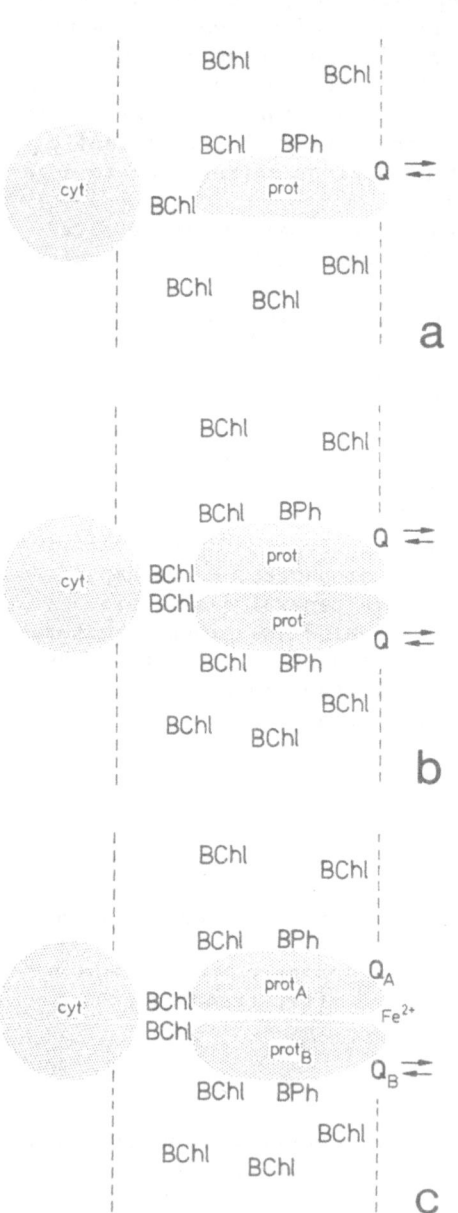

Fig. 8. Possible steps in evolution of bacterial photosynthesis.

a) membrane bound antenna molecules (bacteriochlorophyll or
precursor) and protein that
binds such molecules and loosely binds quinone Q. Chromophores bound to prot act as
weak exciton trap and as pump
of electrons from photoexcited
chromophore to quinone. The
chromophore is regenerated by
accepting an electron from
membrane bound donor cyt.

b) Two electron-transfer systems
of the type shown in Fig. 8a
have joined forming special
pair of bacteriochlorophyll
molecules. The pair is a better exciton trap than the
single molecule (bathochromic
band shift). By evolution of
prot an arrangement of bacteriochlorophyll, bacteriopheophytine and quinone deve-
lops that is optimized for efficient energy storage by
electron transfer.

c) Loss of symmetry by functional
division: Photoinduced electron transfer from cyt to Q_B
via Q_A and Fe^{2+}; Q_A binds more
strongly. Q_B, as before, carries electrons into pool.
Functional division by evolution of prot into $prot_A$ and
$prot_B$. The specific interaction
with these proteins determines
energy levels of chromophores.
Increase in efficiency by
keeping cytochrom in fixed
position.

$$k = \frac{2\pi}{\hbar} S \epsilon^2 \qquad (5)$$

where

$$\epsilon = \langle \psi_{el,D} V \psi_{el,A} \rangle \qquad (6)$$

$$S = \sum_{v,w} e^{-\Delta_{vw}/k_B T} \frac{1}{k_B T} \langle \varphi_{Do} \varphi'_{Dv} \rangle^2 \langle \varphi_{Ao} \varphi'_{Aw} \rangle^2 \qquad (7)$$

V is the perturbation energy of the electron of the donor by the acceptor. This is essentially identical with the usual description [16-19]. The relation

$$\frac{\delta}{k_B T} e^{-\Delta_{vw}/k_B T}$$

is used for the probability that states v and w match (activation energy Δ_{vw}) where δ is the uncertainty in energy. δ cancels in writing the expression for k. It is assumed that the solvent reorganization energy can be neglected in the present case (the chromophores are surrounded. by an essentially rigid protein and the electronic charge, before and after transfer is spread over a large π-electron portion). It is assumed that the effect of charge separation on the medium is reasonably well considered by treating the medium as dielectric with D = 4. Important in the present view is the evaluation by taking the specific electronic structure of donor and acceptor and their relative position into account. The parameter values thus obtained are quite different from those assumed in previous work.

The formalism is valid for small perturbation ($\epsilon < \hbar/t_c$ where t_c is the time to destroy phase relations in electron transfer, i. e. the time between collisions that can lead to energetic match or destroy match). This is the case in the transitions $D' \rightleftharpoons D^+$, $D \rightleftharpoons D'^+$, $W_2^- \rightleftharpoons A$, $A^- \rightleftharpoons W_2$, $W_2^- \rightarrow D^+$, $A^- \rightleftharpoons D^+$. If $\epsilon > \hbar/t_c$ but still small compared to $k_B T$ (this case is given for transition $D^* \rightleftharpoons W_1$, $W_1^- \rightleftharpoons D^+$, $W_1^- \rightleftharpoons W_2$, $W_2^- \rightleftharpoons W_1$) we use the equation [3] [20]

$$k = \frac{1}{t_c} 2\epsilon S \qquad (8)$$

It is obtained by considering that the rate of the transition $\psi_{el,D} \varphi_{Do} \varphi_{Ao} \rightarrow \psi_{el,A} \varphi'_{DV} \varphi'_{Aw}$ is $\frac{1}{t_c}$ times the probability to reach match by a collision, times the probability that the donor, after removing an electron, is in state v (probability $\langle \varphi_{Do} \varphi'_{Dv} \rangle^2$), times the probability that the acceptor, after accepting an electron, is in state w (probability $\langle \varphi_{Ao} \varphi'_{Aw} \rangle^2$) and that the probability to reach match is

$$\frac{\delta}{k_B T} e^{-\Delta_{vw}/k_B T}$$

where $\delta = 2\varepsilon$ in the present case. We use the value $t_c = 10^{-12}$ s through-out the paper. Jortner [17] gives the value $5 \cdot 10^{-12}$ s in ordinary solids based on picosecond spectroscopy, Kenkre and Knox [21] obtained values between 10 fs to 53 ps.

Wave Functions of Electron Before and After Transfer

In the case of the electron transfer from D^* to W_1 a π-electron extend-ing over the special pair D^+ becomes a π-electron extending over W_1. The approximate wave functions of the electron before and after transfer is easily obtained in the picture of the free electron model. It has first 8 and then 10 antinodes in each of the two molecules of the special pair. It is described by atomic orbitals $\psi_{el,i}$:

$$\psi_{el} = \Sigma c_i \psi_{el,i} \tag{9}$$

At small distances r_i of the electron from nucleus i $\psi_{el,i}$ can be represented by a Slater function:

$$\psi_{el,i} = N_{S,i}(\frac{r_i}{a_o}) e^{-Z_{eff,i} r_i/(2a_o)} \frac{z_i}{r_i} \quad (\text{for } r_i \leq r) \tag{10}$$

where $Z_{eff,i} = 3.25$ for C, 3.90 for N, a_o is Bohr's radius. For eva-luating ε the wave function at large distance is of interest, where the electron is effectively in the field of its counter charge e_o in a di-electric of permittivity D (potential energy $- e_o^2 /(Dr_i)$). Since the energy of the electron has a given value $-\varphi$ its wave function at large distance r must fulfill the Schrödinger equation for $V = -e_o^2/(Dr_i)$ for this energy $-\varphi$. This is the case for the function

$$\psi_{el,i} = N_i (\frac{r_i}{a_o})^{n-1} e^{-\alpha r_i} \cdot \frac{z_i}{r_i} \quad (\text{for } r_i \geq r_\pi) \tag{11}$$

$$\alpha = \sqrt{2m\varphi}/\hbar$$

$$n = (\alpha a_o D)^{-1}$$

(m = mass of electron). In the present case (E = -0.83 V; φ = 0.82 eV; D = 4; α = 0.465 $Å^{-1}$; n = 1.02). N_{Si} and N_i are interrelated since $\psi_{el,i}$ must smoothly go over into (10) and (11). For simplicity we apply (10) for $r \leq r_\pi$ and (11) for $r \geq r_\pi$ where r_π = 1.7 Å is the van der

Waals radius of a π-electron. Equalizing (10) and (11) for $r = r_\pi$ gives

$$N_i = N_{S,i} \ (r_\pi/a_o)^{2-n} \ e^{(\alpha - Z_{eff,i}/(2a_o))r_\pi} \tag{12}$$

$N_{S,i}$ is given by normalizing the wave function. Since the electron charge is mainly at $r < r_\pi$ it is reasonable to approximate $N_{S,i}$ by normalizing the Slater function.

$$N_{S,i} = \frac{1}{\sqrt{\pi}} \ (\frac{Z_{eff,i}}{2})^{5/2} \ a_o^{-3/2} \tag{13}$$

In the case of the wave function of the electron after transfer ψ_{el} is obtained accordingly. However it must be taken into account in this case that the electron in W_1^- is bound to a neutral molecule W_1 ($V = o$ at $r > r_\pi$) and to fulfill the Schrödinger equation at large r_i values, n must be taken as zero:

$$\psi_{el,i} = N_i (\frac{r_i}{a_o})^{-1} \ e^{-\alpha r_i} \frac{z_i}{r_i} \quad \text{(for } r_i \geq r_\pi) \tag{14}$$

In the transfer considered here, $D^* \overset{e}{\to} W_1$, the initial wave function has 10 antinodes in each of the two molecules in the special pair [22]. At the end the electron is in W_1 and has again a wave function with 10 antinodes.

The wave functions of the different components indicated in Fig. 9 and 10 are obtained in the same manner from free electron model calculations [22].

Evaluation of ε

ε is obtained from (6) by introducing the wave functions given above. The integral can be easily approximated in all cases of neutral acceptors ($D^* \overset{e}{\to} W_1$; $W_1^- \overset{e}{\to} W_2$, $W_2^- \overset{e}{\to} W_1$, $W_2^- \overset{e}{\to} A$, $A^- \to W_2$ where $V = o$ except at $r_i < r_\pi$. In the region $r_i \leq r_\pi$ V is given by the Slater potential V_i

$$V_i = -\frac{e_o^2}{r_i} \ (Z_{eff,i} - \frac{a_o}{r_i}) + V_o \tag{15}$$

where $Z_{eff,i} = 3.25$ for C, $Z_{eff,i} = 3.90$ for N. The term $V_o = 3.8$ eV $\cdot$ $(1-1/D) = 2.85$ eV is added since the potential energy is considered to be zero in the medium, not in the vacuum. According to (9) and (6):

$$\varepsilon = \sum_{ij} c_i c_j \ \varepsilon_{ij}; \ \varepsilon_{ij} = \langle \psi_{el,Di} \ V_j \ \psi_{el,Aj} \rangle \tag{16}$$

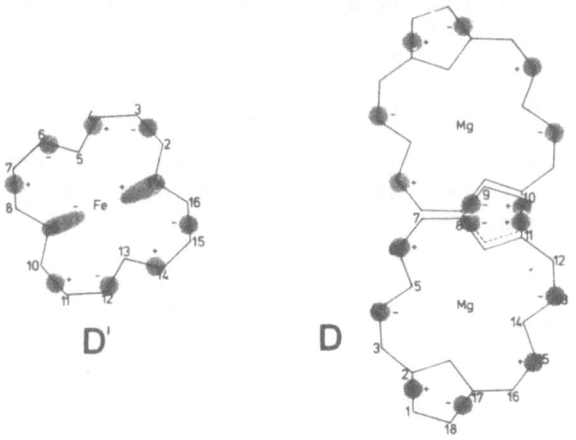

Fig. 9. Upper lobes of wave functions relevant to transition
D' ⇄ D$^+$ indicated. Special pair D and heme D'.

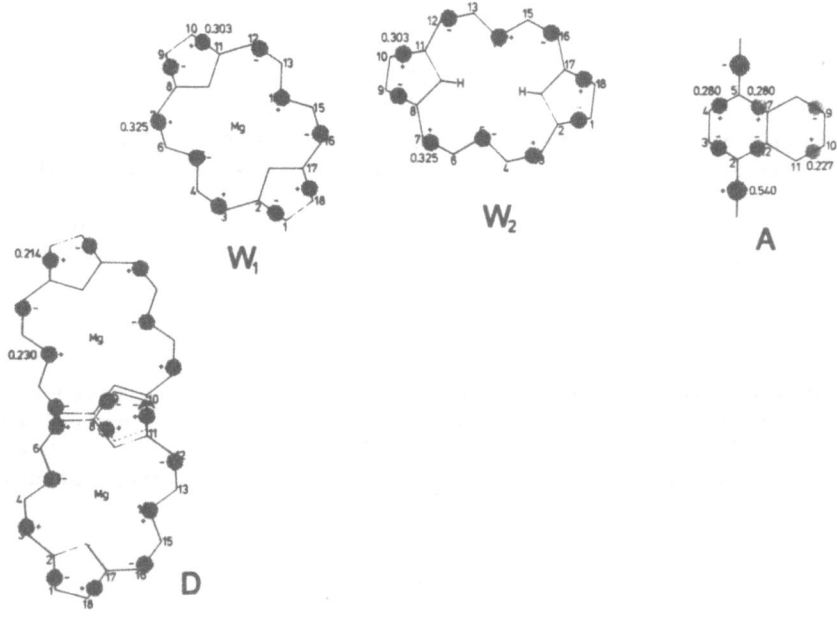

Fig. 10. Wave functions in special pair D, W$_1$, W$_2$ and A
relevant to transitions D* ⇄ W$_1$, W$_1^-$ ⇄ W$_2$, W$_2^-$ ⇄ A.

where the c_i's and c_j''s refer to donor D and acceptor A respectively. In evaluating ε_{ij} the wave functions are divided into components with the orbital axes parallel to the line connecting the atoms (x-axis) in Fig. 11 and perpendicular to this line respectively:

$$\varepsilon_{ij} = \varepsilon_{ij,x} \cos\vartheta_i \cos\vartheta_j + \varepsilon_{ij,yz} \sin\vartheta_i \sin\vartheta_j \cos\alpha_{ij} \tag{17}$$

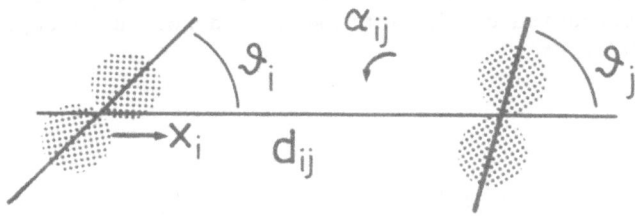

Fig. 11. Angles ϑ_i α_{ij} ϑ_j and distance d_{ij}

The interaction energy resulting from the first component is easily obtained if we set $r_i = x_i$ and expand the exponential in (11)

$$e^{-\alpha r_i} = e^{-\alpha x_i} = e^{-\alpha(d_{ij}+x_j)} = e^{-\alpha d_{ij}}(1-\alpha x_j)$$

The integral obtained by evaluating ε_{ijx} can be solved analytically, and we find

$$\varepsilon_{ij,x} = e^{-\alpha d_{ij}} (d_{ij}/a_o)^{n-1} \text{ const} \tag{18}$$

$$\text{const} = \frac{e_o^2}{2a_o} N_i N_{s,j} a_o^3 \left(\frac{2}{Z_{eff,i}}\right)^3 (\alpha a_o) \frac{80\pi}{3} \tag{19}$$

if the term V_o in (15) is neglected and the integral extended to infinity. In finding the second component r_i and x_i can be taken as constant and we obtain

$$\varepsilon_{ij,yz} = \frac{\varepsilon_{ij,x}}{\alpha d_{ij}} \tag{20}$$

ε can then be evaluated in the five cases $D* \overset{e}{\rightleftharpoons} W_1$, $W_1^- \overset{e}{\rightleftharpoons} W_2$, $W_2^- \overset{e}{\rightleftharpoons} W_1$, $W_2^- \overset{e}{\rightleftharpoons} A$, $A^- \overset{e}{\rightleftharpoons} W_2$ by summation over i and j (eq. 16). In evaluating (20) we have considered each antinode as a pseudo atom of carbon

($Z_{eff,i} = 3.25$). This simplification has little influence on the result. The c_i's ans c_j's are given by the free electron model, see Fig. 9 and 10. The numerical values of the c_i's are indicated at some lobes; all other c_i's are given by these values by symmetry.

In all cases with charged acceptor ($D' \to D^+$, $D \to D'^+$, $W_1^- \lessgtr D^+$, $W_2^- \lessgtr D^+$, $A^- \lessgtr D^+$) an additional contribution besides (18) and (20) results from the Coulomb energy of the electron in the field of the counter charge. It can be estimated easily and turns out to be small in the present cases and is neglected.

The ε-values thus obtained are given in Table 1. In judging the results it should be realized that only a few terms in the sum in (16) are important (the contribution of the antinodes of donor and acceptor that are nearest to each other).

Evaluation of S

Vibronic excitation of acceptor D^+
For calculating S the integrals $<\varphi_{Do} \varphi'_{Dv}>$ and $<\varphi_{Ao} \varphi'_{Aw}>$ must be evaluated. This is illustrated in the case of the electron accepted by the special pair, assuming that the positive charge extends over both molecules. By the increasing charge density due to the extra electron (wave function indicated in Fig. 9) the bonds 1-2, 8-9, 10-11, 17-18 compress and the bonds 1-18, 9-10, 2-3, 7-8, 11-12, 16-17 (where the extra electron has a node and is therefore antibonding) extend. Essentially, the C C valence vibrations of bonds 1-2, 8-9, 10-11, 17-18 (1400 cm^{-1}, $\omega_A = 2.64 \cdot 10^{14}$ s^{-1}) are excited, i. e. a number of normal vibrations of roughly the same frequency. We assume for simplicity that the situation can be described as excitement of a single vibration. This is correct if all the excited normal vibrations have the same frequency [23].

$$<\varphi_{Ao} \varphi'_{Aw}>^2 \text{ is given by the expression}$$

$$<\varphi_{Ao} \varphi'_{Aw}>^2 = e^{-\beta_A} \frac{\beta_A^w}{w!} \text{ with } \beta_A = \frac{\omega_A}{2\hbar} \Sigma M_i \rho_i^2 \tag{21}$$

where M_i is the mass of the nucleus of atom i, ρ_i the distance between equilibrium position of atom i before and after electron transfer. In the present case, as shown below, β_A is small (the change in bond lengths is small): the system that has accepted the electron is most frequently found in the vibrationless state ($w = o$).

To approximate the decrease in length of bonds 1-2, 8-9, 10-11, 17-18, in which the additional π-electron has an antinode the relation

$$\frac{\Delta l}{\overset{\circ}{A}} = -0.37 \cdot \frac{\sigma}{\overset{\circ}{A}^{-1}} \tag{22}$$

is used. Δl is the change in bond length, σ the density of this electron in the center of the bond (measured in number of electrons per unit length) [23]. σ is given by the free electron model. For bond 1-2 we find from ref.[22] taking into account that the electron is spread over 2 molecules:

$$\sigma = \frac{1}{2l} \ (0.3203 \ \sin \frac{2\pi l}{4.419 \ l})^2 = \frac{1}{l} 0.050$$

(where $l = 1.4 \ \overset{o}{A}$ is the bond length) and therefore $\Delta l = -1.3 \cdot 10^{-2} \ \overset{o}{A}$. (The simple perimeter model would give

$$\sigma = \frac{1}{2} \frac{2}{181} \cdot (\sin \frac{4\pi}{9})^2 = \frac{1}{l} \frac{0.9}{18} = \frac{1}{l} \cdot 0.05$$

The resulting shift in equilibrium position of the nuclei 1, 2, 8, 9, 10, 11, 17, 18 and of the corresponding nuclei in the second molecule of special pair D is approximately $(1/2) \cdot 1.3 \cdot 10^{-2} \ \overset{o}{A}$. Similarly, the change in length of bonds 3-4, 6-7, 12-13, 15-16 is

$$\sigma = \frac{1}{2l} \ (0.3006 \ \sin \frac{2\pi(3/2)l}{4.419 \ l})^2 = \frac{1}{l} \cdot 0.032$$

and therefore $\Delta l = -8.2 \cdot 10^{-3} \ \overset{o}{A}$.

Therefore, if the electron extends over both molecules,

$$\Sigma \ M_i \rho_i^2 = 16(2 \cdot 10^{-23} g) \left[(\frac{1}{2} \ 1.3 \cdot 10^{-2} \ \overset{o}{A})^2 + (\frac{1}{2} \cdot 8.2 \cdot 10^{-3} \ \overset{o}{A})^2 \right]$$

According to (21) $\beta_A = 0.24$ and therefore $<\varphi_{Ao}\varphi_{Aw}>^2$ is 0.84 for $w = o$; 0.19 for $w = 1$; 0.02 for $w = 2$. For the present purpose this simple approximation is sufficient. A low frequency lattice vibration was considered but does not appreciably contribute.

In a more rigorous description a normal vibration analysis would be required and it should be taken into account that several normal vibrations k can be excited. $<\varphi_{Ao}\varphi_{Aw}>^2 = e^{-\beta_A} \frac{\beta_A^w}{w!}$ has to be substituted by a product

$$\underset{k}{\Pi} \ e^{-\beta_{Ak}} \ \frac{\beta_{Ak}^{w_k}}{w_k!} \tag{23}$$

where

$$\beta_{Ak} = \frac{\omega_{Ak}}{2\hbar} \frac{(\Sigma_i M_i A_{ki} \rho_i \cos\alpha_{ki})^2}{\Sigma_i M_i A_{ki}^2} \tag{24}$$

A_{ki} is the amplitude of atom i if normal vibration k is excited, α_{ki} the angle between the direction of ρ_i and A_{ki}. In ref. [23] such an analysis was made in considering the electronic excitation of some π-electron systems and the simplified model considering only one normal vibration was found to be well justified.

Vibronic excitation of donor D'

In removing the electron from D' the four Fe-N bonds extend since the wave function in Fig. 9 has antinodes in bonds Fe-1 and Fe-9, the complementary wave function of the two-fold degenerate set has antinodes in bonds Fe-5 and Fe-13. The breathing-like mode then is essentially excited. From Resonance Raman Spectroscopy this mode is at ω_D = 1372 cm$^{-1}\cdot 2\pi c$ and atoms 1, 5, 9, 13 on one hand, atoms 2, 4; 6, 8; 10, 12; 14, 16 on the other hand, are displaced by about 0.01 Å. Then $\Sigma M_i \rho_i^2$ is approximately $12(2\cdot 10^{-23}$ g$)(0.01$ Å$)^2$ and therefore (according to eq. 21) β_D = 0.3; $\langle \varphi_{Do} \varphi'_{Dv} \rangle^2$ = 0.74 for v = o; 0.22 for v = 1; 0.03 for v = 2.

Energetic match of D' and D$^+$

The level of acceptor D$^+$ is by $-\Delta G^o$ = 0.13 eV below the level of D' (difference between mid-point potentials of D'/D'$^+$ and D/D$^+$ multiplied by e) and therefore the following states of D' and D$^+$ almost coincide (activation energy $\Delta_{vw} = |(v\omega_D + w\omega_A) \hbar + \Delta G^o|$ small)

$$v = o \qquad w = 1 \qquad \Delta_{01} = 0.044 \text{ eV}$$

$$v = 1 \qquad w = o \qquad \Delta_{10} = 0.040 \text{ eV}$$

In all other cases the activation energy is much larger and the corresponding contribution to S can be neglected. Therefore

$$S = \frac{1}{k_B T} e^{-\beta_D -\beta_A} \cdot \left[e^{-\Delta_{01}/k_B T} \cdot \beta_A + e^{-\Delta_{10}/k_B T} \cdot \beta_D \right]$$

with β_D = 0.3; β_A = 0.24. Thus S = 2.4 eV^{-1}. All other cases can be evaluated accordingly.

It is of interest to compare the present approach with approaches by Hopfield, Jortner and Marcus. In the present approach, in the case of the electron transfer from cytochrom c to the special pair, S = 2.4 eV^{-1}, and ε = $1.0\cdot 10^{-5}$ eV for d = 20 Å (d = d_1 in Fig.7b) and thus k = $\frac{2\pi}{\hbar}$ S ε^2 = $2\cdot 10^6$ s^{-1} in agreement with experiment. For various d values ε can be approximated by ε = 18 eV$\cdot$exp $(-\alpha d)$ with α = 0.72 Å^{-1} and thus

$$k = 0.74\cdot 10^{19} \text{ s}^{-1} e^{-1.44 \text{ Å}^{-1}\cdot d} \tag{25}$$

Hopfield's and Jortner's approaches are based on considerably smaller S values. Hopfield [24] calculated S from the semiclassical expression S = $(2\pi\sigma^2)^{-1/2}$ exp $[-(E_D-E_A -\Delta)^2/2\sigma^2]$ assuming the values

TABLE 1

	E_D/V	E_A/V	$\Delta G^0/eV$	λ/eV	$\alpha/\text{Å}^{-1}$	n	S/eV^{-1}	ε/eV	k/s^{-1}
a) $D' \rightleftarrows D^+$	0.33	0.46	−0.13	1.98	0.722	0.65	2.4	$5.4\cdot10^{-5}$ $4.6\cdot10^{-6}$ $4.1\cdot10^{-7}$	$6\cdot10^7(d_1=18\text{ Å})$ $5\cdot10^5(d_1=21\text{ Å})$ $4\cdot10^3(d_1=24\text{ Å})$ $2.6\cdot10^6$ (exper)
b) $D \rightleftarrows D'^+$	0.46	0.33	+0.13	2.11	0.745	0.63	$1.6\cdot10^{-2}$	$3.6\cdot10^{-6}$	$2\cdot10^3(d_1=21\text{ Å})$
c) $D^* \rightleftarrows W_1$	−0.83	−0.83	0	0.82	0.465	1.02	16.6	$3.7\cdot10^{-2}$ $4.8\cdot10^{-3}$ $5.8\cdot10^{-4}$	$10^{12}(d_2=10\text{ Å})$ $10^{11}(d_2=13\text{ Å})$ $10^{10}(d_2=16\text{ Å})$ $>2\cdot10^{11}$ (exper)
d) $W_1^- \rightleftarrows W_2$	−0.83	−0.63	−0.2	0.82	0.465	0	5.2	$1.2\cdot10^{-1}$ $5.1\cdot10^{-3}$ $5.6\cdot10^{-5}$	$10^{12}(d_3=8\text{ Å})$ $10^{11}(d_3=11\text{ Å})$ $10^8(d_3=14\text{ Å})$ 10^{11} (exper)
e)α) $W_1^- \rightleftarrows D^+$ (excited singlet)	−0.83	−.083	0	0.82	0.465	0	16.6	$1.7\cdot10^{-3}$	$10^{11}(d_2=13\text{ Å})$
e)β) $W_1^- \rightleftarrows D^+$ (triplet)	−0.83	−0.43	0.4	0.82	0.465	0	0.86	$1.7\cdot10^{-3}$	$10^{10}(d_2=13\text{ Å})$
e)γ) $W_1^- \rightleftarrows D^+$ (ground state)	−0.83	+0.46	−1.29	0.82	0.465	0	$1.3\cdot10^{-4}$	$5\cdot10^{-4}$	$10^5(d_2=13\text{ Å})$
f) $W_2^- \rightleftarrows W_1$	−0.63	−0.83	+0.2	1.02	0.518	0	$2.4\cdot10^{-3}$	$6.5\cdot10^{-3}$	$10^7(d_3=11\text{ Å})$
g)α) $W_2^- \rightarrow D^+$ (excited singlet)	−0.63	−0.83	+0.2	1.02	0.518	0	$7.6\cdot10^{-3}$	$6.1\cdot10^{-6}$	$3\cdot10^3(d_3=17\text{ Å})$
g)β) $W_2^- \rightleftarrows D^+$ (triplet)	−0.63	+0.43	−0.2	1.02	0.518	0	5.1	$8.3\cdot10^{-5}$ $6.1\cdot10^{-6}$ $1.5\cdot10^{-6}$	$3\cdot10^8(d_4=14\text{ Å})$ $2\cdot10^6(d_4=17\text{ Å})$ $1\cdot10^5(d_4=20\text{ Å})$ $4.6\cdot10^7$ (exper)
g)γ) $W_2^- \rightleftarrows D^+$ (ground state)	−0.63	+0.46	−1.09	1.02	0.518	0	$1.1\cdot10^{-3}$	$3.9\cdot10^{-5}$	$1\cdot10^4(d_4=17\text{ Å})$
h) $W_2^- \rightleftarrows A$	−0.63	−0.165	−0.465	1.02	0.518	0	0.7	$1.1\cdot10^{-3}$ $1.4\cdot10^{-4}$ $1.4\cdot10^{-5}$	$2\cdot10^9(d_5=10\text{ Å})$ $2\cdot10^8(d_5=13\text{ Å})$ $2\cdot10^6(d_5=16\text{ Å})$ $3\cdot10^9$ (exper)
i) $A^- \rightleftarrows W_2$	−0.165	−0.63	+0.465	1.485	0.625	0	$1.1\cdot10^{-8}$	$1.1\cdot10^{-4}$	1 $(d_5=13\text{ Å})$
k)α) $A^- \rightarrow D^+$ (triplet)	−0.165	−.043	+0.265	1.485	0.625	0	$3.7\cdot10^{-5}$	$1.3\cdot10^{-8}$	$10^{-6}(d_6=27\text{ Å})$
k)β) $A^- \rightarrow D^+$ (ground state)	−0.165	+0.46	−0.625	1.485	0.625	0.75	0.19	$1.5\cdot10^{-7}$ $1.5\cdot10^{-8}$ $1.5\cdot10^{-9}$	40 $(d_6=24\text{ Å})$ 0.4 $(d_6=27\text{ Å})$ $4\cdot10^{-3}(d_6=30\text{ Å})$ $1\cdot10^2$ (exper)

$\Delta = 1$ eV, $E_D - E_A = 0.05$ eV and $\sigma^2 = 0.06$ (eV)^2 (for 300 K) and obtained the value $S = 10^{-3}$ eV^{-1}. Jortner [18] used a quantum mechanical model, similar to the one used here, but assumed a value $\beta_D + \beta_A = 46$ which is much larger than the value $\beta_D + \beta_A = 0.54$ obtained here and thus the factor $e^{-(\beta_D + \beta_A)}$ in the expression for S then is much smaller. Therefore, in Hopfield's and Jortner's approach the distance between donor and acceptor calculated from the known value of the rate constant is much smaller than the actual distance. Jortner considered the possibility of superexchange type coupling to explain the discrepancy. In the present view it is due to the particular choice of parameters.

Marcus and Sutin [16] assumed that the rate is given by the expression

$$k = 10^{13} \text{ s}^{-1} e^{-\beta d'} e^{-\Delta G_r^*/RT}$$

with $\beta = 1.2$ Å^{-1} and $\Delta G_r^* = 7.1$ kJ mol^{-1}.

d' is the separation of the closest C-atoms of the two reactants (15 Å according to the X-ray analysis of Deisenhofer et al., i. e. 6 Å smaller than the center to center distance d) minus an amount of 3 Å to allow for the extension of the π-electron orbital beyond the carbon nucleus. Thus $d' = d - 9$Å. The value of β and the preexponential factor 10^{13} s^{-1} were obtained by adjustment to experimental rate constants in nonbiological systems such as monolayer assemblies. If we substitute d for d' we obtain for T = 300 K

$$k = 10^{13} \text{ s}^{-1} e^{-\beta d} e^{\beta 9 \text{Å}} 0.059$$

$$= 2.9 \cdot 10^{16} \text{ s}^{-1} \cdot e^{-1.2 \text{ Å}^{-1} \cdot d} \tag{26}$$

In the relevant case d = 20 Å both equations (25) and (26) give a similar value $k \cong 10^6$ s^{-1}. In judging the result it should be taken into account that the constants in the Marcus-Sutin's approach are extracted from experimental rate constants while this is not the case in the present approach.

REFERENCES:

1) D. Möbius, R. Ahuja and G. Debuch, in preparation.

2) H. Kuhn, J. Photochem. 10, 111 (1979).

3) D. Möbius and H. Kuhn, J. Chem. 18, 375 (1979);
   T. L. Penner and D. Möbius, J. Am. Chem. Soc. 104, 7407 (1982).

4) H. Bücher, H. Kuhn, B. Mann, D. Möbius, L. v. Szentpàly and
   P. Tillmann, Photogr. Sci. Eng. 11, 233 (1967);
   L. v. Szentpàly, D. Möbius and H. Kuhn, J. Chem. Phys. 52, 4618
   (1970); R. Steiger, H. Hediger, P. Junod, H. Kuhn and D. Möbius,
   Photogr. Sci. Eng. 24, 185 (1980).

5) R. Steiger, Photogr. Sci. Eng. 28, 35 (1984).

6) D. Möbius, Acc. Chem. Res. 14, 63 (1981).

7) V. Vogel and D. Möbius, Thin Sol. Films submitted.

8) U. Schoeler, K. H. Tews and H. Kuhn, J. Chem. Phys. 61, 5009 (1974).

9) E. E. Polymeropoulos, D. Möbius and H. Kuhn, J. Chem. Phys. 68, 3918 (1978).

10) D. Möbius in: Photochemical Conversion and Storage of Solar Energy, Part A, ed. J. Rabani, The Weizmann Press of Israel 1982, p. 139.

11) H. Kuhn, Chem. Phys. Lipids 8, 401 (1972).

12) K. P. Seefeld, D. Möbius and H. Kuhn, Helv. Chim. Acta 60, 2608 (1977).

13) J. Deisenhofer, O. Epp, K. Miki, R. Huber and H. Michel, J. Mol. Biol. 180, 385 (1984).

14) D. Holten, M. W. Windsor, W. W. Parson, J. P. Thornber, Biochim. Biophys. Acta 501, 112 (1978); P. L. Dutton, J. S. Leigh, R. C. Prince and D. M. Tiede in: Light-Induced Charge Separation in Biology and Chemistry, eds. H. Gerischer and J. J. Katz, Berlin: Dahlem Konferenzen 1978, Verlag Chemie, Weinheim, New York 1979, p. 411.

15) H. Kuhn, submitted.

16) R. A. Marcus and N. Sutin, Biochim. Biophys. Acta 811, 265 (1985).

17) E. Buhks and J. Jortner, FEBS Lett. 109, 117 (1980).

18) J. Jortner, Biochim. Biophys. Acta 594, 193 (1980).

19) J. Ulstrup, Charge Transfer Processes in Condensed Media in Lecture Notes in Chemistry, Springer Verlag Berlin, Heidelberg, New York 1979.

20) H. Kuhn in: Light-Induced Charge Separation in Biology and Chemistry, eds. H. Gerischer and J. J. Katz, Berlin: Dahlem Konferenzen 1978, p. 151, Verlag Chemie Weinheim 1979.

21) V. M. Kenkre and R. S. Knox, J. Luminescence 12, 187 (1976).

22) W. Huber and H. Kuhn, Helv. Chem. Acta 42, 363 (1959).

23) F. Bär, W. Huber and H. Kuhn, Z. Elektrochem. 64, 551 (1960); H. Kuhn, Chimia 15, 53 (1961).

24) J. J. Hopfield, Proc. Natl. Acad. Sci. USA $\underline{71}$, 3640 (1974).

Structure and Dynamics of Metal-Nucleic Acid Interactions

T. THEOPHANIDES* and H.A. TAJMIR-RIAHI
Department of Chemistry
Université de Montréal, C.P. 6128, Succ. A
Montréal, Québec, Canada, H3C 3J7

ABSTRACT. Fourier transform infrared spectra of complexes isolated from DNA and RNA interactions with Mg(II), Cu(II), $Co(NH_3)_6^{3+}$, $Co(NH_3)_5Cl^{2+}$, cis- and trans-$Pt(NH_3)_2^{2+}$ were obtained and correlations between the spectral changes, coordination sites and conformational transitions are suggested. Spectroscopic evidence indicates that there is a direct metal interaction with the N_7-site of the guanine base moiety and an indirect metal-phosphate binding through coordinated H_2O or NH_3 molecules for all the metal cation species employed, except for $Co(NH_3)_6^{3+}$. In the latter case there is an indirect metal-NH_3....N7 and metal-NH_3...$O_6 = C_6$, and an indirect metal-NH_3...$^-$O-P=O interaction via the coordinated NH_3 molecules. The B to Z conformational transition was readily observed for the DNA molecule upon binding to the metal ions, whereas RNA conserves its A conformation in all the metal-RNA adducts. The marker bands for the B conformer (C_2'-endo/anti) were observed at 825 (phosphodiester mode) and 690 cm^{-1} (guanine breathing mode). The marker bands were observed at 810 and 675 cm^{-1} for the A conformer (C_3'-endo/anti). The B to Z conformational transition (C_2'-endo/anti to C_3'-endo/syn) for the DNA molecule was accompanied with the bands at 810 cm^{-1} and 625-600 cm^{-1}. The carbonyl band at 1701 in natural DNA shifts to lower frequencies upon a B to Z transition and the bands at 1485, 1371, 1285 and 1265 cm^{-1} also change considerably upon a B to Z transition.

INTRODUCTION

Since the discovery of the Z-DNA structure by Wang et al (1), numerous efforts have been made to explain the physical nature of the B to Z conformational transition. It has been demonstrated that high concentration of metal ions facilitates the B to Z transition in DNA (2,3), but the role of the metal cation in this transition is not fully understood. Rich et al (4) have suggested that since the phosphate groups in Z-DNA are packed closer than in the B-DNA, therefore a high concentration of metal cations is required to shield the phosphate repulsion and to stabilize the Z-conformation.

551

P. M. Rentzepis and C. Capellos (eds.), Advances in Chemical Reaction Dynamics, 551–563.

It seems that metal ion interaction with DNA, either electrostatically with the phosphate negative charges or by direct coordination
to the bases would facilitate the B to Z transition (5). It should
be noted that the binding of a large atom or a bulky organic
substance such as Br(6) or 2-(acetylamino)fluorene/(carcionogen)
(7)at the C8-position of the guanine base also brings about a B toZ
conformational transition. In a recent communication (8) we have
reported the effects of several metal cations such as, Mg(II),
Cu(II) or cationic species, $Co(NH_3)_6^{3+}$, $Co(NH_3)_5^{2+}$ and trans-
$Pt(NH_3)_2Cl_2(DDP)$ on the infrared spectra of DNA and RNA. The DNA
molecule underwent a B to Z transition in all its metal complexes
(8), whereas the RNA molecule remained in its A-form.

In the present work we wish to examine the FT-IR spectra of the
free DNA, RNA and their metal adducts, in the region 1800-500 cm^{-1}
in order to detect the characteristic features of each structural
type of complexes synthesised and possibly to establish a correlation between the spectral changes and the conformation of nucleic
acids. Furthermore, the effect of metal ions on the rigid structure
of RNA and on the B to Z conformational transition of DNA is investigated and the results are compared with previously published
infrared data (8,9) on DNA and RNA.

The interaction of the heavy metal cation with DNA is predomi-
nently through the N7 site of the guanine molecule. The metal
cation species is first hydrated then a metal-water bond is substituted by the ligand (L) to form complexes as follows:

$$M^{2+} + 6H_2O \qquad M(H_2O)_6^{2+}$$
$$M(H_2O)_6^{2+} + L \qquad M(H_2O)_5L^{2+} + H_2O$$

where M:Mg^{2+}, $Co(NH_3)_6^{3+}$, $Co(NH_3)_5Cl^{2+}$, Cu^{2+}, $cis-Pt(NH_3)_2^{2+}$ and
$trans-Pt(NH_3)_2^{2+}$. These metal cations, when hydrated give the following hydrated species: $Mg(H_2O)_6^{2+}$, $Co(NH_3)_5(H_2O)^{3+}$, $Cu(H_2O)_4^{2+}$,
$cis-Pt(NH_3)_2(H_2O)_2^{2+}$ and $trans-Pt(NH_3)_2(H_2O)_2^{2+}$ the substitution
reaction takes place only with the solvent water, the metal-ammonia
bond being quite strong and inert. The hydrated-ammoniated metal
cations interact with DNA electrostatically through their charges
and chemically by forming a direct chemical covalent bond with the
N7 site of guanine. In addition, the coordinated water and ammonia
molecules may form hydrogen bonds and stabilize certain molecular
conformations of the polynucleotide.

EXPERIMENTAL

Materials. Calf thymus DNA was from Sigma Chemical Company. The RNA
yeast was from Calbiochem and both DNA and RNA were used without
further purification. Cobalt hexaammine and cobalt pentaammine
chloride were prepared and recrystallized by standard methods (10).
Cis- and trans-$Pt(NH_3)_2CL_2$ were prepared from K_2PtCl_4 (11) and were
purified as reported (12). All other chemicals were reagent grade
and were used as supplied.

Preparation of metal-nucleic acid adducts

The DNA and RNA metal adducts were prepared by similar procedures. The DNA and RNA sodium salts were dissolved in a solution of NaCl (4×10^{-3} mol) and kept in the refrigerator at 5C° for a period of 5 days with occasional gentle stirring. After complete dissolution and filtration of the nucleic acid solution a solution of 0.05 mol of metal cation ($MgCl_2 \cdot 6H_2O$, $Cu(NO_3)_2 \cdot 3H_2O$, $Co(NH_3)_6^{3+}$, $Co(NH_3)_5^{2+}$, cis- and trans-DDP) was added to it and the final solution was left for a period of one week. The precipitate formed was isolated and washed with water and air dried. The Mg(II) compounds were precipitated by addition of ethanol.

Any contamination of the DNA and RNA commercial preparations by metal ions was shown to be negligible.

Physical measurements

FT-IR spectra were recorded on a DIGILAB FTS-15C/D Fourier Transform Infrared Interferometer equipped with a HgCdTe detector (Infrared Associates, New Brunswick, NJ), a KBr beam splitter and a Globar source. The spectra were taken as KCl pellets with a resolution of 4 to 2 cm^{-1}. The triangular apodization and the smoothing procedures were those used in our previous report (13).

DATA AND DISCUSSION

The FT-IR spectra of DNA and RNA have been recorded in the region of 1800–500 cm^{-1} and compared with the infrared data reported in the literature (9,14). The results are in good agreement with those of the naturally occuring nucleic acids and there has been no trace of denaturated products in the samples. We have also examined the infrared spectra of several DNA and RNA metal-adducts and analyzed the metal binding and any conformational changes.

Base Binding

In the double bond region (1800–1500 cm^{-1}) there is the coupling between the carbonyl stretching frequency and the NH_2 and NH bending modes. It is difficult to separate these vibrational frequencies. However, the two strong absorption bands at 1701 and 1655 cm^{-1} are mainly assigned to the $C_6=O$ of the guanine moiety and $C_2=O$ of the cytidine residue, respectively (15). The two absorption bands shift towards lower frequencies in the spectra of all metal–DNA adducts (Fig. 1 and Table 1). The C=O stretching vibrations of hydrogen bonded GC pairs can be affected by interbase vibrational coupling and thus the shifts of these carbonyl vibrational frequencies would depend on the rearrangement of the hydrogen bonding caused by a B to Z DNA conformational transition upon metal complexation (16). It should be also noted that similar absorption bands in the RNA spectrum at 1691 and 1647 cm^{-1} did not show considerable changes upon base metalation (Fig. 2 and Table 2) and thus, the RNA secondary structure is not perturbed by metal coordination. The absorption bands at about 1605, 1570 and 1530 cm^{-1} in the spectra of

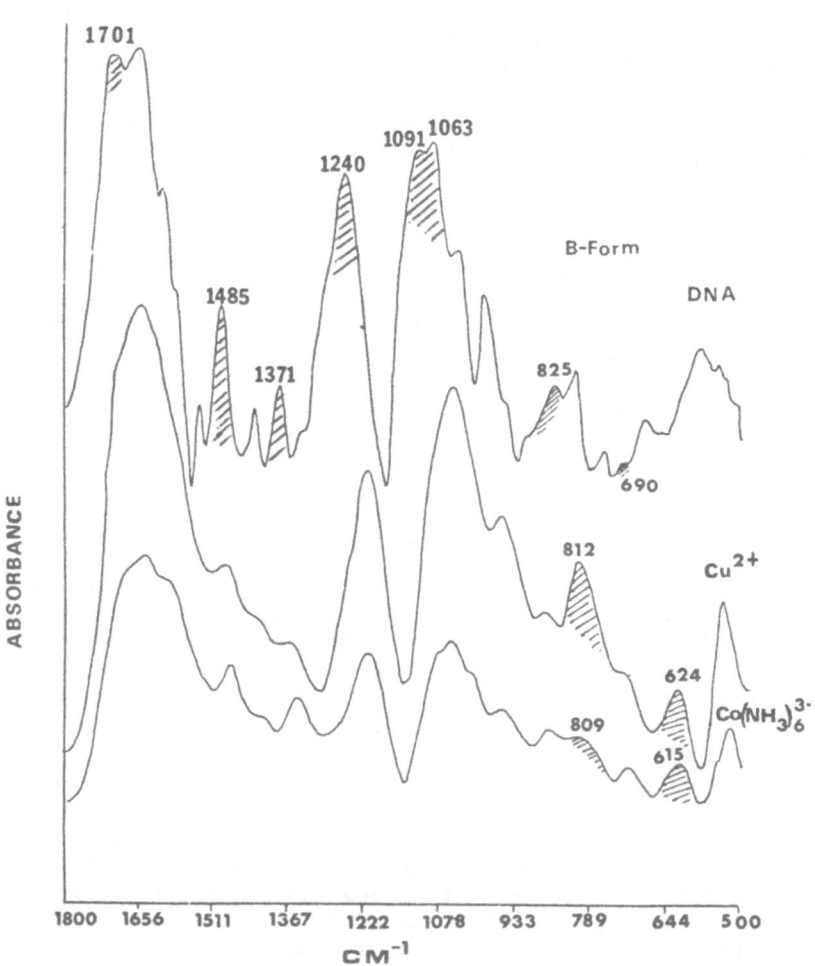

Fig. 1　　FT–IR spectra of DNA and its metal adducts
in the region of 1800–500 cm^{-1}.　The spectra were taken as KCl pellets.

Table 1 FT-IR Absorption Bands(cm^{-1}) for DNA and its Metal Adducts in the Region 1800-500cm^{-1} with possible Band Assignments and Conformational Arrangements.

DNA	Mg-DNA	Cu-DNA	Co(NH$_3$)$_6^{3+}$ DNA	Co(NH$_3$)$_5^{2+}$ DNA	trans-DDP DNA	cis-DDP DNA	Assignments and conformation
1701vs	1678s	1692sh	1691s	1692sh	1693sh	1690sh	Base (C=O) stretch
1655vs	1649s	1648vs	1648vs	1643vs	1650vs	1648vs	A T G
1605s	1600s	1600sh	1600sh	1600sh	1600sh	1587s	A C
1570sh	1575w	1575sh	1570sh	1570sh	1572sh	1570sh	A G
1531w	1531w	1520vw	1530sh	1525sh	1525sh	1520sh	ν NH+δCH
1485s	1480m	1487m	1480s	1481m	1495sh	1502m	νNH+δCH
1420w	1416m	1420w	1415w	1416w	1421w	1420w	δCH$_2$
1371w	1366m	1370m	1353m	1355sh	1358m	1349m	
1329vw	-	-	-	1310m	-	-	C T
1285sh	-	-	-	-	-	1279vw	C T
1265sh	1260sh	-	-	-	-	-	
1240vs	1240bs	1214vs	1217vs	1223s	1222vs	1224vs	PO$_2^-$assymetric
-	1216s	-	-	-	-	-	
1091vs	1097s	1095sh	1092vs	1099sh	1095sh	1095sh	PO$_2^-$symmetric
1063vs	1058vs	1051vs	1055vs	1065vs	1062vs	1059vs	C-O deoxyribose
1015s	1026s	-	1016sh	1010m	1016m	1020sh	
962vs	960vs	962vs	958vs	958vs	958vs	958vs	νC-C+δCCO
922sh	925sh	-	-	-	924sh	-	deoxyribose
883s	878m	880s	869s	-	880s	886w	
850sh	860m	-	-	850bs	-	845s	
825s	-	-	-	-	-	-	
-	811sh	812vs	809s	812sh	809s	-	ν-P-O-sugar B-Form
797m	-	-	-	-	-	790sh	ν-P-O-sugar A-Form
784s	775s	780sh	780s	775s	766s	767s	C
727m	714s	730s	719s	725m	724m	730sh	A
690m	-	-	-	-	-	-	
-	610vs	624vs	615vs	607vs	605vs	597vs	G B-Form
649s	-	-	-	-	-	-	
622m	-	-	-	-	-	-	
538s	555s	537vs	524vs	530vs	530vs	537vs	G Z-Form

s, strong; sh, shoulder, w, weak; m, medium, b, broad; v, very

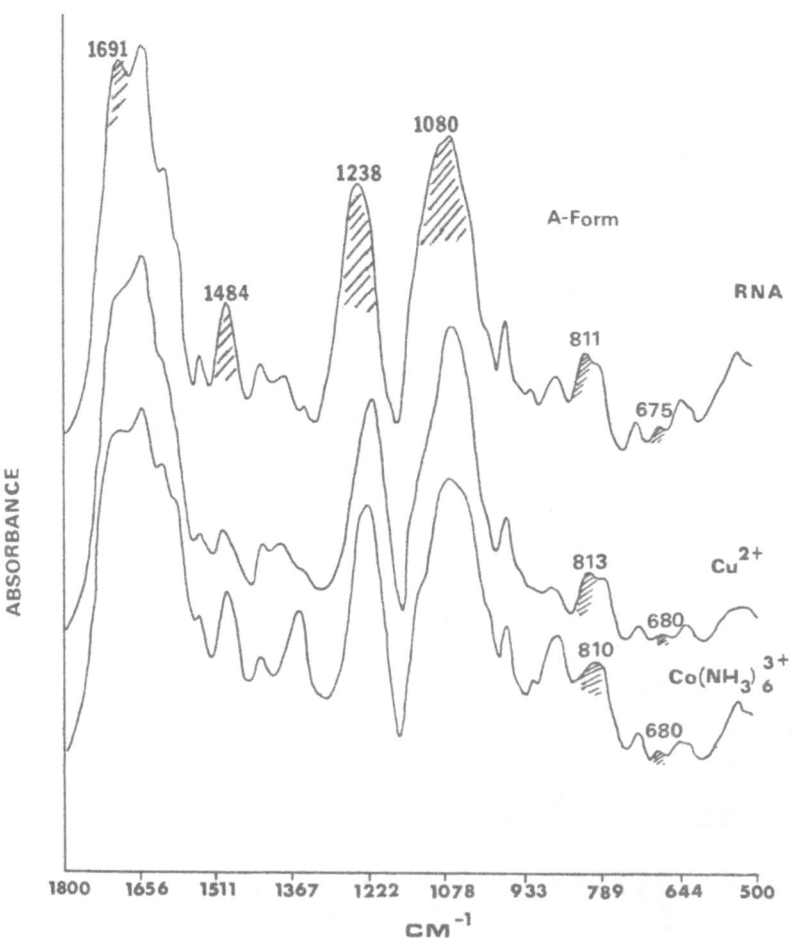

Fig. 2   FT-IR spectra of RNA and its metal adducts
in the region of 1800-500 cm^{-1}. The spectra were taken as KCl pellets.

Table 2   FT-IR Absorption Bands (cm⁻¹) for RNA and its Metal Adducts in the Region 1800-50 cm⁻¹ with Possible Band Assignments and Conformational Arrangements.

RNA A-Form	Mg-RNA	Cu-RNA	Co(NH₃)₆-RNA	Co(NH₃)₅-RNA	Assignments and conformation
1691s	1692zh	1695sh	1688s	1685s	$C_6=O$
1647vs	1640 vs	1646vs	1642vs	1641vs	$C_2=O$
1604s	1600sh	1600sh	1600s	1599s	
1575sh	1570sh	1575sh	1570sh	1570sh	G+A
1530w	1532w	1531w	1529w	1531sh	C
1484m	1486m	1491m	1481m	1481m	G+A
1419w	1410w	1419w	1417w	1419w	A+G
1375w	1367vw	1392w	1359m	1363vw	A+G
1331vw	1330vw	1330vw	-	1331vw	A+G
1238vs	1255bs	1216bs	1217bs	1213bs	PO_2^- assymmetric
-	1220sh	-			
1080s	1116vs	1100sh	1100sh	1100sh	PO_2^- symmetric
1070vs	1063m	1064bs	1062bs	1060bs	C-O ribose
1018sh	1000sh	1005sh	1010sh	1010sh	
962vs	969vs	964vs	960vs	960vs	C-C+CCO ribose
940sh	940sh	944sh	944sh		
913s	912m	915m	910s	910s	
867vs	885s	879vs	865vs	855vs	
811vs	815s	813vs	810sh	809s	O-P-O-sugar A-Form
797vs	794s	795s	795vs	796s	C+U
784s	786s	781s	782s	782s	C
720s	719s	721s	720s	719s	A
675m	680m	681m	680m	680m	
640s	634s	638s	640m	638m	
620sh	622m		620m	622m	G A-Form
566sh	562m	567sh	560sh	560sh	
536bs	552m	543vs	536vs	534s	
515sh	520m	517sh	516m	515m	

s,strong; sh, shoulder; w,weak; m, medium; b,broad; v, very

DNA and RNA which are assigned to the cytidine, guanine and adenine
vibrational frequencies (15), shifted slightly upon metalation (Tables 1
and 2). Since these absorption bands are mainly related to the skeletal
vibrations of the ring systems, metalation of the base moities may cause
small perturbation on these frequencies. Similar observations were made
in the spectra of several mononucleotides upon metal complexation (17-
21). The strong band at 1485 cm⁻ in the spectra of DNA and RNA which
is assigned to the guanine and adenine vibrations (15) exhibited intensity changes and shifting in the spectra of the metal-adducts (Tables 1
and 2). According to previous assignment (22) that absorption band is
assigned to the N -C strethching and C -H bending vibrations of the
guanine moiety in DNA and therefore the changes observed for this
absorption band could be the result of metal-N guanine coordination.
Similar spectral changes occurred for the corresponding absorption band
at 1488 cm⁻ in the spectra of 5'-GMP upon N -metalation (17). Drastic
spectral changes were observed in the bands at 1371, 1329, 1285 and
1265 cm⁻ in the spectra of DNA on complexation (Table 1). Similar
spectral changes were observed in the B to Z conformational transition
in poly(dG-dC). poly(dGdC) (16). On the other hand, the changes were
not significant in the spectra of the A form RNA-metal adducts (Table
2). It is interesting to note that the spectral changes observed in the
direct N -metal interactions (1,23) are very similar to those of an
indirect N -metal cation interaction via NH groups as was found in the
crystal structure of the hexaammine cobalt-tRNA complex (24).

Phosphate Binding

The two strong and broad absorption bands at about 1240, 1090 cm⁻
in the spectra of DNA and RNA are assigned to the PO ⁻ asymmetric and
symmetric stretching vibrations, respectively (14). The band at 1240
cm⁻ shifts towards lower frequencies, while the band at 1090 cm⁻
shifts towards higher frequencies upon nucleic acid metalation (Tables 1
and 2). The shifting of the PO ⁻ vibrational frequencies upon complexation most likely is due to the indirect metal-PO ⁻ interaction via H O
molecules (1) or NH groups (24) M-OH ....⁻O-P-O, or M-NH ...⁻O-P-O
Similar behaviour was observed in the spectra of several structurally
known metal-GMP complexes (GMP = guanosine-5'-monophosphate), where an
indirect metal-PO ⁻ interaction via water molecules was suggested
(17). It is notworthy that the bands at 1240 and 1090 cm⁻ exhibited
major intensity changes, shifting and splitting in the spectra of the
Mg-DNA and Mg-RNA complexes (Tables 1 and 2). These spectral changes
occurred in the PO ⁻ absorption bands could be related to a direct
metal-phosphate interaction in these metal complexes.

Sugar Pucker Conformational Transitions

DNA-Metal Adducts

Free DNA shows a B conformation (C '-endo/anti) with a characteristic infrared absorption bands at 825 cm⁻ , mainly a sugar vibration

and at 690 cm^{-1}, the guanine ring breathing mode (25). Upon coordination of Mg(II), Cu(II), Co(NH$_3$)$_5$$^{2+}$, Co(NH$_3$)$_6$$^{3+}$ and trans-DDP, the absorption band at 825 cm^{-1} disappeared and a new absorption band was observed at about 810 cm^{-1} in all the spectra of the metal-DNA complexes (Table 1 and Fig 1). The guanine band at 690 cm^{-1} also disappeared and a new band at about 600-626 cm^{-1} was observed (Table 1 and Fig.1). The spectral changes occurred are characteristic of a B to Z conformational transition in DNA. Recent Raman spectroscopic studies of calf-thymus DNA showed (26) similar spectral changes for the said vibrational frequencies upon a B to Z conformational transition. The B to Z transition in the polynucleotide poly(dG-dC).poly(dG-dC) was observed (27) in the presence of high salt concentration by shifting of the Raman lines from 830 to 810 cm^{-1} and that of 680 to 625 cm^{-1}. It is found that the infrared spectra of the trans-DDP adduct of DNA was similar to that of the corresponding magnesium, copper and cobalt hexaammine chloride compounds in the region 1000-500 cm^{-1}, showing the absorption band at 809 cm^{-1} and the guanine vibration at 605 cm^{-1} (Fig. 1) which is consistent with a DNA being in the Z-conformation. It is interesting that the major adduct of cis-DDP with DNA showed spectral similarities with that of the cobalt pentaammine-DNA adduct, distinct from the spectra of the other metal-DNA complexes studied here (Table 1). The cis-DDP and cobalt pentaammine-DNA adducts exhibited strong infrared bands at about 850 and 600 cm^{-1} not observed in the spectra of the B or Z conformation of DNA. On the other hand, the dissimilarities observed for cis- and trans-DDP-DNA compounds could be related to different conformational transitions imposed on DNA by the cis- and trans-DDP binding and this may be responsible for the antitumor activity of cis-platinum (28). It should be noted that cis-DDP is a potent anticancer drug (28), whereas the trans-isomer is inactive against cancers. The spectral differences observed here for the two isomers are due to a different binding with DNA .

$$\begin{array}{c}
NH_3 \\ \\
Cl
\end{array} > Pt < \begin{array}{c}
Cl \\ \\
NH_3
\end{array}$$

Trans – DDP

$$\begin{array}{c}
MH_3 \\ \\
MH_3
\end{array} > Pt < \begin{array}{c}
Cl \\ \\
Cl
\end{array}$$

Cis – DDP

The cis-isomer binds to DNA through the N$_7$N$_7$-atoms of the guanine bases, whereas the trans-isomer bindings is not specific. The major binding site of the cis-compound is the N$_7$N$_7$-chelation with guanine bases, whereas the trans-isomer binding is not that specific (29). It is worth mentioning that the direct metal-N$_7$ coordination of Mg(II), Cu(II) and trans-DDP produces a similar conformational transition in the DNA molecule as in the case of the indirect metal-N$_7$ binding found for

the cobalt hexaammine chloride, Co(NH) $^{+}$. The indirect hydrogen bonding of cobalt hexaammine with the N —and O —site is via the NH groups
(See Fig.1 and ref. (24)).

RNA—Metal adducts

     Free RNA has an A-conformation (C '-endo/anti) (30,31). The marker
infrared bands for this conformation are the strong band at 811 cm$^-$,
assigned to the sugar vibration (14,32-34) and the medium intensity band
at 675 cm$^-$ which was assigned to the guanine ring vibration (25). Upon
complexation with Mg(II), Cu(II), cobalt pentaammine and cobalt hexaammine chloride, we have observed slight changes in the region 1000-500
cm$^-$ of the spectra (Fig.2 and Table 2). This, indicates the rigidity
of the RNA secondary structure and that of A-conformation in the RNA-
metal adducts (8) (Table 2). The X-ray structural determination of
tRNA-adducts with cobalt hexaammine (24), ruthenium pentaammine chloride, cis- and trans-DDP have been reported (23). The cobalt hexaammine
binds via NH molecules through hydrogen bonds indirectly to the N — and
to O —sites of the two adjacent guanine bases, whereas ruthenium pentam-
mine and cisDDP bind directly to the N —atoms of the guanine moieties
and hydrogen bonded to the O —site and the phosphate anions through the
ammine groups (23). The trans-DDP binds to the N atom of the adenine
bases and to the N —site of the guanine moieties (23). Since the spectra of all the metal-RNA adducts exhibited marked spectral similarities
with that of free RNA (Table 2 and Fig. 2) it seems, that the characteristic absorption bands in the region of 1000-500 cm$^-$ are mainly related to the sugar and sugar-phosphate (back bone) vibrational modes and
are characteristic of the backbone conformational transition. (Table
3).

CONCLUSION

The sugar-phosphate and base vibrational frequencies in the region of
900-600 cm$^-$, are sensitive to sugar and base conformational transitions
(8,35). Marker bands have been found to be at about 825 cm$^-$ and 690
cm$^-$ for the C '-endo/anti (B-form), whereas the shift of these vibrations to 810 and 625-600 cm$^-$, respectively is diagnostic of a B to Z
conformational transition (8). Extending this investigation to the
region of 1800-500 cm$^-$ for DNA, RNA and their metal adducts the following conclusion can be made:
(1) The carbonyl stretching vibration at 1701 cm$^-$ in natural DNA is
     shifted towards lower frequencies upon a B to Z conformational
     transition, while the carbonyl group in RNA at 1691 cm$^-$ showed no
     significant changes on metal complexation; the shift of 1701cm$^-$
     band, however cannot be considered characteristic of the Z-DNA;
(2) The absorption bands at 1485, 1371, 1329, 1285 and 1265 cm$^-$ (36,
     37) exhibited considerable changes upon DNA conformational transi-
     tion from B to Z upon metalation;
(3) The assymmetric and symmetric PO $^-$ stretching vibrations at about

TABLE 3

Infrared Diagnostic Frequencies and Assignments for DNA and RNA and their Metal-Adducts

	DNA	Metal–DNA	RNA	Metal–RNA
	B-Form	Z-Form	A-Form	A-Form
base (conformer)				
G(C$_2$'-endo/anti)	690			
G(C$_3$'-endo/syn)		624–605		
G(C$_3$'-endo/anti)			675	680
Sugar (conformer)				
G(C$_2$'-endo/anti)	825	800–812		
G(C$_3$'-endo/syn)				
G(C$_3$'-endo/anti)			811	810–815

1240 and 1090 cm^{-1} in the infrared spectra of both DNA and RNA showed considerable changes upon nucleobase metalation (36,37). These changes are most likely caused by an indirect NH_3 or H_2Ometal-phosphate interaction through hydrogen bonding and an electrostatic interaction between the positive charges on the metal and the negative charges on the phosphate groups;

ACKNOWLEDGEMENTS

This work was supported by the National Research Council of Canada.

REFERENCES

1. Wang, A.J.J.; Quigly, G.T.; Kolpak, F.J.; Crawford, J.L.; Van Boom, J.H.; Van der Marel, G. and Rich, A. (1979) Nature, 282 , 680–686.
2. Pohl, F.M. and Jovin, T.M. (1972) J. Mol. Biol., 67 ,375–396.
3. Thomas, T,; Lord, R.C.; Wang, A.H.J. and Rich, A. (1981) Nucleic Acid Res., 9 ,5443–5457.
4. Rich, A.; Quigley, G.J. and Wang, A.H.J. (1981) Biomolecular Stereodynamics, Vol. 1 ,Ed. R.H. Sarma, Adenine Press, New York, pp. 35–52.
5. Ho Chen, H.; Behe, M.J. and Rau, D.C. (1984) Nucleic Acid Res., 12 ,2381–2389.
6. Evans, F.E. and Kaplan, N.O. (1979) FEBS Lett., 105 ,11–14.
7. Evans, F.E.; Miller, D.W. and Levin, R.A. (1984) J. Amer. Chem. Soc., 106 ,396–401.
8. Theophanides, T. and Tajmir-Riahi, J.A. (1985) J. of Biomolecular Structure and Dynamics, 2 ,995–1004
9. Nishimura, Y.; Morikawa, K. and Tsuboi, M. (1974) Bull. Chem. Soc. Jpn, 47, 1043–1044; Sutherland, G.B.B.M. and Tsuboi, M. (1957) Proc. Roy. Soc., 239 ,446–463.
10. Palmer, W.G. (1959) Experimental Inorganic Chemistry, University Press, pp. 530–544.
11. Kauffman, G.B. and Cowan, D.O. (1963) Inorg. Syntheses, 7 , 239–244.
12. Raudashi, G.; Lippert, B. and Hoeschele, I.D. (1983) Inorg. Chim. Acta, 78 ,L43–L44.
13. Tajmir-Riahi, H.A. Theophanides, T. (1984) Can. J. Chem., 62 , 1429–1440.
14. Tsuboi, M. (1974) Infrared and Raman Spectroscopy in Basic Principles in Nucleic Acid Chemistry, Vol. 1 ,Academic Press 399–450 and (1964) Adv. Chem. Phys., 7, 435–497.
15. Prescott, B.; Steinmetz, W. and Thomas, Jr., G.J. (1984) Biopolymers, 23 ,235–255.
16. Benevides, J.M., Want, A.H.J., Van der Marel, G.A., Van Boom, J.H., Rich, A. and Thomas, Jr., G.J. (1984) Nucleic Acids Research, 12 , 5913–5925.

17. Tajmir-Riahi, H.A. and Theophanides, T. (1983) Can. J. Chem., 61 , 1813-1822.
18. Tajmir-Riahi, H.A. and Theophanides, T. (1983) Inorg. Chim. Acta, 80 ,183-190.
19. Tajmir-Riahi, H.A. and Theophanides, T. (1983) Inorg. Chim. Acta, 80 ,223-230.
20. Tajmir-Riahi, H.A. Theophanides, T. (1984) Can. J. Chem., 62 , 223-230.
21. Sherer, E., Tajmir-Riahi, H.A. and Theophanides, T. (1984) Inorg. Chim. Acta, 92 ,285-292.
22. Mansy, S. and Peticolas, W.L. (1976) Biochem., 15 ,2650-2655.
23. Rubin, J.R.; Sabat, M. and Sundaralingam, M. (1983) Nucleic Acid Res., 11 ,6571-6586.
24. Hingerty, B.E.; Brown, R.S. and Klug, A. (1982) Biochim. Biophys. Acta, 698 ,78-82.
25. Nishimura, Y.; Tsuboi, M.; Nakano, T.; Higuchi, S.; Sato, T.; Shida, T.; Uesugi, S.; Ohsuka, E. and Ikehera, M. (1983) Nucleic Acid Res., 11 ,1579-1588.
26. Martin, J.C. and Wartel, R.M. (1982) Biopolymers, 21 ,499-512.
27. Wartel, R.M.; Harell, J.T.; Zacharias, W. and Wells, R.D. (1983) J. Biomolecular Structure and Dynamics, 1 ,83-96.
28. Rosenberg, B.; Van Camp, L.; Trsoko, J.E. and Mansour, V.H. (1969) Nature, 222 ,385-386.
29. Wing, B.; Drew, H.; Takano, T.; Broka, C.; Tanaka, S.; Itakura, K. and Dickerson, R.E. 1980) Nature, 287 ,755-758.
30. Arnott, S.; Jukins, D.W.L. and Dover, S.D. (1972) Biochim. Biophys. Res. Commun., 47 ,1504-1509 and (1972) 48, 1392-1399.
31. Quigley, G.J. and Rich, A. (1976) Science, 194 ,796-806.
32. Thomson, Jr., G.J.; Chen, M.C. and Harman, K.A. (1973) Biochim. Biophys. Acta, 324 ,37-49 and (1974) Biopolymer, 13, 615-626.
33. Erfurth, S.C. and Peticolas, W.L. (1975) Biopolymers, 14 ,247-264 and 1259-1271.
34. Lu, K.C.; Prohofsky, E.W. and Van Zandt, L.L. (1977) Biopolymers, 16 ,2491-2506.
35. Theophanides, T. and Tajmir-Riahi, H.A.(1985) Structure and Motion: Membranes, Nucleic Acids and Proteins (Eds., E. Clementi, G. Corongiu,, M.H. Sarma and R.H. Sarma, 521-530.
36. Theophanides, T. (Ed.), (1984) in Fourier Transform Infrared Spectroscopy, 105-124, by D. Reidel Publishing Company; ibid. (1979) in Infrared and Raman Spectroscopy of Biological Molecules, 205-223, by D. Reidel Publishing Company.
37. Theophanides, T. and Tajmir-Riahi, H.A. (1984) in Spectroscopy of Biological Molecules, C. Sandorfy and T. Theophanides (Eds.) 137-152, by D. Reidel Publishing Company.

DETECTION AND CHARACTERIZATION OF SHORT-LIVED SPECIES BY NANOSECOND TIME-RESOLVED SPECTROSCOPIC TECHNIQUES

J. Aubard,[*] J.J. Meyer, R. Dubest, A. Adenier and P. Levoir

Institut de Topologie et de Dynamique des Systèmes de l'Université Paris VII, associé au CNRS, 1, rue Guy de la Brosse, 75005 Paris – France.

During the last decade considerable interest has been shown in time-resolved spectroscopic techniques because they make it possible to detect and characterize transient species in the picosecond or nanosecond time range, even when these species are highly diluted.

In this contribution we briefly describe the experimental techniques we have developed to perturb or initiate a chemical (biological) reaction in solution on the nanosecond time scale but we focus our attention on detection systems, which can be used to follow the extent of the reaction after the perturbation (U.V.-Visible absorption, resonance Raman scattering). Applications concerning reaction dynamics, involving biological molecules and photochromic species are presented to demonstrate the efficiency of these detection systems in the identification and characterization of reaction transients.

NANOSECOND LASER T-JUMP CHEMICAL RELAXATION TECHNIQUE

Chemical relaxation methods are the best tool available today to study very fast chemical or biological equilibria in solution. The basic principle of these methods[1] is to perturb a chemical system at equilibrium by a rapid change (transient or periodic) of an external parameter such as temperature, pressure, electric field[2].... Following the rapid variation of the reactant's concentrations to their new equilibrium value allows the determination of the reaction rates. In transient chemical relaxation techniques, such as temperature-jump (T-Jump), the concentration change following the perturbation (i.e. the relaxation process) is usually monitored by means of absorption, emission or diffusion of the light. However, today, time-resolved optical absorption spectroscopy is the most popular detection method used for monitoring the relaxation.

To obtain T-Jump relaxation measurements in the nanosecond time range, one of the most efficient ways is the Raman laser T-Jump technique.[3] In this technique short T-Jump perturbations are obtained from a pulsed Neodymium-Glass laser, whose wavelength has been shifted,

P. M. Rentzepis and C. Capellos (eds.), Advances in Chemical Reaction Dynamics, 565–570.
© *1986 by D. Reidel Publishing Company.*

by stimulated Raman effect, to a spectral region where the usual
solvents have a marked absorbance. Using liquid nitrogen as the Raman
active medium, the laser wavelength is shifted from 1.06 µm to 1.41 µm
where water and hydroxylic solvents absorb strongly. Thus, by this way,
it is possible to carry out, in water, temperature-jumps up to 8°C in
20 ns, in a small volume of a few hundred µls. As stated above, we use
an optical detection to monitor the relaxation. However, for these
measurements, a high detection bandwidth is required (ca. 20 MHz) and
led to an important shot noise. Therefore when firing the laser the
detection light source is pulsed for 1 ms to increase the intensity,[4]
leading to an important improvement in the signal-to-noise ratio.

    This apparatus was first developed to study fast proton transfers
in solution, particularly prototropic transformations in purine and
pyrimidine bases aqueous solutions.[5] Recently, we have undertaken a
program to investigate the dynamics of aggregation processes of active
biological dyes in solution.

    A large number of dyes form dimer aggregates in aqueous solution
and we know today that this behaviour must play a role in their biological activity.[6] For the aminoacridine studied, the reaction considered,
in neutral aqueous solution, is the dimerization of the cationic species
as it is shown on the Fig. 1.

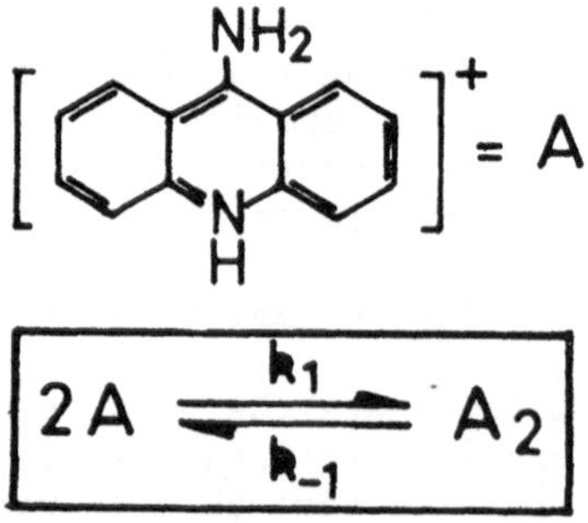

$$\left[\text{NH}_2\text{-acridine} \right]^+ = A$$

$$2A \underset{k_{-1}}{\overset{k_1}{\rightleftharpoons}} A_2$$

$$C_o = \overline{[A]} + 2\overline{[A_2]} \qquad K = k_1/k_{-1}$$

Fig. 1. 9-aminoacridine : Dimerization process in neutral aqueous
solution.

    Even at relatively diluted concentrations (10^{-4}–10^{-3} M) this
reaction is known to be very rapid, because the forward rate (k_1) is
almost diffusion controlled,[7] and the Raman laser T-Jump spectrometer
was required for this study. On the other hand, at these concentrations
the dimer only represents a small percent of the species in the solution
and therefore its detection is not very easy. Fig. 2 shows the results
of the relaxation experiments. Since the concentration dependence of
the relaxation time, for the dimerization reaction (Fig. 1) is given by

$$\tau^{-2} = 8\ k_1 k_{-1}\ Co + k_{-1}^2 \qquad (1)$$

where Co is the total concentration of dissolved aminoacridine, the rate constants k_1 and k_{-1} are easily derived from Fig. 2 and reported in the frame of this figure. The ratio of k_1 and k_{-1} gives an equilibrium constant of ca. 100 which agrees with recent NMR data.[7]

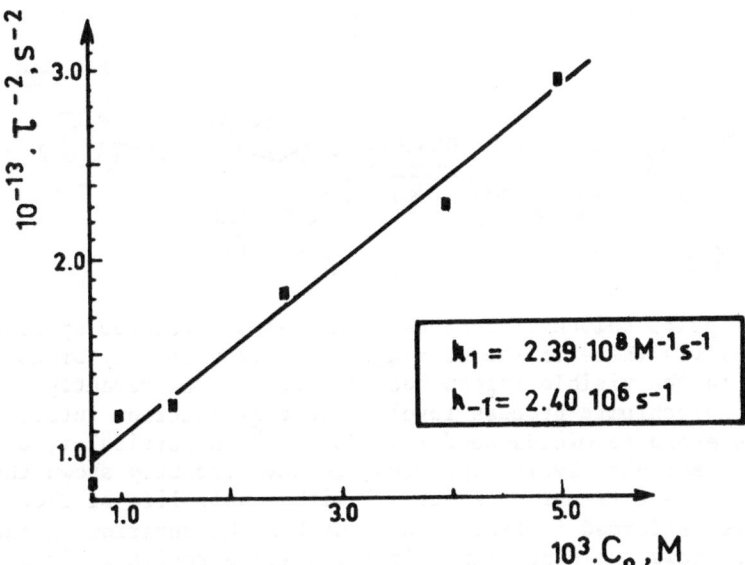

Fig. 2. Plot of Raman laser T-Jump data (square of reciprocical relaxation times) as a function of total concentration for 9-aminoacridine dimerization in aqueous solution. Experimental conditions : observation wavelength λ = 423 nm ; final temperature T = 25°C ; pH 6.86. The full line is the best fit of the experimental data according with equation (1).

Thus the Raman laser T-Jump technique allows to detect nanosecond transient species, even diluted in solution, and provides kinetic and thermodynamic data with accuracy. Moreover this technique only requires small quantities of compounds which is very convenient for samples of biological interest.

TIME-RESOLVED RAMAN SPECTROSCOPY

The analysis of the kinetics of very fast reactions in solution is only one of the aspects in reaction dynamics. Another major aspect concerns the identification and structural characterization of transient species involved in the course of the reaction. Unfortunately, monitoring relaxation phenomena by optical absorption or emission spectroscopy suffers from certain limitations. Indeed such spectroscopies do not provide information on the molecular structure. This can be improved, in principle, by studying Raman spectra. In the following of this

contribution we will discuss, through a typical experiment, the system
we use to obtain time-resolved Raman spectra of transient species, up
to nanosecond time scale.

During the last years we have studied the photochemical formation
of photomerocyanines. As shown on the scheme, photomerocyanines (M)
are produced under U.V. irradiation by the photochromic transformation
of spiropyrans (SPP).

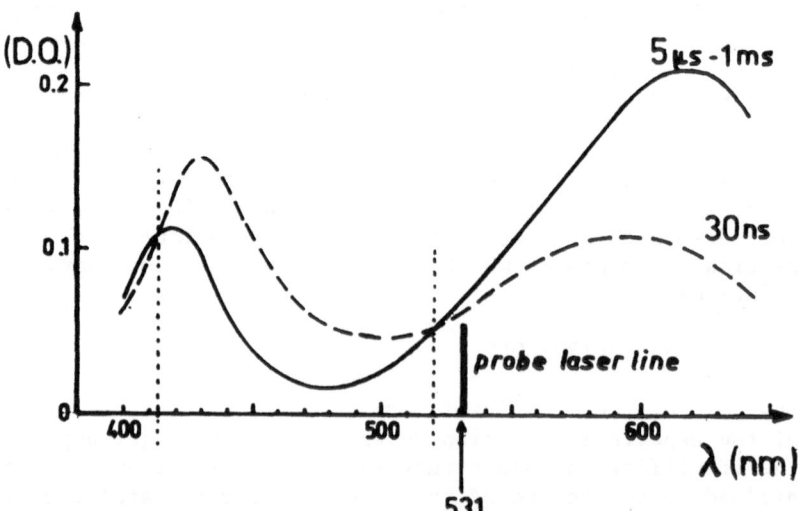

In non polar solvents, photomerocyanines are colored species, with
short half-lives (here in the ms range), and they display broad absorption bands in the visible region near 600 nm. It was recently established
that this photochromic process involves several reaction intermediates
in the picosecond to microsecond time domain.[8] In particular, using a
nanosecond laser photolysis apparatus, we have recently shown that this
reaction implies a transient species I with a half-life of about 500 ns.[9]
This species is formed instantaneously within the duration of the laser
(30 ns) and shows two broad bands in the visible region as it appears
on the Fig. 3. On the same figure, the spectrum of the photomerocyanine
M obtained in the ms range has been reported.

Fig. 3. Transient absorption spectra measured at 30 ns, 5 μs and 1 ms,
for 2x10⁻³ M spiropyran SPP in toluene at 25°C, irradiated by a 354 nm,
30 ns laser pulse.

To characterize more precisely the I and M species we have recorded their Raman spectra. The technical basis of the multichannel Raman spectrometer used in this study has been reported in details.[9-11] It consists of a spectrograph coupled to a multichannel detection which is composed of an image intensifier unit and a sensitive camera. The T.V. output of the camera is directly interfaced to a minicomputer allowing each spectrum to be digitized over 500 points. In the present experiments, photomerocyanines M are produced by irradiating toluene solutions of spiropyrans with a UV flash lasting 1 ms. Their Raman spectra are then obtained by a 30 ns probe laser pulse at 531 nm (second harmonic of Nd-glass laser), synchronized with the UV flash perturbation. To obtain the Raman spectrum of the short-lived intermediate I, the toluene solution of spiropyran is now irradiated by the third harmonic of the laser (354 nm) and the second harmonic is always used for Raman scattering. Since the second and third harmonics are delivered simultaneously, it is then possible to obtain the Raman spectrum of the species produced within the 30 ns duration of the laser pulse.

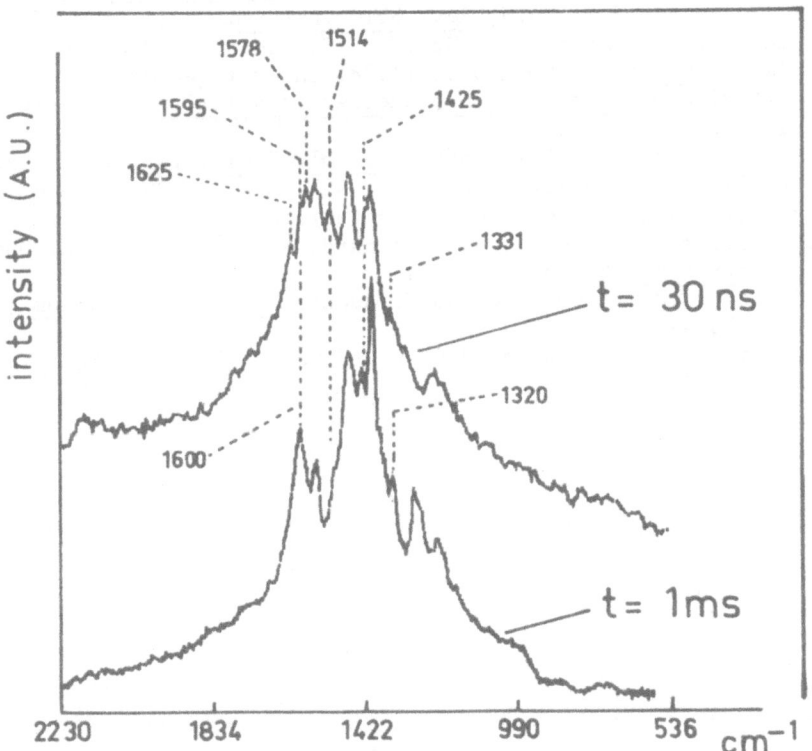

Fig. 4. Time-resolved resonance Raman spectra, excited by a 30 ns probe 531 nm laser pulse, of a toluene solution of spiropyran (SPP) irradiated, by a 354 nm laser pulse (t = 30 ns) , by a UV flash (t = 1 ms).

Fig. 4 gives pulsed resonance Raman spectra obtained from a
10^{-3} M toluene solution of spiropyran, irradiated either by a UV flash
(t = 1 ms) or by the 354 nm laser line (t = 30 ns) and probed by the
531 nm second harmonic line of the laser. Since this laser wavelength
falls within an absorption band of both the photomerocyanine M and of
the intermediate I (Fig. 3), selective enhancement of the Raman light
occurs, allowing the detection and characterization of these species
even highly diluted in solution. From the detailed analysis of the Raman
spectra obtained it has been possible, we believe for the first time, to
investigate the structure of transient photoproducts involved in the
photochromic reaction up to the ns time range.[9]

REFERENCES

1. M. Eigen and L. De Maeyer, "Techniques of Organic Chemistry", 8,
    Part. 2, 895, Wiley-Interscience, (1963).
2. A. Persoons and L. Hellemans (this book).
3. J. Aubard, J.J. Meyer and J.E. Dubois, Chem. Instrum., 8, 1, (1977).
4. J.J. Meyer and J. Aubard, Rev. Sci. Instrum., 48, 695, (1977).
5. O. Bensaude, J. Aubard, M. Dreyfus, G. Dodin and J.E. Dubois,
    J. Am. Chem. Soc., 100, 2823, (1978) and references cited therein.
6. V. Vitagliano, "Aggregation processes in solution", Elsevier Sci.
    Publ. Co., 271, (1983).
7. P.R. Young and N.R. Kallenbach, J. Mol. Biol., 145, 785, (1981).
8. Y. Kalisky, T.E. Orlowski and D.J. Williams, J. Phys. Chem., 87.
    5333, (1983) and references cited therein.
9. J.L. Albert, J.P. Bertigny, J. Aubard, R. Dubest and J.E. Dubois,
    J. Chim. Phys., 82, 521, (1985).
10. J.J. Meyer, J.C. Fontaine and J. Aubard, Rev. Sci. Instrum., 55,
    1879, (1984).
11. J.J. Meyer, J. Aubard and J.C. Fontaine, J. Phys. E. Sci. Instrum.,
    18, 430, (1985).

FORMATION, DECAY, AND SPECTRAL CHARACTERIZATION OF SOME SULFONYL RADICALS

C. Chatgilialoglu
Istituto dei composti del carbonio
Consiglio Nazionale delle Ricerche
40064 Ozzano Emilia (Bologna)
Italy

ABSTRACT. The absolute rate constants for the reactions of a variety of radicals with a number of sulfonyl halides have been measured in solution by using laser flash photolysis techniques. The reactivities cover a wide range, e.g., the rate constants for the reaction of n-butyl, phenyl, and triethylsilyl radicals with α-toluenesulfonyl chloride are 1.3×10^6, 1.0×10^8, and 5.7×10^9 $M^{-1}s^{-1}$ respectively at 298K. A comparison of the Arrhenius parameters determined for a few representative substrates suggests the involvement of charge-transfer interactions in these reactions. Sulfonyl radicals, $RSO_2\cdot$, show a weak absorption band with a maximum around 340nm. Evidences are presented suggesting that such a transition places a nonbonding oxygen electron in the single occupied molecular orbital. Additional kinetic data are also given for the unimolecular and bimolecular decay of sulfonyl radicals.

INTRODUCTION

Sulfonyl radicals, $RSO_2\cdot$, have been postulated in the past as intermediates in many reactions of sulfur containing compounds. In recent years their structures have been investigated in details by EPR spectroscopy and molecular orbital calculations(1-3). Thus, sulfonyl radicals have a σ ground state with the unpaired electron localized on the SO_2 moiety and a pyramidal center at sulfur, pyramidality depending on the electronegativity of the substituents. A detailed conformational analysis has also been presented.

X = alkane, amino, alkoxy, alkene, arene

Despite the significant body of informations now available for these species, absolute kinetic data for the reactions involving sulfonyl

P. M. Rentzepis and C. Capellos (eds.), Advances in Chemical Reaction Dynamics, 571–579.
© *1986 by D. Reidel Publishing Company.*

radicals are very scarce. One of the purposes of this paper, is to report our recent quantitative rate data for this important class of reactions obtained by using different spectroscopic techniques together with the limited data which appeared in literature in the last years.

Although the electronic spectra, either in organic solvents or in aqueous solution, of several sulfonyl radicals have been reported(4,5), the data altogether are still somehow misleading and in none of these studies was there any attempt to rationalize them. We have now extended our experimental study to some of the hitherto unreported optical absorption spectra of sulfonyl radicals; in particular we tried to rationalize some of the unusual properties observed.

RESULTS AND DISCUSSION

Optical Absorption Spectra

In the present study we employed modulation spectroscopy, laser flash photolysis and pulse radiolysis techniques. Sulfonyl radicals were generated by the sequence of reactions 1-3 in modulation spectroscopy and

$$Me_3COOCMe_3 \xrightarrow{h\nu} 2\ Me_3CO\cdot \tag{1}$$

$$Me_3CO\cdot + Et_3SiH \longrightarrow Me_3COH + Et_3Si\cdot \tag{2}$$

$$Et_3Si\cdot + RSO_2Cl \longrightarrow Et_3SiCl + RSO_2\cdot \tag{3}$$

laser flash photolysis (see *infra*), while in pulse radiolysis the radicals were generated by the following fast dissociative electron-capture reaction, *vis.*,

$$RSO_2Cl + e_{aq}^- \longrightarrow RSO_2\cdot + Cl^- \tag{4}$$

The optical absorption spectra of sulfonyl radicals obtained in the region 300 to 500nm show a continuous broad band with λ_{max} at around 340nm. All the data obtained(6) together with those available in the literature (4,5) have been summarized in table I.

TABLE I. Summary of the Optical Absorption Spectra of Sulfonyl Radicals

Radicals	Solvent	λ_{max},nm	ε_{max},$M^{-1}cm^{-1}$
$RSO_2\cdot$	hydrocarbons	355	~1000
	H_2O	332	
$ArSO_2\cdot$	hydrocarbons	330	~1000
	H_2O	315	

In table I the extinction coefficients at λ_{max} obtained by three different methods are also reported. The values of $ca.$ 1000 $M^{-1}cm^{-1}$ which
should be associated, to first approximation, with low oscillator strenghs,
suggest that the transition is forbidden.

These results can be rationalized by constructing a qualitative
molecular orbital diagram for sulfonyl radicals (see figure 1). Thus, on
the left side of figure 1 are reported the four highly occupied molecular

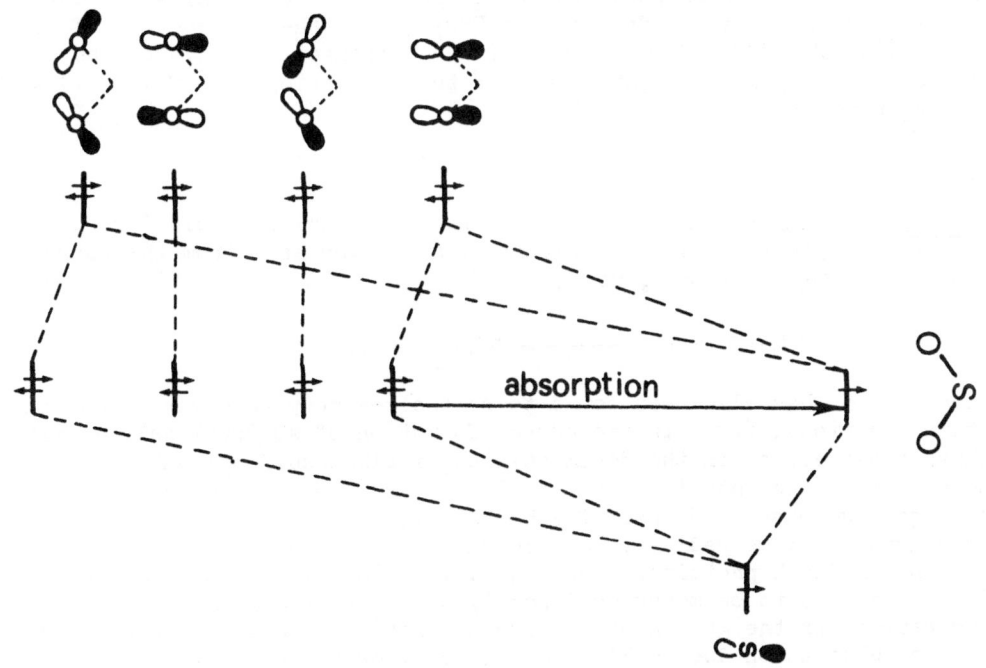

Figure 1. Qualitative MO diagram for sulfonyl radicals

orbitals derived by combination of four oxygen 2p, one σ-bonding, and two
sulfur 3d orbitals similar to the ones reported for the parent diamag-
netic molecule based on the photoelectron spectroscopic data (for simplicity only the four oxygen 2p orbitals have been reported)(7). Furthermore, for symmetry reasons the unpaired electron on sulfur should interact
with the highest and lower of these four orbitals, as shown in figure 1,
to give the orbital sequence in sulfonyl radicals.

Figure 1 shows the kind of transition for the observed band, $i.e.$

a nonbonding oxygen electron is placed in the SOMO (Single Occupied Molecular Orbital). This assignment is in accord with our preliminary *ab initio* calculations(6). Moreover, these transitions occuring within the SO_2 moiety are also in agreement with the σ character of the sulfonyl radicals derived from EPR spectroscopic data(1-3).

From table I two observations can be made. The first concerns the blue shift of 22 and 15nm in aqueous solution relative to hydrocarbon as a solvent for alkanesulfonyl and arenesulfonyl radicals respectively; in fact, in aqueous solution the four highest doubly occupied molecular orbitals in sulfonyl radicals will be stabilized by intermolecular hydrogen bonding and consequently a blue shift should be observed. The second observation is the red shift of $RSO_2 \cdot$ with respect to $ArSO_2 \cdot$; the explanation stands on the fact that the HOMO (Highest Occupied Molecular Orbital) of the aryl fragment in $ArSO_2 \cdot$ is higher in energy with respect to the HOMO of the alkyl fragment in $RSO_2 \cdot$, thus the interaction of the former with the SOMO of SO_2 moiety will be stronger relative to the latter. That is, the energetic gap for the absorption (see figure 1) will be larger in $ArSO_2 \cdot$ than in $RSO_2 \cdot$.

Kinetic Data

Formation. The most common way to generate sulfonyl radicals for spectroscopic studies is the abstraction of a halogen atom from the corresponding sulfonyl halides, *viz.*,

$$R'SO_2X + R \cdot \longrightarrow R'SO_2 \cdot + RX \qquad (5)$$

Reaction 5 has also been found to be the key step in several radical chain reactions, *i.e.*, in the adduct formation of sulfonyl halides with alkenes as well as in the decomposition of alkanesulfonyl halides(8). However, kinetic data for reaction 5 are very limited. In order to fill this gap, we have used laser flash photolysis techniques to examine the reactivity of some radicals towards sulfonyl halides.

Laser flash photolysis technique, in principle, allows rate constants for reaction 5 to be measured directly by monitoring either the decay of the reagent or the growth of the product radicals. However, transient optical absorption due to R· (alkyl, phenyl or triethylsilyl)(9,10) or $R'SO_2 \cdot$ radicals are too weak to be used for accurate kinetic studies, particularly if one considers that in order to minimize the occurence of second-order processes, it is usually necessary to considerably attenuate the excitation dose. Therefore we have preferred to use two indirect methods(11,12):
(A) The rates of the reactions of triethylsilyl radicals with various sulfonyl halides were measured by using the technique described previously, in which benzil is used as a probe to monitor the $Et_3Si \cdot$ radical concentration(11).

$$Et_3Si \cdot + RSO_2X \longrightarrow Et_3SiX + RSO_2 \cdot \qquad (6)$$

$$Et_3Si \cdot + PhCOCOPh \longrightarrow Ph\dot{C}(OSiEt_3)COPh \qquad (7)$$

In this system (see eqs 1,2,6,7), the kinetics of formation of $Ph\overset{.}{C}(OSiEt_3)$ COPh is monitored for a series of samples with a constant concentration of benzil and variable concentrations of sulfonyl halides. The time profile for formation of $PhC(OSiEt_3)COPh$ (monitored at 378nm) leads to experimental pseudo-first-order rate constant, k_{expt}, which is related to the elementary step of interest according to equation 8

$$k_{expt} = k_0 + k_7 \left[PhCOCOPh \right] + k_6 \left[RSO_2X \right] \qquad (8)$$

where k_0 includes all pseudo-first-order modes of decay of $Et_3Si\cdot$ other than reactions with benzil and sulfonyl halides.

(B) The rates of reactions of alkyl and phenyl radicals with α-toluene-sulfonyl chloride, $PhCH_2SO_2Cl$, were measured by using the following system(12):

$$\xrightarrow{\text{fast}} R\cdot$$

$$PhCH_2SO_2Cl + R\cdot \longrightarrow PhCH_2SO_2\cdot + RCl \qquad (9)$$

$$PhCH_2SO_2\cdot \longrightarrow PhCH_2\cdot + SO_2 \qquad (10)$$

In a system like this, where reaction 10 is a very fast process(see *infra*), when radical $R\cdot$ can be generated "instantaneously" the time profile for the formation of $PhCH_2\cdot$ (monitored at 317nm) leads to experimental pseudo-first-order rate constant, k'_{expt}, which is related to the elementary step of interest according to equation 11

$$k'_{expt} = k'_0 + k_9 \left[PhCH_2SO_2Cl \right] \qquad (11)$$

where k'_0 is a rate constant that includes all first- and pseudo-first-order processes that alkyl or phenyl radicals undergo in the absence of substrate $PhCH_2SO_2Cl$ and which are not affected by $PhCH_2SO_2Cl$.

The values of k_6 and k_9 obtained by these two ways are absolute rate constants and measure the overall (or molecular) reactivity of the substrate. The results are summarized in table II together with data of Horowitz which had been measured previously(13,14).

The rate constants, measured at *ca.* 298K, show that reaction 6 is essentially a diffusion-controlled process for both alkane- and arene-sulfonyl chlorides. While for the sulfonyl fluorides the rate constants are considerably lower than diffusion-controlled ones, still they are very similar for both alkane- and arenesulfonyl fluorides. We therefore conclude that arenesulfonyl radicals are not appreciably stabilized relative to alkanesulfonyl radicals in agreement with the σ-type structure derived from our EPR data(1,2) but in contrast with Benson's thermodynamic arguments which suggest that the $PhSO_2\cdot$ has a stabilization energy of 13 kcal/mol(15).

The Arrhenius parameters listed in table II show that in the Cl-transfer reaction between $PhCH_2SO_2Cl$ and primary or secondary alkyl radicals the reactivity trends, to a large extend, are determined by the abnormally high preexponential factor (*n.b.* a "normal" preexponential factor for free-radical abstraction is considered to be $10^{8.5\pm0.5}M^{-1}s^{-1}$(16).

TABLE II. Absolute Kinetic Data for Halogen Abstraction by Some Radicals from Sulfonyl Halides

Substrate	Radical	$\log A/M^{-1}s^{-1}$	E_a,kcal/mol	k,$M^{-1}s^{-1}$ at 298K
CH_3SO_2F	$(CH_3CH_2)_3Si\cdot$			1.3×10^7
$p\text{-}CH_3C_6H_4SO_2F$	$(CH_3CH_2)_3Si\cdot$			8.9×10^6
CH_3SO_2Cl	$c\text{-}C_6H_{11}\cdot{}^a$	9.1	4.9	2.9×10^5
	$(CH_3CH_2)_3Si\cdot$			3.2×10^9
$CH_3CH_2CH_2SO_2Cl$	$c\text{-}C_6H_{11}\cdot{}^b$			1.5×10^{6c}
$C_6H_5CH_2SO_2Cl$	$CH_3CH_2CH_2CH_2\cdot$	10.1	5.5	1.3×10^6
	$CH_3CH_2\dot{C}HCH_3$	9.6	4.8	1.2×10^6
	$C_6H_5\cdot$			1.0×10^8
	$(CH_3CH_2)_3Si\cdot$			5.7×10^9
$C_6H_5SO_2Cl$	$(CH_3CH_2)_3Si\cdot$			4.6×10^9

[a]From ref. 13   [b]From ref. 14   [c]At 393K

We attribute the high preexponential factors to the great importance of polar contribution to the transition state(17).

$$PhCH_2SO_2Cl \ \cdot R \longleftrightarrow PhCH_2SO_2\cdot \ Cl^- \ {}^+R \longleftrightarrow PhCH_2SO_2^- \ \dot{C}l \ {}^+R \longrightarrow$$

$$(I) \qquad\qquad (II) \qquad\qquad (III)$$

$$\longrightarrow PhCH_2SO_2\cdot + RCl \qquad\qquad\qquad (12)$$

That is, the great electron affinity of sulfonyl chlorides(18) will in- crease the contribution of the canonical structures with charge separa- tion to the transition state. So far, it has been recognized that charge- transfer interactions to the transition state enhance the reaction rate by reducing the activation energy, this being in excellent agreement with the present results. However, the charge-transfer may exert a greater influence on the preexponential factor than on the activation energy: When there is no polar contribution to the transition state the reaction occurs only when C, Cl, and S atoms are colinear or nearly so, whereas a strong polar contribution to the transition state will relax this restriction (see scheme I). In the limiting case of complete electron transfer the resultant ion·pair would not be subject to any restriction

SCHEME I

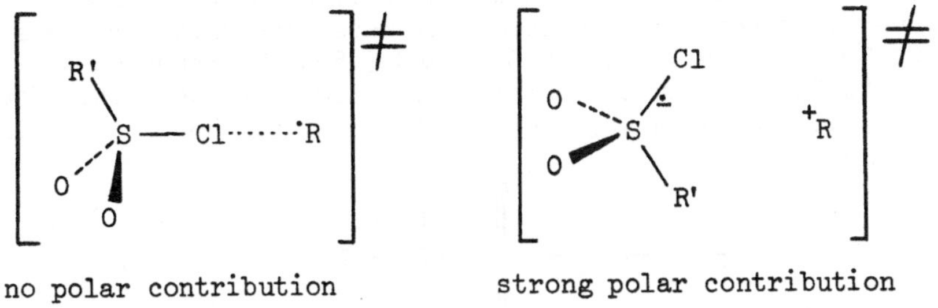

no polar contribution   strong polar contribution

in their relative rotational motion. This gain of two rotational degree$_2$ of freedom in the transition state would increase the A-factor by $ca.10^2$. The fact that the A-factor of reaction 9 is at least one order of magnitude greater than a "normal" one means that electron transfer is extensive in the reaction transition state. In a similar way, the increase in the reactivity from CH_3SO_2Cl to $PhCH_2SO_2Cl$ is largely due to an increase in the preexponential factror (see table II), which we attribute to the increased importance of polar effects; that is, the electron-withdrawing phenyl group will stabilize the sulfinate anion of the resonance form (III) in eq 12.

Decay. One of the peculiar features of sulfonyl radicals is the α-scission process, $viz.$,

$$RSO_2{\cdot} \longrightarrow R{\cdot} + SO_2 \qquad (13)$$

For comparison purposes activation energies for reaction 13, in both gas and liquid phase, are collected in table III. The value of α-toluenesulfonyl radical is in line with the activation energies of the other $RSO_2{\cdot}$ desulfonylation, provided allowance is made for differences in the stabilization energies of the organic radicals that are produced. However, electrostatic repulsions within the radical molecule and phase effect may influence considerably activation energies.

TABLE III. Activation Energies for the Decomposition of Some Sulfonyl Radicals

Radical	E_a,kcal/mol	Phase	Ref.
$CH_3SO_2{\cdot}$	22.4	gas	(19)
$CH_3CH_2SO_2{\cdot}$	19.9	gas	(20)
$CH_3CH(Cl)SO_2{\cdot}$	13.4	liquid	(21)
$C_6H_5CH_2SO_2{\cdot}$	5	liquid	(22)

A characteristic reaction of free-radicals is the bimolecular self-reaction which in many cases proceeds at the diffusion-controlled limit. We have measured by kinetic EPR spectroscopy rate constants for the self-reaction of $CH_3SO_2\cdot$, $CH_3CH_2SO_2\cdot$, $C_6H_5SO_2\cdot$ and 2,5-dichlorobenzenesulfonyl radical(22). At 233K and in cyclopropane as a solvent the values found for $2k_{14}$ were in the range $(4.5\pm1.5)\times10^9$ $M^{-1}s^{-1}$ for all $RSO_2\cdot$ radicals.

$$2\ RSO_2\cdot \xrightarrow{\ 2k_{14}\ }
\begin{cases}
\ \ \overset{O}{\underset{C}{\overset{\|}{R-S}}}-\overset{O}{\underset{O}{\overset{\|}{S}}}-R \\[2em]
\ \ \overset{O}{\underset{O}{\overset{\|}{R-S}}}-O-\overset{}{\underset{O}{\overset{}{S}}}-R
\end{cases} \tag{14}$$

Previous measurements in solution of $2k_{14}/\mathcal{E}_\lambda$, where $\mathcal{E}_\lambda$ is the extinction coefficient at the monitoring wavelenght, by time-resolved optical spectroscopy have given rate constants also at the diffusion-controlled limit but slightly lower than our ones (ca. 5×10^8 $M^{-1}s^{-1}$)(5,23). This discrepancy may arise from the uncertainties in the values used for extinction coefficient for the sulfonyl radicals in the optical kinetic work. It is worth mentioning that the most interesting feature of the bimolecular self-reaction of sulfonyl radicals is that they can combine to form either an intermediate sulfonyl sulfinyl "anhydride" the decomposition of which is expected to yield $RSO\cdot$ and $RSO_3\cdot$ radicals or an α-disulfone (see eq 14)(24). Finally, the rate constants for the radical-radical reactions of c-$C_6H_{11}SO_2\cdot$ with c-$C_6H_{11}OO\cdot$ and $CH_3(CH_2)_{12}OO\cdot$ have been recently found to be 1.5×10^8 and 3.0×10^8 $M^{-1}s^{-1}$ respectively at 293K(25).

REFERENCES

(1) C. Chatgilialoglu, B.C. Gilbert, and R.O.C. Norman, J. Chem. Soc., Perkin Trans. 2, 770 (1979).
(2) C. Chatgilialoglu, B.C. Gilbert, and R.O.C. Norman, J. Chem. Soc., Perkin Trans. 2, 1429 (1980).
(3) C. Chatgilialoglu, B. C. Gilbert, R.O.C. Norman, and M.C.R. Symons, J. Chem. Res., (S)185, (M)2610 (1980).
(4) T.E. Eriksen, and J. Lind, Radiochem. Radioanal. Lett., 25, 11 (1976).
(5) H.H. Thoi, O. Ito, M. Iino, and M. Matsuda, J. Phys. Chem., 82, 314 (1978).
(6) C. Chatgilialoglu, D. Griller, and M. Guerra, manuscript in preparation.
(7) B. Solouki, H. Bock, and R. Appel, Angew. Chem. Internat. Ed., 11, 927 (1972).
(8) E. Block, Reactions of Organosulfur Compounds, Academic Press: New York, 1978.

(9) C. Chatgilialoglu, J.C. Scaiano, and K.U. Ingold, *Organometallics*, 1, 466 (1982).

(10) C. Chatgilialoglu, K.U. Ingold, J. Lusztyk, A.S. Nazran, and J.C. Scaiano, *Organometallics*, 2, 1332 (1983).

(11) C. Chatgilialoglu, K.U. Ingold, and J.C. Scaiano, *J. Am. Chem. Soc.*, 104, 5119 (1982)

(12) C. Chatgilialoglu, *J. Org. Chem.*, submitted for pubblication.

(13) A. Horowitz, and L.A. Rajbenbach, *J. Am. Chem. Soc.*, 97, 10 (1975).

(14) A. Horowitz, *Int. J. Chem. Kinet.*, 8, 709 (1976).

(15) S.W. Benson, *Chem. Rev.*, 78, 23 (1978).

(16) S.W. Benson, *Thermochemical Kinetics*, 2nd ed., Wiley: New York, 1976.

(17) For general discussions of polar effects in radical reactions see: F. Minisci, and A. Citterio, *Adv. Free-radical Chem.*, 6, 65 (1980).

(18) G. Barbarella, C. Chatgilialoglu, S. Rossini, and V. Tugnoli, *Tetrahedron Lett.*, submitted for pubblication.

(19) A. Good, and J.C.J. Thynne, *Trans. Faraday Soc.*, 63, 2708 (1967).

(20) A. Good, and J.C.J. Thynne, *Trans. Faraday Soc.*, 63, 2720 (1967).

(21) H.H.Thoi, M. Iino, and M. Matsuda, *Macromolecules*, 12, 338 (1979).

(22) C. Chatgilialoglu, L. Lunazzi, and K.U. Ingold, *J. Org. Chem.*, 48, 3588 (1983).

(23) V.D. Komissarov, and P.L. Safiullin, *React. Kinet. Catal. Lett.*, 14, 67 (1980).

(24) C. Chatgilialoglu, B.C. Gilbert, B. Gill, and M.D. Sexton, *J. Chem. Soc.*, *Perkin Trans. 2*, 1141 (1980).

(25) A.I. Nicolaev, R.L. Safiullin, and V.D. Komissarov, *Khim. Fiz.*, 3, 257 (1984).

DENSITY AND CONCENTRATION FLUCTUATIONS IN MACROMOLECULAR SYSTEMS AS PROBED BY DYNAMIC LASER LIGHT SCATTERING.

G. Fytas
Department of Chemistry, University of Crete and Research Center of Crete, 711 10 Iraklion, Crete, Greece.

ABSTRACT: A potential information on space and time resolved properties of polymeric systems may be obtained by employing dynamic light scattering. For undiluted polymers, the light is scattered by local thermal density fluctuations which are related to dynamic mechanical properties. For polymer mixtures, concentration fluctuations dominate the polarized Rayleigh spectrum which yields insight into the mutual diffusion of the polymer chains. Examples of application are given from recent studies of amorphous bulk poly(phenylmethylsiloxane), poly(styrene) and their compatible mixture as well.

1. INTRODUCTION

The quantities measured in the light scattering spectroscopy are either the autocorrelation function of the scattered intensity $G(t)$ or the spectral density of the scattered electric field $I(\omega)$. Both quantities are determined by the fluctuations [1]:

$$\delta\alpha(q,t) = \Sigma\alpha_{if}(j,t) \exp iqr(j,t) \qquad (1)$$

where $\alpha_{if}(j,t)$ is the projection of the polarizability tensor $\alpha(j,t)$ of the i-th segment onto the initial (i) and final (f) polarization of the electromagnetic wave, $r(j,t)$ is the position of the center of mass and q is the scattering wave vector.

P. M. Rentzepis and C. Capellos (eds.), Advances in Chemical Reaction Dynamics, 581–587.
© 1986 by D. Reidel Publishing Company.

In the VV scattering geometry (V denotes vertical polarization with respect to the scattering plane) the major part of the scattering intensity for undiluted polymers arises from local thermal density fluctuations . The correlation functions G(t) exhibits a highly nonexponential shape and the average relaxation time is found to be q-independent in the $q \to 0$ limit [2]. Alternatively for compatible polymer mixtures, concentration fluctuations should dominate the VV-intensity. For non-entangled chains, G(t) decays exponentially with a characteristic q^2 - dependent relaxation rate [3]. In this short paper, we demonstrate the applicability of the photon correlation spectroscopy to the study of molecular motion in polymers. We report results from recent investigations of low molecular weight (MW) undiluted poly(phenylmethylsiloxane) (PPMS), [4], polystyrene (PS) [5] and their compatible mixture [6] above the glass transition temperature Tg.

2. SLOW DENSITY FLUCTUATIONS IN BULK POLYMERS

Near and above Tg the density fluctuations possess characteristic decay times falling into the time range $(10-10^{-7}s)$ of the photon correlation spectroscopy. Under self beating conditions the desired relaxation function g(t) can be obtained from the measured quantity G(t) through the relation:

$$G(t) = A (1 + b |g(t)|^2) \qquad (2)$$

where A is the base line computed or measured at long delay times t. The contrast factor b which is treated as a fitting parameter, is proportional to the fraction of the total scattered intensity arising from slow nonpropagating density fluctuations [7].

$$b^{1/2} \sim (\frac{\partial n}{\partial \rho})^2_T \frac{\chi_0 - \chi_\infty}{\chi_\infty} \qquad (3)$$

In Eq.(3), χ_0, χ_∞ are the limiting compressibilities at low and high frequencies respectively and n is the refractive index.

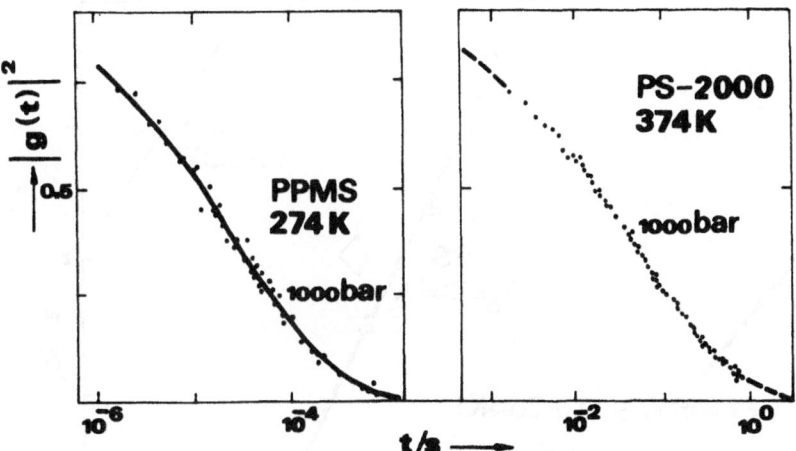

Fig. 1 Normalized density correlation functions (dots) in bulk poly(phenylmethylsiloxane) (PPMS) [4] and polystyrene (PS) of molecular weight 2000 [5]. Solid lines represent fits of eq. (4) to the experimental data with β=0.44 for both polymers.

Figure 1 shows two normalized correlation functions $(g(t))^2$ for the undiluted PPMS (MW = 2600, Tg = 229 K) and PS (MW = 2000, Tg=325 K) at 1000 bar and temperatures 274 and 374 K respectively. The experimental G(t) were measured with a 96-channel Malvern correlator.

The functions g(t) do not conform to a single exponential form but can be well described by

$$g(t) = \exp\left|-(t/\tau^*)^\beta\right| \qquad (4)$$

where β (0<β<1) is a measure of the width of the distribution of relaxation times. In most case β>0.3 and a meaningful average time $\tau=(\tau^*/\beta).\Gamma(1/\beta)$ ($\Gamma(1/\beta)$ being the gamma function) can be obtained. The time τ exhibits a strong temperature dependence (Fig. 2) which can be described by the free volume equation .

$$\tau = \tau_\infty \exp\left|\frac{E}{R(T-T_0)}\right| \qquad (5)$$

The activation parameter E* and the time τ* as well can be discussed in the framework of a generalized coupling model of relaxations reported recently [8].

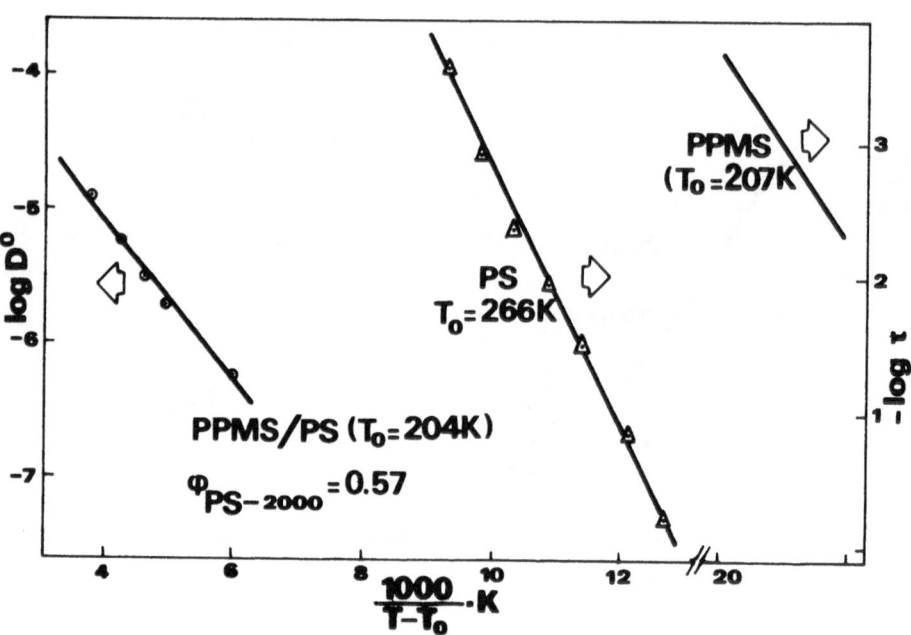

Fig. 2 Semilog plot of the monomeric diffusion coefficient D^O (cm^2/s) of the mixture PPMS/PS and the structural relaxation time τ for the undiluted components vs $1/(T-T_O)$. The solid lines represent the least squares fit of eq. (5) to the data. For PPMS, the solid line was calculated from high pressure data [4].

The model predicts the fractional exponential eq (4) with a coupling parameter $n=1-\beta$ and the relation $\tau^* \sim \tau_0^{1/(1-n)}$ between the apparent time τ^* and a primitive time τ_O. The latter is proportional to the monomeric friction coefficient j_O, which obeys eq (5) but with an activation energy $E_O = E^* \cdot (1-n)$. Thus the temperature dependence of τ^* can be used to infer the temperature dependence of j_O. This possibility was tested in the case of PPMS, where $n=0.56$ and E^*, E_O amount to 13.8 and 5.8 kJ/mol respectively [9]. For PS , less flexible chain $n = 0.6$ and $E^* = 18.4$ kJ/mol [5] . With these values, the activation energy E_O for PS is computed to be 7.4 kJ/mol. To provide a test of this estimate, we need to measure E either for the promitive relaxation time τ_O or the microscopic coefficient j_O for sufficient short PS chains ($n \sim 0$). This will be discussed below when the diffusion results are represented.

Finally, a practical application of the relaxation functions g(t) is the computation of the retardation spectrum and hence the prediction of important viscoelastic functions of amorphous bulk polymers near Tg [10] .

3. CONCENTRATION FLUCTUATIONS IN COMPATIBLE POLYMER MIXTURES

Recently, considerable theoretical interest was focused on the dynamics of concentration fluctuations in polymer mixtures [3,11-13]. For non-entangled chains in the regime $q \cdot R_O \ll 1$ (R_O being the radius of gyration) the relaxation function $g(q,t)$ decays exponentially with a relaxation rate $1/\tau_m = D_m q^2$ where the mutual diffusion D_m is given by

$$D_m = D^O \; \varphi \; (1-\varphi) \; /S(0) \qquad (6)$$

$S(0)$ in the limit $q \to 0$ is the static scattering intensity due to concentration fluctuation and φ is the volume fraction. The monomeric diffusion D^O depends on the composition and the microscopic Rouse diffusivities D_{AB}^O, D_{BA}^O in the limit $\varphi_A \to 0$ and $\varphi_B \to 0$ respectively. While in the molecular theories [3,10] , D^O is a geometrical average of the individual diffusivities, in the phenomenological theory of interdiffusion [12, 13], D^O is an arithmetical mean of D_{AB}^O and D_{BA}^O.

To verify these rather strong theoretical predictions we have conducted a photon correlation study of $g(q,t)$ of the compatible mixture PPMS/PS for different fractions of PS over a wide temperature range [6]. Figure 3 shows a semilog plot of a net correlation function $b\,g(q,t)$ (eq. (2)) at 168 °C. The solid line on the plot represents the least squares fit of eq. (4) to the experimental data taken with a 28 log-spaced delay channel Malvern (K 7027) correlator. The amplitude b is proportional to the total scattered intensity arising from concentration fluctuations $\delta\varphi(q)$

$$b^{1/2} (\partial n/\partial \varphi)^2 \quad < |\delta\varphi(q)|^2 > \qquad (7)$$

where $(\partial n/\partial \varphi)$ is the refractive index increment. The value of the distribution parameter β amounts to 0.86. This should be contrasted with the corresponding value of $\beta(\sim 0.4)$ in the pure components where the laser light is scattered by density fluctuations.

The strong temperature dependence of D^O which is mainly dictated by the monomeric friction j_O is depicted in Fig. 2. Figure 2 includes for comparison the structural relaxation

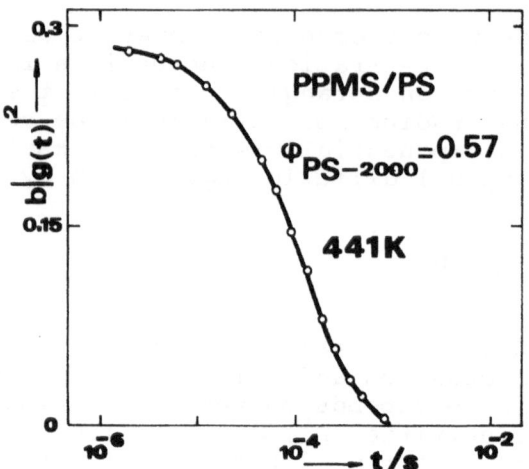

Fig. 3 Net concentration auto correlation function (dots) vs logt for the mixture PPMS/PS. The solid line represents the fit of eq.(4) with =0.86 to the experimental function.

time τ for the pure components. Equation (5) describes well the temperature dependence of D with T_O= 204.3 K and E_O= 11 kJ/mol. The values of E^O and c_2 = $(Tg-To= 59K)$ should be compared with those of the friction coefficient in pure components. For PPMS [4], the reported values are 5.8 kJ/mol and 20 K whereas the corresponding values for pure PS are 13±2 kJ/mol and 55±7 K [14]. The experimental finding that the values of E^O and c_2 of the mixture resemble to those of pure PS and not of PPMS, is in favor of the molecular theory [3,11], which predicts a dominance of the less mobile chain.

If we assume, that the temperature dependence of the mutual diffusion D_m is that of the monomeric friction of PS chain then the ratio E^O/E^* is larger than the predicted value 1-n of the coupling theory. This quantitative discrepancy may stem from the uncertainty in the value of Tg for the mixture and the temperature dependence of the prefactor connecting τ_O with j_O. To clarify the latter point, as well as the concentration dependence of D^O further studies are in course.

REFERENCES

1. B.J. Berne and R. Pecora : Dynamic light scattering, Wiley Interscience, New York 1976

2. C.H. Wang, G. Fytas, D. Lilge and Th. Dorfmüller, Macromolecules 18, 1363 (1981)

3. F. Brochard and P.G. de Gennes Physica 118A, 289 (1983)

4. G. Fytas, Th. Dorfmüller and B. Chu, J. Polym. Sci., Polym. Phys. Ed. 22, 1471 (1984)

5. U. Mittag and G. Fytas, Unpublished data

6. U. Murschall, E.W. Fischer, Ch. Herkt-Maetzky and G. Fytas, J. Polym. Sci. Polym. Lett. Ed. (1986)

7. C.H. Wang and E.W. Fischer, J. Chem. Phys. 82, 632 (1985)

8. A.K. Rajagopal, K.L. Ngai and S. Teitler, J. Phys. C17, 6611 (1984) and references cited herein

9. K. L. Ngai and G. Fytas, J. Polym. Sci. Polym. Phys. Ed. (1986)

10. G. Fytas, C.H. Wang, G. Meier, and E.W. Fischer, Macromolecules, 18, 1492 (1982)

11. K. Binder, J. Chem. Phys. 79. 6387 (1983)

12. Krammer, P.Green and C. Palmstrom, Polym.25, 473 (1984)

13. H.Sillescu, Makromol. Chem., Rapid Comm. 5, 519 (1984)

14. D.J. Plazek, V.M.O Rourke, J. Polym. Sci.A2, 209 (1971.

LIST OF PARTICIPANTS

LECTURERS

Prof. George **ATKINSON**
Head, Chemistry Dept
University of Arizona
Tucson, AZ 85721

Dr. P. **ARGYRAKIS**
Dept of Physics
University of Crete
Iraklion, Crete
Greece

Dr. Ronald **ARMSTRONG**
Dept of Mechanical Engineering
University of Maryland
College Park, MD 20742

Dr. Jean **AUBARD**
C.N.R.S.
I.T.O.D.Y.S.
1 Rue Guy de la Brosse
75005 Paris
France

Dr. Richard **BERSOHN**
Dept of Chemistry
Columbia University
New York, NY 10027

Dr. Martin **BRAITHWAITE**
Detonation Research Group
ICI PLC
Ayrshire KA20 3LN
United Kingdom

Prof. W. Byers **BROWN**
Dept of Chemistry
University of Manchester, U.K.
Manchester, M12 9PL
United Kingdom

Dr. Chris **CAPELLOS***
Energetic Materials Division
Building 3022
LCWSL, ARRADCOM
Dover, NJ 07801-5001

Dr. Chris **CHATGILIALOGLU**
Italian NRC
Instituto Dei Composti del Carbonio
Consiglio Nazionale Delle Ricerche
400064 022ANO
Emilia Bologna, Italy

Prof. Mostafa A. **EL SAYED**
Department of Chemistry & Biochemistry
University of California, Los Angeles
Los Angeles, CA 90024

Prof. Edward R. **GRANT**
Dept. of Chemistry
Corness University
Ithaca, NY 14853

Dr. Jan **GRYKO**
Institute of Physical Chemistry
01-224 Warsaw
Kasprazaka 44/52
Poland

Prof. Bryan **HENRY**
Chemistry Department
University of Manitoba
Winnipeg, R3T 2N2
Canada

Dr. George **HIDA**
Technion - Israel Institute of Technology
Dept of Mineral Engineering
Technion City
Haifa 32000, Israel

Dr. J.B. **HOPKINS**
AT&T Bell Laboratories
Physical & Inorganic Chemistry
  Research Departent
Murray Hill, NJ 07974

*Co-director

Dr. Daniel **HUPPERT**
ChemistryDepartment
Tel Aviv University
Tel Aviv, Israel

Prof. J. **JORTNER**
Dept of Chemistry
Tel Aviv University
Ramat Aviv
Tel Aviv, Israel

Dr. Joyce J. **KAUFMAN**
Dept of Chemistry
The Johns Hopkins University
North Charles and 34th Streets
Baltimore, MD 21218

Prof. Walter **KOSKI**
Chemistry Department
The Johns Hopkins University
North Charles and 34th Streets
Baltimore, MD 21218

Dr. Hans **KUHN**
Max-Planck-Institut fur
  biophysikalisch Chemie am Fassberg
D-3400 Gottingen-Nikolausberg
West Germany

Prof. Peter **LAMBROPOULOUS**
University of Crete
PO Box 470
Iraklion, Crete
Greece

Prof J.H. **LEE**
Mechanical Engineering Department
McGill University
817 Sherbrooke Street West
Montreal PQ H3A 2K6, Canada

Dr. M. **LEEUW**
Prins Maurits Lab TNO
Postbus 45
2280 AA Rijswijk
The Netherlands

Dr. Constantine **MAVROYANNIS**
Div. of Chemistry
Natl Res. Council of Canada
Ottawa, Ontario
K1A 0RG, Canada

Dr. A. **METROPOULOS**
Theoretical Chem. Inst.
N.H.R.F.
48 Vas. Constantinou Avenue
Athens 11635, Greece

Dr. C.A. **NICOLAIDES**
Theoretical Chemistry Inst.
National Hellenic Reserch Fdn.
48, Vassileos Constantinous Ave.
Athens, Greece

Prof. Simone **ODIOT**
Universite Pet M Curie
3 quai de la tournelle
Paris 75005, France

Prof. P. **PAPAGIANAKOPOULOS**
University of Crete
Crete, Greece

Prof. A. **PERSOONS**
Dept of Chemistry
University of Leuven
Celestijnenlaan 200D
B-3030 Heverlee, Belgium

Dr. C. **PFEIL**
Fraunhofer-Gesellsch
Fraunhofer-Institut for Treib
  und Explosivstoffe
Postfach 1240, Pfinztal 1
D-75057, West Germany

Dr. H. **REISS**
Department of Chemistry
University of California, Los Angeles
Los Angeles, CA 90024

Dr. P.M. **RENTZEPIS****
Physical and Inorganic Chemistry
  Research Department
AT&T Bell Laboratories
Murray Hill, NJ 07974

Dr. James **RITCHIE**
MS B214
Los Alamos Natl Laboratory
Los Alamos, NM 87545

Dr. Stephen **SCHMIDT**
Los Alamos Natl Lab
PO Box 1663
MS J970
Los Alamos, NM 87545

Dr. Scott **SHACKELFORD**
EOARD
Box 14
FPO New York 09510

Dr. David **SQUIRE**
U.S. Army ERO
223 Old Marylebone Road
London NW1 5th, England

Dr. K. David **STRAUB**
John L. McClellan
Memorial Veterans Hospital
4300 West 7th Street
Little Rock, AR 72205

Dr. Theophile **THEOPHANIDES**
University of Montreal
2900 Edovard Monpetit
Montreal, Quebec
Canada H3C3V1

Dr. Bruno **VAN WONTERTHEM**
Lab Chem and Biol Dyn
University of Leuven
Celestijnenlaan 200D
B-3030 Leuven, Belgium

****Director**

PARTICIPANTS

Dr. L.G. **ARNAUT**
Centro de Quimica
Dept de Quimica
Universidade de Coimbra
3049 Coimbra Codex
portugal

Dr. Francis **BANKS**
College of the Bahamas
NASCI, COB
Box N4912
Nassau, Bahamas

Dr. Sotiris **CAPELLOS**
University of Athens
Athens

Dr. Thomas C. **CASTORINA**
TCC Consulting Service
72 Cliffside Trail
Denville, NJ   07834

Ms. Natalia **CHESTNOY**
AT&T Bell Laboratories
Murray Hill, NJ   07974

Mr. Loen **CLAYS**
Lab of Chem and Biol Dyn
University of Leuven
Celestijnenlaan 200D
B-3030 Leuven, Belgium

Dr. O. **DIMOPOULOU-RADEMANN**
Institute fur Physikalische Chemie
University of Marburg
Auf den Lahnbergen Hans-Meerwern-Strasse
D-3550 Marburg/Lahn
West Germany

Dr. S. **DURMAZ**
Chemistry Dept
Ondoluzmayis Uuniversitesi
Fed-Ed Fakutesi
Kimya Bolumu
Samsun, Turkey

Dr. J.L. **FIGUEIREDO**
University of Porto-Portugal
Faculdade de Engenharia
4099 Porto Codex
Portugal

Dr. Guy **FLEURY**
Faculty of Pharmacie
Rue du Prof Laguelle
59045 Lille Cedex
France

Dr. Kostas **FOTAKIS**
Dept of Chemistry
University of Crete
Iraklion, Crete
Greece

Dr. Leonard **GAMSS**
ARDC
Dover, NJ   07801-5001

Dr. Chris **HADJIKOSTAS**
Faculty of Chem Dept
Dept of General & Inorganic Chem
University of Thessaloniki
Thessaloniki   54006
Greece

Dr. Peter John **HASKINS**
Ministry of Defense
XM1 Division, RARDE
Fort Halstead
Sevenoaks, Kent   TN147BP
England

Dr. Elias **HONTZOUPOULOS**
15 N. Pazilsi St
N. Psichiko   15451
Athens, Greece

Dr. Stefanos **KARAGEORGIOU**
Fac of Chemistry
Dept of General & Inorganic Chem
Univeristy of Thessaloniki
Thessaloniki   54006
Greece

Dr. Miltiades **KARAYANNIS**
Chemistry Department
University of Ioannina
Ioannina, Greece

Dr. Nikos **KATSAROS**
NRC "Demokritos"
Chemistry Department
Aghia Paraskevi Attikis
Athens, 153-10
Greece

J.V. Richard **KAUFMANN**
Battelle - Res Tri Park
Eaglenest at Hifields
Washington Depot, CT   06794

Dr. Y. **KONSTANTATOS**
NRC "Demokritos"
Chemistry Department
Aghia Paraskevi Attikis
Athens, 153-10, Greece

Dr. George **MANOUSAKIS**
Faculty of Chemistry
Dept of General & Inorganic Chem
University of Thessaloniki
Thessaloniki   54006
Greece

Dr. Louis **MOREIRA**
University of Porto-Portugal
Faculdade de Engenharia
4099 Porto Cedex
Portugal

Dr. Giancarlo **MARCONI**
Italian Council for Research
Istituto FRAE-CNR
via Castagnolo 1
40126 Bologna
Italy

Dr. Vladimiro **MUJICA**
Uppsala University
Quantum Chemistry Group
Box 518, S-751
Uppsala, Sweden

Dr. Cemil **OGRETIR**
University of Anatolia
Chemistry Dept
Faculty of Arts & Sciences
Eskisehir, Turkey

Dr. Mine **PEHLIVAN**
Middle East Tech University
Turkey

Dr. K. **RADEMANN**
Institute fur Physikalische Chemie
Auf den Lahnbergen Hans-Meerwern-Strasse
D-3550 Marburg/Lahn
West Germany

Dr. Michael **SCHIEFERSTEIN**
Max-Planck-Institue fur Strahlenchemie
Stifstr 34-36
D-4300Muhlheim a.d. Ruhr 1
West Germany

Dr. P. **SISKOS**
University of Athens
Chemistry Dept
Lab of Analyt Chem
104 Solonos Str
10680 Athens
Greece

Dr. A. **SOLOMONIVICI**
RAFAELK ARDA
Dept 24
PO Box 2250
HAIFA 31021
Israel

Dr. Chris **TSIAMIS**
Faculty of Chemistry
Dept of General & Inorganic Chem
University of Thessaloniki
Thessaloniki   54006
Greece

Dr. S. **TZOUWARA-KARAYANNI**
Univ of Ioannina
Chem Dept
Ioannina, Greece

Dr. J. **VAN DOOREN**
Catholic University of Lourain
Lab de Physico-Chimie de la Combustion
1, pl. La Pasteur
1348 Louvain-la-Neuve
Belgium

Dr. Maria **VRETTOU**
Faculty of Chemistry
Dept of General & Inorganic Chem
University of Thessaloniki
Thessaloniki   54006
Greece

Dr. Raymond F. **WALKER**
ARDC
Dover, NJ   07801-5001